Petroleum Geochemistry and Source Rock Potential of Carbonate Rocks

AAPG Studies in Geology #18

Geological Society
Withdrawn from the
Library

REFERENCE ONLY

Petroleum Geochemistry and Source Rock Potential of Carbonate Rocks

Edited by
James G. Palacas

AAPG

REFERENCE COPY
DONATED BY THE PUBLISHER

Published by
The American Association of Petroleum Geologists
Tulsa, Oklahoma 74101, U.S.A.

Geological Society
Withdrawn from the
Library

Copyright © 1984
The American Association of Petroleum Geologists
All Rights Reserved
Printed in the United States of America

Petroleum geochemistry and source rock potential of carbonate rocks.

(AAPG studies in geology; #18)
Bibliography: p.
Includes index.
1. Petroleum—Geology—Congresses. 2. Gas, Natural—Geology—Congresses. 3. Organic geochemistry—Congresses. 4. Rocks, Carbonate—Congresses. I. Palacas, J. G. II. Series: AAPG studies in geology; no. 18.
TN870.5.P478 1985 553.2'82 84-45746
ISBN 0-89181-024-2

AAPG grants permission for a single photocopy of any article herein for research or non-commercial educational purposes. Other photocopying not covered by the copyright law as Fair Use is prohibited. For permission to photocopy more than one copy of any article, or parts thereof, contact: Permissions Editor, AAPG, P.O. Box 979, Tulsa, Oklahoma 74101.

Association Editor: Richard Steinmetz
Science Director: Edward A. Beaumont
Project Editor: Dougas A. White
Special Editor: William D. Rose
Design and Production: S. Wally Powell
Typographers: Eula Matheny and Tricia Kinion

Contents

Special Abstracts:

INTRODUCTION

Carbonate rocks have diverse characteristics. They can be excellent reservoirs as well as prolific source rocks for oil. Oils from carbonate rocks commonly have distinctive bulk chemical and molecular characteristics that reveal their origin. The lack of widespread appreciation for these facts in the geological community was one reason that a symposium entitled "Petroleum Geochemistry and Source Rock Potential of Carbonate Rocks" was organized and held at the Geological Society of America annual meeting in Atlanta, Georgia, in October 1980. The symposium was sponsored by the Organic Geochemistry Division of the Geochemical Society during my term as chairman of the division. Of the 18 papers given in the symposium, 12 papers and four abstracts are included herein. Also included in this volume are two papers that were prepared later.

I hope that this collection of original papers, which synthesize data from about 20 different sedimentary basins, will help to correct any lingering misconceptions concerning the effectiveness of carbonate rocks as major sources of petroleum. I also believe that the information presented herein, including the references, will serve as a valuable resource for evaluating petroleum occurrence in other carbonate sequences and for locating petroleum reserves in unexplored, partially explored, and even maturely explored basins where possible carbonate-generated oil and gas may have been overlooked.

The first 11 papers, arranged in geochronological order, are descriptions and interpretations (that is, case histories) of specific carbonate source rocks that range in age from Precambrian to Miocene. Some of the highlights of these papers are summarized below.

The paper by Fu Jia Mo, Dai Yong Ding, Liu De Han, and Jia Rong Fen, in addition to describing the geochemistry of petroleum accumulations and source rocks ranging in age from Precambrian to Triassic, points out some interesting differences in thermal histories of Precambrian carbonate-rock sequences in eastern China. In one basin, Precambrian carbonate rocks are surprisingly thermally immature and have yielded heavy oils and asphalts. In another basin, on the other hand, Precambrian carbonate rocks are definitely overmature and have generated methane-rich gas.

The paper by McKirdy, Kantsler, Emmett, and Aldridge on the Eastern Officer basin, South Australia, includes the first reported examples of *nonmarine* carbonate rocks and oils of Cambrian age that are similar to those of the Eocene Green River Formation, Utah.

In their study of crude oils in the Michigan basin, Gardner and Bray indicate that the interreef, laminated carbonate rocks of Silurian age are the primary source of commercial oil accumulations in the Silurian pinnacle reefs. One interesting aspect of this finding is that it provides an example of a proposed source-rock facies that is characterized by a relatively low organic-carbon content (average about 0.3%)—a value considered by some geochemists to be the lowest limit for source-rock requirement.

Powell's paper presents strong evidence that in western Canada thermally immature, organic-rich carbonate rocks of Devonian age have yielded heavy oils and bitumens, whereas comparable carbonate facies in a mature setting are the probable source of lighter, conventional oils.

In four papers, detailed geochemical data support previously held concepts, which were based primarily on geologic evidence, that three well-known carbonate rock units of Mesozoic age are the source of "giant" oil accumulations. The three prolific carbonate units are (1) the lower part of the Jurassic Smackover Formation in the Gulf Coast region (Oehler), (2) the Upper Cretaceous La Luna Formation in Colombia (equivalent to the La Luna Formation, Venezuela) (Zumberge), and (3) the Upper Cretaceous Austin Chalk in Texas (Grabowski; Hunt and McNichol).

The paper by Palacas, Anders, and King demonstrates the use of relatively new but powerful techniques for correlation and interpretation of maturity, using high-molecular-weight biological-marker distributions (for example, steranes and triterpanes), to indicate the specific carbonate source-rock facies for Lower Cretaceous Sunniland oil in south Florida.

Although carbonate source rocks are generally oil prone, the report by Rice outlines the occurrence and origin of early diagenetic (biogenic) methane-rich gas in shallow, thermally immature, organic-rich chalks of the Upper Cretaceous Niobrara Formation in the eastern Denver basin of Colorado, Kansas, and

Nebraska. Gas reserves in that area are estimated to be 1.2 tcf (0.34×10^{12} m^3). In the past, there was little exploration for such gas accumulations in the Denver basin, and similar occurrences may have been overlooked in other areas.

In the last of the case-history papers, Demaison and Bourgeois briefly document, by means of geochemical oil–source-rock correlations, an interesting set of geologic conditions. They show that at the Casablanca field, Tarragona basin, offshore Spain, overlying younger carbonate source facies (middle Miocene Alcanar Formation) generated the oil that charged the underlying older (Jurassic) limestone reservoirs. The emphasis of the report, however, is on the unique integration of sedimentological, paleontological, and geochemical analyses of organic facies to initially determine the paleoenvironmental conditions and source potential and, ultimately, to predict the stratigraphic distribution of source beds in a given area.

In the first of three topical studies, Jones compares geological and geochemical features of carbonate and shale source beds and in that context discusses the general nature of carbonate source rocks. (Two other papers listed above—Hunt and McNichol, and Oehler—also contain significant contributions concerning the general nature of carbonate source rocks. These two papers should also be consulted for a more complete synthesis of geological and geochemical source-rock characteristics.) Included in Jones' paper is a detailed discussion of the concept of organic facies and a critical reexamination of factors such as organic-carbon requirements, catalytic effects, and expulsion–migration mechanisms that affect carbonate source rocks. The sometimes controversial interpretations in this paper should stimulate further discussion and research, which may help to resolve possible conflicting opinions.

Hughes demonstrates the use of organic-sulfur-containing compounds (thiophenes) for distinguishing oils derived from carbonate rocks from oils derived from siliciclastic rocks. Abundant sulfur compounds are characteristic of many carbonate source rocks and carbonate-derived oils, and an understanding of their relative abundances may lead to improved oil and source-rock fingerprinting techniques.

In the last complete paper, Shinn, Robbin, and Claypool address the problem of expulsion and migration mechanisms in carbonate rocks—processes that still are far from being fully understood. They conducted both compaction experiments, at pressures equivalent to as much as 13,500 ft (4,000 m) of overburden, and low-heating (100°–200°C or 212°–392°F) experiments on cores of Holocene carbonate sediment from Florida Bay. The preliminary results indicated that both mechanical and chemical (that is, pressure dissolution) compaction could provide the "driving force" for expulsion of hydrocarbon-bearing fluids into carrier beds or directly into reservoirs.

The third section of the volume includes four abstracts of papers that were given at the carbonate symposium of the annual meeting of the Geological Society of America in 1980. The abstracts provide additional geographic coverage of carbonate source-rock studies and support most of the conclusions in the present collection of complete papers.

I thank all of the authors for their contributions to this volume. Whatever benefits may accrue from this collection of reports should be credited to their efforts. Recognition and appreciation are extended to the following referees, whose conscientious reviews and helpful comments improved the papers: C. Barker, J. R. Castaño, G. E. Claypool, J. L. Clayton, H. Dembicki, T. D. Fouch, W. E. Harrison, J. M. Hunt, K. A. Kvenvolden, S. R. Larter, C. F. Jordan, M. D. Lewan, D. Leythaeuser, P. Müller, W. L. Orr, P. L. Parker, K. E. Peters, T. G. Powell, L. M. Pratt, M. A. Rogers, L. M. Ross, E. A. Shinn, C. P. Summerhayes, B. E. Torkelson, J. A. Williams, J. C. Winters, A. Young, and J. E. Zumberge.

Thanks are also due V. E. Swanson for his meticulous and helpful editorial review of several manuscripts. I appreciate the assistance of W. D. Rose, the project editor for this volume. I am deeply grateful to D. R. Malone for her efforts in the voluminous correspondence and in the typing of some of the manuscripts.

Finally I wish to express my appreciation to the American Association of Petroleum Geologists for publishing this collection of papers.

James G. Palacas
U.S. Geological Survey
Denver, Colorado

Distribution and Origin of Hydrocarbons in Carbonate Rocks (Precambrian to Triassic) in China

Fu Jia Mo
Liu De Han
Jia Rong Fen
Institute of Geochemistry, Academia Sinica, Guiyang, China

Dai Yong Ding
Institute of Geology, Academia Sinica, Beijing, China

Carbonate rocks range in age from Sinian (upper Proterozoic) to Triassic and are widely distributed in China. In the northern, southwestern, and southeastern parts of eastern China, these rocks contain abundant oil seeps and disseminated asphaltic substances. Large gas and small oil accumulations also have been discovered in these strata. Geologic and geochemical evidence suggests that the carbonate rocks are at least partially the source of these petroleum accumulations.

In the eastern part of China, hydrocarbons in carbonate rocks have generally undergone heating to high levels and have reached a much higher stage of thermochemical evolution. A late-stage evolution profile has been set up for evaluating hydrocarbon prospects in carbonate-rock formations. Although solid asphalts and kerogen exhibit a high degree of maturity at some places, commercial quantities of natural gas have been generated and preserved in subsurface reservoir rocks.

INTRODUCTION

Carbonate rocks are extensively distributed in China, especially in the eastern part (Fig. 1). In the southern part of eastern China, hereafter referred to as South China, a well-developed carbonate-rock sequence, 5,000 m (16,400 ft) in thickness, of Precambrian (late Proterozoic) to Triassic in age ($1.95–12 \times 10^8$yr), contains abundant oil seeps and disseminated asphaltic substances (Fig. 2). Some extensive asphalt deposits are commercially mined for fuel. Large gas fields and small oil fields also have been developed in these rocks in the southwestern part of eastern China. In the northern part of eastern China (referred to as North China), most of the carbonate rocks, of Sinian to Ordovician age, underlie thick sequences of Paleozoic to Cenozoic nonmarine formations, especially Tertiary nonmarine shales. In these Sinian to early Paleozoic carbonate rocks, numerous oil seeps and disseminated asphalt deposits occur in both outcrops and subsurface rocks. Figure 3 illustrates a representative area (the Yan Shan Mountain area) that is underlain by Sinian and lower Paleozoic carbonate rocks (Wang Tieguan, 1981). In addition to the indigenous oil seeps and asphalt deposits, commercial oil accumulations believed to have been derived largely from Tertiary nonmarine shales have been discovered in the Sinian carbonate rocks.

SOUTH SICHUAN AREA, SOUTHWEST CHINA

Southern Sichuan is one of the most extensively explored and geologically studied areas in China. Medium to large gas fields and small oil fields have been discovered in this area. Most are developed in upper Sinian dolomites and lower Permian (P_1) and Triassic (Tc and Tr) carbonates (Fig. 4). Geological and geochemical evidence indicates that gas, especially in P_1 rocks, and oil, in Tc and Tr rocks, probably are derived from local carbonate source rocks. Recently, annual production of natural gas from this area was 5.8 billion m^3 (205 billion ft^3), half of which is produced from lower Permian (P_1) carbonates.

In southern Sichuan, lower Permian (P_1) strata consist mainly of bioclastic limestones, overlain by a sequence of upper Permian (P_2) coals and basalts and lower Triassic (Tf) shales (Fig. 4). Triassic carbonate formations (Tc, Tr) are composed of dolomitic limestones and dolomites, interlayered with evaporites. These Triassic formations are overlain by thick nonmarine shale and sandstone sequences of Jurassic age.

Table 1 shows $^{40}Ar/^{36}Ar$ age determinations of natural-gas samples from rock strata of Sinian to Jurassic age in southern Sichuan Province. Results of $^{40}Ar/^{36}Ar$ dating of gas samples indicate that the P_1 gases are indigenous to P_1 rocks; i.e., the gases were generated from and accumulated in P_1 carbonate rocks (Shen Ping et al, in press). Most Triassic (Tc) gases also were shown to have similar ages as P_1 gases, suggesting that the Tc gases also were derived from P_1 carbonate source rocks. Hydrocarbon compositions (indicated by ratios of C_2 to C_5 alkanes/C_1) of Tc and P_1 gases (ratios ranging approximately from 0.005 to 0.05) are significantly different from those of the coal-mine gases in P_2 coal beds (ratios <0.001), strongly supporting the conclusion from the $^{40}Ar/^{36}Ar$ data that neither the P_1 nor the Tc gases were derived from the P_2 coal beds (see Fig. 10).

If the Permian (P_1) natural gas is indeed indigenous, then the question might be asked whether lower Permian limestones contain adequate source rocks to have produced enough hydrocarbons to account for the large gas accumulations.

Geologic studies, including detailed microscopic examination of outcrop samples and drill cuttings of P_1 carbonates from southern Sichuan (Dai Yong Ding et al, 1978), indicate the presence of

Figure 1 — Index map showing region referred to as eastern China and the two specific areas studied, South China (see Fig. 2) and North China (see Fig. 3).

Table 1

Stratigraphic Interval	$^{40}Ar/^{36}Ar$ Range	$^{40}Ar/^{36}Ar$ Average	No. of Samples	Ar–Ar Age (10^6 years)
Jurassic	445–686	566	2	172–180
Tc^5, Tc^4, Tc^3, Tc^2, Tc^1 + Tf	435–1371	875	35	Most samples show similar ages to that of Permian gases (P_1)
P_2, P_1	630–1189	925	79	290
Sinian	4440–9255	7009	3	2900

Table 1 — $^{40}Ar/^{36}Ar$ ratio and Ar–Ar ages of Sinian to Jurassic strata, South Sichuan Province.

favorable source rock and reservoir facies. The bioclastic limestones of the Moukou Formation (P_1) contain an abundant and diverse suite of marine fossils that are thought to have played an important role in the generation and accumulation of oil and gas. Some of the principal carbonate facies that are considered source and reservoir rocks in the Moukou Formation include red algal sparite, green algal micrite, foraminiferal (*Glomospira*) sparite and micrite, shelly micrite, and skeletal (coral) micrite (Plates 1 through 8, end of paper). These carbonates reflect deposition in shallow carbonate-bank environments rich in algal debris, nearshore lagoonal environments rich in shelly and skeletal debris, and marginal sea environments where foraminiferal limestones accumulated (Fig. 5). The algal and foraminiferal limestones deposited in the carbonate-bank and marginal sea environments are good reservoir facies, while the shelly and skeletal micrites deposited in the lagoonal environments are considered to be important source rocks for oil and gas.

Analyses of cores and cuttings from the Moukou and Qixia (P_1) Formations indicate that the organic-carbon contents of the purer limestones range from 0.05 to 1.86% and average 0.35% (126 samples). The extractable-bitumen content averages 120 ppm (61 samples). In contrast, marlstones and clayey limestones have a higher average organic-carbon content of 0.93% (61 samples). The amounts of organic carbon in most of the limestones, clayey limestones, and marlstones are

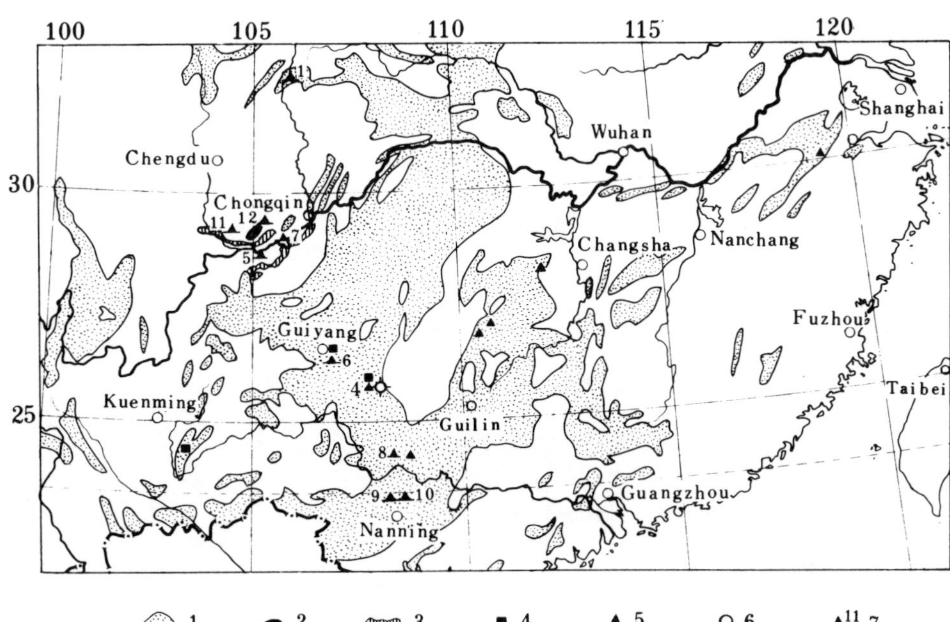

Figure 2 — Distribution of oil fields, gas fields, oil seeps, and asphalts in Sinian–Triassic carbonate rocks in South China. Key: 1, outcrops of carbonate rocks; 2, oil fields; 3, gas fields; 4, oil seeps; 5, asphalts; 6, wells; 7, location and number of samples analyzed (see Table 4).

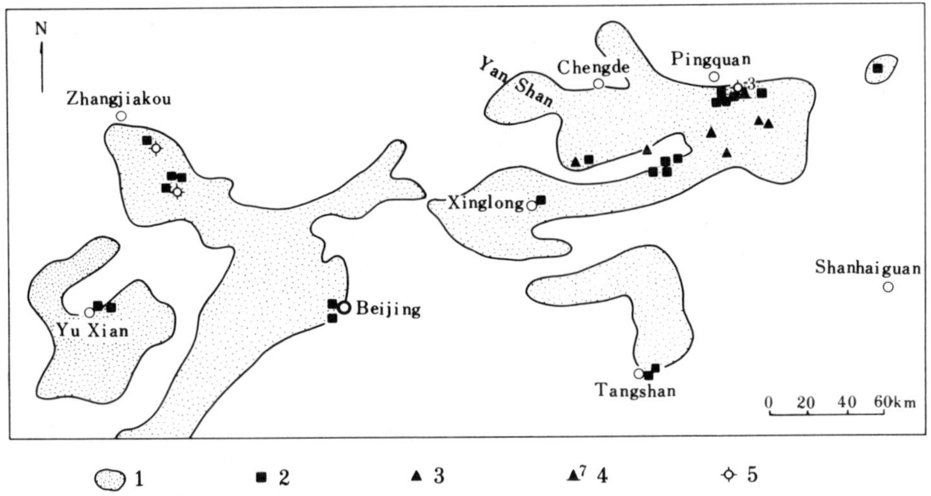

Figure 3 — Distribution of oil seeps and asphalts in Sinian–Ordovician carbonate rocks in North China. Key: 1, outcrops of carbonate rocks; 2, oil seeps; 3, asphalts; 4, location and number of samples analyzed (see Table 4); 5, wells.

COLUMN OF STRATA	STRAT. INTERVAL	THICKNESS (m)	LITHOLOGY	GAS, OIL ETC.
	Jh	514	Dark gray sandstones claystones and coal beds	Gas, Small amount of oil
	Tr	160	Gray dolomite	Gas, Light oil Source rock
	Tc	650	Bioclastic limestones with some evaporites	
	Tf	400	Purple graywacke and shales	Cover rock
	P_2	172	Coal-bearing formation	
	P_1	360	Mainly bioclastic gray limestones	Gas Source rock
	S	150	Mainly purple silts and shales	Cover rock
	O	350	Marl and limestone	
	Є	806	Dolomitic limestones, gypsum and claystone Basal black shale	
	Zb	150	Light gray algal dolomites	Source rock Gas

Figure 4 — Sedimentary column for South Sichuan Province, South China. See text for amplification.

great enough for these rocks to be considered potential sources of petroleum. By comparison with the shaly Tertiary source rocks of eastern China, the carbonate source rocks of southern Sichuan are richer in alkane and aromatic hydrocarbons (Fig. 6). Carbonate source rocks are estimated to make up 140 m (460 ft) of the 350-m (1,150-ft) total thickness of the Moukou and Qixia Formations in the Gu Song area.

The wide distribution of potential carbonate source rocks and closely associated reservoir carbonates in the Permian strata is probably a main cause of the many large gas deposits that have been discovered in the South Sichuan Basin. Total natural-gas production from these fields each year ranges up to 3 billion m^3 (106 ft^3).

YAN SHAN MOUNTAIN AREA, NORTH CHINA

Hydrocarbon seeps, including oils and asphalts, have been found in Sinian to Ordovician carbonate rocks in the North China region. Seeps are most common in the Sinian formations of the Yan Shan Mountain area. Sinian rocks, 5,000 to 9,130 m (16,400–29,950 ft) thick and 0.8 to 1.9 b.y. old, are composed predominantly of carbonate rocks. The major types of seeps in these carbonates are black to yellow fluid-like oils, black semi-fluid-like oils, and solid asphalts that can be partially dissolved in organic solvents. Gas seeps are not common in this area.

Geologic and organic-geochemical evidence indicates that most of the hydrocarbon seeps are derived from Sinian carbonate source rocks. Solid asphaltic substances are distributed paral-

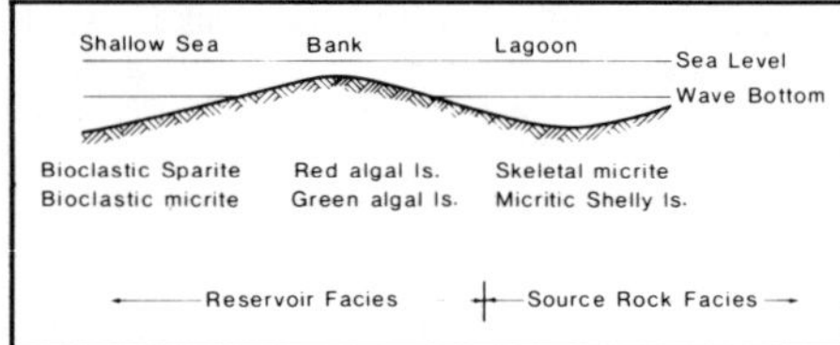

Figure 5 — Schematic distribution of lithofacies of Moukou Formation carbonates, South Sichuan Province.

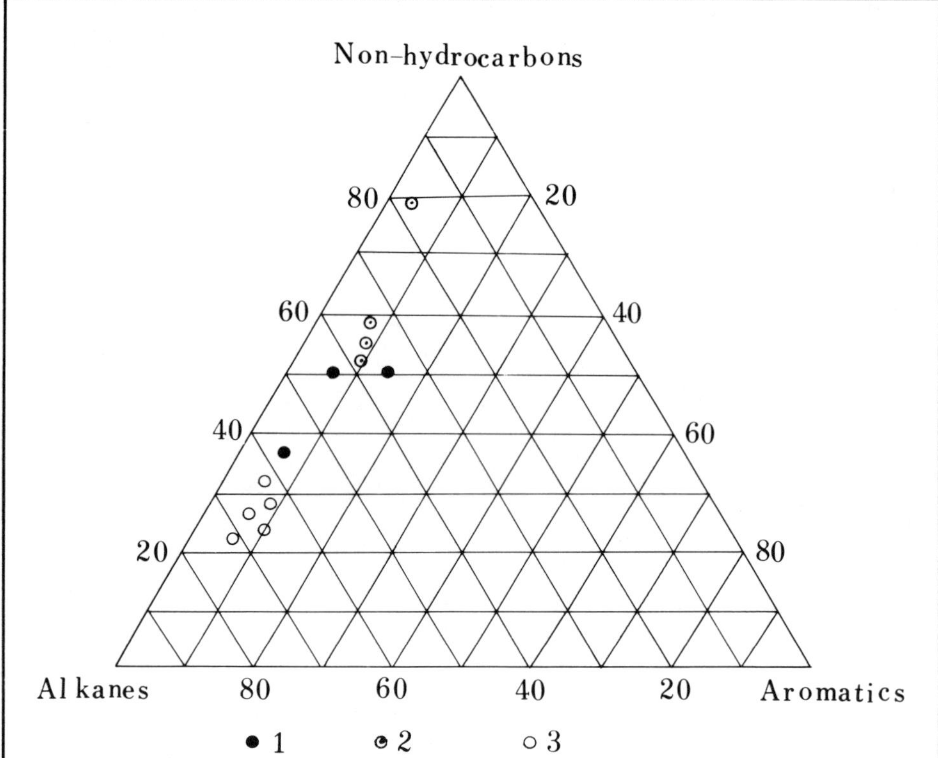

Figure 6 — Composition of extractable organic matter of carbonate rocks, South Sichuan Province. Key: 1, lower Permian (P_1^3) limestones; 2, Tertiary shales; 3, Triassic (Tc) dolomitic limestones.

Table 2

Well No.	Sample	Depth (m)	$\frac{5\alpha\text{-}C_{27}}{5\alpha\text{-}C_{29}}$	$\frac{5\beta\text{-}C_{27}}{5\alpha\text{-}C_{27}}$	$\frac{5\alpha\text{-}20S\text{-}C_{29}}{5\alpha\text{-}20R\text{-}C_{29}}$	$\frac{\text{Rearranged Sterane}}{5\alpha(C_{27}+C_{28}+C_{29})}$
Jin 30	oil	2001–2037	1.84	1.8	1.8	1.1
An 29	shale	2835	1.11	0.6	0.3	0.4
An 29	shale	3515–3520	1.64	1.4	1.0	1.0

Table 2 — Ratios of steranes in an Ordovician oil and Tertiary shales.

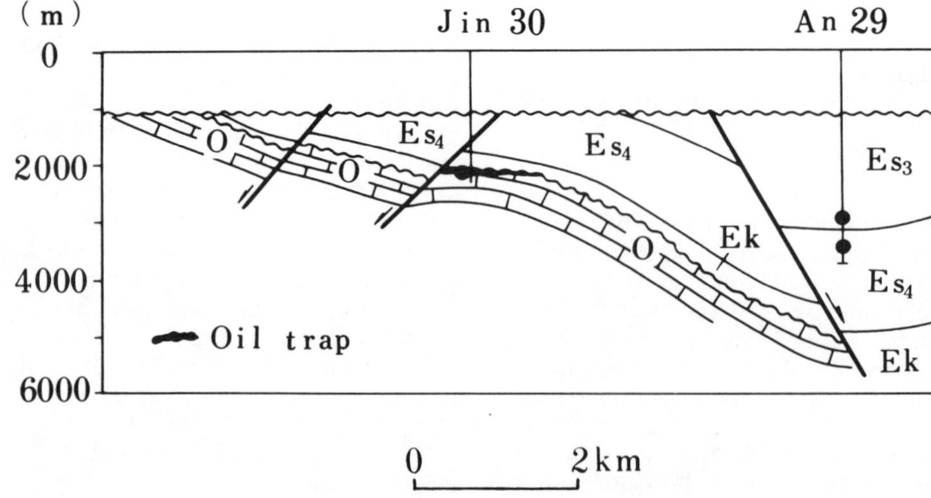

Figure 7 — Cross section of Langu oil field, south of Beijing, North China. Key: Ek, Oligocene shale and sandstone formation; ES, Eocene shale and sandstone formation; O, Middle Ordovician limestone.

lel to sedimentary stratification, and oils are finely dispersed or concentrated in small voids and fractures in the carbonate rocks. The composition of the seepage oils is markedly different from that of extracts from the nearby Tertiary nonmarine shales. The seepage oils are rich in alkanes, including alkanes of low molecular weight and aromatic hydrocarbons.

Those Sinian carbonates designated as source rocks are dark-gray to black dolomites and limestones that contain 0.3 to 0.6% organic carbon and 100–400 ppm bitumen.

Just south of the Yan Shan Mountain area, oil has accumulated in both Sinian carbonate and Tertiary sandstone reservoirs. These hydrocarbons are considered to have been derived mainly from Tertiary shales, although there is some evidence that the Sinian carbonates also have contributed to these oils. Langu oil field south of Beijing lies in the Langu deep subsidence basin of Tertiary age. As shown in Figure 7, oil in well Jin 30 of this field comes from a trap in an Ordovician limestone, which is different from the other oil occurrences in this Tertiary basin.

By means of GC–MS analyses of ste-

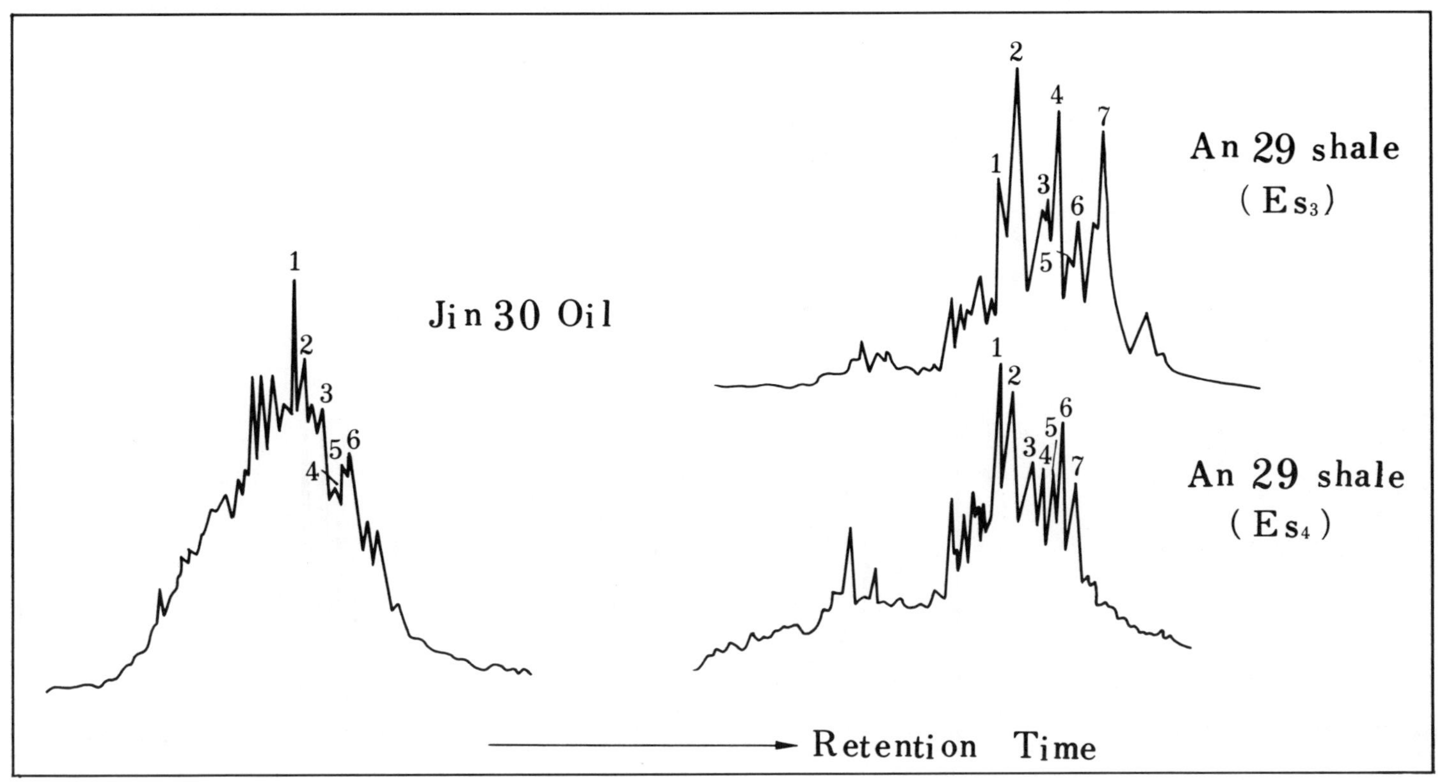

Figure 8 — Mass fragmentograms of m/z 217, showing that the Ordovician oil in Figure 7 does not correlate with possible Tertiary shale source rocks. Key: 1, 5β-C_{27} sterane; 2, 5α–C_{27} sterane; 3, 5β14βC_{28} sterane + 5β14βC_{29} sterane; 4, 5α–C_{28} sterane; 5, 5α–C_{29} (20S) sterane; 6, 5β14βC_{29} sterane; 7, 5α–C_{29} (20R) sterane.

ranes and terpanes, good correlations have been achieved between the Tertiary oils and their probable Tertiary source rocks. The oil in well Jin 30, however, does not appear to correlate with the oil extracts from the Tertiary shales (Fig. 8). Using m/z 217 mass spectra of steranes, four sterane ratios were chosen to characterize the oils (Zhang Zhencai et al, 1981). These ratios for oil from well Jin 30 and for shales from well An 29 are given in Table 2 and indicate that the Jin 30 oil has attained the highest degree of maturation of the samples from this basin (Mackenzie et al, 1980). Therefore, we believe that the oil in Jin 30 may be derived from sources within the Ordovician limestones or from sources deeper in the basin than those sampled. More evidence is needed to confirm the source of the oil in well Jin 30.

LATE-STAGE EVOLUTION OF HYDROCARBONS IN CARBONATE ROCKS

As indicated by many petroleum geologists and geochemists (e.g., Hunt, 1967; Zhou Zhongyi and Jia Rong Fen, 1974), and as discussed earlier in this article, both the abundance of organic matter and the environment of sedimentation are important factors that produce carbonate source rocks. Thermal maturation is also important for the development of carbonate source rocks. In eastern China, hydrocarbons in carbonate rocks generally have experienced an extensive thermal history and have reached a high stage of evolution. The carbonate formations of Sinian and Paleozoic age in South Sichuan can be cited as excellent examples of the late-stage evolution of hydrocarbons. Based on field surveys and laboratory investigations, we have systematically characterized the evolution of hydrocarbons and also have defined the exact lower limits of the disappearance of crude oil and the generation and accumulation of commercial gas deposits. It is obvious that the compositional characteristics of hydrocarbons during the late stage of thermal evolution can be used as an aid in the exploration for crude oil and natural gas in carbonate rocks as well as an aid for identification of possible source rocks in eastern China.

There are many evolutionary characteristics of late-stage hydrocarbons. In general, the ratio of heavy to light hydrocarbons is low, and, therefore, much more natural gas is discovered than crude oil. Almost all of the crude oil and oil seepages found in carbonate rocks are made up of light hydrocarbons. Most of the dispersed and concentrated solid asphalts are of thermal metamorphic origin. Particularly important is the complete disappearance of some biological marker compounds and distinct n-alkane distributions that can reflect the nature of the precursor organic matter. All kinds of rocks in South Sichuan, including carbonates and coal-bearing strata rich in nonmarine organisms, show a great similarity to one another in their composition of normal alkanes. They are composed essentially of light hydrocarbons with only traces of odd–even predominances (Fig. 9). The average odd–even predominance (OEP) for samples of 90 different rock types is 1.01, with a range of 0.87 to 1.16 OEP; the main n-alkane peak is at C_{20} (Wang Benshan et al, 1975).

In addition to the above evolutionary characteristics of hydrocarbons from late-

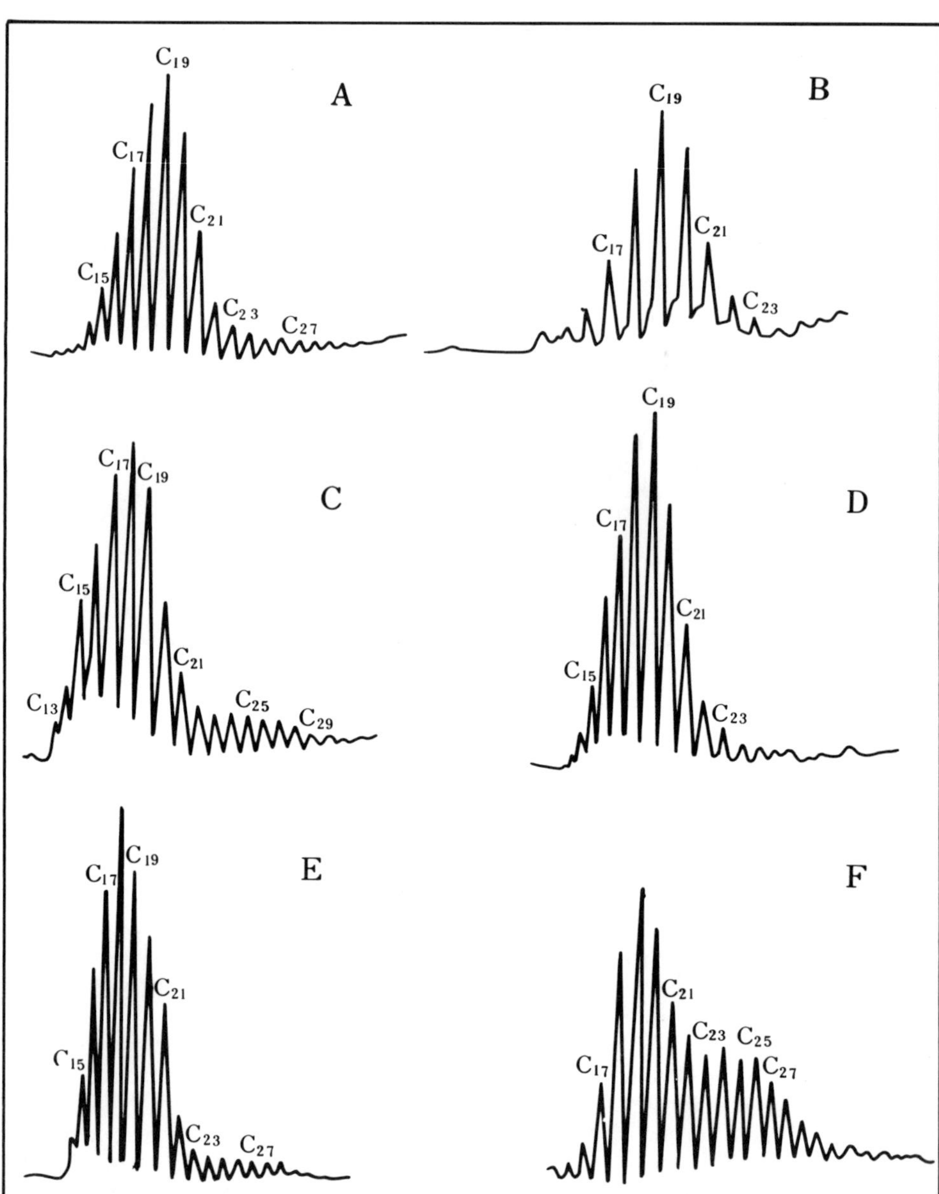

Figure 9 — Gas chromatograms of n-alkane distribution from rock samples of South Sichuan. Key: A, limestone, Tc^1; B, nonmarine shale, P_2^1; C, limestone, P_1^3; D, limestone, O_2; E, limestone, C_1^3; F, dolomite, Zn.

Table 3

	H/C Atomic Ratio	Reflectance (R_o)*	X-Ray Diffraction Equivalent Spacing
Light oil and condensate	0.8	≥1.5	≥3.53
Methane gas	<0.3–0.4	≥4.5	≤3.44

*Reflectance measured in oil.

Table 3 — Indicator values for lower limits of commercial oil and gas in South Sichuan area.

stage thermal maturation, there are also specific differences in the evolution of hydrocarbons in carbonate rocks. With increasing geologic age and burial depth, a series of changes in the organic matter takes place, such as: (1) absorbed hydrocarbons in the rocks (such as ethane and other heavier gases) decrease gradually, with only methane surviving (Fig. 10) (Zhang Lijie and Wang Wenlin, 1978); (2) the ratio of bitumens to organic carbon decreases markedly, almost reaching zero at the bottom of the profile shown in Figure 11 (the same relationship holds for the ratio of hydrocarbons to organic carbon with depth); (3) the reflectance (R_o) of kerogen and solid asphalts changes from 0.59 to greater than 4.5 (Table 4), and the atomic H/C ratio in these substances changes from about 1 to about 0.3 (Fig. 13); (4) the solid asphaltenes are metamorphosed into anthraxolite (Plates 9, 10 and 11, end of paper). An effective way to study the maturity of asphaltic substances is by x-ray diffraction. Using this method, we observed a change in equivalent spacing (d_{002}) from 3.91 Å to 3.44 Å (Table 4), which can be used as an indicator of thermal maturity.

An excellent evolution profile of late-stage hydrocarbons formed in carbonate rocks has been well documented in South Sichuan. There are a few commercially

Table 4

Sample Number–Location	Strat. Interval	H/C Atomic Ratio	Reflectance R_o max (%)	X-Ray Diffraction Equivalent Spacing d_{002} (Å)	d_{100} (Å)	Comments
1 N. Sichuan	S	1.18	0.59	3.9069		Crude oil and natural gas
2 (Coal) Liaoning	Pg	1.00	0.62	3.8567		
3 Hebei	Zn		0.59–1.00			
4 Guizhou	O	0.90	1.09	3.5442		
5 S. Sichuan	Tc		1.51			
6	P_1		1.65			Natural gas only
7 S. Sichuan	P_1	0.73	1.85	3.5147		
8 Kuangsi	D_2	0.19	3.85	3.5065	2.0891	
9 Kuangsi	D_2	0.27	3.84	3.5092	2.0939	
10 Kuangsi	D_2	0.19	4.8	3.4849	2.0822	
11 S. Sichuan	Z_n		4.5	3.4703		
12 Sichuan	Z_n	0.37	>4.5	3.4434	2.0385	
13 (Asphaltite) N. Sichuan	S	0.63	4.01	3.4930		Simulate experiment conditions: 400°C/500 atm. 400 hours
14 (Coal) Liaoning	Pg	0.48	1.84	3.5366		

Table 4 — Maturity indicators of selected asphalt samples from carbonate rocks, eastern China.

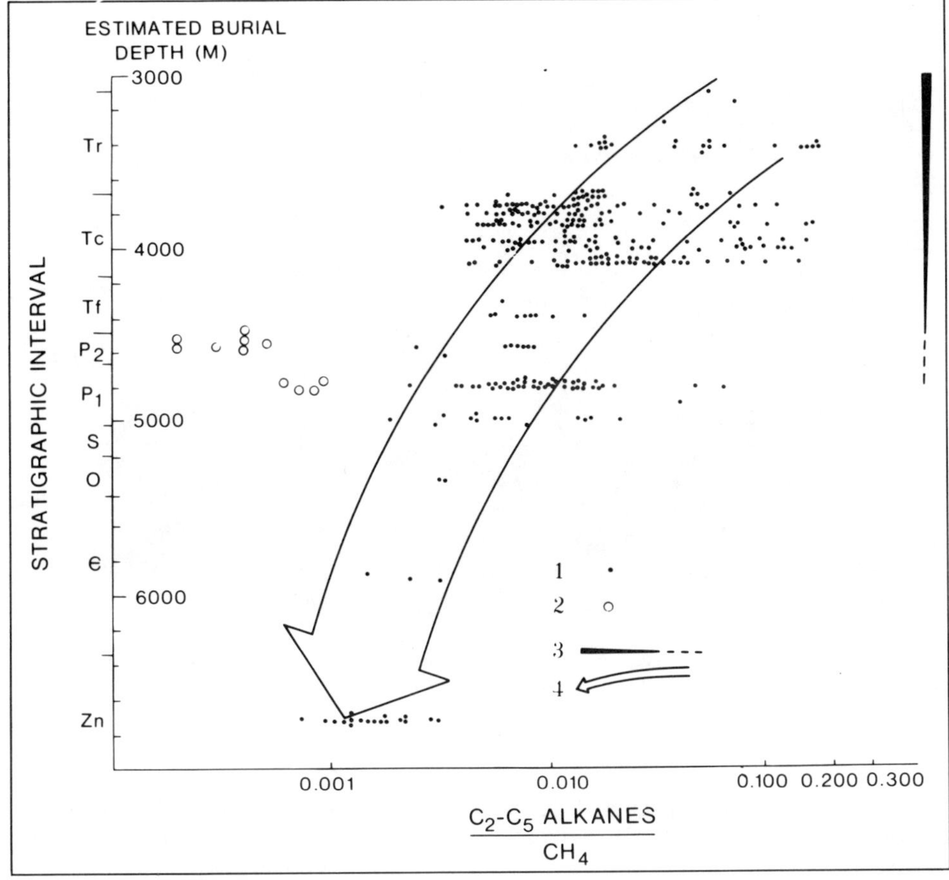

Figure 10 — Maturation (conversion of natural gas to methane) in Sinian to Jurassic rocks of South Sichuan Province. Key: 1, natural gas occurring in carbonate rocks; 2, natural gas occurring in coal deposits; 3, crude-oil occurrences; 4, direction of maturation (conversion of natural gas to methane) tendency. The stratigraphic depth at the ordinate Y shows that the present burial depth of Sinian natural-gas reservoir rocks ranges from 4,000 to 5,000 m (13,000–16,400 ft), and the surrounding cover loads are in the range of about 3,000 m (9,850 ft; i.e., the restored maximum depth is 7,000 m or 23,000 ft).

productive strata in this area. It is concluded from geologic and geochemical data that the crude oil and part of the natural gas in Triassic reservoirs came from Triassic source rocks and that the natural gas found in Permian strata came from Permian source rocks. The natural gas in Sinian rocks could have come from Sinian rocks or from more deeply buried strata. Based on a large amount of exploration data and numerous comprehensive analyses, we found some very interesting changes in composition of crude oils and natural gases with increasing geologic age and burial depth. It is evident that crude oil exists in strata overlying Tf (Tc_1, Tc_3, Tc_5, Tr, Jh, etc.; see Fig. 4) and some within Tf, but there is neither preservation of crude oil nor any sign of oil accumulation in the strata underlying Tf. As previously mentioned, natural gas found in Permian rocks is the driest (containing 95–98% C_1), followed by gas in Triassic rocks (86–98% C_1), and then gas in Sinian rocks (80–90% C_1). Using the ratio of heavier gaseous hydrocarbons to methane (C_2–C_5/C_1), determined from more than 170 natural-gas samples, a distinct evolutionary trend of natural gases was observed. As shown by Figure 10, this ratio decreases (i.e., methane increases) with increasing geologic age (Fu Jia Mo and Shi Jiyang, 1977). The range of variation found in this gas ratio

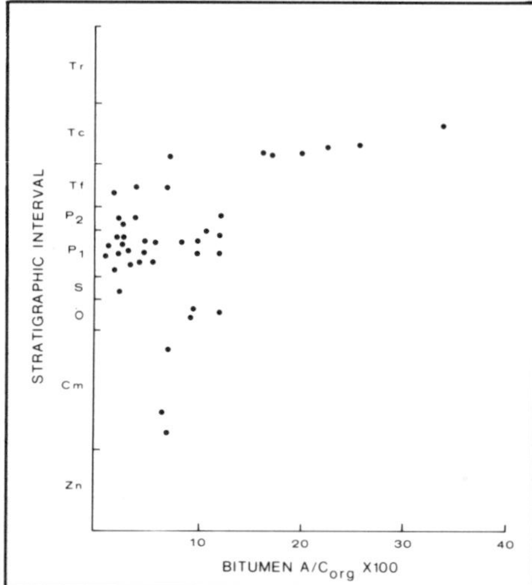

Figure 11 — Changes of ratio of bitumen A/ organic carbon with stratigraphic interval (geologic age).

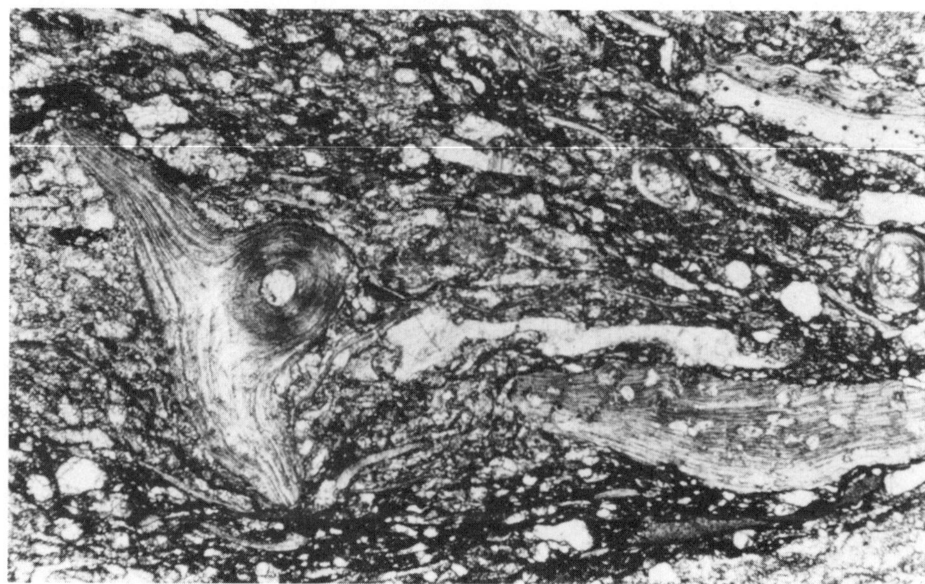

Plate 1 — Shelly micrite. Includes fragments of brachiopods, ostracodes, foraminifers, and algae. Matrix is micrite. Rock has fine laminae with small amounts of asphalt and dolomite. Source rock. Permian (P_1^m unit), ×18.

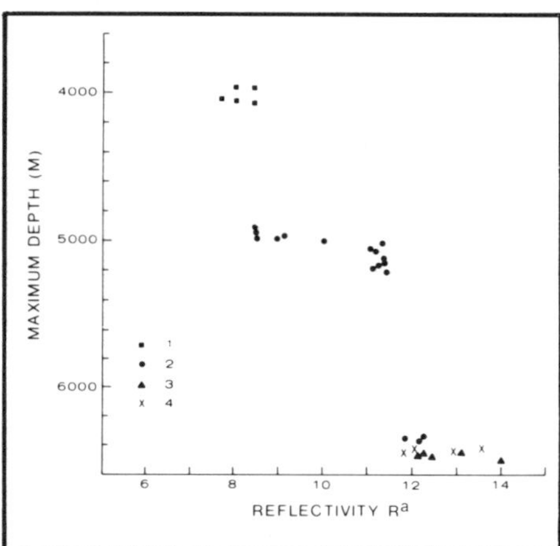

Figure 12 — Changes of reflectance of organic matter with maximum depth of burial (South Sichuan). Key: 1, vitrinite in shales; 2, kerogen in carbonate rocks; 3, pyrobitumen in carbonate rocks; 4, pyrobitumen in quartz crystals.

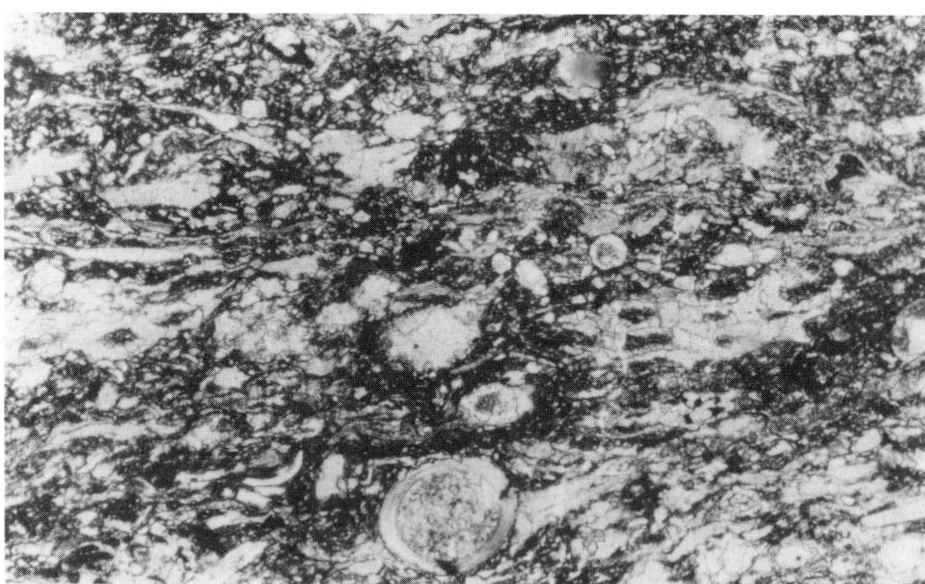

Plate 2 — Algal shelly micrite. Includes fragments of green algae, ostracodes, brachiopods, and echinoderms. Matrix is micrite impregnated with asphalt. Source rock. Permian (P_1^m unit), ×37.

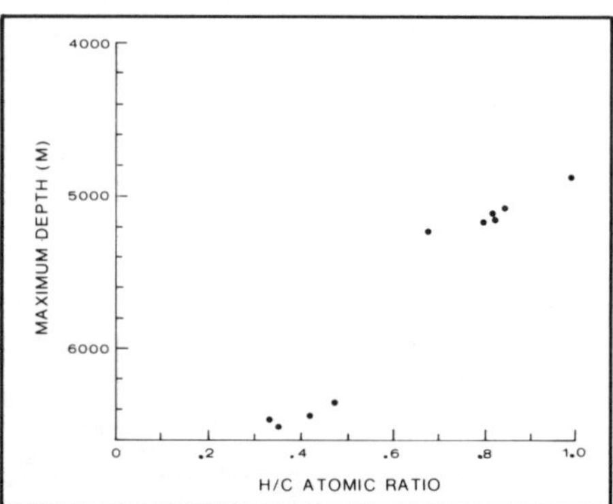

Figure 13 — Changes of H/C atomic ratio of kerogen in carbonate rocks with maximum depth of burial (South Sichuan).

is as follows: Triassic (Tc), 0.010–0.120; Permian ($P_1^2 + P_1^3$), 0.005–0.058; Sinian (Zn_2), 0.001–0.003.

Kerogen and asphalts from the South Sichuan profile also were analyzed, and these data were used to compile a thermal-evolution diagram for kerogen (Figs. 12, 13). By comparing Figure 10 with Figures 12 and 13, indicator values for the depth extinction of commercial oil and gas deposits are proposed (Table 3). These values may be useful in evaluation of oil and gas prospects in areas underlain by carbonate rocks.

Maturation parameters of some representative asphalt samples from carbonate rocks in eastern China are given in Table 4. The organic matter in upper Sinian dolomites from South Sichuan is some of the most thermally mature organic mat-

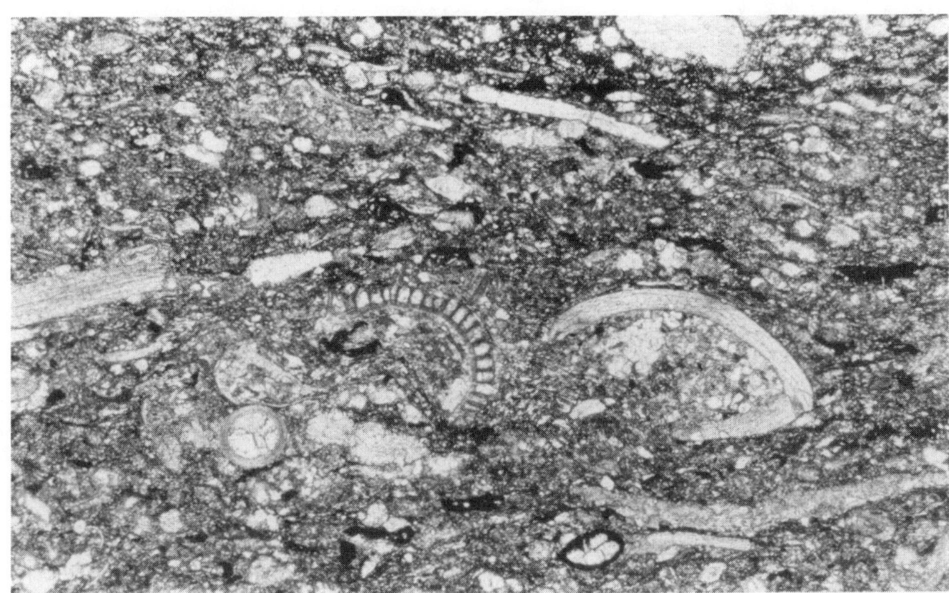

Plate 3 — Algal shelly calcisiltite. Rock contains fragments of ostracodes, brachiopods, algae, and foraminifers. Matrix is micrite. Asphalts occur locally. Permian (P_1^m unit), ×20.

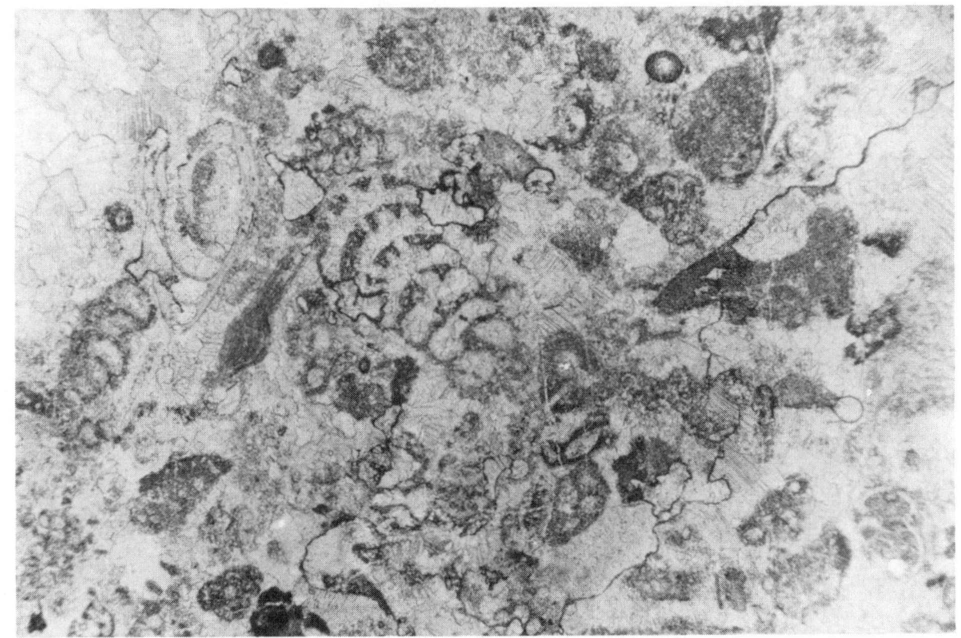

Plate 4 — Foraminifer sparite. Mainly composed of skeletons of foraminifers and fusulinids, with sparite as matrix. Reservoir rock. Permian (P_1^m unit), ×37.

ter in eastern China (Table 4, samples 11, 12). The maximum burial depth of these 650–700-m.y.-old dolomites is nearly 7,000 m (23,000 ft). Quartz ores containing numerous gaseous, liquid, and solid inclusions within the quartz crystals have been discovered in the Sinian dolomites. Trace gaseous hydrocarbons are dominated by methane. Homogenization temperatures in the range of 160°–250°C (320°–482°F) were measured on liquid inclusions and indicate that homogenization formation temperatures had upper limits of 290°–300°C (554°–572°F), and the maximum temperatures exerted on organic matter in the carbonate rocks did not exceed 300°–350°C (572°–662°F). This temperature range is in accordance with the results of thermal-maturation-simulation experiments (Table 4, sample 14; Plate 12, end of paper).

It is interesting to note that one of the most thermally mature samples from eastern China was collected from a large natural-gas field producing from a Sinian dolomite (Table 4, sample 12). This indicates that even in carbonate rocks that have high levels of thermal maturation and in which all the solid organic matter has been metamorphosed to anthraxo-

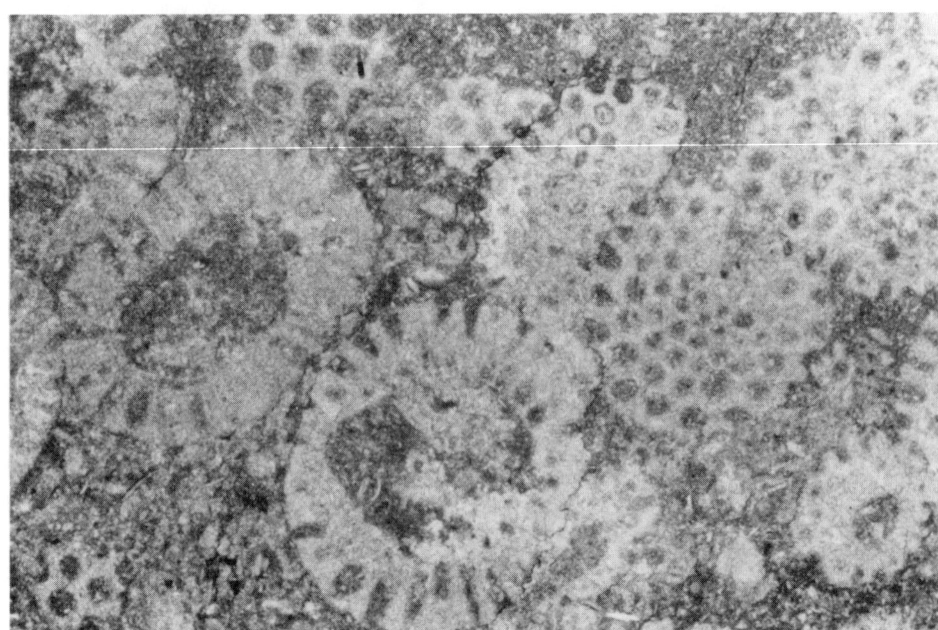

Plate 5 — Green algal micrite. Green algae are mainly fragments of *Mizzia*. Asphalts exist along suture line. Reservoir rock. Permian (P_1^m unit), ×11.

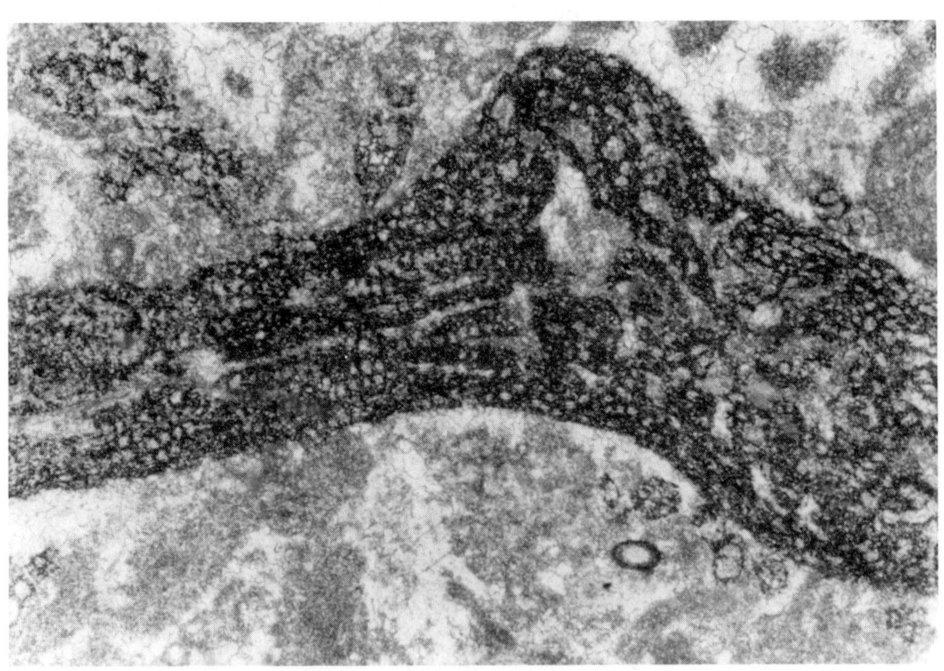

Plate 6 — Red algal sparite. Red algal-cell wall is filled with asphalt. Permian (P_1^m unit), ×16.

lite, large natural-gas fields can still occur. In contrast, organic mater in lower Sinian dolomites (as old as 1.1 b.y.) from the Yan Shan Mountain area is still in an early stage of maturation (Table 4, sample 3). This is probably a result of shallow burial depth caused by crustal movement and discontinuity and low geothermal gradient. These facts indicate that these old dolomites may not only contain many indigenous oil seeps but also might make some contributions to commercial oil deposits.

SUMMARY

1. In eastern China well-developed and thick carbonate sections of Sinian (Precambrian) to Triassic age contain abundant oil seepages, disseminated asphaltic substances, and gaseous and liquid hydrocarbons. Some of these hydrocarbon deposits are believed to have their sources in carbonate rocks.

2. Some medium to large gas fields in the South Sichuan Basin produce a total of 5.8 billion m^3 (205 billion ft^3) of natural

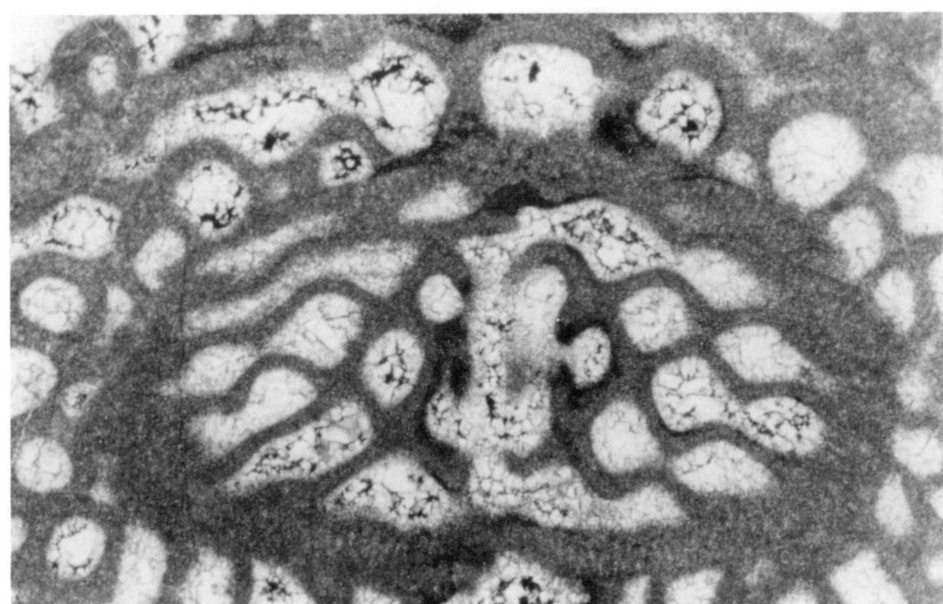

Plate 7 — Asphalts exist in intercrystalline pores of fusulinids. Permian (P_1^m unit), ×37.

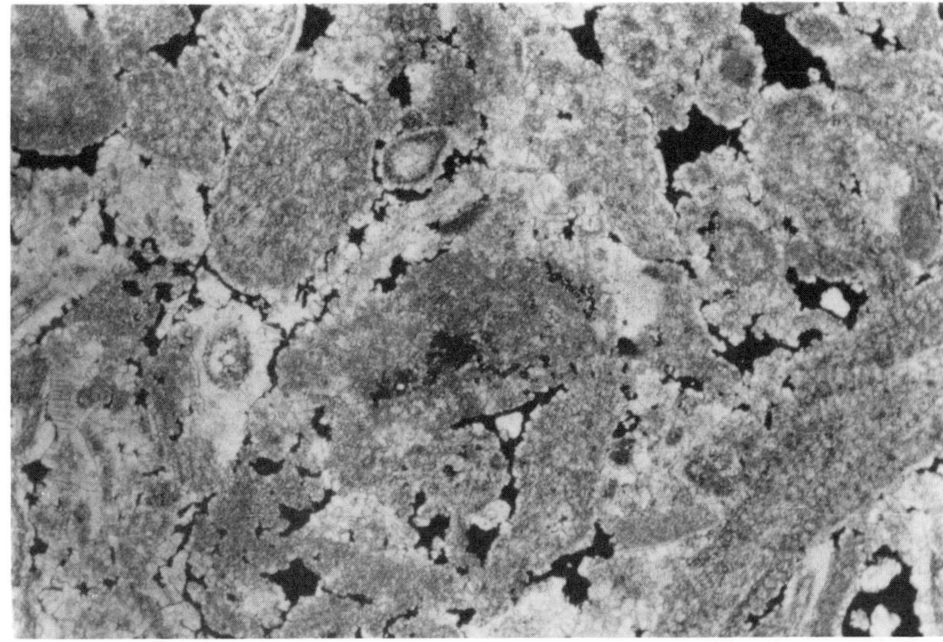

Plate 8 — Asphalts exist in intercrystalline fissures of red algae. Red algal sparite. Permian (P_1^m unit), ×37.

gas annually. One-half of this natural gas is produced from lower Permian (P_1) carbonates. Sedimentological and geochemical lines of evidence suggest that most commercial gas, especially P_1 natural gas, originated from carbonate source rocks.

3. In eastern China, the hydrocarbons in carbonate rocks generally have experienced an extensive thermal history and reached a much higher stage of evolution than carbonates in other parts of China. Geochemical characteristics of organic matter in these carbonates have been discussed. A profile of late-stage evolution of hydrocarbons in carbonate rocks has been set up, whch may be used to evaluate prospective hydrocarbon occurrences in carbonate rocks.

4. One of the most thermally mature sequences of carbonate rocks was found to be upper Sinian dolomites in the South Sichuan Basin. The maximum burial depth of these strata was nearly 7,000 m (23,000 ft), and their maximum temperature was about 300°C (572°F). The remaining solid organic matter has been

Plate 9 — Anthraxolite, plumelike texture. Polished section. Devonian (D_2 unit); crossed nicols, × 200.

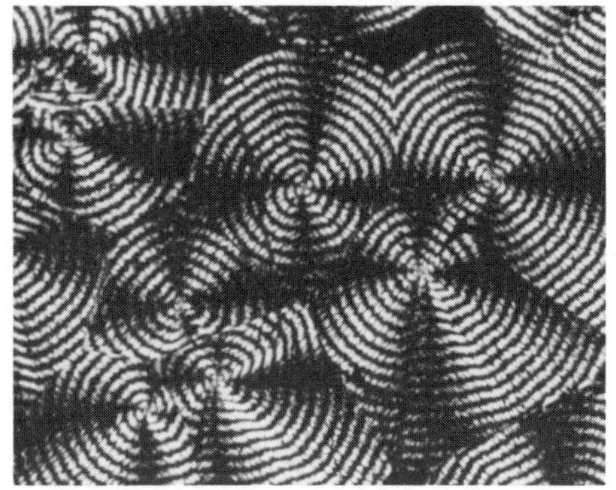

Plate 10 — Anthraxolite, columnar structure, concentric-symmetrical texture. Polished section. Devonian (D_2 unit); crossed nicols, × 50.

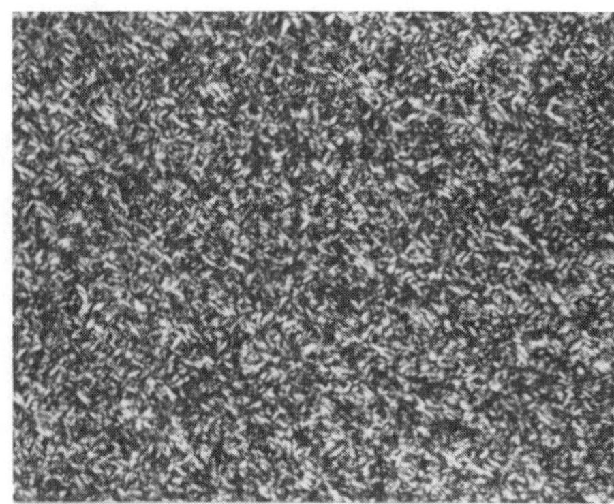

Plate 11 — Anthraxolite, fine mosaic texture. Polished section. Sinian (Zn); crossed nicols, × 500.

metamorphosed to anthraxolite. Even in such thermally mature strata, natural gas (mainly methane) has not been destroyed, and large gas fields are still being discovered.

5. Some of the organic matter found in other Sinian dolomites, such as in the Yan Shan Mountain area of North China, still remains in an early stage of thermal evolution. These dolomites, as old as 1.1 b.y., contain numerous oil seeps and asphalt deposits. The presence of these liquid hydrocarbons is attributed to shallow burial depths caused by crustal movement and discontinuity and to a lower geothermal gradient.

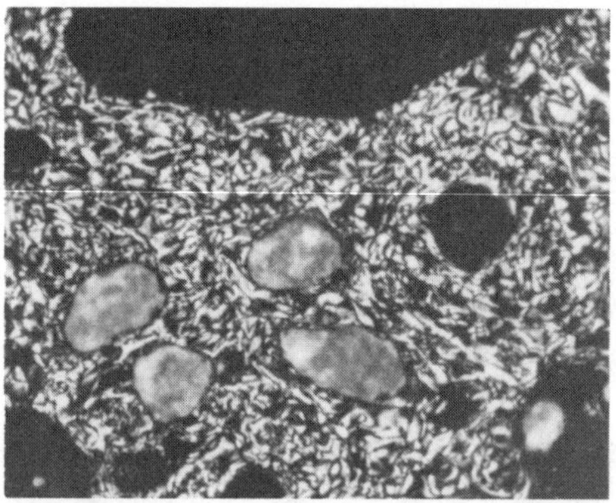

Plate 12 — Asphalt of lower maturity, simulated under 400°C (752°F)/500 atm. for 100 hours. Fine mosaic texture with gas voids. Polished section. Crossed nicols, × 500.

ACKNOWLEDGMENTS

We thank all of our colleagues at the Institute of Geochemistry, Academia Sinica, who provided us with data obtained in their work during recent years. We are grateful also to Dr. J. G. Palacas, U.S. Geological Survey, Denver Federal Center, Denver, Colorado, for inviting two of us to attend a symposium on "Petroleum Geochemistry and Source Rock Potential of Carbonate Rocks," as well as for assistance in preparation of this paper.

REFERENCES CITED

Dai Yong Ding et al, 1978, Petrography of carbonate rocks from the Permian Moukou Formation and their reservoir properties in southern Sichuan Basin: Scientia Geologica Sinica, no. 3, p. 203–219.

Fu Jia Mo and Shi Jiyang, 1977, Theory and practice on petroleum evolution (2)—The practical model of petroleum evolution and its significance: Geochemica, no. 2, p. 87–104.

Hunt, J.M., 1967, The origin of petroleum in carbonate rocks, *in* G.V. Chilingar, H.J. Bissell, and R.W. Fairbridge, eds., Carbonate rocks: New York, Elsevier, p. 225–251.

Mackenzie, A.S., et al, 1980, Molecular parameters of maturation in the Toarcian shales, Paris Basin, France: I. Changes in the configurations of acyclic isoprenoid alkanes, steranes and triterpanes: Geochimica et Cosmochimica Acta, v. 44, p. 1709–1721.

Shen Ping et al, in press, Study of isotope and adsorbed gas hydrocarbon, as well as correlation of oil/source rocks, *in* Ye Lianjun et al, eds., Proceedings of Symposium on Sedimentology and Organic Geochemistry, Beijing, 1979.

Wang Benshan et al, 1975, The role of n-alkanes in petroleum geochemistry: Geochemica, no. 4, p. 258–272.

Wang Tieguan, in press, Indigenous origin of Sinian oil seeps, Yan Shan Mountain area, China, and its significance, *in* Ye Lianjun et al, eds., Proceedings of Symposium on Sedimentology and Organic Geochemistry, Beijing, 1979.

Zhang Lijie and Wang Wenlin, 1978, The relationship between oil–gas and absorbed hydrocarbons in rocks: Geochemica, no. 4, p. 261–269.

Zhang Zhencai, in press, Steranes and terpanes found in oils and source rock, Beijing–Trianjing area, China: Acta Petrolei Sinica.

Zhou Zhongyi and Jia Rong Fen, 1974, Organic geochemical and petrological characteristics of carbonate rocks as a source rock for petroleum: Geochemica, no. 4, p. 278–298.

Hydrocarbon Genesis and Organic Facies in Cambrian Carbonates of the Eastern Officer Basin, South Australia

David M. McKirdy
South Australian Department of Mines and Energy, Eastwood, S.A. 5063, Australia[1]

Agu J. Kantsler
Wollongong University, Wollongong, N.S.W. 2500, Australia[2]

John K. Emmett
Australian Mineral Development Laboratories, Frewville, S.A. 5063, Australia[3]

Alan K. Aldridge
Masspec Analytical, Stroud, Gloucestershire GL5 3JA, England

In mid-1979, a stratigraphic well drilled by the South Australian Department of Mines and Energy (SADME), Byilkaoora–1, encountered shows of aromatic-naphthenic and naphthenic oils while coring Cambrian carbonates in the northeastern Officer basin, South Australia. The oils occupy vugs and partly healed fractures in an alkaline playa-lake sequence of algal-plate dolomite mudstone, dolomite mudstone with Magadi-type chert, and organic-rich dolomitic argillaceous mudstone containing calcite pseudomorphs of sodium carbonate–bicarbonate minerals, notably trona and shortite. Oil–source-rock correlations confirm that the oils originated within those facies drilled. The Byilkaoora oils are thus the first reported examples of *nonmarine* Cambrian petroleum. They are unusually rich in the C_{15}–C_{25} and C_{30} acyclic isoprenoid alkanes, except where severely biodegraded. Sterane and hopane distributions indicate that the oils were expelled from marginally mature source rocks. On the southeastern flank of the Officer basin, micritic carbonates deposited in a *marine*-sabkha environment are marginally mature to mature and have good oil-source potential, although minor staining is the only evidence of oil generation and migration in these rocks.

The primary organic facies of the Officer basin carbonates determines whether they are oil or gas prone. Kerogen in the oil-prone carbonates is either Type I (deposited in an alkaline playa-lacustrine environment) or Type II (deposited in a marine-sabkha or lacustrine environment). Marine lagoonal carbonates contain gas-prone Type III kerogen derived from algal (including cyanobacterial) mucilage. Petrographically, the major components of the oil-prone kerogens are lamellar alginite and bituminite. Likely precursors include the lipids of cyanobacteria (blue-green algae) and various heterotrophic bacteria. The sesterterpanes and squalane present in high concentrations in the Byilkaoora oils and their source rocks may be biological markers of halophilic and/or methanogenic archaebacteria.

INTRODUCTION

The intracratonic Officer basin spans the border between South Australia and Western Australia. The South Australian portion of the basin covers an area of some 100,000 km^2 (38,600 mi^2) and contains sedimentary rocks of Late Proterozoic to Devonian age. The basin is poorly explored, as shown by the sparse well coverage (Fig. 1). However, recent mapping and stratigraphic drilling by the South Australian Department of Mines and Energy (SADME) along the eastern edge of the basin has established the widespread presence of Early(?) Cambrian carbonate rocks (Pitt, Benbow, and Youngs, 1980). These carbonates are assigned to the Observatory Hill Beds (Fig. 2).

The present study is based on data from continuously cored carbonate sections at four well localities: Byilkaoora–1, where substantial shows of oil and insoluble pyrobitumen were discovered (McKirdy and Kantsler, 1980a, 1980b; White and Youngs, 1980); Marla–1A, 1B; Wilkinson–1, in which oil-stained fractures were noted in micritic carbonate (McKirdy and Kantsler, 1980a, 1980b); and Wallira West–1 (Fig. 2). The carbonates were deposited in environments which ranged from nonmarine alkaline playa-lacustrine at Byilkaoora, through marginal-marine sabkha at Wilkinson and Wallira West, to perisabkha grading upward to open-lagoonal marine at Marla (Pitt, Benbow, and Youngs, 1980; White and Youngs, 1980; Youngs, 1980; Fig. 3). Differences in the type and concentration of the organic matter preserved in these contrasting environments are reflected in source-rock quality. This paper reviews the geochemistry and origin of the unusual Cambrian oils at Byilkaoora (McKirdy and Kantsler, 1980a, 1980b) in the light of new data on their isoprenoid alkanes, triterpanes, and steranes, and

[1]Present address: Australian Mineral Development Laboratories, P.O. Box 114, Eastwood, S.A. 5063, Australia
[2]Present address: Shell Development (Australia) Pty., Ltd., G.P.O. Box W2050, Perth, W.A. 6000, Australia
[3]Present address: Esso Australia, Ltd., G.P.O. Box 4047, Sydney, N.S.W. 2001, Australia

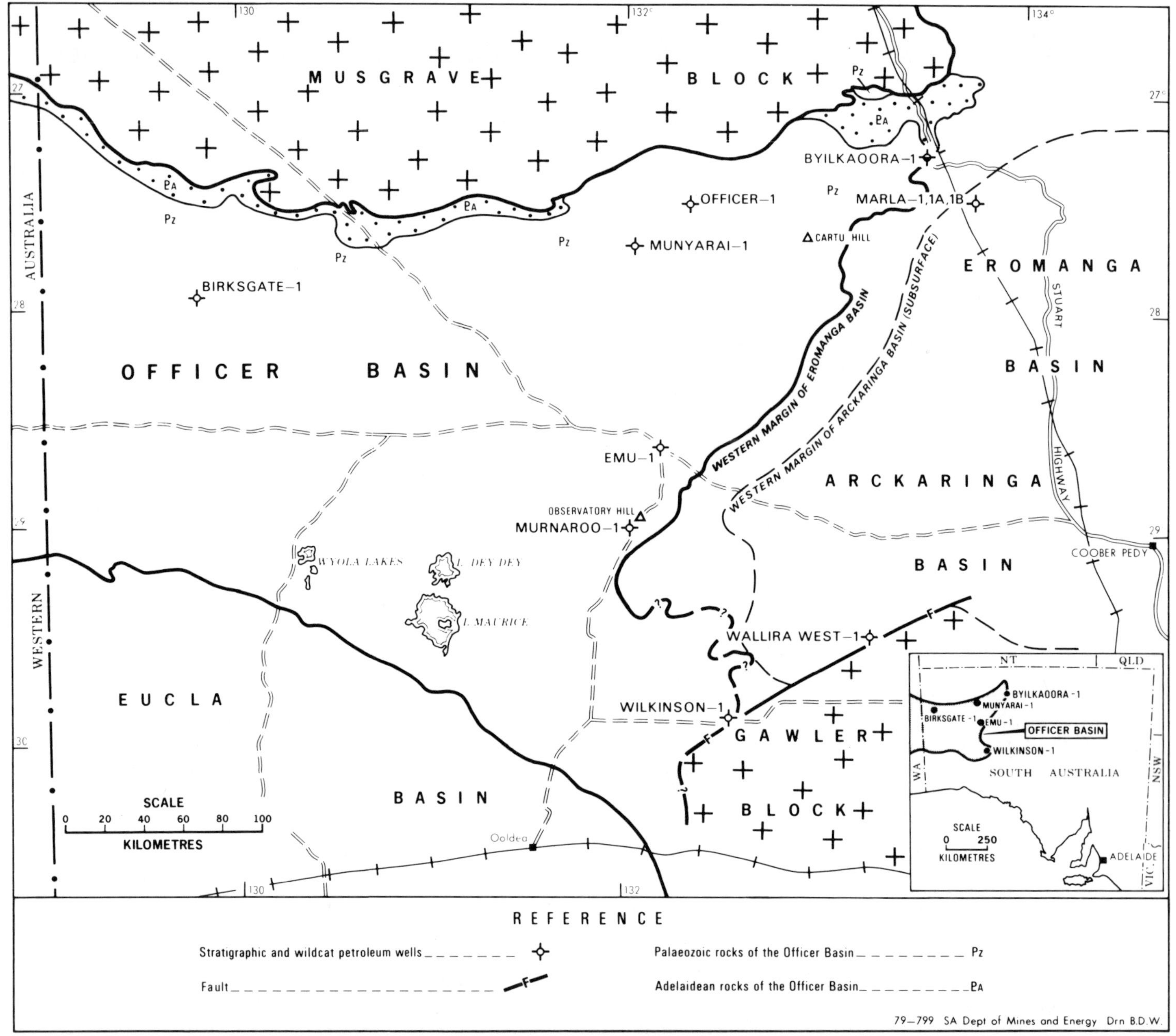

Figure 1 — Map showing outcrop limits of the Officer basin and well locations in western South Australia.

compares the organic geochemistry and organic petrology of the *nonmarine* carbonate source facies with that of the *marine* carbonates at Wilkinson, Wallira West, and Marla.

ANALYTICAL METHODS

A combination of organic-geochemical and microscopic techniques was used to evaluate the hydrocarbon-generating potential and thermal maturity of the Officer basin carbonates. Cambrian sediments, by virtue of their pre–Middle Ordovician age, do not contain woody-herbaceous organic matter (Gray, Massa, and Boucot, 1982). Thus, vitrinite reflectance cannot be used to establish degree of thermal alteration; neither can vascular land-plant precursors be invoked for C_{23+} hydrocarbons and Type III kerogens (Tissot and Welte, 1978).

Sample Preparation

Of the 10 oils from Byilkaoora–1 that were analyzed, only one (from 293.17-m or 961.84-ft depth) was recoverable in sufficient quantity for measurement of API gravity. This sample and one other (from 317.10-m or 1,040.35-ft depth) were scraped from the vugs in which they occurred and transferred to glass vials. Each of the remaining oils was isolated by a 5-minute ultrasonic washing of the core in benzene/methanol (60:40), followed by rotary evaporation of the solvent. The oil then was fractionated and analyzed in much the same manner

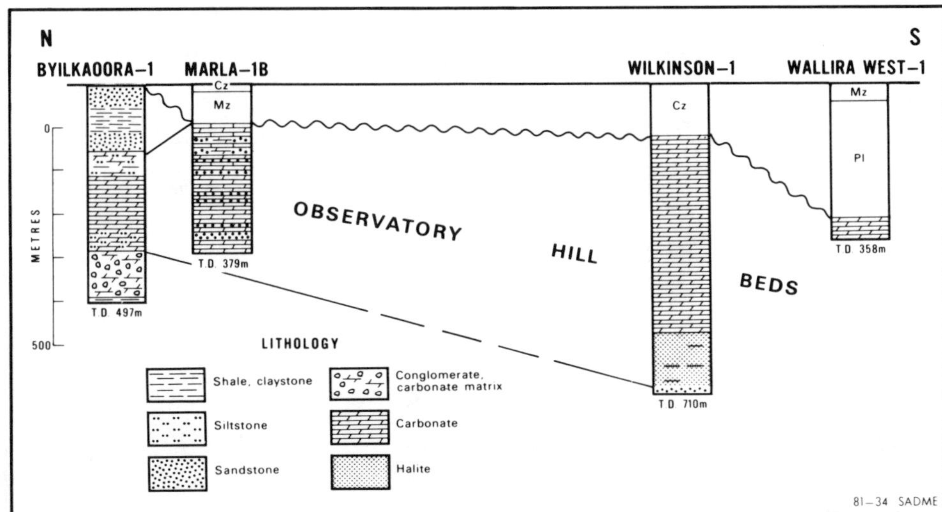

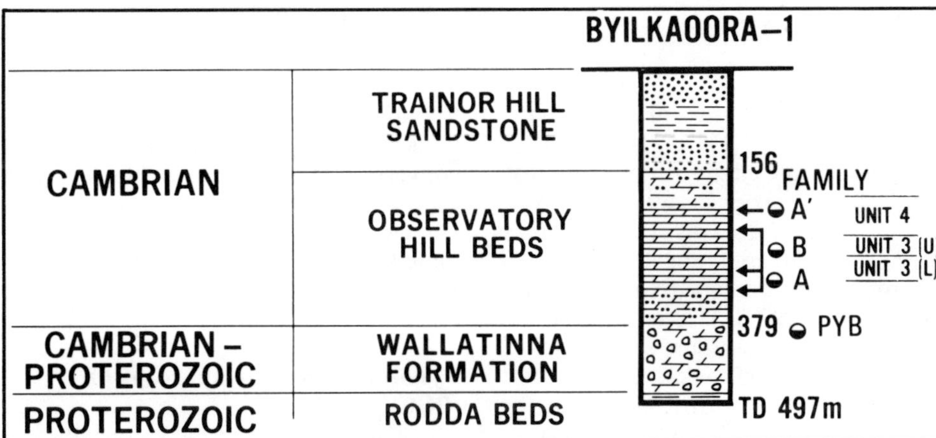

Figure 2 — Correlation of the Observatory Hill Beds in the four SADME stratigraphic wells examined in this study. The summary lithologic log for SADME Byilkaoora–1 illustrates the stratigraphic zonation of oil shows and pyrobitumen (PYB).

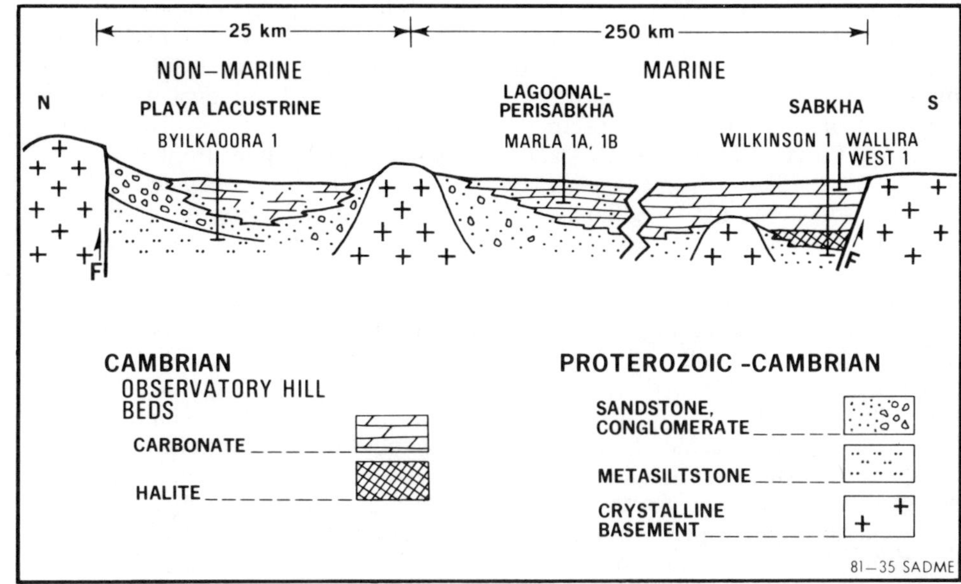

Figure 3 — Diagrammatic section through the eastern Officer basin, showing rock relationships and paleoenvironments toward the end of deposition of the Observatory Hill Beds (modified from Youngs, 1980).

as a rock extract (see below). Lubricants and mud additives used during the drilling of Byilkaoora were screened for hydrocarbons, and deemed acceptable, prior to spudding of the well.

All rock samples analyzed were parts of slabbed or broken core. To ensure removal of any surficial hydrocarbons, each core sample was washed by brief sonication in solvent prior to crushing in a Siebtechnik mill. Cores from Byilkaoora–1 selected for source-rock analysis were free of visible oil staining and lacked obvious vuggy or fracture porosity.

Organic Carbon and C_{15+} Extract

Total organic carbon (TOC) was determined by digestion in dilute HCl to remove carbonate minerals, followed by combustion in a Leco induction furnace and measurement of organic carbon as CO_2. Powdered core (100–150 g) was Soxhlet-extracted for 24 hours with benzene/methanol (60:40). In the case of samples from the Marla wells, the solvent was azeotropic chloroform/methanol (87:13). Evaporation of the solvent yielded the extractable organic matter (EOM, nominally C_{15+}) as a brown gum.

Liquid Chromatography

After precipitation of asphaltenes from the EOM (IP method 143/57), the asphaltene-free extract was fractionated into saturated hydrocarbons (alkanes), aromatic hydrocarbons, and polar (i.e., O-, N-, and S-bearing) compounds by liquid chromatography on Merck Grade 1 neutral alumina (alumina:sample ratio = 300:1), eluting successively with petroleum ether, benzene, and methanol. Where necessary, elemental sulfur was removed from the petroleum ether eluate by treatment with colloidal copper. This chromatographic procedure later was modified, following the discovery of C_{26+} n-alkanes in the aromatic-hydrocarbon fraction of some extracts. Thus, the asphaltene-free Byilkaoora and Marla extracts were chromatographed on a column comprising 80 parts silica gel (Grace Grade 12, 28–200 mesh, activated at 250°C or 482°F for 24 hours) over 20

Table 1

Depth (m)	C_{15+} Composition Sat. (%)	Arom. (%)	ONS (%)	Asph. (%)	$\delta^{13}C_{PDB}$ ‰	Alkane Parameters* Pr/n-C_{17}	Ph/n-C_{18}	Pr/Ph	i-C_{25}/n-C_{22}	Sq/n-C_{26}	Type**
Family A[1]											
219.52	24.0	4.4	71.1	0.5	−30.16	1.21	1.38	1.10	0.57	0.43	A-N
220.00	25.1	2.2	61.7	11.0	n.d.	1.16	1.31	1.04	0.42	0.29	A-N
Family B											
277.00	24.9	7.7	67.1	0.2	n.d.	featureless naphthene hump					A-N
278.55	58.7	3.8	37.0	0.5	n.d.	2.22	2.92	0.89	0.95	0.88	N
278.77	53.4	3.5	42.6	0.5	n.d.	2.28	2.98	0.94	0.98	0.96	N
Family A											
292.94	23.5	4.0	60.1	12.4	n.d.	3.06	4.05	0.82	1.96	3.00	A-N
293.17	23.7	6.4	53.6	16.3	−28.75	2.72	3.35	0.92	1.56	2.23	A-N
314.95	35.6	8.5	40.3	15.6	n.d.	n.d.					A-N
317.10	37.5	7.7	47.9	6.9	−30.64	3.10	3.92	0.82	2.36	1.31	A-N
330.05	25.7	6.7	63.1	4.5	n.d.	n.d.					A-N

*Pr = pristane; Ph = phytane; i-C_{25} = C_{25} regular isoprenoid; Sq = squalane.
**A–N = aromatic-naphthenic; N = naphthenic.
n.d. = not determined.

Table 1 — Geochemistry of oil shows, Observatory Hill Beds, Byilkaoora–1, Officer basin.

parts alumina (adsorbent:sample ratio = 100:1), using petroleum ether, petroleum ether/benzene (85:15), and methanol as the eluting solvents. In the case of the Byilkaoora deasphalted oils, preliminary fractionation into hydrocarbons and nonhydrocarbons was achieved by chromatography on a silica-gel/alumina column (eluting respectively with benzene and methanol) before proceeding to final separation of the saturated and aromatic hydrocarbons as above.

Gas Chromatography (GC)

Alkanes were analyzed by GC, using OV-101 and SP-2100 SCOT glass columns (45 m × 0.5 mm i.d.) and N_2 as carrier gas. Preliminary identification of the C_{15}–C_{25} regular and C_{30} irregular acyclic isoprenoid alkanes was achieved by comparison of retention indices with published values (e.g., Shlyakhov, Koreshkova, and Telkova, 1975) and, in the case of pristane (C_{19}), phytane (C_{20}), and squalane (C_{30}), also by coinjection of standards. In selected samples from Byilkaoora–1, the identity of the isoprenoids was confirmed by mass spectrometry.

Gas Chromatography–Mass Spectrometry (GC–MS)

GC–MS analysis of the alkanes in five oils and one rock extract from Byilkaoora–1 was performed using a Finnigan 3200 quadrupole GCMS interfaced to an INCOS 6100 data system. The glass GC column (20 m × 0.3 mm OV-1 WCOT) was temperature-programmed from 80° to 270°C (176°–518°F) at 4°C (7°F) per minute after a 1-minute hold at ambient. The carrier gas was He (12 psi). The ionizing voltage was 70 eV, and mass spectra (50–450 amu range) were recorded every 2 seconds. In addition to total-ion-current (TIC) traces, the following mass fragmentograms were examined: m/e 85 (acyclic alkanes), m/e 183 (acyclic isoprenoids), m/e 177 (desmethyl hopanes), m/e 191 (hopanes), m/e 205 (3-methyl hopanes), m/e 217, 218 (steranes), m/e 231 (4-methyl steranes), and m/e 259 (rearranged steranes). The structures of the hopanes and steranes were assigned by comparison of their relative retention times with literature data (Seifert and Moldowan, 1979; Mackenzie, 1980) and the use of molecular-ion mass chromatography. Background-subtracted mass spectra of individual isoprenoids were compared with published spectra (Haug and Curry, 1974; Holzer, Oro, and Tornabene, 1979) and the mass spectra of authentic squalane and C_{25} regular isoprenoid standards.

Kerogen Isolation and Analysis

Kerogen was isolated using the procedure of Powell, Cook, and McKirdy (1975) and analyzed for carbon, hydrogen, nitrogen, sulfur, and ash by the Australian Microanalytical Service, Melbourne. Complete removal of pyrite from the kerogen concentrates rarely was achieved. Accordingly, pyritic iron was determined by atomic-absorption spectrophotometry, and appropriate corrections made for pyritic sulfur and the conversion of pyrite to hematite during ashing. Stable-carbon-isotope ratios of the kerogens (and of selected oils) were determined by the Chemistry Department, Western Australian Institute of Technology.

Pyrolysis-Gas Chromatography (PGC)

PGC analysis of selected kerogens, asphaltene fractions, and pyrobitumen was undertaken with a modified Chemical Data Systems Pyroprobe 120 solids pyrolyzer coupled to a Perkin Elmer Sigma 3 gas chromatograph equipped with a flame-ionization detector. The sample (ca.1 mg) was pyrolyzed under He at 700°C (1292°F) for 20 seconds, after preheating for 1 minute at 300°C (572°F) to ensure complete removal of any residual EOM from the kerogen. The pyrolyzate was trapped on the front of the GC column (45 m × 0.5 mm glass SE-30 SCOT) by a liquid-nitrogen-cooled cold block. The column-temperature program was as follows: 35°C (95°F) for 4 minutes; 35° to 260°C (95°–500°F) at 2.5°C (4.5°F) per minute; isothermal at 260°C (500°F) until all peaks eluted. The flow rate of carrier gas through the column was a nominal 2.5 mL per minute. Key compounds in the pyrolyzate have been tentatively identified by comparison of the PGC traces with published examples (e.g., Larter et al, 1979; Van de Meent et al, 1980). In the case of the C_{21}–C_{25} isopre-

Table 2

Sample	Depth (m)	Steranes		Hopanes		
		$C_{29}\frac{\alpha\alpha 20S}{\alpha\alpha 20R}$	C_{27} rearr $\frac{20S}{20R}$	$C_{27}\frac{T_m}{T_s}$	$C_{32}\frac{22S}{22R}$	$C_{29}+C_{30}\frac{\text{Moretanes}}{17\alpha(H)\text{ Hopanes}}$
Family B oil	277.00	>2*	0.90	2.5	1.3	0.24
Family B oil	278.71	1.3	0.91	2.8	1.4	0.25
Stained source rock	289.77	1.3	1.1	3.4	1.6	0.16
Family A oil	293.17	1.1	0.91	3.6	1.4	0.21
Family A oil	314.95	>2*	0.95	2.6	1.3	0.18
Parameter**		1	2	3	4	5

* >50% 20R epimer removed by extensive biodegradation.
**Details of each parameter given elsewhere, as follows: (1) Seifert and Moldowan (1981); (2) Mackenzie et al (1980); (3) Seifert and Moldowan (1978); (4) Seifert and Moldowan (1980); (5) Mackenzie et al (1980).

Table 2 — Biomarker indicators of maturity in four oils and an oil-stained source rock, Byilkaoora–1.

noids (sesterterpanes) and squalane, peak assignments were made solely on the basis of literature retention indices and have yet to be validated by pyrolysis–GCMS.

Organic Petrography

Dispersed organic matter (DOM) was studied in polished sections of core by ordinary and fluorescence-mode reflected-light microscopy as described in McKirdy and Kantsler (1980b).

RESULTS AND DISCUSSION

The geochemistry of a selection of oil shows from Byilkaoora–1 is summarized in Tables 1 and 2. Organic carbon and C_{15+} extract and hydrocarbon data for each of the four wells studied are listed in Table 3. The elemental and carbon-isotopic compositions of kerogens isolated from selected carbonate-rock samples are given in Table 4.

Oils and Pyrobitumen, Byilkaoora–1

The oil shows in Byilkaoora–1 can be grouped into two families, family A and family B, on the basis of their stratigraphic position; the type of porosity they occupy; and aspects of their composition, including their asphaltene and saturated-hydrocarbon contents, and the distribution and relative proportions of paraffins and naphthenes in the saturate fraction (McKirdy and Kantsler, 1980a). Secondary migration of one of the main oil types has resulted in a third, stratigraphically separate type, family A[1]. The stratigraphic zonation of oil types in the Observatory Hill Beds at Byilkaoora–1 is illustrated in Figure 2. Family A and family B oils overlap by at least 5 m (16 ft). Asphaltic pyrobitumen infills fracture and matrix porosity within a 25-m-thick zone of brecciated, bleached, sulfide-bearing conglomerate at the top of the underlying Wallatinna Formation.

Family A Oils

The aromatic-naphthenic oils of family A are black, viscous, heavy crudes. The sample from 293.17-m (961.84-ft) depth (Table 1) has an API gravity of 14.4° and a low sulfur content (0.3%). The alkanes (Fig. 4) are predominantly naphthenic in composition. The prominent bimodal naphthene hump is skewed toward the sterane–triterpane region. The presence of steranes and hopanoid triterpanes was confirmed by GC–MS (Figs. 5, 6). These oils are also unusually rich in acyclic isoprenoid alkanes, including the C_{21}–C_{25} regular sesterterpanes, and squalane, a C_{30} irregular acyclic triterpane (Fig. 7). Individual isoprenoids typically are present in higher concentration than the adjacent n-alkanes (Fig. 4).

Family B Oils

The naphthenic oils of family B have a much lower asphaltene content than do the oils of family A and therefore are less viscous and leave a paler stain on the core. Again, the C_{15}–C_{25} and C_{30} acyclic isoprenoids are strikingly abundant in the saturate fraction (Fig. 4), although their concentration relative to the n-alkanes is somewhat lower than in family A oils. Other contrasts with the oils of family A include a less dominant high-molecular-weight mode in the bimodal naphthene hump and a lower relative abundance of C_{33}–C_{35} hopanes (cf. Figs. 5, 9).

Family A[1] Oils

Family A[1] oils are associated with partly healed fractures in red beds within the upper part of the oil-bearing zone of the Observatory Hill Beds at Byilkaoora (Fig. 2). The color of their stain ranges from pale to dark, depending on asphaltene content. Oils of this group are interpreted to be remobilized aromatic-naphthenic family A oils that underwent secondary migration during and/or after the development of fracture porosity and, in so doing, became depleted in asphaltenes, high-molecular-weight naphthenes, and possibly some acyclic isoprenoids (Fig. 4; McKirdy and Kantsler, 1980a, 1980b).

Pyrobitumen

Asphaltic pyrobitumen occurs in fractured and brecciated conglomerate of the upper Wallatinna Formation. Its reflectance (R_o = 1.25%) and elemental composition (atomic H/C = 0.85, N + S/O = 0.082) are characteristic of epi-impsonite (Jacob, 1975; Hunt, 1978). Although not illustrated in this paper, the carbon-isotopic composition ($\delta^{13}C_{PDB}$ = −30.2 ‰) and PGC trace of the pyrobitumen are almost identical with those of the asphaltene fraction of a family A oil. The bleached character of the host rock and the ubiquitous presence of sulfides (albeit in trace amounts) suggest that this level of the Wallatinna Formation was invaded by sulfur-bearing hydrothermal fluids. The contact of these hot fluids with oil migrating out of the overlying Observatory Hill Beds is believed to have caused the polymerization of ONS compounds and asphaltenes into insoluble pyrobitumen. A similar origin is proposed by Powell (this volume) for epi-impsonite in the Pine Point ore field.

Table 3

Sample Number	Depth (m)	TOC (wt%)	EOM (ppm)	Hydrocarbons	
				mg/g TOC	%EOM
Byilkaoora–1					
5643/RS59	200.30	0.66	195	6	18.5
5643/RS61	211.60	0.18	566	147	46.8
5643/RS63	225.90	0.21	92	8	17.7
5643/RS65	252.75	0.09	152	28	16.7
5643/RS66	262.95	0.56	1329	79	33.5
A2282/79	278.66	0.50	1128	47	20.9
A2283/79	278.86	0.45	508	42	36.9
5643/RS67	279.50	0.32	745	69	29.8
5643/RS68	285.00	0.70	1877	57	21.1
A2284/79	289.77	0.96	1806	83	44.4
—	289.80	1.30			
5643/RS69	296.25	0.38	1304	129	37.5
5643/RS71	306.15	0.64	1043	49	30.3
5643/RS72	312.30	0.33	267	21	26.4
5643/RS73	313.30	0.42	270	12	19.3
—	315.02	0.40			
5643/RS74	316.50	0.48	753	56	35.9
Wilkinson–1					
A6115/78	288.52	0.15	345	53	22.3
A6117/78	314.47	0.37	127	2	4.3
A6118/78	315.05	0.18	152	14	16.8
A6119/78	333.07	0.29	605	71	34.5
A6121/78	344.54	0.56	299	16	30.9
A6122/78	390.08	0.43	799	43	23.5
A6123/78	461.42	0.68	1638	88	37.6
A6124/78	461.87	0.60	1939	110	34.0
A6125/78	462.17	0.63	1576	93	37.3
A6126/78	463.57	0.18	459	65	25.8
A6128/78	481.62	0.23	341	45	30.3
A6130/78	512.07	0.34	660	78	40.0
A6131/78	513.45	0.22	290	23	17.8
A6133/78	538.41	0.16	106	13	19.3
A6134/78	553.37	0.28	406	40	28.3
Wallira West–1					
A6136/78	330.50	0.54	1321	139	56.5
A6137/78	330.97	0.44	1187	157	57.9
A6138/78	343.63	0.49	1250	139	53.9
A6139/78	344.30	0.45	1337	161	54.2
A6140/78	357.15	0.06	60	23	23.3
Marla–1A, 1B					
5643/RS88A*	127.12	4.56	2400	2	2.6
5643/RS88B**	127.12	0.18	54	5	16.7
5643/RS89	131.50	0.57	86	3	16.4
5643/RS90	139.00	0.21	47	7	25.0
5643/RS91	143.85	0.27	86	8	25.9
5643/RS127	148.40	0.26	157	22	37.1
5643/RS128	149.10	0.33	133	15	37.6
5643/RS129	160.90	0.29	137	9	18.3
5643/RS132	223.47	0.08	80	8	7.7
5643/RS133	244.40	0.10	68	7	10.5

*Shale parting from stylolite.
**Host dolomite.

Table 3 — Organic carbon, C_{15+} extract, and hydrocarbon contents of Cambrian carbonates, Observatory Hill Beds, Officer basin.

Source and Maturity

As a group, the Byilkaoora oils are characterized by uniformly low concentrations of aromatic hydrocarbons (<10%; Table 1) and, except where severely biodegraded, by distinctive C_{15+} alkane distributions in which acyclic isoprenoids are dominant (Fig. 4). Acyclic alkane (m/e 85) and isoprenoid alkane (m/e 183) mass fragmentograms (Fig. 7) reveal a series of regular isoprenoids extending up to C_{25}, and squalane, in a family A oil. The appearance of the latter fragmentogram between scans 1000 and 1200 suggests that C_{30+} isoprenoids also may be present (cf. Albaiges, Borbon, and Salagre, 1978; Moldowan and Seifert, 1979). The assigned structures of the C_{25} regular isoprenoid and squalane, a C_{30} irregular isoprenoid, were confirmed by their mass spectra (Fig. 8). This unusual isoprenoid distribution is source-specific.

Acyclic isoprenoids in the range C_{15} to C_{30} have been identified as major neutral

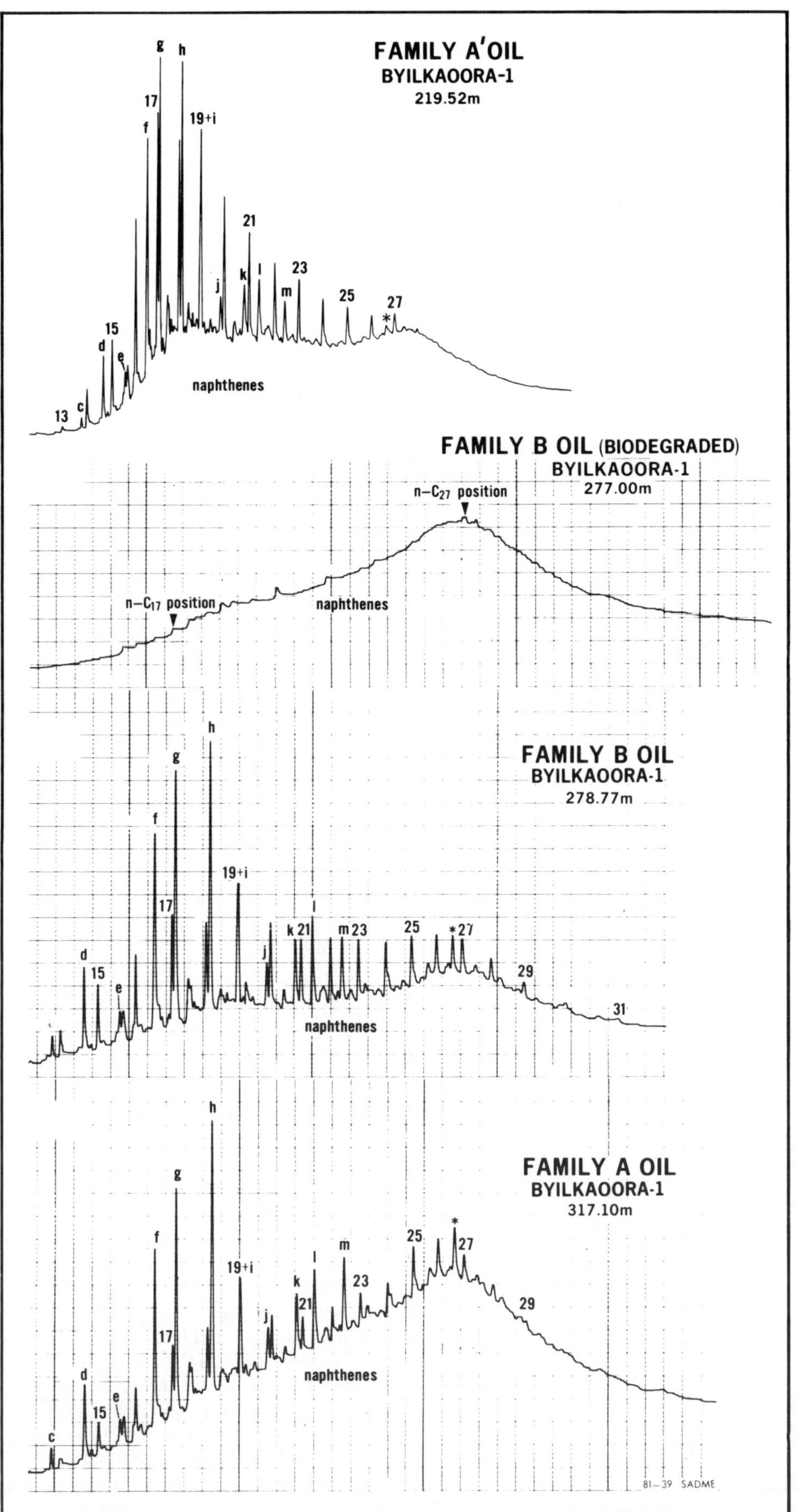

Figure 4 — Gas chromatograms of alkanes from selected oil shows, Observatory Hill Beds, Byilkaoora-1. Numbers refer to carbon number of normal alkanes: a–m are the C_{13}–C_{25} regular acyclic isoprenoids (note: g = pristane, h = phytane); * is squalane.

Table 4

Well	Depth (m)	C	H	N	S	O	Ash (wt%)	H/C	O/C	C/N	Type	$\delta^{13}C_{PDB}$ ‰
		(wt% d.m.m.f.*)						atomic				
Byilkaoora–1												
	200.30	77.75	6.08	2.35	7.8	6.0	28.8	0.93	0.058	39	II	−30.07
	262.95	n.d.					n.d.	0.98	n.d.	42	II	−31.30
	278.66	n.d.					n.d.	1.25	n.d.	59	?	−30.41
	285.00	85.29	9.07	1.42	1.7	2.5	7.6	1.27	0.022	70	I	−29.59
	289.77	86.80	8.70	1.28	1.4	1.8	7.7	1.19	0.015	79	I	−29.00
	296.25	85.40	9.62	1.64	1.6	1.8	3.8	1.34	0.016	61	I	−29.10
	306.15	82.51	8.65	1.30	3.7	3.8	13.4	1.25	0.035	74	I	−28.47
	313.30	n.d.					n.d.	1.25	n.d.	117	I	−29.11
	316.50	n.d.					n.d.	1.24	n.d.	76	I	−29.07
Wilkinson–1												
	314.47	73.56	5.61	2.88	3.7	14.2	16.0	0.91	0.14	30	III	−29.81
	333.07	77.56	6.57	1.71	7.5	6.7	12.5	1.01	0.065	53	II	−29.58
	344.54	79.57	7.10	1.44	3.9	8.0	9.5	1.06	0.075	64	II	−29.10
	390.08	78.71	6.31	2.26	2.8	9.9	10.8	0.96	0.094	41	II	−28.60
	461.42	79.29	7.08	1.79	4.8	7.0	4.0	1.06	0.066	52	II	−31.06
	462.17	78.67	6.36	1.97	4.1	8.9	2.7	0.96	0.085	47	II	−30.70
	481.62	83.26	7.03	1.77	4.1	3.8	26.6	1.01	0.034	55	II	−31.81
	512.07	79.52	6.55	2.17	2.6	9.2	6.7	0.98	0.087	43	II	−30.62
	553.37	80.58	5.84	2.22	3.2	8.2	14.3	0.86	0.076	42	II	−30.23
Wallira West–1												
	330.50	80.68	6.68	1.30	6.1	5.2	8.1	0.99	0.048	72	II	−31.95
	344.30	75.45	6.28	1.70	7.7	8.9	8.4	0.95	0.089	52	II	−30.92
Marla–1A												
	131.50	82.40	4.77	0.91	0.7	11.2	38.1	0.69	0.10	106	III	n.d.
	143.85	82.22	4.63	1.58	0.3	11.3	20.8	0.67	0.10	61	III	n.d.

d.m.m.f. = dry, mineral-matter-free. Oxygen determined by difference.
n.d. = not determined.

Table 4 — Elemental and isotopic composition of kerogens from Cambrian carbonates, Observatory Hill Beds, Officer basin.

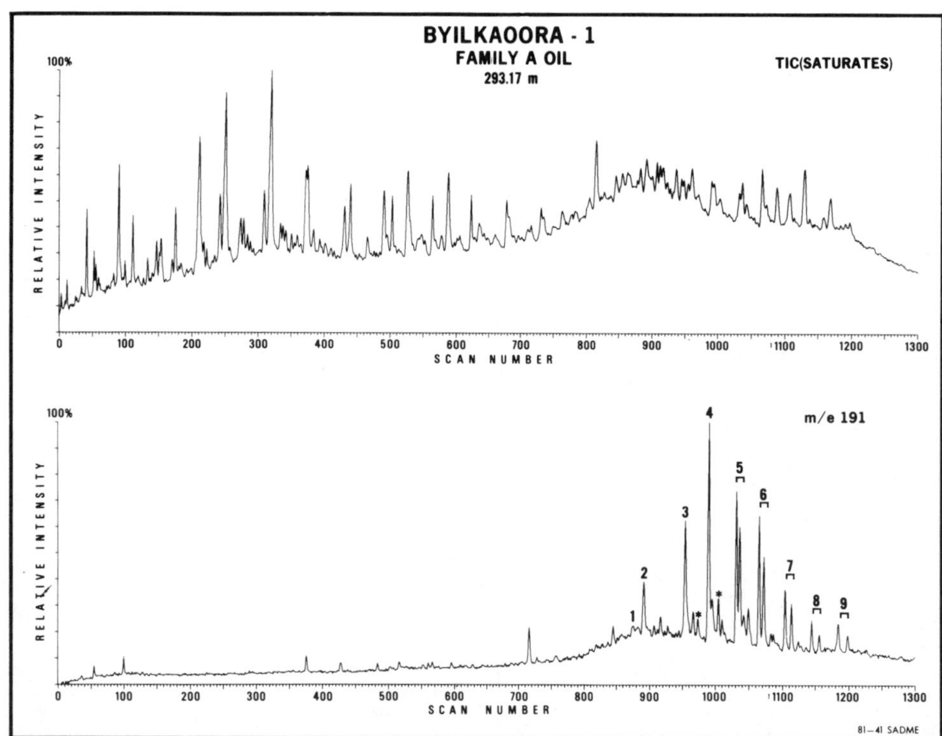

Figure 5 — Saturates total-ion-current (TIC) and terpane (m/e 191) mass fragmentogram of family A oil from Byilkaoora–1. 1 and 2 are T_s and T_m, the C_{27} 18α(H) and 17α(H) trisnorhopanes of Seifert and Moldowan (1978); 3 to 9 are C_{29} through C_{35} 17α(H)21β(H) hopanes; doublets comprise 20S (left) and 20R (right) epimers; * are C_{29} (left) and C_{30} (right) 17β(H)21α(H) hopanes (moretanes).

lipid constituents of methanogenic and thermoacidophilic bacteria (Holzer, Oro, and Tornabene, 1979). In addition, dibiphytanyl (C_{40}) ethers are present in the polar lipid fraction of these bacteria (Tornabene and Langworthy, 1979). Extremely halophilic bacteria contain lipids consisting of phytanyl (C_{20}) diethers, and appreciable concentrations of squalene and related hydrosqualenes (Tornabene, 1978). Of these primitive archaebacteria (Woese, Magrum, and Fox, 1978), both the halophiles and the methanogens are likely precursors of the Byilkaoora oils. Their growth habitat is entirely consistent with the inferred dep-

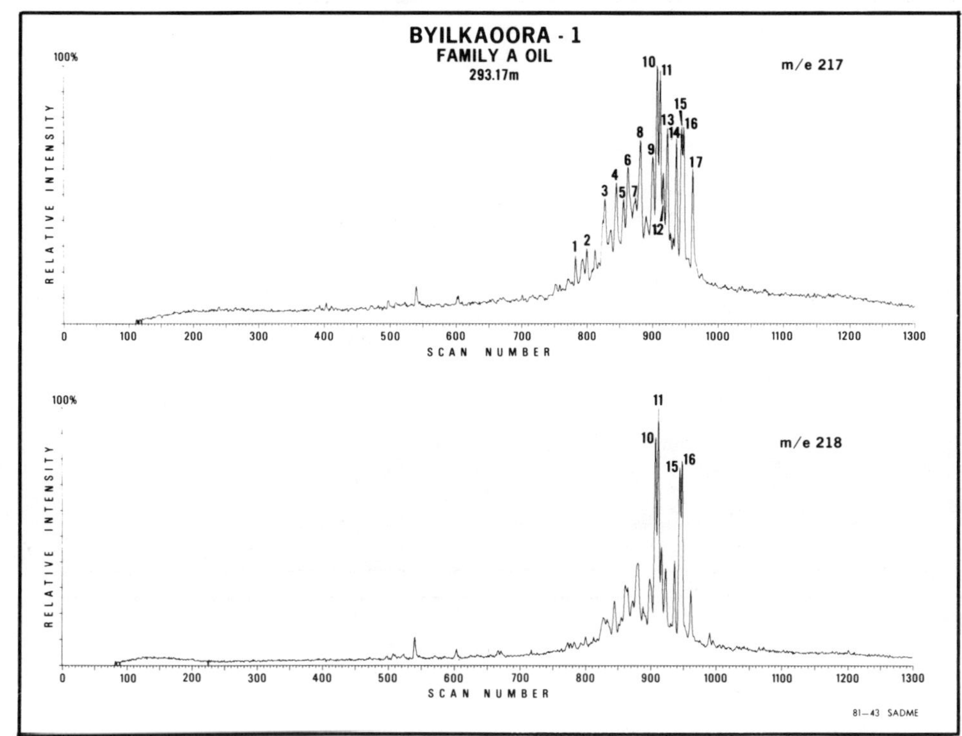

Figure 6 — Sterane (m/e 217, 218) mass fragmentograms of family A oil from Byilkaoora–1. 1 and 2 are C_{27} 20S and 20R diasteranes; 3 and 4 are C_{28} 20S and 20R diasteranes; 6 and 8 are C_{29} 20S and 20R diasteranes; 9 and 13 are C_{28} 14α(H)17α(H) 20S and 20R steranes; 10 and 11 are C_{28} 14β(H)17β(H) 20R and 20S steranes; 14 and 17 are C_{29} 14α(H)17α(H) 20S and 20R steranes; 15 and 16 are C_{29} 14β(H)17β(H) 20R and 20S steranes; 5, 7, and 12 are unidentified steranes.

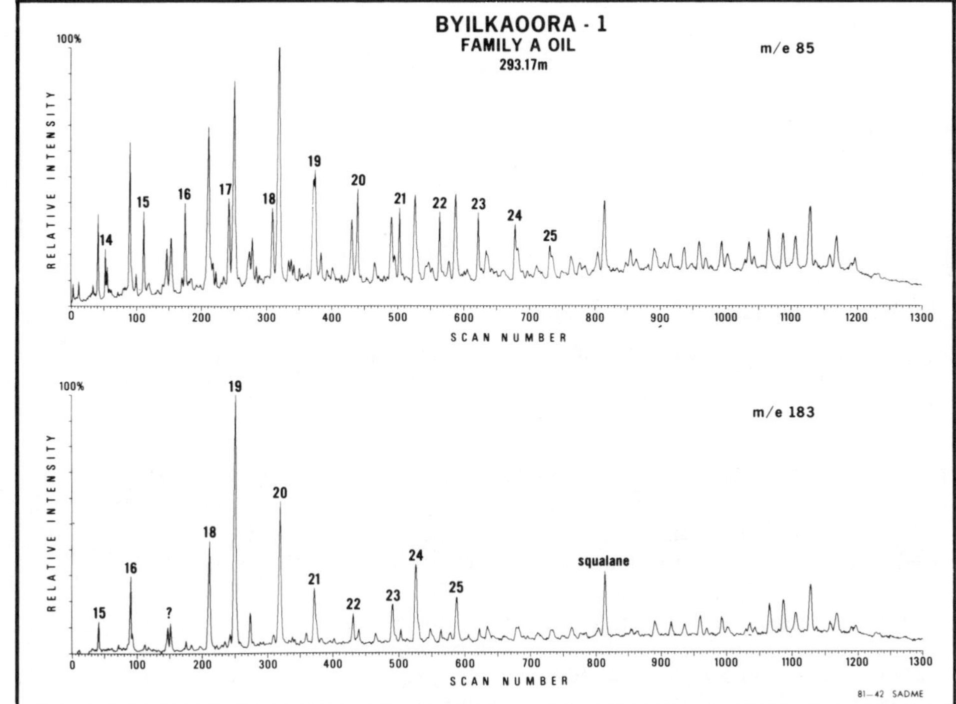

Figure 7 — Acyclic alkane (m/e 85) and isoprenoid alkane (m/e 183) mass fragmentograms of family A oil from Byilkaoora–1. Numbers refer to n-alkane (top) and isoprenoid (bottom) carbon numbers.

ositional environment of the source beds of the oils (see below). The structure of the *irregular* C_{25} isoprenoid alkane, reported to be a specific biological marker of methanogenic bacteria in younger sediments (Brassell et al, 1981; Rowland et al, 1982), differs from that of the *regular* C_{25} isoprenoid found in these Cambrian oils. However, a C_{25} alkene with the same carbon skeleton as the latter compound has recently been isolated from a methanogenic bacterium (S. J. Rowland, personal communication).

A low content of rearranged steranes (cf. Fig. 6 and low signal-to-noise ratio in m/e 259 fragmentogram), a full complement of bacteriohopanes up to and including C_{35} in relatively high concentration (Fig. 5), high values for the ratio of the two C_{27} hopanes ($T_m/T_s >> 1$) (peaks 1 and 2 in Fig. 5), and low pristane/phytane ratios (pr/ph ≤ 1) (Fig. 4) are other features that are consistent with the carbonate habitat (source and reservoir) of the oils (McKirdy, Aldridge, and Ypma, 1983). Their unusual sterane distributions ($C_{28} > C_{29} >> C_{27}$: Fig. 6) are remarkably similar to those of Devonian *lacustrine* sediments from Caithness, Scotland (Hall and Douglas, 1983). Only relatively minor amounts of 4-methyl steranes (C_{28} to ?C_{30}), biomarkers of dinoflagellates and methanotrophic bacteria (Comet et al, 1981), were detected in the family A oil from 293.17-m (961.84-ft) depth.

The maturity of the oils is difficult to assess. High isoprenoid/n-alkane ratios

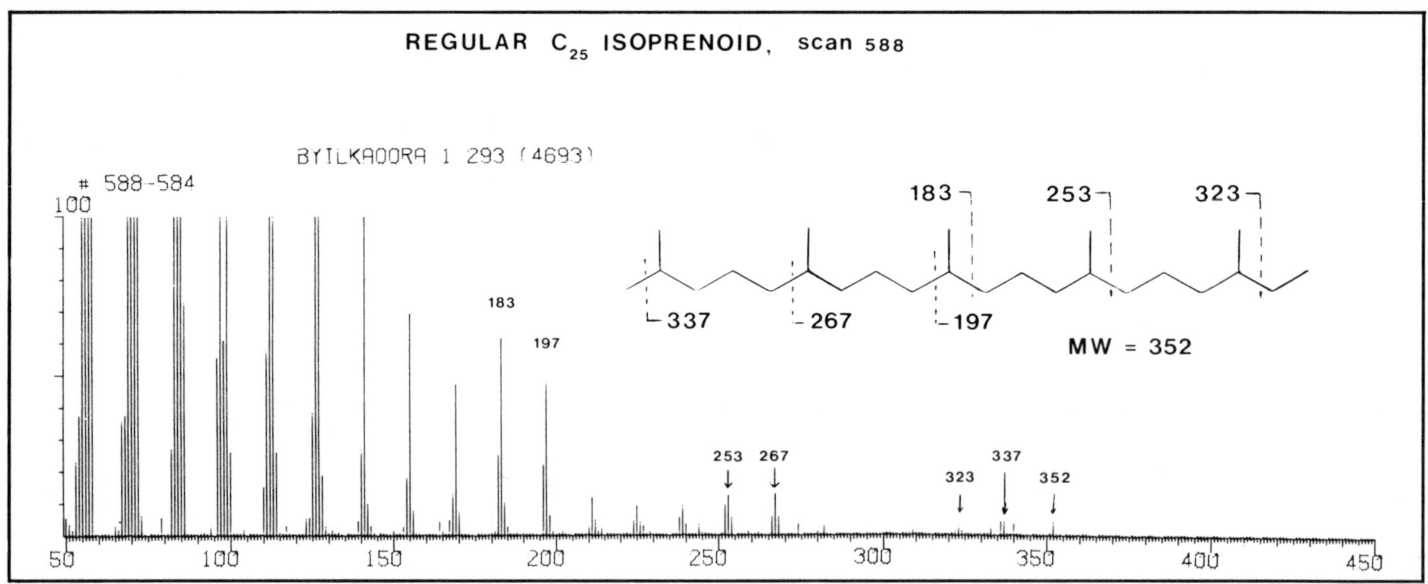

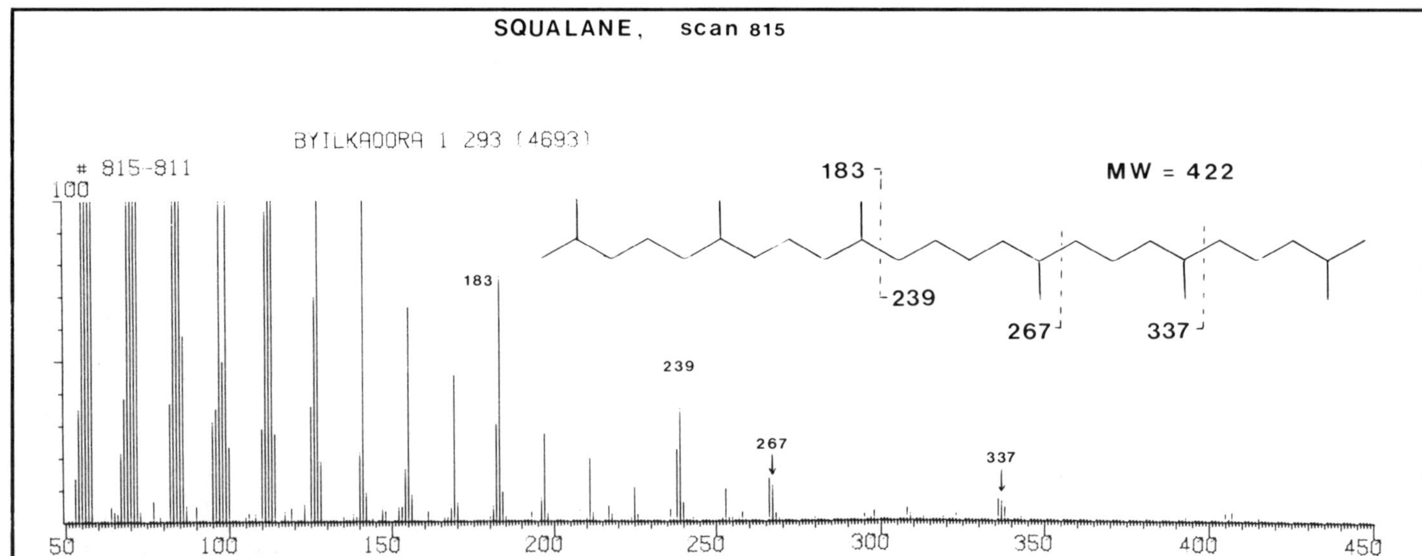

Figure 8 — Mass spectra of the C_{25} regular isoprenoid alkane and squalane (C_{30} tail-to-tail isoprenoid alkane) in family A oil from Byilkaoora-1.

(Table 1) and the distinct naphthene hump in the sterane–triterpane region of the alkane chromatograms (Fig. 4), features commonly associated with immature oils and rock extracts, are in the present case at least partly attributable to biodegradation. Low C_{27} rearranged sterane (diacholestane) 20S/20R values (<1) and high T_m/T_s values (≥ 2.5) (Table 2) suggest immaturity (Seifert and Moldowan, 1978, 1980; Mackenzie et al, 1980), although there is some evidence that these parameters are also *source* dependent. McKirdy, Aldridge, and Ypma (1983) suggested that in carbonate source rocks these two ratios may be less responsive to thermal stress than they are in shales because of the scarcity (or absence) of acidic clays that catalyze the relevant epimerization and isomerization reactions.

Moretane/17α(H) hopane ratios measured on the C_{29} and C_{30} components (0.18–0.21, family A; 0.24–0.25, family B: Table 2) indicate expulsion from marginally mature source rocks (Mackenzie et al, 1980). This observation is consistent with 22S/22R epimer ratios of 1.3–1.4 for the C_{32} hopanes (peak doublet 6 in Fig. 5). The 5α(H)14α(H)17α(H) C_{29} steranes (peaks 14 and 17 in Fig. 6) give 20S/20R epimer ratios (Table 2) that exceed the theoretical equilibrium value (20S/20R = 1) for mature oils. This is attributable to partial removal of the 20R epimer by biodegradation (see below).

Biodegradation

A high proportion of acyclic isoprenoid alkanes in a crude oil may be regarded as a sign of partial biodegradation in the reservoir. However, as is shown below, relatively high concentrations of isoprenoids are present in the source rocks of the oils, both in the EOM and the kerogen pyrolyzate. This makes it unnecessary to invoke substantial post-accumulation biodegradation, although some variable degree of bacterial alteration perhaps would account for the different saturate contents (23–38%: Table 1) of the family A oils.

The effect of moderately severe biodegradation on a family B oil is illustrated in Figure 4. In the case of the oil from 277.0-

m (908.79-ft) depth, complete removal of normal and branched alkanes has left a residual saturate fraction comprising an unresolved complex mixture of cyclic alkanes. In the process, the oil has become heavier and darker, and its overall composition has changed from naphthenic to aromatic-naphthenic. The co-genetic relationship between this oil and a less altered family B oil (from 278.77-m or 914.60-ft depth) is supported by the similarity of their hopane distributions (Fig. 9) and moretane/17α(H) hopane ratios (Table 2). According to Seifert and Moldowan (1979), hopanes are less susceptible to degradation by bacteria than are steranes. Examination of the steranes (m/e 217, 218) in this pair of Byilkaoora oils (McKirdy, Aldridge, and Ypma, 1983, Fig. 5) reveals a novel pattern of biodegradation whereby the 14α(H)17α(H) and 14β(H)17β(H)20R epimers were removed more readily than their 20S counterparts. The degraded oil contains no detectable desmethyl hopanes (m/e 177), suggesting that its degree of bacterial alteration is considerably less severe than the heavily degraded Tertiary oils from California analyzed by Seifert and Moldowan (1979).

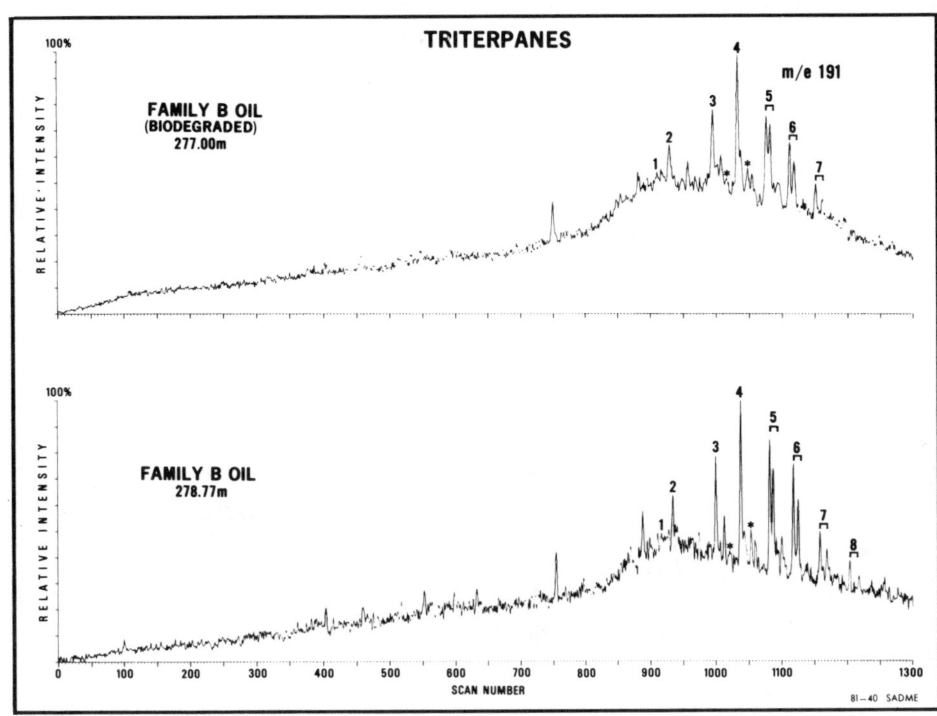

Figure 9 — Terpane (m/e 191) mass fragmentograms of biodegraded (top) and less altered (bottom) family B oils from Byilkaoora–1. See Figure 5 for key.

Nonmarine Carbonates, Byilkaoora–1

The Observatory Hill Beds between 202 m (663 ft) and 322 m (1,056 ft) in depth in Byilkaoora–1 comprise green-gray, pyritic, dolomitic, argillaceous mudstone and algal-laminated dolomite mudstone containing up to 1.3% TOC (McKirdy and Kantsler, 1980a, 1980b; Pitt, Benbow, and Youngs, 1980; Table 3; Fig. 2). The carbonate content of these rocks is highly variable (21–77%, mean 51%; n = 8). The interval has been divided into three stratigraphic units, two of which (lower and upper unit 3) have the characteristics of petroleum source rocks. *Lower unit 3* contains abundant Magadi-type chert and calcite pseudomorphs of the alkaline evaporite minerals, trona and shortite. This sedimentary association is reminiscent of the Wilkins Peak Member of the Green River Formation (Eocene), Wyoming (Eugster and Hardie, 1978). The carbonates of *upper unit 3* contain fewer pseudomorphs of as yet unidentified evaporites and only minor chert. The Early(?) Cambrian carbonate sequence penetrated by Byilkaoora–1 is interpreted as having been deposited in a nonmarine, alkaline

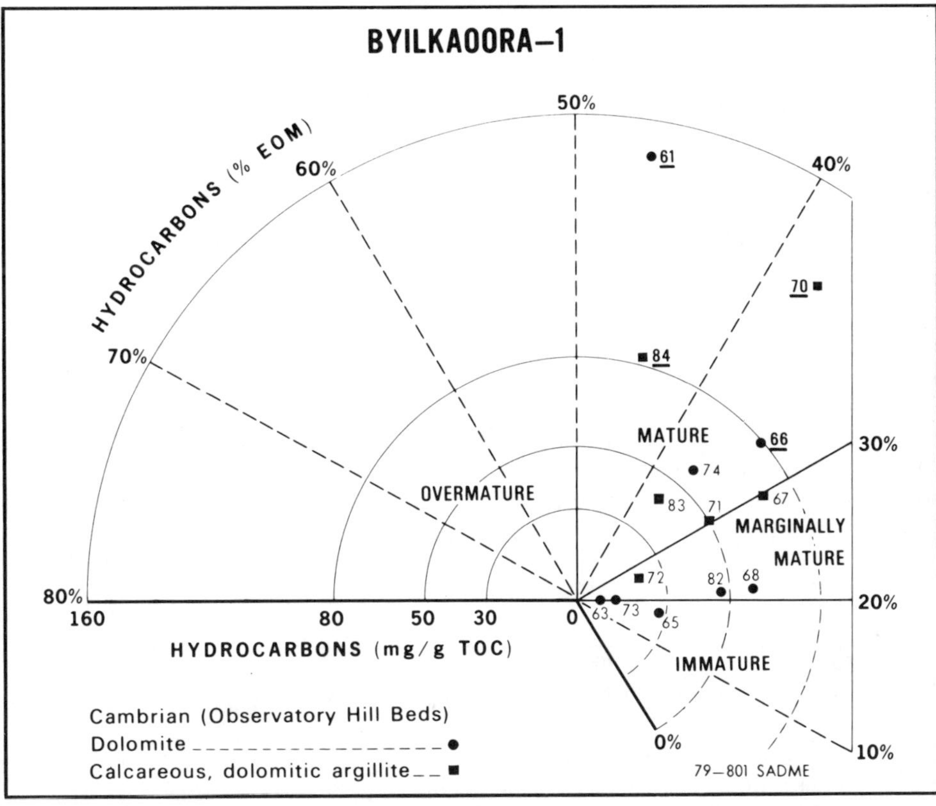

Figure 10 — Maturation state and hydrocarbon potential, Observatory Hill Beds, Byilkaoora–1. Numbers refer to sample numbers listed in Table 3; underlined numbers indicate stained samples. Oil-source-potential scale (mg hydrocarbons/g TOC): <30, none (gas only); 30–50, fair; 50–80, good; >80, excellent (Powell, 1978).

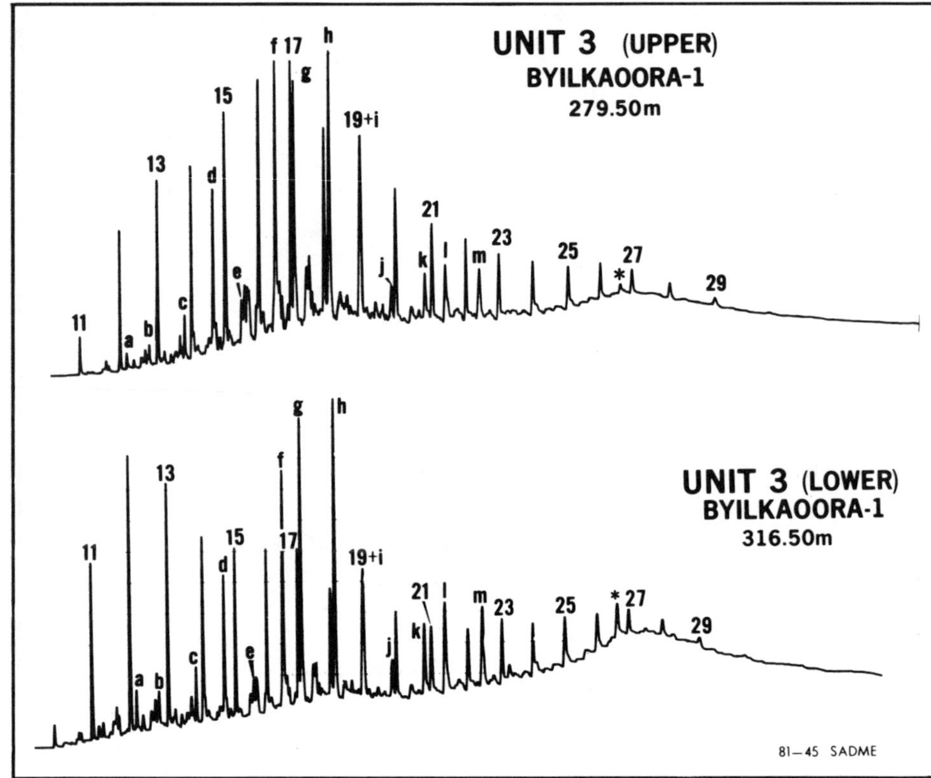

Figure 11 — Gas chromatograms of alkanes from potential oil-source rocks, Observatory Hill Beds, Byilkaoora–1. See Figure 4 for key.

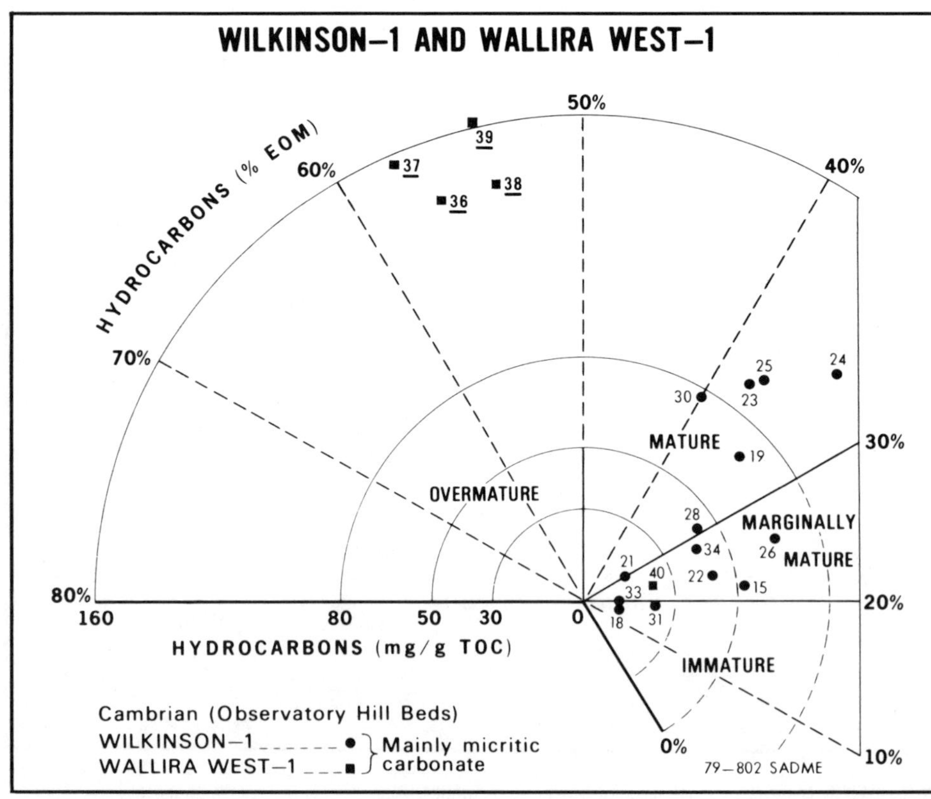

Figure 12 — Maturation state and hydrocarbon potential, Observatory Hill Beds, Wilkinson–1 and Wallira West–1. See Figure 10 for key.

playa-lacustrine environment (White and Youngs, 1980; Fig. 3).

Byilkaoora–1 rock-extract data (Table 3) were plotted on a modified Powell diagram (Fig. 10) to ascertain source-rock quality and maturation state. The considerable spread in data from a relatively narrow depth range (117 m or 384 ft), and hence maturity range, results from migration of hydrocarbons, which has caused depletion or enhancement (through staining) of the indigenous hydrocarbon content. The underlined sample numbers in Figure 10 refer to stained core samples. Most of the analyzed carbonates from unit 3 display fair to good oil-source potential and appear marginally mature to mature. A more definitive assessment of maturity is provided by the elemental composition of the associated kerogen (Fig. 15) and the fluorescence colors of its component lamellar alginite (see below). Both suggest equivalent vitrinite reflectances of $R_o = 0.8 \pm 0.2\%$ (McKirdy and Kantsler, 1980b).

Oil–Source Correlation

Unstained source rocks from *lower unit 3* yield an average of 30 mg saturates per gram TOC. These alkanes (Fig. 11, bottom) have the same relative abundance and carbon-number distribution of isoprenoids as seen in family A oils, and the same skewed naphthene hump in which the high-molecular-weight mode is dominant. Thus, the aromatic-naphthenic oils of *family A* appear to have originated in the evaporite facies of lower unit 3. This correlation is supported by carbon-isotopic evidence (McKirdy, Aldridge, and Ypma, 1983). The kerogen in these sediments is of the Type I variety (Table 4; Figs. 15, 16).

In upper unit 3, unstained carbonates yield an average of 45 mg saturates per gram TOC (i.e., 50% more than the underlying source beds of lower unit 3). The same extended isoprenoid homology and bimodal naphthene distribution are in evidence, but here, as in the family B oils, the isoprenoids occur in somewhat lower concentration relative to the n-alkanes. Also, the low- and high-molecular-weight naphthenes are present in approximately equal proportions (Fig. 11, top). Hence, the naphthenic oils of *family B* may be correlated with carbonates from the upper part of unit 3. The kerogen in this part of the Observatory Hill Beds has a Type II composition

(Table 4; Fig. 15).

The aromatic-naphthenic oils of *family A[1]* probably originated in lower unit 3 and subsequently migrated to their present stratigraphic level after the development of fracture porosity. By contrast, the oils of family A and family B appear to occupy the available porosity (mainly vugs) within their respective source-bed facies, which for the aromatic-naphthenic family A oils is the evaporite-rich lower part of unit 3, and for the naphthenic family B oils is the less markedly evaporitic upper part of the same lacustrine unit.

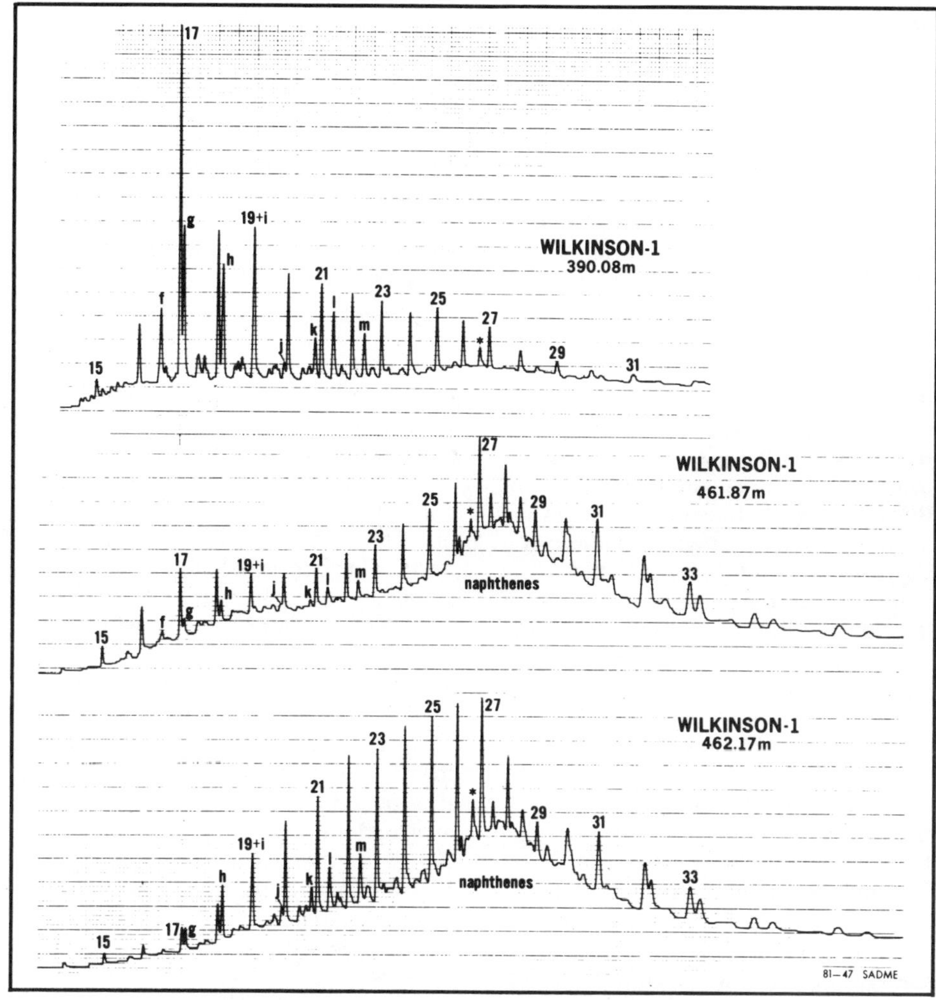

Figure 13 — Gas chromatograms of alkanes from oil-prone micritic carbonates, Observatory Hill Beds, Wilkinson–1. See Figure 4 for key.

Marine Carbonates, Wilkinson–1 and Wallira West–1

The carbonates at Wilkinson–1 and Wallira West–1 (Fig. 2) were deposited in a marginal-marine sabkha environment (Pitt, Benbow, and Youngs, 1980; Fig. 3). They are predominantly fine-grained limestones and dolomites, with disseminated sulfides at several levels and some gypsum veins. At the Wilkinson locality, they overlie thick halite beds. These carbonates contain up to 0.7% TOC, with above-average values (i.e., $> 0.25\%$) being relatively common (Table 3). Their acid-insoluble residue is quite low (8–25%, mean 14%; $n = 15$).

At Wilkinson–1, as at Byilkaoora–1, the carbonates are marginally mature to mature (Figs. 12, 15). Most of the Wilkinson samples show fair to good oil-source potential. The three micritic carbonates that contain >80 mg hydrocarbons per gram TOC are rated as excellent potential oil-source rocks. Although stained, four organic-rich micrites from Wallira West–1 (Table 3) also are considered to be mature, oil-prone, potential source rocks (Figs. 12, 15).

n-Heptadecane is prominent in the alkane distributions of most of these carbonates (Fig. 13, top) and is indicative of a primary cyanobacterial (blue-green-algal) source for the organic matter. Uniformly low pristane/phytane values (<1.5: McKirdy and Kantsler, 1980a, 1980b) suggest that anoxic conditions prevailed during deposition and early diagenesis (Powell and McKirdy, 1973; Didyk et al, 1978). The sesterterpanes and squalane that characterize the *non-marine* Byilkaoora oils and their source rocks are present also in some of the marine sabkha carbonates at Wilkinson and Wallira West but in much lower concentration relative to the n-alkanes.

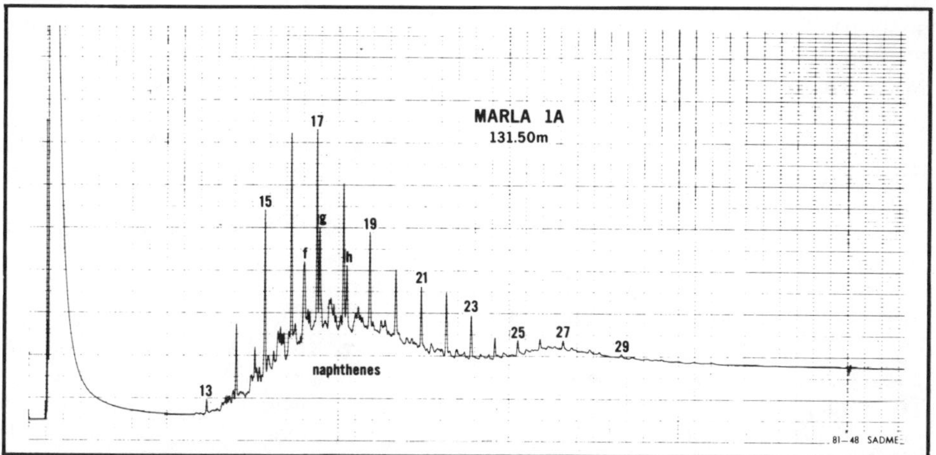

Figure 14 — Gas chromatogram of alkanes from gas-prone micritic carbonate, Observatory Hill Beds, Marla–1A. See Figure 4 for key.

As already discussed, these high-molecular-weight, acyclic isoprenoids are biological markers of the *archaebacteria*, which include the extreme halophiles and the strictly anaerobic methanogens. The evidence suggests that similar bacteria were part of the microbial ecosystem inhabiting hypersaline Cambrian depositional environments—both nonmarine and marine—in the eastern Officer basin. Their organic input to the sediments was particularly high at Byilkaoora.

The Wilkinson–1 carbonates that are rated as excellent potential oil sources (Fig. 13, middle and lower) have a prominent naphthene hump in the sterane–triterpane region and an n-alkane profile that either is bimodal (maxima at n-C_{17} and n-C_{27}) or has a single maximum at n-C_{22} or n-C_{24}. This may reflect intense anoxic reworking (i.e., sapropelization) of the primary algal and/or cyanobacterial debris by anaerobic heterotrophs (Dembicki, Meinschein, and Hattin, 1976), resulting in Type II kerogen and its associated extractable geolipids.

Marine Carbonates, Marla–1A, 1B

At Marla–1A, 1B (Fig. 2), the Observatory Hill Beds consist of a sequence of fine-grained, pale- to dark-gray carbonates comprising interbedded micrites, intraclastic mudstones and packstones, and thin calcareous mudstones (Pitt, Benbow, and Youngs, 1980; Youngs, 1980). Stromatolites occur in the upper part of the section, and evaporites are present in the basal 10 m (33 ft) at Marla–1B. Fluorite and minor pyrite have also been recorded in these carbonates. Organic-rich laminae occur in the mudstones, and organic matter is concentrated in argillaceous partings along subhorizontal stylolites. One such stylolite from 127.12-m (417.06-ft) depth in Marla–1A contains 4.56% TOC, compared with 0.18% TOC for the host dolomite. This carbonate sequence was deposited in a marine environment that ranged from perisabkha (base of sequence) to open lagoonal (Youngs, 1980).

In contrast to the carbonates already discussed, those at Marla have considerably lower TOC contents (mostly <0.3%) and hydrocarbon yields (<30 mg/g TOC) (Table 3) but slightly higher pristane/phytane ratios (0.9–2.3). Conditions during deposition and early diagenesis probably were more oxic than elsewhere. Although their n-alkane profiles (Fig. 14) are indicative of an *algal source*, the Marla carbonates are *gas prone*. The associated kerogen has a humic Type III composition and appears to be of somewhat higher rank than the oil-prone kerogen at Byilkaoora, Wilkinson, and Wallira West (Fig. 15).

Kerogen Composition and Precursors

Kerogen preserved in Early(?) Cambrian carbonates of the Officer basin covers the complete compositional spectrum from lipid-rich Type I to humic-rich Type III organic matter (Table 4). In terms of thermal maturity, all but one of the samples analyzed are well within the oil window (equivalent to vitrinite R_o values of 0.5 to 1.3%: Fig. 15).

Playa-lacustrine, alkaline evaporitic carbonates at Byilkaoora–1 characteristically contain Type I kerogen with high H/C (1.19–1.34) and C/N (61–117) atomic ratios. This oil-prone organic matter ($\delta^{13}C_{PDB} = -28.5$ to $-29.6‰$) is slightly but consistently more enriched in the ^{13}C isotope (by 1–2‰) than is Type II kerogen from evaporite-poor carbonates higher in the sequence (Table 4). The biomarker geochemistry of the Byilkaoora carbonates, in conjunction with their organic petrology (see below) and sedimentology (White and Youngs, 1980), suggests that the principal precursors of their Type I kerogen were the lipids of cyanobacteria and archaebacteria (halophiles, methanogens). A PGC trace of the kerogen (Fig. 16) shows that these lipids included the acyclic isoprenoids of the archaebacteria, as well as microbial n-alkyl moeities up to C_{30}. The abundance of aromatics in the C_6–C_{10} region of the pyrogram agrees with the reported occurrence of alkylbenzenes in extant archaebacteria (Holzer, Oro, and Tornabene, 1979). The pronounced naphthene hump in the sterane–triterpane region is similar to that seen in the family A oils generated from this type of kerogen.

Type II kerogen is dominant in the marine-sabkha carbonates at Wilkinson–1 and Wallira West–1. It has lower atomic H/C ratios (0.86–1.06), and higher S contents (2.8–7.7%), than the Byilkaoora Type I kerogen (Table 4), and a correspondingly lower (but still considerable) oil-source potential. Pyrograms of Type II kerogen (Fig. 17) display lower relative concentrations of both isoprenoids and aromatics and have a less pronounced high-molecular-weight naphthene hump than do those of Type I kerogen. The prominence of n-heptadecane in the Wallira West sample is common to both the kerogen pyrolyzate and the saturate fraction of the extract. Marine photoautotrophs (including cyanobacteria) and heterotrophic bacteria (including sulfate reducers) probably were major sources of the lipid-rich microbial detritus that gave rise to this Type II kerogen. The contribution of archaebacteria was less important than for the Type I organic matter at Byilkaoora. Type II kerogen also occurs at Byilkaoora–1 in lacustrine dolomitic carbonates that lack evidence of appreciable sodium carbonate evaporites (Table 4).

Gas-prone Type III kerogen was isolated from marine-lagoonal carbonates at Marla, and from one sample in the upper Wilkinson carbonate sequence. Its low atomic H/C ratios (0.67–0.69, Table 4) reflect a much more aromatic composition. Alkyl chains in the kerogen structure are shorter and commonly display a marked even or odd carbon-number preference in the C_{12}–C_{20} range (Fig. 18). Mucopolysaccharides comprising the mucilage of benthonic cyanobacteria,

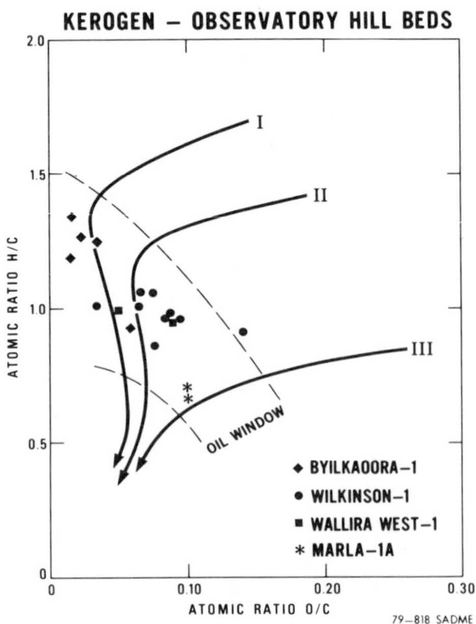

Figure 15 — Modified van Krevelen diagram showing the elemental composition of kerogens isolated from carbonates of the Observatory Hill Beds. Solid lines are the maturation pathways for the principal kerogen types defined by Tissot and Welte (1978). Dashed lines indicate the approximate limits of the oil window (equivalent to vitrinite-reflectance values of 0.5–1.3%).

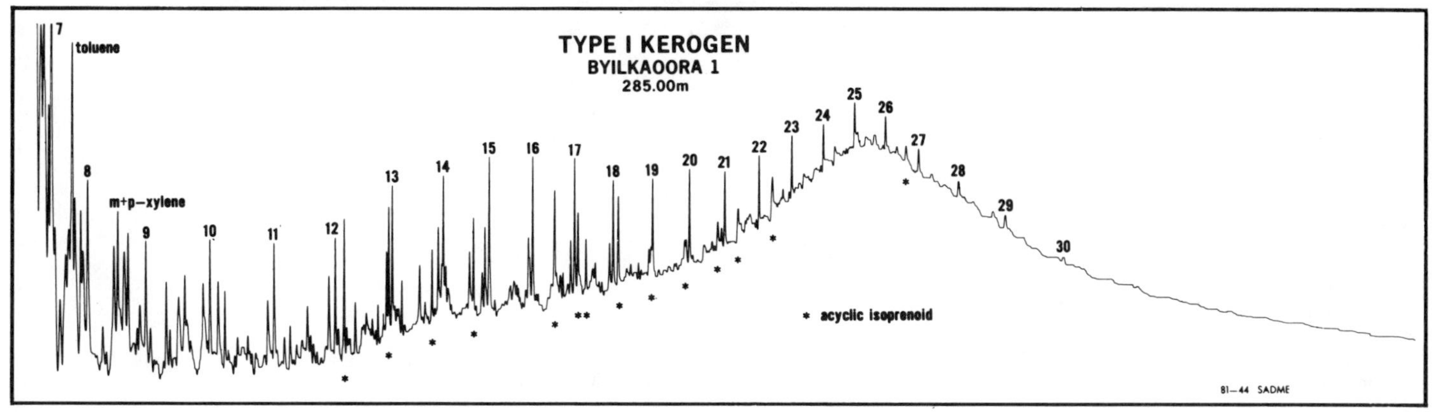

Figure 16 — Pyrolysis–GC trace of Type I kerogen from the Observatory Hill Beds. Numbers refer to carbon number of n-alkene/n-alkane doublets.

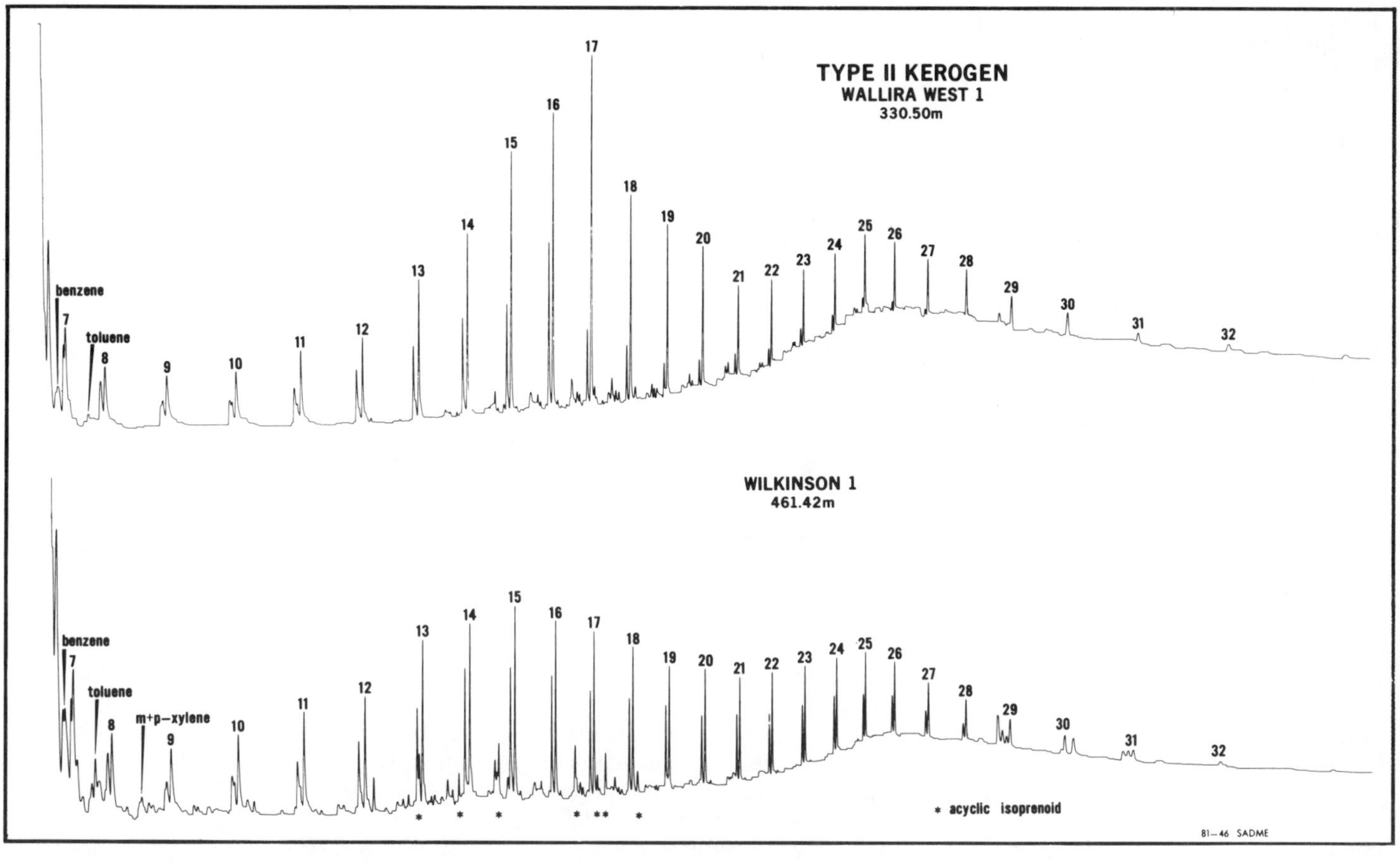

Figure 17 — Pyrolysis–GC traces of Type II kerogens from the Observatory Hill Beds. See Figure 16 for key.

and relatively minor amounts of bacterial lipids, are possible precursors of this microbial kerogen (Shearman and Skipwith, 1965; McKirdy, 1976; McKirdy, McHugh, and Tardif, 1980).

Organic Petrology

A study of the organic petrology of these ancient carbonate rocks was undertaken in parallel with the geochemical work in order to identify individual components of the dispersed organic matter and to obtain an independent check of the thermal maturity of the organic matter by way of fluorescence colors of alginite B and discrete phytoplankton. With the exception of carbonates cored at Marla–1A, 1B, the organic petrology of the Observatory Hill Beds has been reported in detail elsewhere (McKirdy and Kantsler, 1980a, 1980b). Only a brief résumé will be given here.

A varied array of bitumens, ranging in size from <0.01 to 0.2 mm, is present. *Reservoir bitumen* occupies vugs, veins, and cavities between carbonate crystals. Alternatively, in some Byilkaoora core samples, large "cauliflower-shaped" aggregates of bitumen have rounded margins that bear little or no relation to

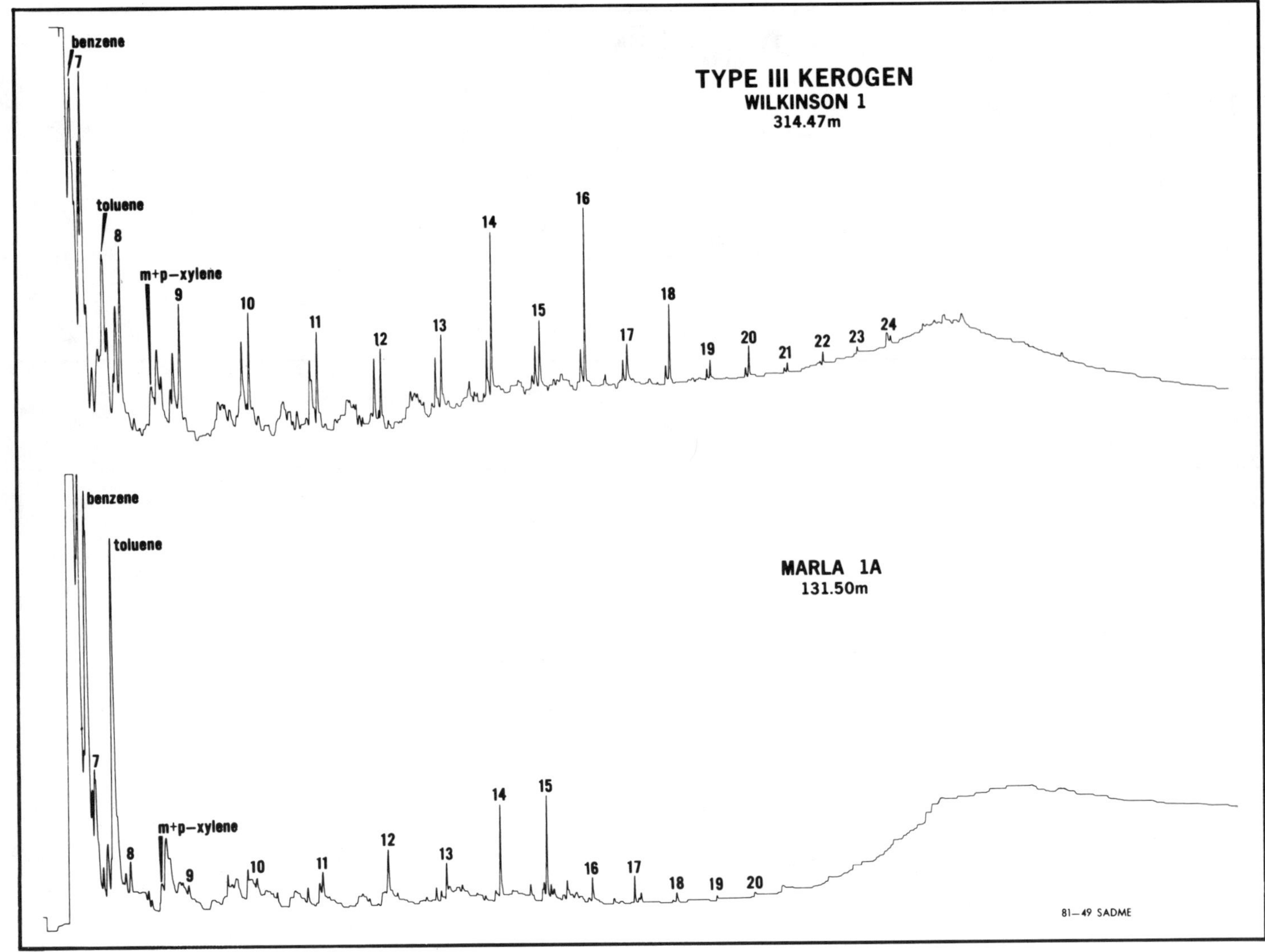

Figure 18 — Pyrolysis–GC traces of Type III kerogens from the Observatory Hill Beds. See Figure 16 for key.

the fabric of the host rock. Their reflectance values (R_o = 0.5–0.8%, but also up to 1.4%), insolubility in immersion oil, lack of fluorescence, and nonporous appearance suggest that these reservoir bitumens may be classified as *grahamite* or *albertite* and, in the case of the higher reflectance varieties, *epi-impsonite* (Jacob, 1975). *Bitumen balls* are spherical aggregates of reflective bitumen (R_o = 0.5–1.1%), some of which have a fluorescent rim. Those with a core of zircon or some other uraniferous mineral are referable to *thucholite*. Possibly because of the diversity of bitumen types represented, bitumen reflectance shows no obvious relationship to other maturation parameters.

Many of the carbonates examined contain an *oil stain* that fluoresces bright green to bright orange. The stain typically occurs in vugs and veins in coarsely crystalline carbonate or in fractures and commonly is seen to originate in lamellar alginite. At Byilkaoora–1, bright-green- or yellow-fluorescing oil stains (?family B) occur mostly in fractures, whereas orange-fluorescing oil stains (?family A or biodegraded family B) occupy vugs. In some oil-filled vugs, very fine sulfide grains are scattered throughout the oil. This intimate association of sulfide and bitumen may be an artifact of *in-situ* biodegradation.

Algal mucilage occurs in short (0.03–0.07 mm) laths of translucent, cream to dark-brown carbonate. The laths are less than 0.01 mm thick and appear to be brecciated laminae of cryptalgal carbonate that is perfused with humic-like organic matter derived from benthic cyanobacteria. Partial replacement of the carbonate by pyrite is common, although in one case the sulfide was identified as sphalerite.

Alginite B is the term introduced by Hutton et al (1980) for the fine lamellar organic matter that constitutes the bulk of the organic content of Tertiary oil shales, such as those of the Green River Formation in Utah, Colorado, and Wyoming. In certain marine oil shales, alginite B is associated with *bituminite* (Sherwood and Cook, 1983). Layers rich in alginite B and/or bituminite are conspicuous in hand specimens of laminated carbonates from Byilkaoora–1 and Wilkinson–1. These layers are 1–5 mm thick and dark gray or brown in color. In polished section under incident light, the organic matter is translucent pale brown or red brown. Its fluorescence colors range from moderate orange to dull brown. Bituminite, the predominant liptinite maceral in the marine carbonates at Wilkinson–1 and Wallira West–1, has a duller fluorescence than alginite B of equivalent rank and typically occurs in thicker laminae or as disseminated amorphous organic-matrix material. The organic matter is finely interwoven with carbonate and argillaceous mineral matter. It tends to follow grain boundaries subparallel with the bedding and commonly forms a web of organic matter typical of an algal mat. Very fine-grained bituminite is recognized by a diffuse yellow fluorescence, suggesting its adsorption onto clays. Both alginite B and bituminite are considered to be the "amorphous sapropel" or structureless organic matter of palynologists and organic geochemists. The association of alginite B with micropyrite in the Byilkaoora carbonates and of bituminite with framboidal pyrite in the marine carbonates is common. Bituminite in these rocks represents photosynthetic algal and cyanobacterial organic matter that has been extensively degraded and resynthesized by anaerobic bacteria.

In the oil-prone Cambrian carbonates at Byilkaoora–1, Wilkinson–1, and Wallira West–1, alginite B, bituminite, and oil stain are the major organic entities, usually making up greater than 99% of the dispersed organic matter. The translucence and fluorescence colors of rare intact acritarchs (phytoplankton) are characteristic of lipid-rich organic matter (kerogen Types I and II) in its initial phase of oil generation. In contrast, at Marla–1A, 1B, algal and/or cyanobacterial mucilage is the dominant variety of dispersed organic matter (Type III kerogen) in association with minor intergranular bitumen and stringers of inertinitic material. Accordingly, these latter carbonates are gas prone.

SUMMARY AND CONCLUSIONS

Detailed study of oil shows and dispersed organic matter in Early(?) Cambrian carbonates from the Observatory Hill Beds in the eastern Officer basin, South Australia, illustrates that pre–Middle Ordovician carbonate rocks may be oil or gas prone, depending on their primary organic facies.

The oil shows in Byilkaoora–1 are the first reported examples of nonmarine Cambrian oils and are among the oldest known crude oils in which steranes, hopanes, and high-molecular-weight (C_{21+}) acyclic isoprenoid alkanes are preserved. Their composition, mode of occurrence, and genetic affinity are similar to certain oils and asphaltic bitumens associated with oil shales and bedded evaporites in the Eocene Green River Formation, Utah (Hunt, Stewart, and Dickey, 1954; Reed and Henderson, 1972). The sesterterpanes and squalane present in high concentrations in the Byilkaoora oils and their carbonate source beds may be biological markers of halophilic and/or methanogenic archaebacteria. Features of the Byilkaoora oils that are common to other oils having carbonate sources include a low pristane/phytane ratio (≤ 1), a low concentration of diasteranes relative to unrearranged steranes, and an anomalously high ratio of the trisnorhopanes, T_m/T_s.

At least three distinct organic facies can be recognized among the various carbonate lithofacies of the Observatory Hill Beds, reflecting significant original differences in microbial input and depositional environment (anoxicity, alkalinity, salinity). The occurrence of oil-prone Type I kerogen in pyritic carbonate mudstones containing early diagenetic calcite pseudomorphs of trona and shortite, and in micritic dolomite with nodular Magadi-type chert, clearly indicates that its precursor organic matter accumulated in a highly alkaline and chemically reducing playa-lacustrine environment. Oil-prone Type II kerogen is associated mainly with micritic carbonates for which a marine-sabkha environment has been inferred, although similar organic matter is found in lacustrine dolomites that lacked abundant penecontemporaneous sodium carbonate–bicarbonate evaporites. Somewhat lower concentrations of gas-prone Type III kerogen are preserved in carbonates that were deposited in a lagoonal to open-marine shelf environment.

Photosynthetic cyanobacteria (blue-green algae) are envisaged as a major progenitor of all three kerogen types. Blooms of lacustrine cyanobacterial, partially reworked during early diagenesis by lipid-rich heterotrophic bacteria, including extreme halophiles and/or anaerobic methanogens, gave rise to Type I kerogen. Cyanobacteria were also part of microbial ecosystems inhabiting the Cambrian marine-sabkha and nonevaporitic lacustrine environments that deposited carbonate sediments in which Type II kerogen was ultimately preserved. In these environments archaebacteria were relatively unimportant, but phytoplankton (as evidenced by acritarchs) and sulfate-reducing bacteria were significant additional sources of sedimentary organic matter. Lamellar alginite and bituminite, respectively, are the dominant petrographic constituents of the two oil-prone kerogen types. Type III kerogen formed where growth conditions favored the production of abundant cyanobacterial mucilage.

NOTE ADDED IN PROOF

Recent geological work in the eastern Officer basin has established that the marine carbonate sequences intersected in Wilkinson–1, Wallira West–1, and Marla–1A, 1B are stratigraphically older than the nonmarine Observatory Hill Beds at Byilkaoora–1. The marine carbonates are now informally assigned to the Lower Cambrian "Wintinna Formation" (Weste et al, 1984).

ACKNOWLEDGMENTS

This paper is published with the permission of the Director-General of Mines and Energy, South Australia, and is a contribution to the IUGS–UNESCO International Geological Correlation Program, Project 157: Early Organic Evolution and Mineral and Energy Resources. Thanks are due Harold Sears (AMDEL) for skilled technical assistance; Steve Rowland (OGU, Bristol University) for coinjection of authentic 2, 6, 10, 14, 18-pentamethyleicosane with the saturate fraction of a crude oil from the Byilkaoora–1 well; the Publications Drafting Section, SADME, for preparation of most of the figures; Conoco, Inc., for access to its word-processing facilities; Jim Palacas, Trevor Powell, and John Zumberge for their helpful critiques of the penultimate draft; and Bev Sellmann for typing the final manuscript. We are especially grateful to Dr. John H. Oehler, Conoco, Inc., who presented the paper on our behalf at the 1980 Organic Geochemistry Division Symposium in Atlanta; and to Graham Pitt and Bridget

Youngs, members of the original SADME Officer Basin Study Group, and Andy White, Comalco, for their interest, encouragement, and advice on matters geological. This work was supported in part by a grant from the National Energy Research Development and Demonstration Council of Australia. Figures 1, 4, 9–13, and 15 (in whole or in part) are reproduced by permission of the Australian Petroleum Exploration Association.

REFERENCES CITED

Albaiges, J., J. Borbon, and P. Salagre, 1978, Identification of a series of C_{25}–C_{40} acyclic isoprenoid hydrocarbons in crude oils: Tetrahedron Letters, no. 6, p. 595–598.

Brassell, S.C., et al, 1981, Specific acyclic isoprenoids as biological markers of methanogenic bacteria in marine sediments: Nature, v. 290, p. 693–696.

Comet, P.A., et al, 1981, Lipids of an Upper Albian limestone, Deep Sea Drilling Project Site 465, Section 465-A-38-3: Initial Reports of Deep Sea Drilling Project, v. 62, p. 923–937.

Dembicki, H., Jr., W.G. Meinschein, and D.E. Hattin, 1976, Possible ecological and environmental significance of the predominance of even-carbon number C_{20}–C_{30} n-alkanes: Geochimica et Cosmochimica Acta, v. 40, p. 203–208.

Didyk, B.M., et al, 1978, Organic geochemical indicators of palaeoenvironmental conditions of sedimentation: Nature, v. 272, p. 216–222.

Eugster, H.P., and L.A. Hardie, 1978, Saline lakes, *in* A. Lerman, ed., Lakes—chemistry, geology and physics: New York, Springer–Verlag, p. 237–293.

Gray, J., D. Massa, and A.J. Boucot, 1982, Caradocian land plant microfossils from Libya: Geology, v. 10, p. 197–201.

Hall, P.B., and A.G. Douglas, 1983, The distribution of cyclic alkanes in two lacustrine deposits, *in* M. Bjoroy et al, eds., Advances in organic geochemistry 1981: Chichester, John Wiley, p. 576–587.

Haug, P., and D.J. Curry, 1974, Isoprenoids in a Costa Rican seep oil: Geochimica et Cosmochimica Acta, v. 38, p. 601–610.

Holzer, G., J. Oro, and T.G. Tornabene, 1979, Gas chromatographic–mass spectrometric analysis of neutral lipids from methanogenic and thermoacidophilic bacteria: Journal of Chromatography, v. 186, p. 795–809.

Hunt, J.M., 1978, Characterization of bitumens and coals: AAPG Bulletin, v. 62, p. 301–303.

——, F. Stewart, and P.A. Dickey, 1954, Origin of hydrocarbons of Uinta basin, Utah: AAPG Bulletin, v. 38, p. 1671–1698.

Hutton, A.C., et al, 1980, Organic matter in oil shales: Australian Petroleum Exploration Association Journal, v. 20, no. 1, p. 44–67.

Jacob, H., 1975, Mikroskop-photometrische Analyse natürlicher fester Erdölbitumina, *in* B. Alpern, ed., Pétrographie de la matière organique des sédiments, relations avec la paleotempérature et le potentiel pétrolier: Paris, Éditions du Centre National de la Recherche Scientifique, p. 103–113.

Larter, S.R., et al, 1979, Occurrence and significance of prist-l-ene in kerogen pyrolysates: Nature, v. 279, p. 405–408.

Mackenzie, A.S., 1980, Applications of biological marker compounds to subsurface geological processes: Bristol, England, University of Bristol unpublished Ph.D. thesis.

——, et al, 1980, Molecular parameters of maturation in the Toarcian shales, Paris Basin, France: I. Changes in the configurations of acyclic isoprenoid alkanes, steranes and triterpanes: Geochimica et Cosmochimica Acta, v. 44, p. 1709–1721.

McKirdy, D.M., 1976, Biochemical markers in stromatolites, *in* M.R. Walter, ed., Stromatolites: Amsterdam, Elsevier, p. 163–191.

—— and A.J. Kantsler, 1980a, Oil geochemistry and potential source rocks of the Officer Basin, South Australia: Australian Petroleum Exploration Association Journal, v. 20, no. 1, p. 68–86.

—— and ——, 1980b, Oil geochemistry and assessment of hydrocarbon-source potential of Cambrian and Devonian rocks, Officer Basin, South Australia: South Australian Department of Mines and Energy and Report Book 80/10 (unpublished), 104 p.

McKirdy, D.M., D.J. McHugh, and J.W. Tardif, 1980, Comparative analysis of stromatolitic and other microbial kerogens by pyrolysis–hydrogenation–gas chromatography (PHGC), *in* P.A. Trudinger, M.R. Walter, and B.J. Ralph, eds., Biogeochemistry of ancient and modern environments: Berlin, Springer–Verlag, p. 187–200.

McKirdy, D.M., A.K. Aldridge, and P.J.M. Ypma, 1983, A geochemical comparison of some crude oils from pre-Ordovician carbonate rocks, *in* M. Bjorøy et al, eds., Advances in organic geochemistry 1981: Chichester, John Wiley, p. 99–107.

Moldowan, J.M., and W.K. Seifert, 1979, Head-to-head linked isoprenoid hydrocarbons in petroleum: Science, v. 204, p. 169–171.

Pitt, G.M., M.C. Benbow, and B.C. Youngs, 1980, A review of recent geological work in the Officer Basin, South Australia: Australian Petroleum Exploration Association Journal, v. 80, no. 1, p. 209–220.

Powell, T.G., 1978, An assessment of the hydrocarbon source rock potential of the Canadian Arctic Islands: Geological Survey of Canada Paper 78–12, 82 p.

—— and D.M. McKirdy, 1973, Relationship between ratio of pristane to phytane, crude oil composition and geological environment in Australia: Nature, Physical Science, v. 243, p. 37–39.

Powell, T.G., P.J. Cook, and D.M. McKirdy, 1975, Organic geochemistry of phosphorites—relevance to petroleum genesis: AAPG Bulletin, v. 59, p. 618–632.

Reed, W.E., and W. Henderson, 1972, Proposed stratigraphic controls on the composition of crude oils reservoired in the Green River Formation, Uinta Basin, Utah, *in* H.R. von Gaertner and H. Wehner, eds., Advances in organic geochemistry 1971: Oxford, Pergamon Press, p. 499–515.

Rowland, S.J., et al, 1982, Confirmation of 2, 6, 10, 15, 19-pentamethyleicosane in methanogenic bacteria and sediments: Tetrahedron Letters, v. 23, no. 1, p. 101–104.

Seifert, W.K., and J.M. Moldowan, 1978, Applications of steranes, terpanes and monoaromatics to the maturation, migration and source of crude oils: Geochimica et Cosmochimica Acta, v. 42, p. 77–95.

—— and ——, 1979, The effect of biodegradation on steranes and terpanes in crude oils: Geochimica et

Cosmochimica Acta, v. 43, p. 111–126.

——— and ———, 1980, The effect of thermal stress on source-rock quality as measured by hopane stereochemistry, *in* A.G. Douglas and J.R. Maxwell, eds., Advances in organic geochemistry 1979: Oxford, Pergamon Press, p. 229–237.

——— and ———, 1981, Paleoreconstruction by biological markers: Geochimica et Cosmochimica Acta, v. 45, p. 783–794.

Shearman, D.J., and P.A.d'E. Skipwith, 1965, Organic matter in recent and ancient limestones, and its role in their diagenesis: Nature, v. 208, p. 1310–1311.

Sherwood, N.R., and A.C. Cook, 1983, Petrology of organic matter in the Toolebuc Formation oil shales: Proceedings of First Australian Workshop on Oil Shale, p. 35–37.

Shlyakhov, A.F., R.I. Koreshkova, and M.S. Telkova, 1975, Gas chromatography of isoprenoid alkanes: Journal of Chromatography, v. 104, p. 337–349.

Tissot, B., and D.H. Welte, 1978, Petroleum formation and occurrence: Berlin, Springer–Verlag, 538 p.

Tornabene, T.G., 1978, Non-aerated cultivation of *Halobacterium cutirubrum* and its effect on cellular squalenes: Journal of Molecular Evolution, v. 11, p. 253–257.

——— and T.A. Langworthy, 1979, Diphytanyl and dibiphytanyl glycerol ether lipids of methanogenic archaebacteria: Science, v. 203, p. 51–53.

Van de Meent, D., et al, 1980, Pyrolysis–high resolution gas chromatography and pyrolysis gas chromatography–mass spectrometry of kerogens and kerogen precursors: Geochimica et Cosmochimica Acta, v. 44, p. 999–1013.

Weste, G., et al, 1984, Cambrian palaeoenvironments and source rocks of the eastern Officer Basin: Geological Society of Australia Abstract Series, no. 12, p. 542–544.

White, A.H., and B.C. Youngs, 1980, Cambrian alkali playa–lacustrine sequence in the northeastern Officer Basin, South Australia: Journal of Sedimentary Petrology, v. 50, p. 1279–1286.

Woese, C.R., L.J. Magrum, and G.E. Fox, 1978, Archaebacteria: Journal of Molecular Evolution, v. 11, p. 245–252.

Youngs, B.C., 1980, Lithologies and interpretations of the Observatory Hill Beds, Marla-1A and 1B: South Australian Department of Mines and Energy, Report Book 80/79 (unpublished), 82 p.

Oils and Source Rocks of Niagaran Reefs (Silurian) in the Michigan Basin

W. C. Gardner[1] and E. E. Bray[2]
Mobil Research & Development Corporation, Dallas, Texas

Potential source rocks were selected from some 300 Silurian core samples by measurements of total organic carbon. Characterization of 60 Michigan basin oils by use of infrared-absorption spectra and carbon-isotope data indicates that the group of oils produced from Niagaran reefs are unique and distinct from other oils in the Michigan basin. Correlation of the oils with potential source-rock kerogen by use of carbon-isotope data indicates that carbonaceous, laminated, interreef Salina A-1 carbonates are the principal source rocks, with the upper algal or "Brown Niagaran" facies of the reef itself constituting a subordinate source. Both facies were deposited in a mesosaline environment, i.e., in waters more saline than normal marine but insufficiently concentrated to precipitate calcium sulfate.

INTRODUCTION

The Michigan basin is a Paleozoic intracratonic basin, with virtually all petroleum production having been derived from carbonate rocks. Substantial gas production from Mississippian sandstones during the 1940's is a principal exception. Niagaran-reef production occurs in an inner pinnacle-reef belt of Silurian age. Niagaran reefs have been known in the subsurface of southwestern Ontario since the early 1900's, when production was obtained by shallow drilling. Three areas of principal Niagaran production in Michigan are recognized for purposes of this discussion: (1) southeastern Michigan (marked "SE" in Fig. 1), where earliest exploration activity dates back to the 1950's; (2) south-central Michigan ("SC" in Fig. 1), where Mobil Corp. was active in the early 1970's; and (3) northern Michigan ("N" in Fig. 1), where Shell Oil Co. and Amoco Production Co., in particular, were successful throughout the 1970's.

Silurian strata crop out on the flanks of the arches that surround the basin (Fig. 1). Niagaran strata reach depths of 3,000–5,000 ft (915–1,525 m) in the pinnacle-reef belt and about 10,000 ft (3,050 m) in the basin center (Fig. 2). The total Silurian thickness increases from 1,000 ft (305 m) on the basin periphery to 4,000 ft (1,200 m) in the basin center. The Salina Group thickens basinward as a complement to the basinward thinning of the Niagaran sequence. Pinnacle reefs near the basinward edge of a barrier-reef complex produce oil and gas and have been buried in halite, anhydrite, and carbonate muds of the Salina. The controversial temporal relationship between Niagaran reef growth and Salina evaporite deposition was discussed and clarified by Shaver (1974) and colleagues at the Indiana Geological Survey and need not be elaborated upon except to state that the terms Niagaran and Salina, as applied herein, are used in a lithostratigraphic sense only.

THERMAL HISTORY IN THE MICHIGAN BASIN

Present-day temperature data indicate that the pinnacle-reef belt is now relatively cool (Fig. 3). Unadjusted temperature versus depth plots of more than 100 electric-log temperatures in the Michigan basin show an average gradient of about 1°F/100 ft (1.82°C/100 m). The AAPG geothermal-gradient map of North America (Shelton, 1976) incorporates a correction for borehole-fluid equilibration and shows 1.6°F/100 ft (2.9°C/100 m) gradient values over most of the basin, with a few small localities reaching 1.8°F/100 ft (3.28°C/100 m). Even at maximum gradient conditions along the pinnacle-reef belt, Niagaran temperatures should not significantly exceed 150°F (65°C). In view of the long, relatively stable epeirogenic history of the basin, it is difficult to show that Niagaran strata have ever been much deeper in the past than they are now. On the other hand, the roughly 420-m.y. age of the Niagaran strata may supplement the effect of temperature in the maturation process, with the rock-matrix composition playing an unknown role.

Palynological profiles show mature thermal-alteration indices (TAI) of about 3 to 3.5 for Silurian strata, using Staplin's (1969) scale (Fig. 4). The values for the Silurian and older parts of the curve are based on acritarchs and chitinozoans rather than on spores and pollen, as land plants did not appear until mid-Silurian time. Vitrinite studies are not possible for the same reason. Palynomorphs encountered in several profiles such as this showed a mature yellow-brown color in the Traverse limestone, the Salina Group, and the Niagaran dolomite, in contrast to those in the Trenton limestone, which were black and metamorphosed. The TAI values for the Coldwater, Utica, and Glenwood shales also appear to fit nicely with carbonate rocks on the profile. These rough data indicate sufficient maturity to generate petroleum in Silurian strata in the Michigan basin.

LOCAL SOURCE-ROCK AND RESERVOIR GEOLOGY

Figure 5 shows a thickness for Niagaran rocks within the basin of some 100 ft (30 m), which contrasts sharply with thicknesses of up to 500 ft (150 m) for the barrier-reef complex. The pinnacle reefs lie along the belt of maximum isopach gradient. Three cross sections, CC′, DD′, and EE′, were examined for stratigraphic details in the three principal productive areas along the reef belt. Only section

[1]Present address: Mobil Oil Indonesia, Post Bag 400, Jakarta Pusat, Jakarta, Indonesia.
[2]Present address (retired): P.O. Box 277, Cedar Hill, Texas 75104

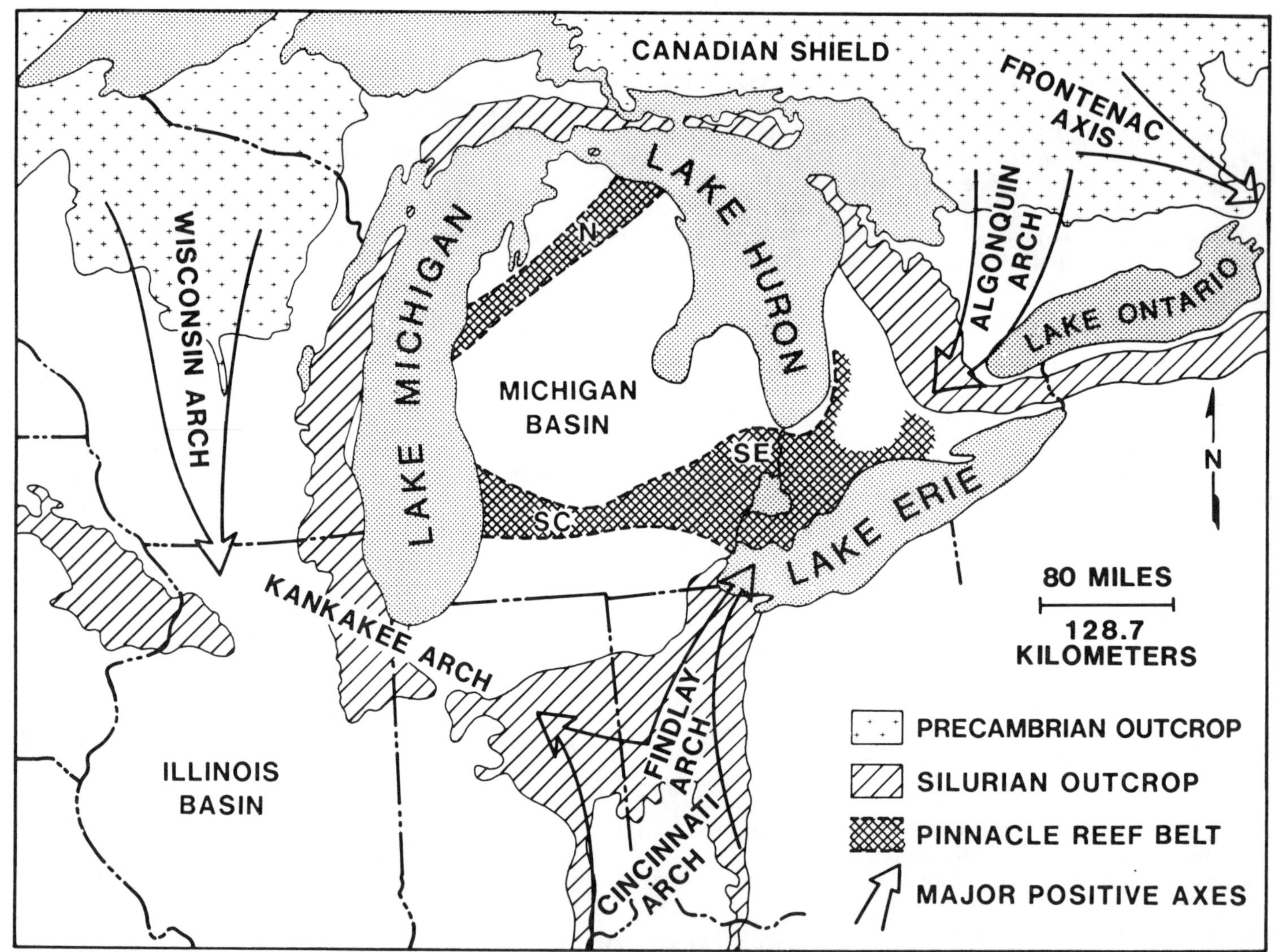

Figure 1 — Tectonic framework of Michigan basin region. SE = southeastern Michigan; SC = south-central Michigan; N = northern Michigan.

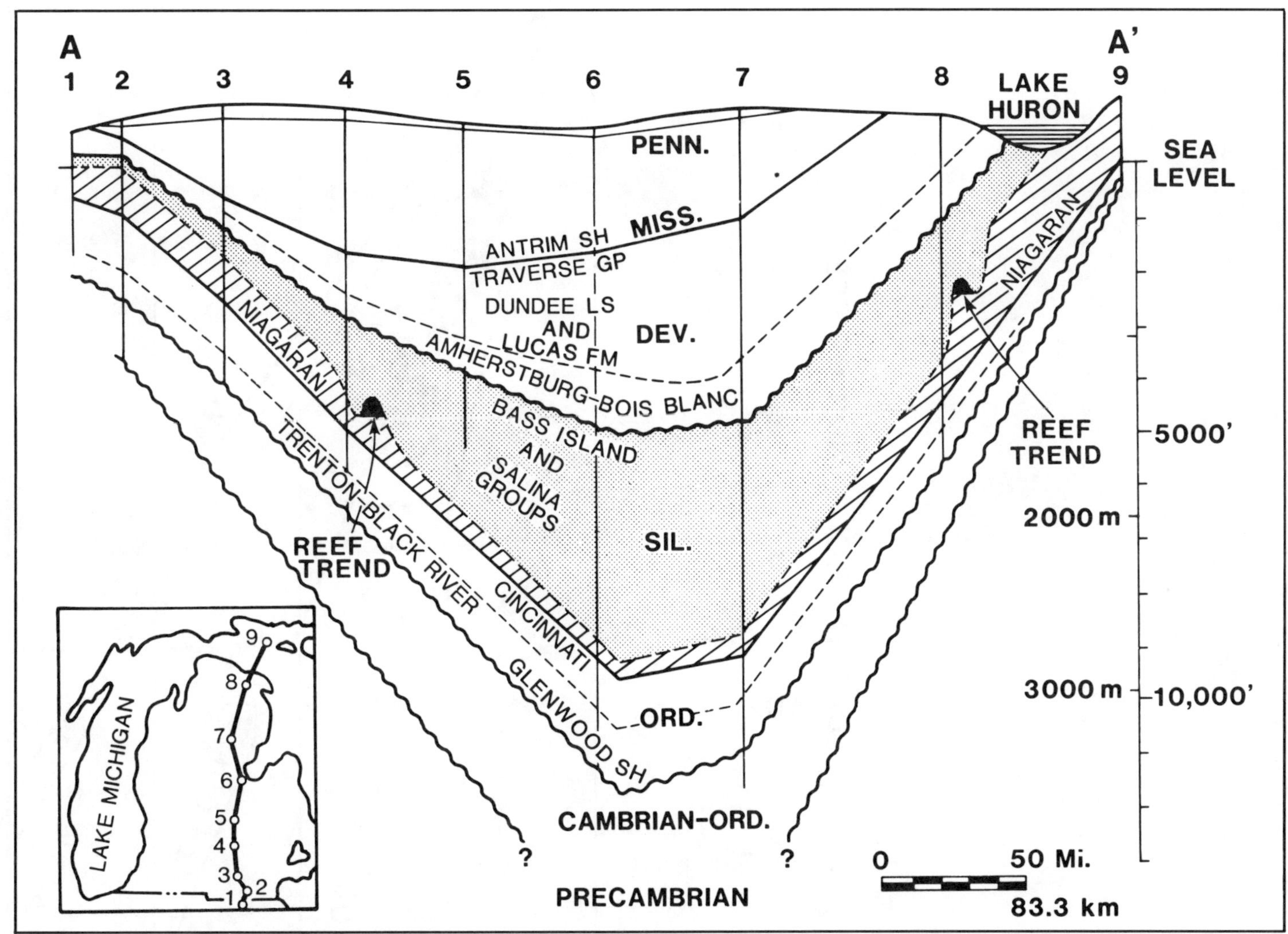

Figure 2 — Structural cross section across Michigan basin.

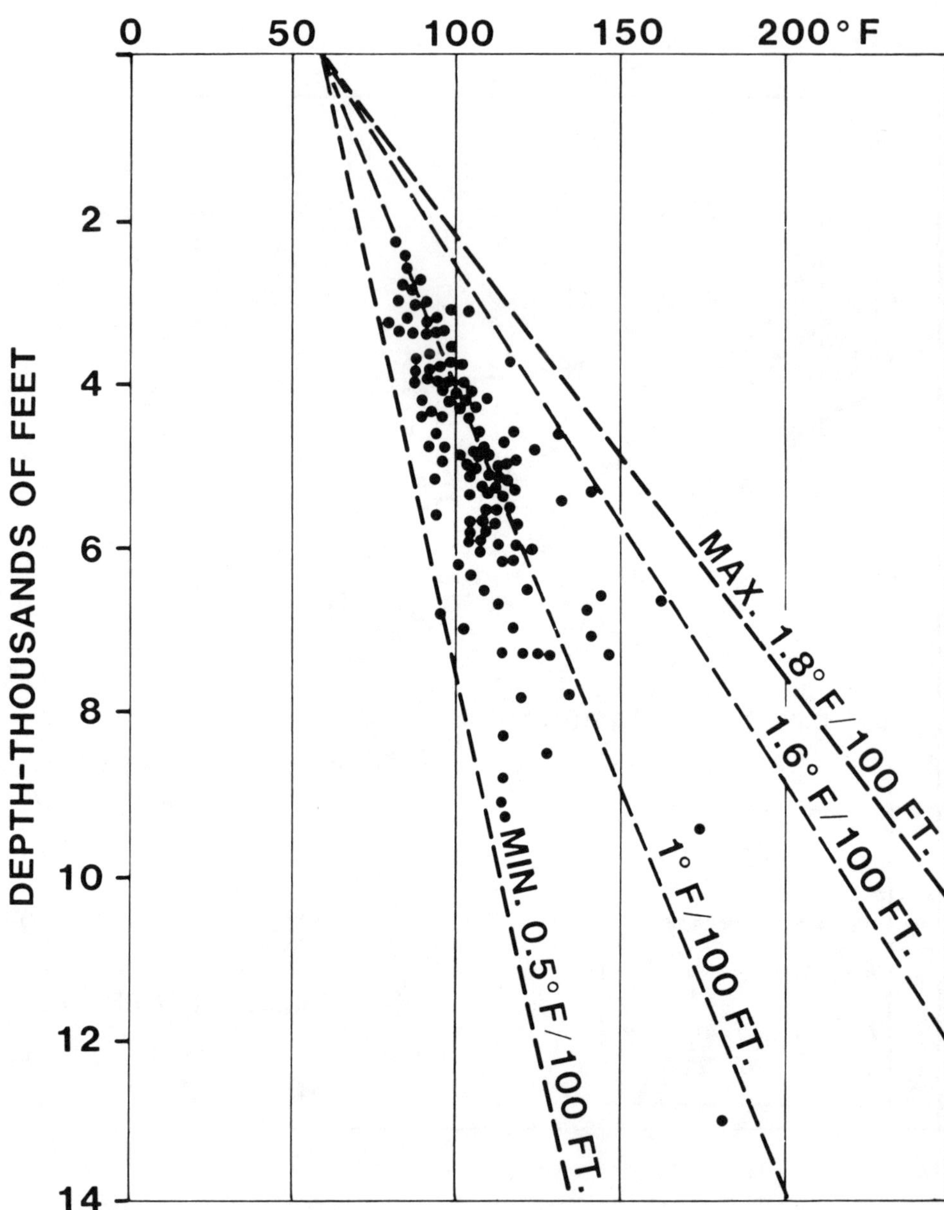

Figure 3 — Michigan basin temperature profile.

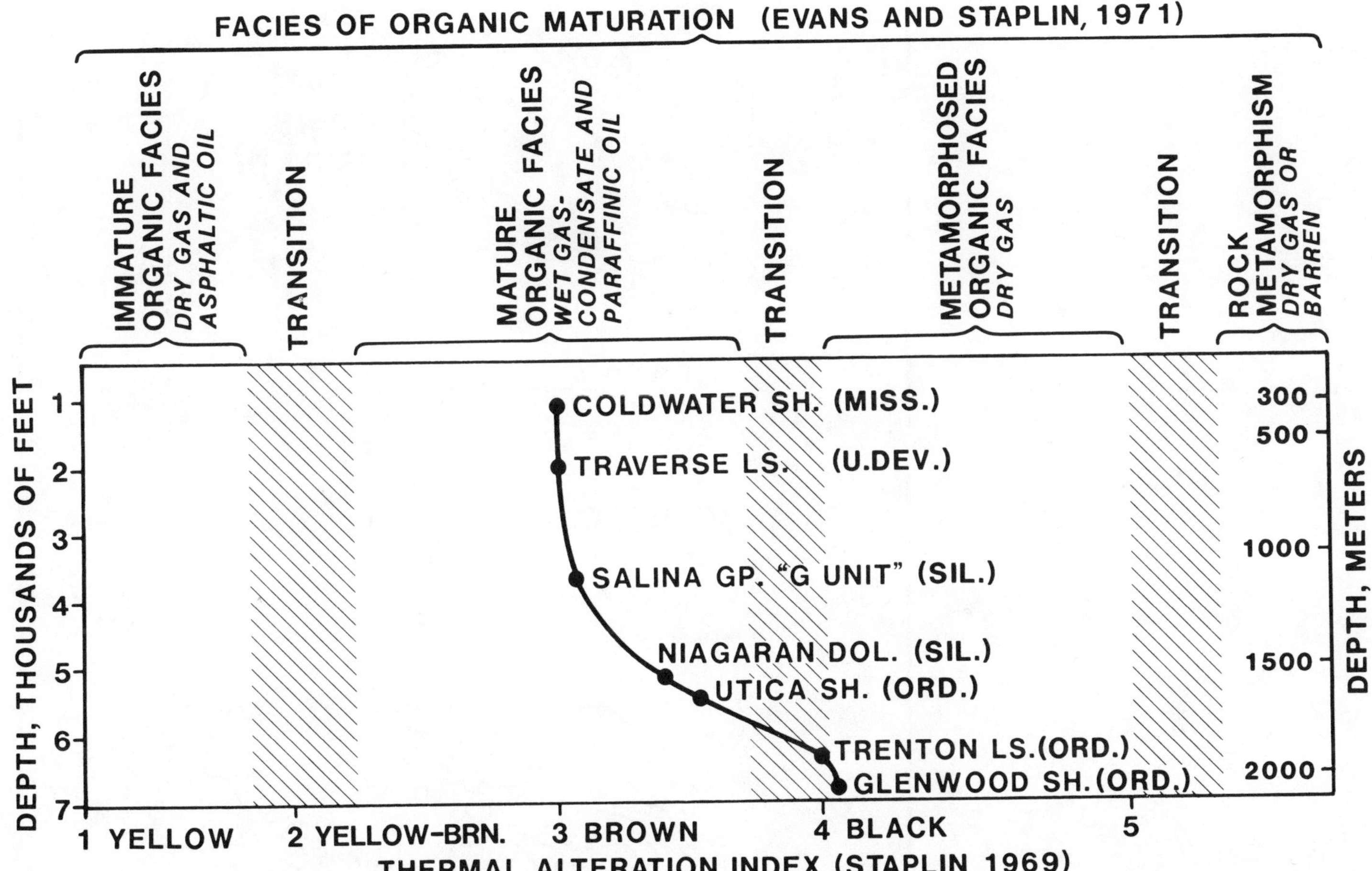

Figure 4 — Palynomorph carbonization, based on samples from Mobil 1 Jelinek–Ferris, Sec. 5, T5N, R2E, Shiawassee County, Michigan. (Staplin's scale is for spores and pollen, which are not present in the Silurian. However, experience at Mobil indicates that coloration of chitinozoans and acritarchs may be similarly applied as an approximation.)

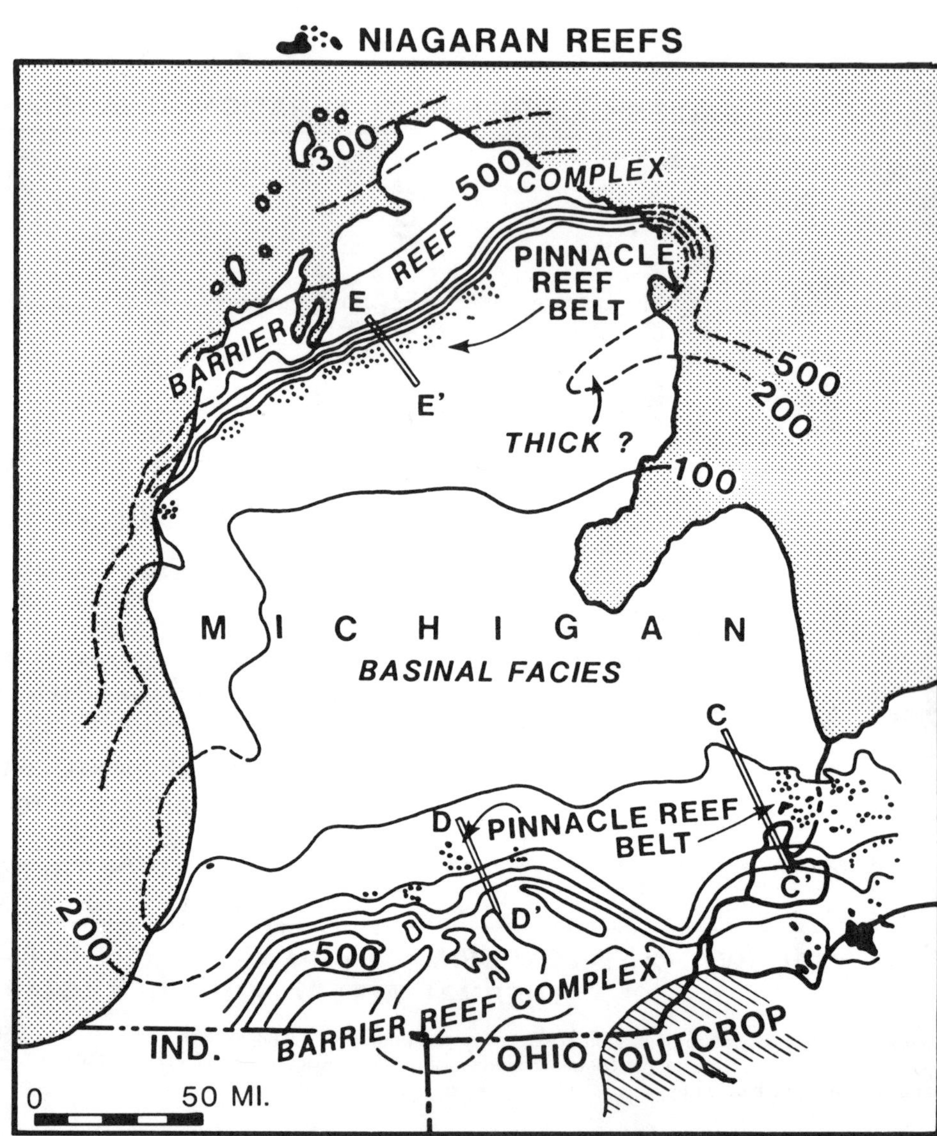

Figure 5 — Isopach map of Niagaran rocks in Michigan basin; contour interval, 100 ft (30.5 m).

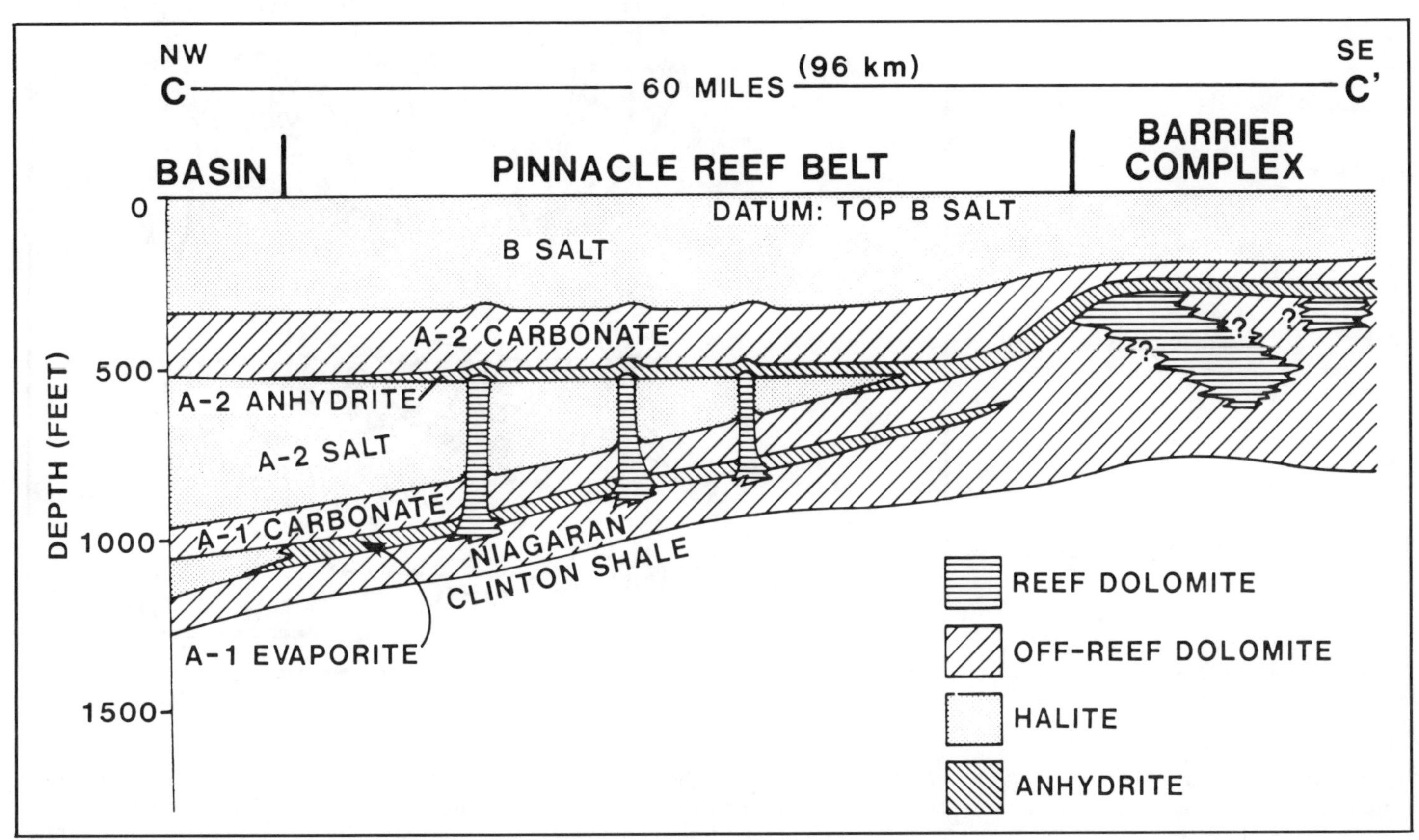

Figure 6 — Stratigraphic section across pinnacle-reef belt, southeastern Michigan. Line of section shown in Figure 5.

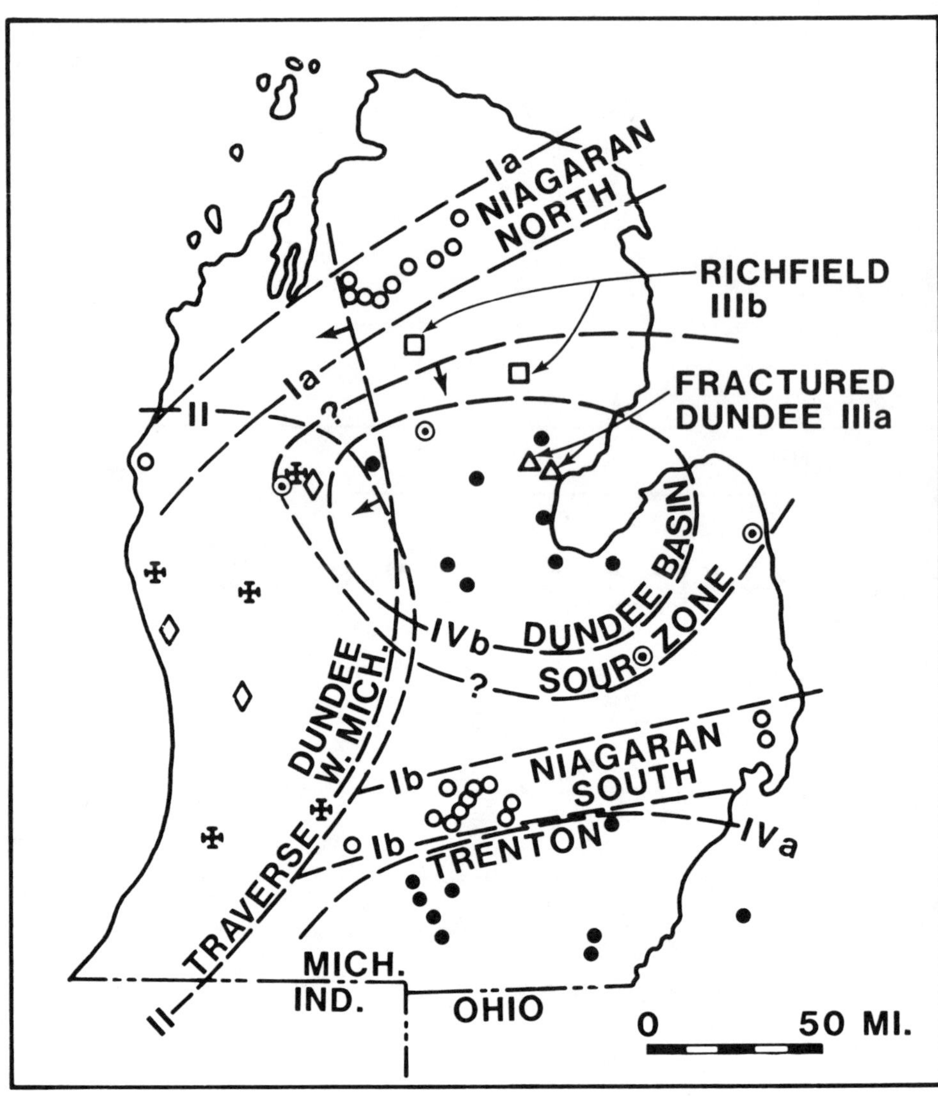

Figure 7 — Map of Michigan basin showing geographic distribution of oil groups.

CC′ across southeastern Michigan (Fig. 6) is discussed, since all three sections show essentially identical stratigraphic relationships.

Vertical exaggerations in geologic cross sections of nearly 100, such as in Figure 6, are commonly used to illustrate the "pinnacle" reefs, which are several hundred feet in height and about one-half mile to a mile in diameter. The reefs are capped and sealed by impermeable anhydrite and are encased mostly in Salina A-2 halite (Fig. 6). The Salina A-1 carbonate muds reach a thickness of about 100 ft (30 m) along the reef belt. They appear to be the most likely source of the petroleum in the reef because of their strong petroliferous odor, carbonaceous appearance, and gas and oil shows. In contrast, the basal Niagaran or platform strata are impermeable and generally lack the petroliferous aspects of the Salina A-1 beds. The underlying maroon and varicolored strata of the Clinton Shale are devoid of significant organic content and are clearly non-source rocks.

The coral/stromatoporoid skeletal reef core ("Gray Niagaran") grew in shallow to moderate water depths, far from land and terrestrial influx, in a marine environment that was intermittently sediment starved (Mesolella et al, 1974). Rapid halite sedimentation, represented by the A-2 salt, buried and sealed the reef edges. A brown, carbonaceous, petroliferous, laminated dolomite facies ("Brown Niagaran") on the flanks and in the upper parts of the reefs reflects increased salinities prior to the final capping and sealing of the reefs by anhydrite. The brown, micritic, dolomite facies does not contain the normal-marine fauna of the "Gray Niagaran" but rather exhibits a restricted algal character. The "Brown Niagaran" is a subordinate source of petroleum in the Niagaran reefs. It, like the A-1, was deposited in a mesosaline environment, i.e., in water saltier than normal-marine waters but insufficiently concentrated to precipitate calcium sulfate (Kirkland and Evans, 1981).

GEOCHEMICAL EVIDENCE FOR SOURCE–RESERVOIR RELATIONSHIP

Some 300 Silurian core samples were collected, lithologically described, and analyzed for total organic carbon (TOC). The more promising samples were analyzed further for extractable organic matter (EOM).

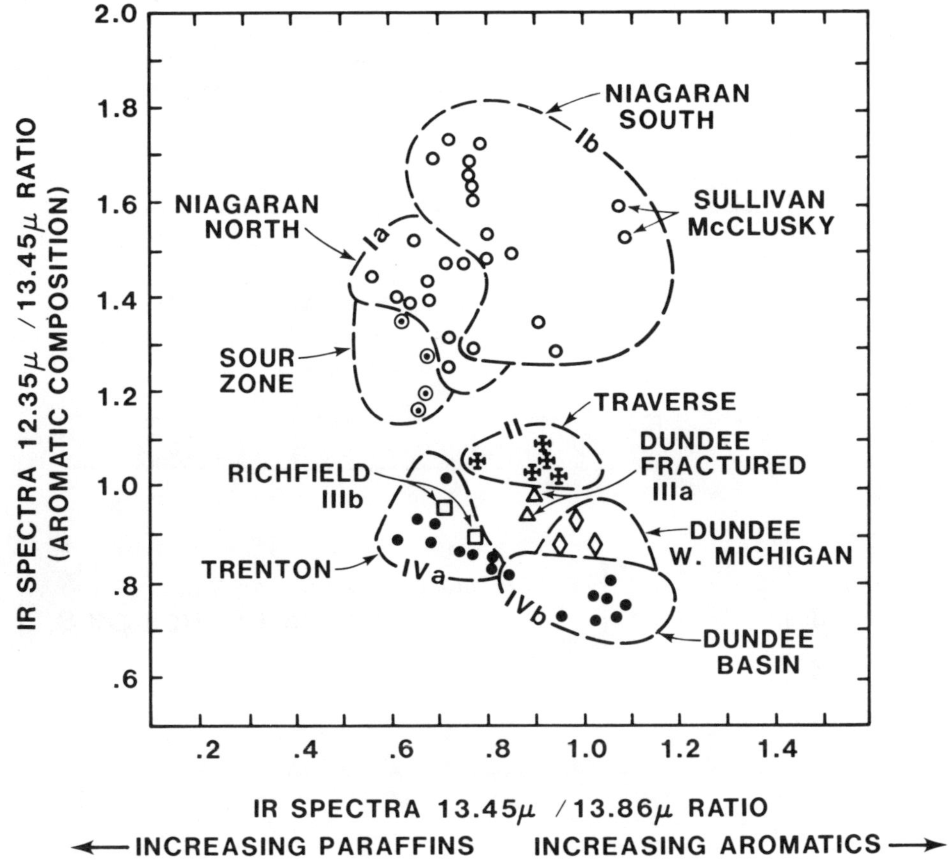

Figure 8 — Michigan basin oil groups.

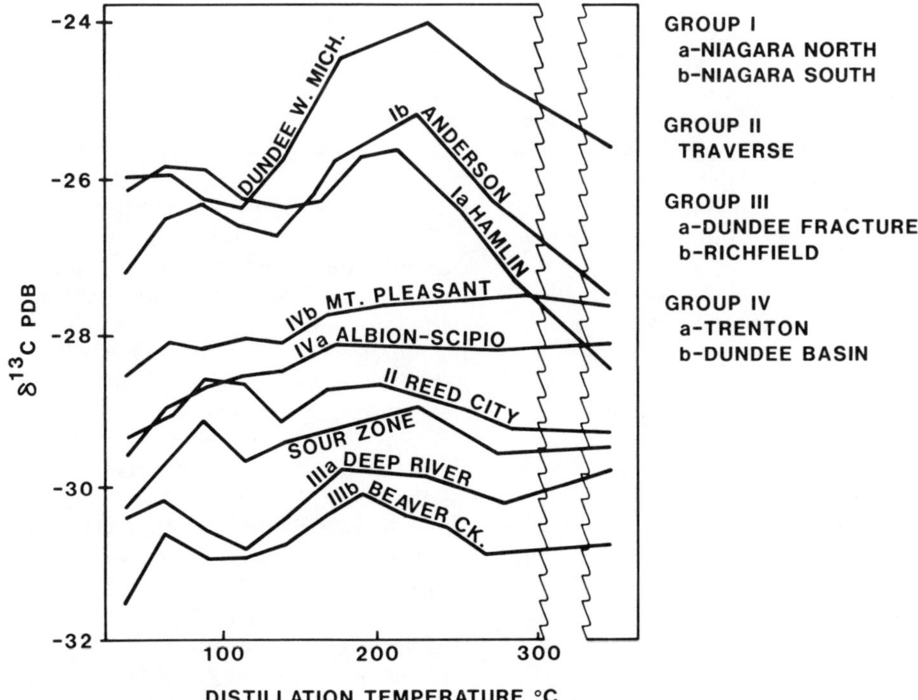

Figure 9 — Typical carbon-isotope profiles of Michigan basin oil groups.

Fifteen anhydrite and salt samples were analyzed to determine TOC values for those lithologies that contain black, carbonaceous laminae and reticulations. On the basis of the their extremely low TOC values (average 0.04%) and impermeable nature, anhydrite and salt, which make up such a large part of the Silurian section, are unlikely source rocks for petroleum trapped in the reefs.

In contrast, the average TOC for 50 Salina A-1 carbonate samples was 0.28%, with 40% of them having values equal to and greater than 0.3% and ranging up to 0.6%. Salina A-2 carbonate averaged 0.17% TOC, based on 29 samples. One hundred samples of Niagaran strata that served as platform material for the reefs (Fig. 6) averaged 0.12% TOC, with 99% of the samples having less than 0.3%. On the other hand, a brown-dolomite facies ("Brown Niagaran") on the flanks and in the upper parts of the reefs contained an average of 0.27% TOC, with 36% of the samples having 0.3–0.5%. The Salina A-1 carbonate and the "Brown Niagaran" dolomite have the highest organic content and are also the strata that have the strong petroliferous odor, carbonaceous appearance, and gas and oil shows.

Standards of TOC and EOM commonly applied to shale source rocks do not appear appropriate for carbonates. For example, average TOC's are low for carbonates; the world TOC average is about 0.25% (Hunt, 1961). Hunt (1961) and Gehman (1962) reported that carbonate rocks, in contrast to shales, yield a higher amount of hydrocarbons per unit of organic carbon. Palacas (1978) came to the same conclusion in his assessment of source rock potential of Cretaceous and Lower Tertiary carbonates in south Florida. He further pointed out that, in many of the carbonate samples, 80% or more of the organic matter is amorphous, sapropelic, hydrogen-rich kerogen. Kerogens of the Silurian samples from the Michigan basin are of similar form and composition.

CORRELATION OF THE OILS

Sixty Ordovician, Silurian, and Devonian crude oils were analyzed to determine composition and geochemical characteristics critical to the understanding of the generation, migration, and accumulation history of oils in the Michigan basin. The principal geochemical

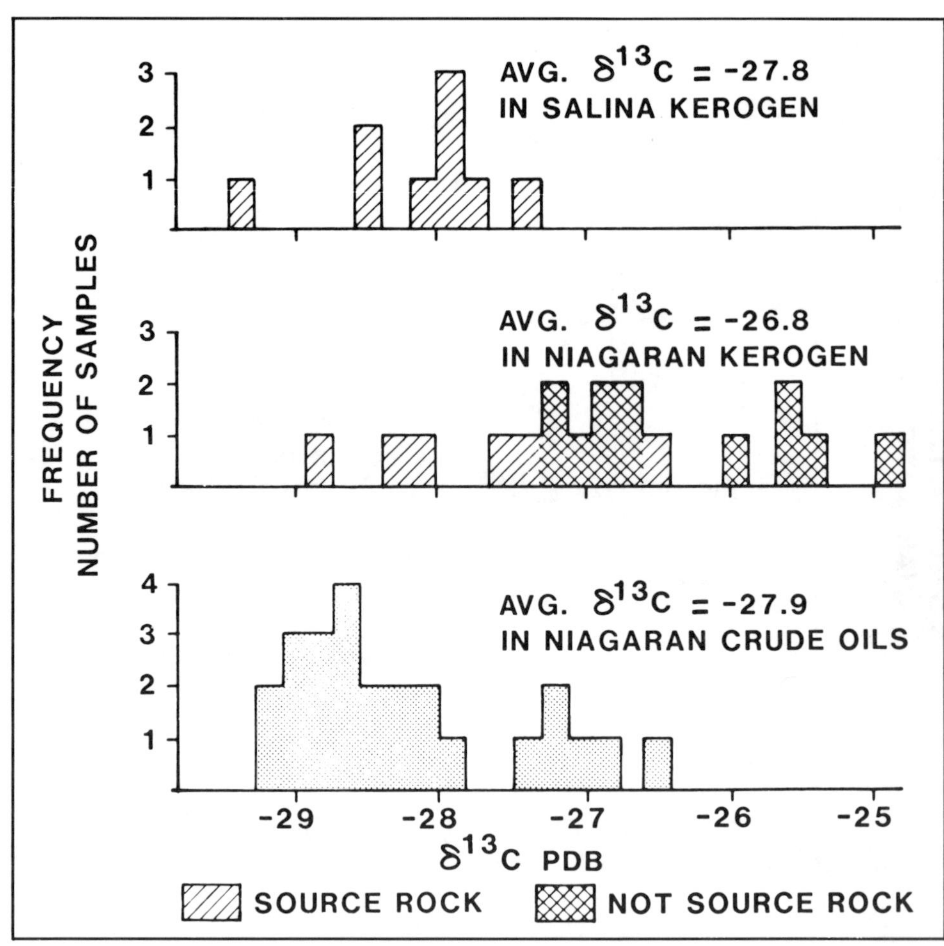

Figure 10 — Silurian ^{13}C distribution in kerogens and oils from Michigan basin.

methods utilized for correlation purposes were comparisons of carbon-isotope data, infrared spectrograms of C_{18+} aromatics, and infrared spectrograms of the total crude oil stabilized by evaporation at 40°C (104°F).

The 60 oils were classified into four principal groups, based on a visual best fit of the IR spectrograms of the C_{18+} aromatic fraction of each oil. The four groups (I–IV) have been further divided into subgroups on the basis of more subtle differences. The geographic distribution of these groups and subgroups (shown in Fig. 7) includes: Ia, Niagaran north (Silurian); Ib, Niagaran south (Silurian); II, Traverse of western Michigan (Devonian); IIIa, fractured Dundee (Devonian) of basin center; IIIb, Richfield (Devonian) of basin center; IVa, Trenton (Ordovician); IVb, Dundee (Devonian), basin center. The Devonian "Sour Zone" and the Dundee western Michigan groups stand apart, as shown later from isotope data.

The similarities between group IVa Trenton and IVb Dundee basin-center oils are surprising unless similarity of source lithology and depositional environment is considered. Also, the group IIIa oil in fractured Dundee rocks appears to have migrated from underlying evaporitic source rocks of the Lucas (Devonian), which are environmentally similar to the source rocks for the IIIb Richfield oil.

Figure 8 illustrates the discrimination of the subgroups, based on ratios of infrared-absorption intensity of the aromatic fraction (ordinate) and whole oil (abscissa). The 12.35/13.45 μ ratio, plotted on the ordinate, is a function of the character of the aromatic fraction. The 13.45/13.86 μ ratio on the abscissa reflects the relative amounts of paraffin and aromatic components present in the total stabilized (40°C, 104°F) crude. Points situated to the left are more paraffinic; those to the right are increasingly aromatic.

Niagaran oils (group I) cluster nicely within the upper part of the diagram, owing to their distinctive aromatic character. The oils may be further subdivided into Niagaran north (Ia) and Niagaran south (Ib). The Devonian "Sour Zone" oils from Reed City, Headquarters, Otter Lake, and Volmering fields, in the central part of Michigan, are more similar to IR group I. As the A-1 carbonate, the source of Niagaran oils, and the "Sour Zone" carbonate muds were both deposited in similar mesosaline environments, the partial similarity of their oils is not surprising. However, the "Sour Zone" oils are different isotopically.

An example of variation in the Niagaran oils is seen in two samples from Sullivan 1 McClusky, a Niagaran reef well in southwestern Michigan. These oils are situated to the right of the other Niagaran oils in Figure 8, being less paraffinic in composition. The deficiency in paraffins, together with the relatively shallow depths (2,700 ft or 825 m) and 15–20° API gravity, suggest biodegradation.

Carbon-isotope profiles of 36 oils generally support the infrared-absorption

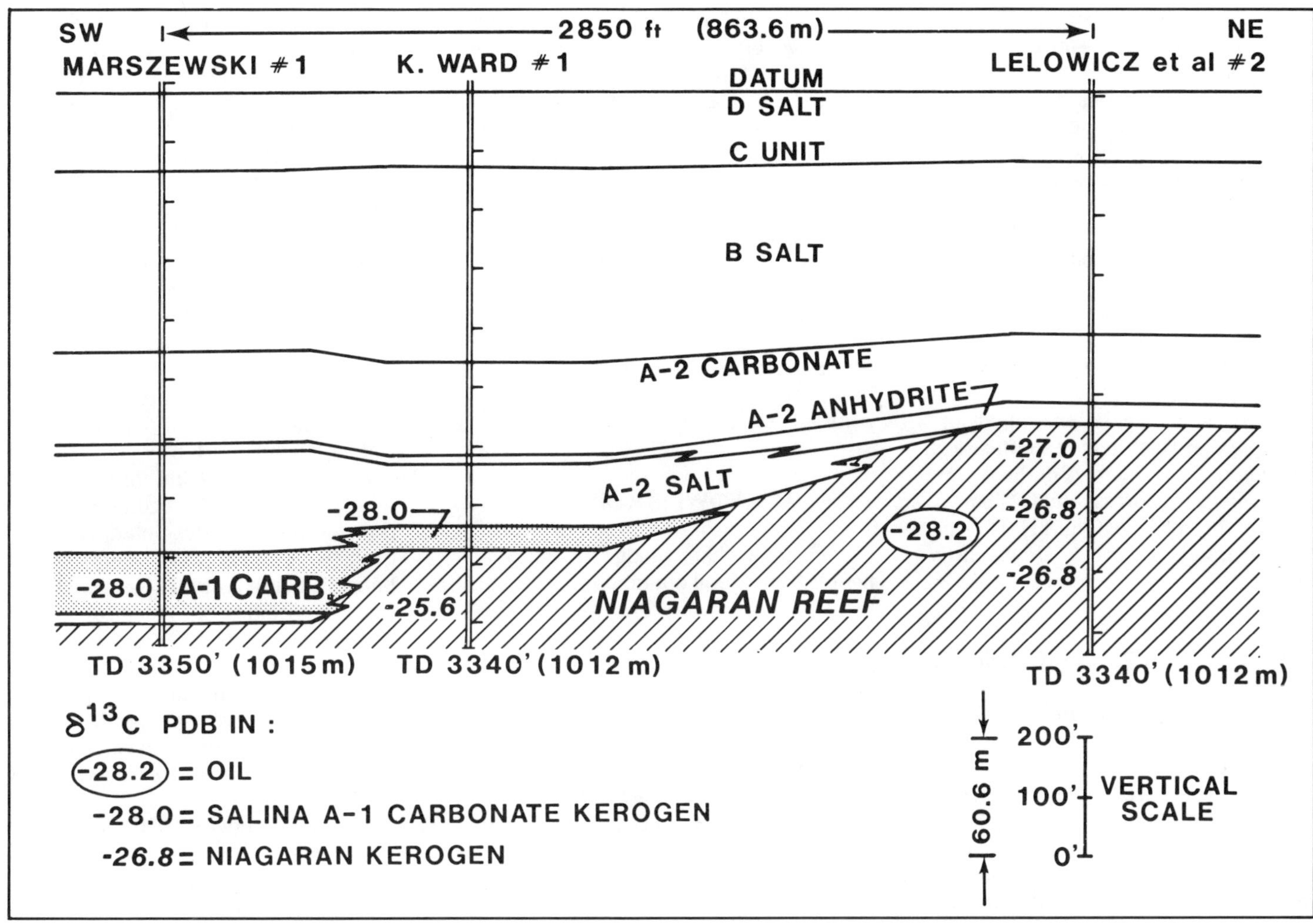

Figure 11 — Cross section through Tiger field, St. Clair County, Michigan.

interpretation. The isotope data are restricted to those compounds that have less than 18 carbon atoms per molecule. By comparison, the aromatic fraction, used for infrared characterization, includes only those compounds with more than 18 carbon atoms per molecule. In a graphic representation of the isotope profile, $\delta^{13}C$ is plotted on the ordinate, and distillation temperatures for 11 distillation fractions to 300°C (572°F) plus a residuum fraction are plotted along the abscissa (Fig. 9). The nine profiles shown are intended only to indicate typical shapes of the profiles for oil groups of the Michigan basin.

All of the Niagaran oils have a characteristic bimodal profile and relatively lower ^{13}C concentrations in the high boiling residuum. The Devonian "Sour Zone" oils, so similar on infrared spectrograms, appear distinctly different from the Niagaran oils in the carbon-isotope profile and are generally lower in ^{13}C concentration. The $\delta^{13}C$ profiles distinguish the Niagaran oils from all other oils in the basin and suggest that there has been no migration from other sources outside the Silurian into the Niagaran reef reservoirs.

COMPARISON OF THE OILS WITH SOURCE-ROCK KEROGEN

Carbonate-source-rock evaluation presents some important differences from the interpretation of terrigenous-clastic source rocks. With uncontaminated shale samples, there is no problem of hydrocarbon remigration into an unfractured shale matrix. In contrast, carbonate rocks serve as both reservoir and source rock. Except for dense, completely impermeable samples, the question commonly arises as to whether the oil is indigenous or whether it migrated from another source. Niagaran oils exhibit the property of having heavy ends and residuum with isotopically lighter $\delta^{13}C$ values than those found in the light and intermediate fractions, as shown by Figure 9. This suggests that the source kerogen, normally isotopically heavier than the oil it generates, has a $\delta^{13}C$ value closer to the oil in this case, making a correlation of the $\delta^{13}C$ values of Niagaran oil with the $\delta^{13}C$ value of its source kerogen more valid.

The $\delta^{13}C$ values of 27 Niagaran and lower Salina A-1 carbonate kerogens were measured to compare them with $\delta^{13}C$ values in oils from Niagaran reservoirs. The $\delta^{13}C$ values for the oils are shown in the lowermost histogram of Figure 10. The average $\delta^{13}C$ value of the 25 Niagaran oils is -27.9. The average of

18 Niagaran kerogen samples is −26.8, while the average value for the 9 Salina A-1 kerogen samples is very close to the −27.9 value for the oils. The "Brown Niagaran" algal-laminated facies also have $\delta^{13}C$ values in the same range as the oils. These data further support the hypothesis that the Salina A-1 kerogen is the likely principal oil source, but that "Brown Niagaran" kerogen may also have contributed to a lesser degree. This correlation of kerogen and oil by $\delta^{13}C$ values goes a step beyond inference regarding the source–reservoir relationship.

When a single field is examined, for example, the Tiger field of southeastern Michigan, the $\delta^{13}C$ values in the A-1 carbonate kerogen correspond to the $\delta^{13}C$ value of the crude oil (Fig. 11). The values for Niagaran kerogen $\delta^{13}C$ are distinctly higher.

SUMMARY AND CONCLUSIONS

Total-organic-carbon and correlation data determined from Silurian core samples and oils indicate that Salina A-1 carbonaceous, carbonate muds surrounding and encasing the lower part of reefs are the principal source for the Niagaran oil. The A-1 carbonate was deposited in a mesosaline environment. "Brown Niagaran" facies of the upper and peripheral parts of the reef also reflect deposition in mesosaline waters and constitute a subordinate source of petroleum.

Oil-correlation studies show the Niagaran oils to be unique in the Michigan basin and help provide a basis for the conclusion that most of the Niagaran oils are indigenous to the Silurian rocks. The $\delta^{13}C$ values of the kerogen are consistent with the A-1 carbonate as the source of Niagaran oil. Indigenous oil in the Silurian is not surprising in view of the numerous thick, impermeable salt and anhydrite units overlying and the impermeable limestone and shale underlying the pinnacle reefs. Where sealing strata are fractured or pervious, some vertical migration may be possible. Thermal-alteration indices of chitinozoans and acritarchs indicate that the organic matter is of a mature facies and has not undergone excessive thermal history.

ACKNOWLEDGMENTS

We want to acknowledge the many people from Mobil operations and research groups who contributed to an extensive internal study of source-to-reservoir relationships in the Michigan basin. This paper is a summary of a part of that work. We also gratefully acknowledge the late Dr. Louis I. Briggs and the Subsurface Geological Laboratory at the University of Michigan for making well cores available for sampling purposes.

REFERENCES CITED

Evans, C.R., and F.L. Staplin, 1971, Regional facies of organic metamorphism in geochemical exploration: Montreal, Canadian Institute of Mining and Metallurgy Special Volume 11, Third International Geochemical Exploration Symposium Proceedings, p. 517–520.

Gehman, H.M., Jr., 1962, Organic matter in limestones: Geochimica et Cosmochimica Acta, v. 26, p. 885–897.

Hunt, J.M., 1961, Distribution of hydrocarbons in sedimentary rocks: Geochimica et Cosmochimica Acta, v. 22, p. 37–49.

Kirkland, D.W., and R. Evans, 1981, Source-rock potential of evaporitic environments: AAPG Bulletin, v. 65, p. 181–190.

Mesolella, K.J., et al, 1974, Cyclic deposition of Silurian carbonates and evaporites in Michigan basin: AAPG Bulletin, v. 58, p. 34–62.

Palacas, J.G., 1978, Preliminary assessment of organic carbon content and petroleum source rock potential of Cretaceous and Lower Tertiary carbonates, South Florida Basin: Gulf Coast Association of Geological Societies Transactions, v. 28, p. 357–381.

Shaver, R.H., 1974, Structural evolution of northern Indiana during Silurian time, *in* Silurian reef–evaporite relationships: Michigan Basin Geological Society Field Conference Guidebook, p. 55–77.

Shelton, J.W., compiler, 1976, Geothermal gradient map of North America: AAPG and U.S. Geological Survey, scale 1:5,000,000.

Staplin, F.L., 1969, Sedimentary organic matter, organic metamorphism, and oil and gas occurrence: Bulletin of Canadian Petroleum Geology, v. 17, p. 47–66.

Some Aspects of the Hydrocarbon Geochemistry of a Middle Devonian Barrier-Reef Complex, Western Canada

T. G. Powell

Geological Survey of Canada, Institute of Sedimentary & Petroleum Geology, Calgary, Alberta[1]

Organic geochemical data have been obtained on potential source rocks, bitumens, and oils from within and in the vicinity of a Middle Devonian carbonate-barrier complex in western Canada. Organic-carbon values in carbonate rocks vary widely (0.11 to 30.1%) according to facies type. Extract yields from immature samples from the basinal facies can be extremely high and appear to be a consequence of the highly reducing conditions in which the organic matter was deposited. Characteristic features of these high-yield extracts are a low proportion of hydrocarbon, low saturate-to-aromatic ratios, pristane-to-phytane ratios less than 1.0, a high content of acyclic isoprenoids relative to n-alkanes, and an even-to-odd predominance in the n-alkanes. The hydrocarbon yields at low levels of maturity are sufficiently high for these carbonates to be source rocks. The main migration product is heavy oil/bitumen, which shows all the characteristics of the parent-rock extracts. Locally these bitumens have been altered by heat during dolomitization to form an insoluble hydrogen-rich pyrobitumen classified as epi-impsonite. This pyrobitumen probably was formed by polymerization of the NSO/asphaltenes fraction. With increasing depth the normal rock extracts show a decline in NSO (nitrogen, sulfur, and oxygen) compounds and asphaltenes yield owing to polymerization under normal conditions of maturation but with essentially no change in hydrocarbon yield. The oils in pinnacle-reef reservoirs in the Rainbow and Zama fields south of the barrier complex also show features inherited from a source rock deposited under highly reducing conditions. These oils are more mature than heavy oils encountered in the low-maturity area.

INTRODUCTION

A carbonate-barrier complex divides Middle Devonian rocks of western Canada into two broad depositional areas. Throughout Alberta and Saskatchewan a predominantly evaporite basin exists, whereas in the Northwest Territories shales replace evaporites as the dominant basinal facies (Basset and Stout, 1967; Fig. 1). The barrier complex (Figs. 1, 2) crops out south of Great Slave Lake in the Northwest Territories and dips gently southwestward following the regional dip. It can be traced in the subsurface into northern British Columbia. In the outcrop area the barrier complex is the site of a major Mississippi Valley–type lead–zinc deposit, the Pine Point ore field (Skall, 1975) (Figs. 1, 2). South of the barrier in northern Alberta, major oil accumulations are present in pinnacle reefs in the Rainbow and Zama oil fields (Fig. 1) (Hriskevich, 1970). These pinnacles are equivalent in age to the barrier complex.

Oil shows and bitumen occurrences are ubiquitous throughout the barrier complex, including the ore field. An organic geochemical study has been undertaken in the area of Pine Point ore field with the following objectives: (1) to examine depositional controls on organic-matter composition in the various facies of the barrier complex at low-maturation levels, (2) to provide information on the nature of potential carbonate source beds in this area, (3) to examine the relationship between indigenous organic matter and bitumen/oil occurrence in these shallow rocks, (4) to provide preliminary information on potential sources for Rainbow–Zama oils, and (5) to examine the role (if any) of organic matter and/or bitumen in the formation of the lead–zinc deposit. The latter aspect is not the concern of this paper, however, and is dealt with elsewhere (Macqueen and Powell, 1983). This study was facilitated by the availability of a detailed facies study in the Pine Point area (Skall, 1975) and of samples from closely spaced diamond-drill holes and ore-body pits. In order to place the results from the Pine Point area in their regional context, samples of rocks of corresponding age were obtained from petroleum-exploration wells in the Pine Point area and to the west (Fig. 1).

CORRELATION AND NOMENCLATURE

The regional stratigraphic relationships of the barrier complex are illustrated in Figure 2. The middle Paleozoic rocks of the Pine Point region dip gently to the southwest from the exposed Precambrian rocks of the Canadian shield. The barrier complex is considered to be Givetian in age and is underlain by the Chinchaga Formation, an evaporite unit, 90–110 m (295–360 ft) thick, of Eifelian age. The Givetian units making up the barrier complex were studied in the Pine Point area by Skall (1975), who recognized a distinctive suite of rock facies that serve to outline both the detailed stratigraphy of the area and to determine the evolution of the barrier (Fig. 3). Skall (1975) recognized four main units, the Keg River Formation (facies A), the Pine Point Group (facies B through K), the Watt Mountain Formation (facies L), and the Slave Point Formation (facies M through P).

The Keg River Formation (facies A)

[1]Now with Bureau of Mineral Resources, Geology and Geophysics, Canberra, ACT, Australia.

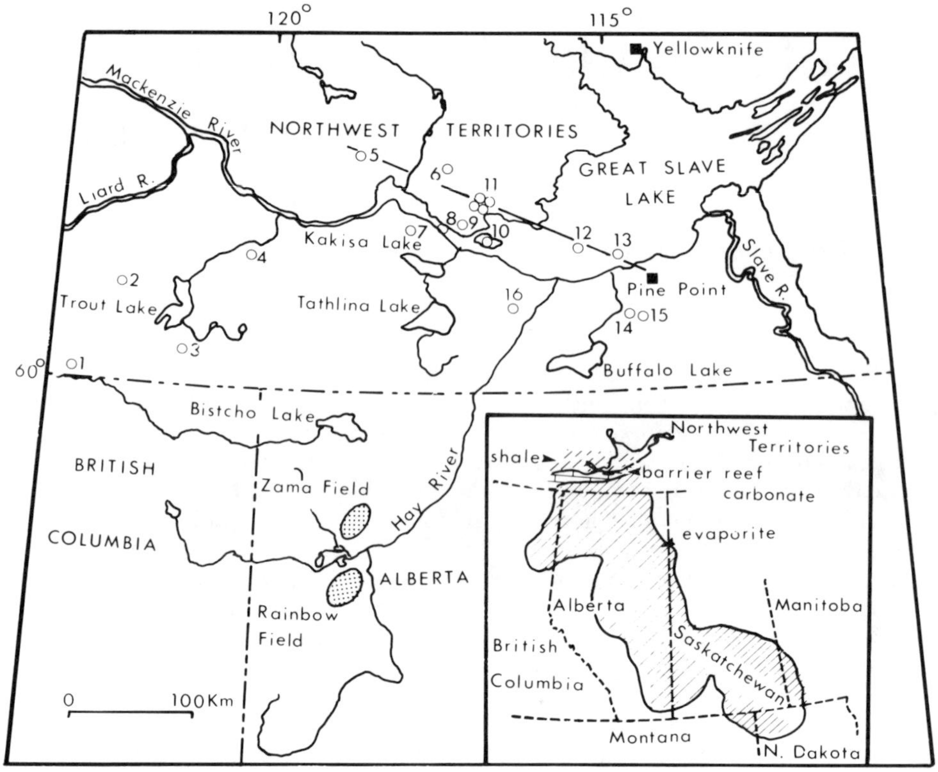

Figure 1 — Location map for barrier-reef complex at Pine Point, Rainbow and Zama oil fields, and sampled petroleum-exploration wells. Inset after Skall (1975). Line of section is that for Figure 2: (1) Texaco Bovie Lake J-72; (2) Imp. Sun Arrowhead I-46; (3) Imp. Island River 1; (4) Briggs Trout River 4; (5) C.S. Laferte River M-16; (6) IOE Providence K-45; (7) Central Del Rio Mills Lake A-70; (8) Central Del Rio Mills Lake B-75; (9) NWT Deep Bay 2; (10) NWT Big Island 1; (11) Punch Deep Bay 2, 3, and 6; (12) McDermott et al Hay River I-47; (13) McDermott et al Sulphur Point 0-7; (14) Central Del Rio Wood Buffalo C-74; (15) Central Del Rio Wood Buffalo C-03; (16) oil sample from Slave Point Formation.

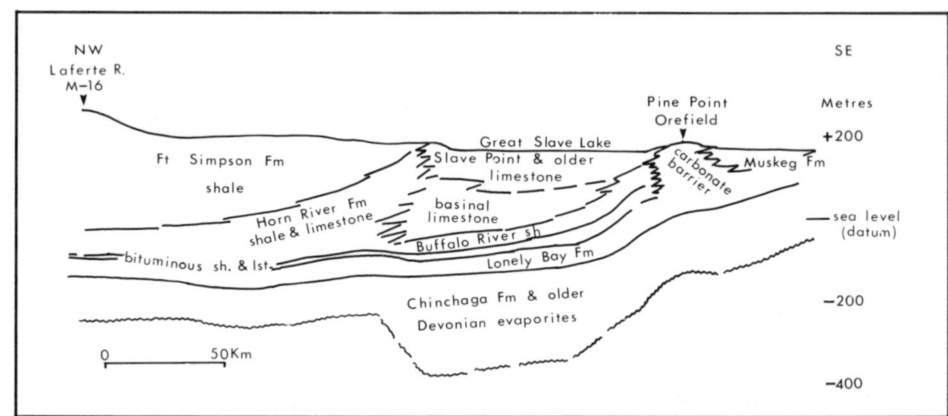

Figure 2 — Cross section showing stratigraphic relationships of barrier-reef complex in Pine Point ore-field area. Line of section is that shown in Figure 1. (After Macqueen et al, 1975.)

constitutes the carbonate platform upon which reef growth developed. In the Pine Point area it is 65–75 m (213–246 ft) thick and is probably equivalent to the lower Keg River recognized in the Rainbow–Zama oil-field area. The Keg River Formation is approximately synonymous with the Lonely Bay Formation north of the barrier (Fig. 2) and to the Nahanni Formation to the west.

The Pine Point Group (facies B through K) is 75 to 150 m (246–492 ft) thick in the Pine Point area and constitutes the barrier complex. The depositional settings for the various facies of the Pine Point Group are as follows (Skall, 1975) (Fig. 3):

Facies B (off-reef facies): open-marine fore reef in a transitional position relative to organic-barrier facies to the south and basinal facies to the north.

Facies C (shallow fore-reef facies): shallow proximal fore-reef environment.

Facies D (organic-barrier facies): agitated-water conditions with organic-reef growth and cementation of skeletal debris.

Facies E (clean arenite facies): agitated-water conditions in the fore reef with sediment derived by attrition of skeletal debris.

Facies F (*Tentaculites* facies): quiet-water conditions in basinal setting.

Facies G (Buffalo River Shale): basinal shale replaced by facies G to the south.

Facies H and I (gastropod and *Amphiphora* facies): lagoonal to tidal-flat conditions.

Facies J (south flank back-reef facies): tidal-flat to hypersaline-lagoonal conditions.

Facies K (Presquile facies): diagenetic facies (see below).

The lower part of the Pine Point Group is equivalent to the upper Keg River Formation of the Rainbow–Zama area and represents the phase of the pinnacle-

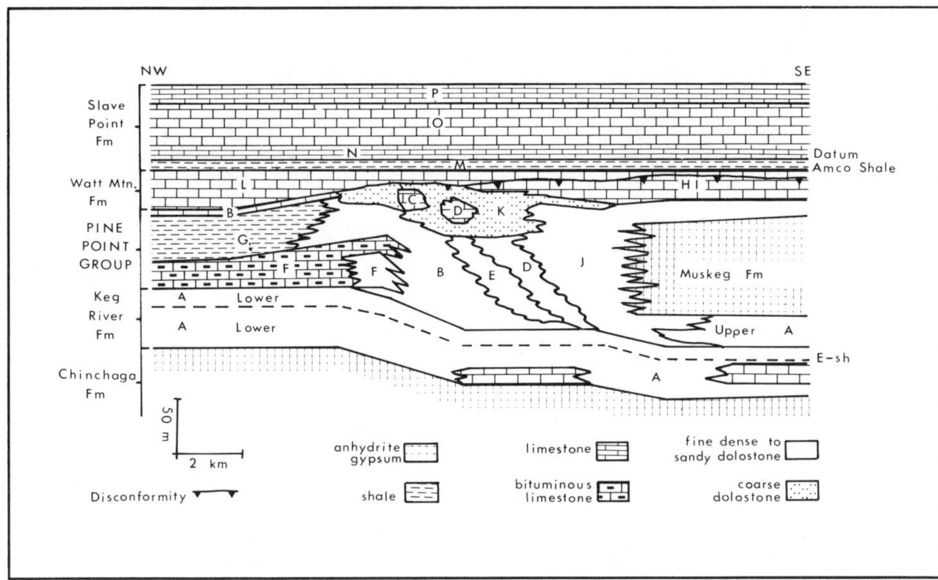

Figure 3 — Diagrammatic cross section through barrier-reef complex at Pine Point, showing facies relationships (after Kyle, 1980, but based on Skall, 1975): A = marine platform; B = off-reef facies; C = shallow fore-reef facies; D = organic barrier facies; E = clean arenite facies; F = *Tentaculites* facies; G = Buffalo River Shale; H and I = gastropod and *Amphiphora* facies; J = back-reef facies; K = Presquile facies (diagenetic); L = green shaly facies; M = Amco Member; N = tidal-flat member; O = shallow platform member; P = deep platform member.

reef growth. South of the barrier the period of reef growth was succeeded by deposition of the Muskeg evaporites, which buried the pinnacle reefs. These evaporites are not represented over the barrier itself. The upper part of Skall's Pine Point Group (facies J, H, and I, Fig. 3) probably correlates with most of the Muskeg evaporites and Sulphur Point carbonates of the evaporite (Elk Point) basin. The bituminous limestone and shale unit in Figure 2 correlates with facies F (*Tentaculites* facies) of Skall (1975). This unit is thick and contains abundant carbonate detritus adjacent to the second edge of the barrier but is thin and consists entirely of mud rock elsewhere.

The Watt Mountain Formation constitutes facies L (green shaly facies) of Skall (1975). It is a fine-micritic limestone with light-green clay layers and was deposited in tidal-flat to lagoonal conditions.

The Slave Point Formation is divided into four facies. Facies M (Amco Member), comprising shale, was deposited under open-marine conditions. Facies N, O, and P are carbonates representing tidal-flat, shallow-platform, and deep-platform environments, respectively. The combined thickness of the Watt Mountain and Slave Point Formations ranges from 50 m (164 ft) over the barrier carbonate to 115 m (377 ft) in the fore-barrier belt.

The Horn River Formation contains basinal shale and limestone and occurs north of the barrier complex (Fig. 2).

Two types of dolomitization have affected the barrier complex (Skall, 1975). The first dolomite is finely crystalline and dense to sandy textured. It was probably formed by brines refluxing through the reef complex from the evaporitic environment to the south (Skall, 1975). The second dolomite is coarsely crystalline (up to several millimeters in crystal size) with only ghost textures and outlines of fossils representing the original fabric of the rock. This facies was previously assigned formation status (Presquile Formation) but is now regarded as a diagenetic rock type that may occur in a variety of facies. This Presquile facies is characterized by its vugs, which range in diameter from a few millimeters to several centimeters.

SAMPLES AND METHODS

Samples

Three types of organic matter are found in the barrier-reef complex: (1) dispersed, indigenous organic matter consisting of insoluble kerogen and associated soluble components (extract); (2) accumulations of soft, brown, sticky semi-fluid bitumen; and (3) accumulations of black, brittle and glassy or dull powdery-textured bitumen. A total of 125 samples were collected from 22 diamond-drill holes and five open pits in the ore-field area. Fifty-seven samples were selected for analysis. In order to facilitate analysis the selection was made in favor of those samples that apparently contained large amounts of organic matter. Of the selected samples (Tables 1, 2), 22 contained indigenous organic matter, and 35 were bitumens.

The barrier complex and the region to the north contain a number of petroleum-exploration boreholes that penetrate Middle Devonian strata. Because of the westerly dip, strata at the level of the barrier complex occur at depths ranging to about 3,000 m (9,850 ft) below the surface at Texaco Bovie Lake J-72, located in the Liard River area about 450 km (280 mi) west of Pine Point (Fig. 1). Some 28 core samples were obtained from 16 boreholes (Table 1), representing the bituminous-shale and limestone unit of the Pine Point Formation, the Horn River Shale, and the Nahanni and Lonely Bay platform carbonates.

Thirteen oil samples have been examined in order to compare their geochemical characteristics with the rocks of the Pine Point barrier. Eleven of these samples came from the Rainbow field, one from the Zama field, and one from a diamond-drill hole in the Slave Point Formation below 400 m (1,300 ft) and west of the ore-field area (Fig. 1, location 16). The Rainbow and Zama oils are produced from Keg River pinnacle reefs south of the barrier reef (Fig. 1).

Analytical Methods

Analytical procedures for extraction, fractionation, and gas chromatography were essentially those used by Powell (1978). An extraction mixture of chloroform/methanol (87/13) was substituted for benzene/methanol (60/40), and a pentane/dichloromethane (50/50)

Table 1 Sample No.	Well Formation/Facies	Depth (m)	C_{org} (%)	Extract mg per g C_{org}	Hydrocarbons mg per g C_{org}	Hydrocarbons % In Extract	Saturates / Aromatics
Ore-Field Area							
1	Pine Point Gp/B	16.7–19.8	0.82	183.6	29.8	16.2	0.2
2	Pine Point Gp/B	6.1–7.6	0.32	387.8	46.2	11.9	0.2
3	Pine Point Gp/B	23.2	0.43	360.2	77.3	21.5	0.2
4	Pine Point Gp/B	36.3	1.81	527.4	168.7	32.0	0.1
5	Pine Point Gp/F	50.6	0.80	601.5	205.6	34.2	0.3
6	Pine Point Gp/F	62.8	3.86	97.8	32.6	33.3	0.2
7	Pine Point Gp/F	106.7	3.02	453.0	119.5	26.4	0.2
8	Pine Point Gp/B	132.2	0.15	610.4	375.2	61.5	2.6
9	Keg River Gp/A	138.0	0.88	364.3	95.6	26.0	0.3
10	Pine Point Gp/F	39.3	3.59	867.1	314.1	36.2	0.1
11	Pine Point Gp/F	62.5	5.13	563.7	228.5	40.5	0.2
12	Pine Point Gp/I	86.6	0.18	252.2	93.8	37.1	1.8
13	Pine Point Gp/K	41.1	0.13	1,686.3	813.2	48.2	0.8
14	Pine Point Gp/J	61.1	0.16	259.2	48.6	18.7	0.4
64	Pine Point Gp/F	116.4	3.60	298.9	64.4	21.5	0.18
74	Pine Point Gp/F	117.4	7.51	149.3	27.0	18.1	0.18
75	Pine Point Gp/B	139.3	0.95	617.5	133.1	21.5	0.15
76	Pine Point Gp/F	147.8	2.47	850.1	340.2	40.0	0.20
104	Keg River/A	41.8	nd	nd	nd	10.9	0.17
115	Pine Point Gp/F	157.3	nd	nd	nd	24.7	0.14
137	Pine Point Gp/K	0.0	0.49	157.3	38.5	24.4	0.17
151	Pine Point GP/J	7.0	0.27	165.5	26.5	18.9	0.32
Petroleum-Exploration Wells							
	McDermott el al Sulphur Pt 0-7						
20	Bit. Lst. & Sh.	231.6	10.9	31.6	22.4	71.0	0.9
	McDermott el al Hay River I-27						
21	Bit. Lst. & Sh.	246.9	7.30	66.9	48.8	73.0	1.1
22	Bit. Lst. & Sh.	248.1	6.00	52.2	34.7	65.0	0.8
	IOE Providence K-45						
23	Bit. Lst. & Sh.	277.8	4.70	155.0	77.5	50.0	1.3
	Central Del Rio Mills Lake B-75						
24	Horn River Sh.	444.1	30.10	64.0	55.5	87.0	0.6
25	Bit. Lst. & Sh.	640.4	13.00	97.0	49.5	51.5	0.7
	Central Del Rio Mills Lake A-70						
26	Bit. Lst. & Sh.	640.9	1.30	246.4	96.4	39.1	0.8
27	Lonely Bay	648.9	3.18	191.3	73.5	38.4	0.4
28	Lonely Bay	649.2	0.95	522.7	246.0	47.0	0.5
	Central Del Rio Wood Buffalo C-03						
29	Keg River	338.9	0.38	180.4	39.0	21.6	0.9
	Central Del Rio Wood Buffalo C-74						
30	Keg River	309.0	0.85	310.6	75.5	24.4	1.0
31	Keg River	325.2	0.64	172.4	74.3	44.1	2.1
	N•W•T• Big Island 1						
32	Lonely Bay	387.1	0.48	198.0	52.4	25.8	0.8
	N•W•T• Deep Bay 2						
33	Lonely Bay	319.7	1.02	87.0	27.5	31.6	0.8
34	Lonely Bay	320.0	1.00	266.4	102.7	38.6	0.7
	Punch Deep Bay 2						
35	Bit. Lst. & Sh.	298.7	0.43	403.0	186.4	46.2	1.3
36	Lonely Bay	315.4	0.19	195.6	65.2	33.4	2.6
	Punch Deep Bay 3						
37	Bit. Lst. & Sh.	248.1	3.82	131.2	45.0	34.2	1.0
38	Lonely Bay	302.0	0.20	351.4	148.6	42.3	2.4
	Punch Deep Bay 6						
39	Lonely Bay	318.5	0.23	136.3	31.4	23.4	0.8
	Briggs Trout River 4						
40	Lonely Bay	1,037.8	0.15	46.4	nd	nd	nd
41	Lonely Bay	1,039.0	0.62	52.1	23.6	45.3	2.1
	Imp. Sun Arrowhead I-46						
42	Loney Bay (equiv)	2,012.5	0.32	16.2	nd	nd	nd
43	Loney Bay (equiv)	2,033.6	0.68	8.9	nd	nd	nd
	Imp. Island River 1						
44	Pine Point Gp. (equiv)	2,250.3	0.11	75.0	45.9	61.3	1.7
45	Lonely Bay (equiv)	2,356.4	0.24	61.4	36.4	59.3	4.2
	Texaco Bovie Lake J-72						
46	Nahanni	2,969.9	0.45	50.0	37.1	74.2	2.7
47	Nahanni	2,973.9	0.44	39.7	27.9	70.3	2.6

equiv = equivalent. Bit. Lst. & Sh. = bituminous limestone and shale.

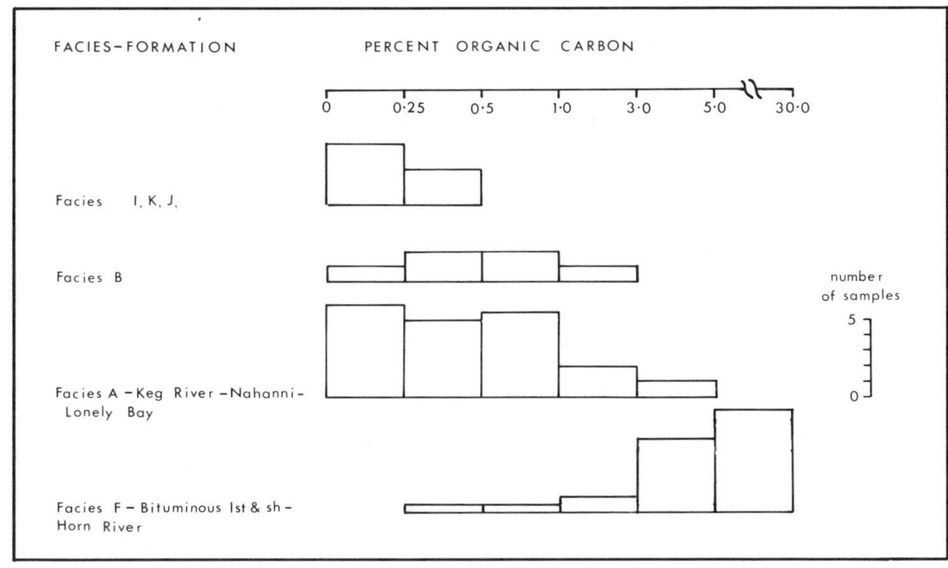

Figure 4 — Distribution of organic-carbon content according to facies type and formation.

mixture was substituted for benzene when isolating aromatics from the chromatographic column. Analysis of saturated hydrocarbon utilizing a salt eutectic column (Snowdon and Peake, 1978) was supplemented by analysis utilizing a 30-m (98.5-ft) glass capillary column coated with SP2100 (Supelco, Inc.).

Organic-carbon contents were determined on unextracted rocks after acid treatment with hot 6N HCl to remove carbonates. Care was taken to ensure that no organic carbon was lost during washing of the samples prior to carbon analysis. A Leco induction furnace was used in conjunction with a Leco WR 12 carbon determinator for organic-carbon analysis.

Bitumen concentrates were prepared by prying pieces of bitumen from the rock using a needle and extracted as described above.

Kerogen-isolation procedures are essentially those used by Powell, Cook, and McKirdy (1975), although treatment with sodium borohydride to remove pyrite was not attempted. Elemental analyses for carbon, hydrogen, and nitrogen on kerogen concentrates and bitumen residues were carried out on a Perkin Elmer 240 Elemental Analyzer.

Oil samples were separated into saturates, aromatics, NSO (nitrogen, sulfur, and oxygen compounds), and asphaltenes as described by Powell (1978) and modified as described above. Analysis of gasoline-range components was carried out on a 100-m (328-ft) stainless steel capillary column coated with squalene after helium stripping (Thompson, 1979).

Procedures for preparation of samples for microscopic examination and procedures used for microscopy were those described by Creaney (1980).

RESULTS AND DISCUSSION

Organic Carbon and Microscopy

In Figure 4 the organic-carbon data are plotted in terms of a facies type. Hunt (1972) summarized the available data concerning the organic-carbon content of carbonate rocks. He gives an average carbon content of 0.33% for carbonates from the continents, shelf, and slope and 0.28% for oceanic carbonates. By comparison, many of the samples in this study are extremely rich in organic carbon (Fig. 4; Table 1). However, this sample set is biased in favor of those rocks that are richer in organic matter. Thus rocks from lagoonal or tidal-flat deposits (facies I and J) and the diagenetic facies (facies K) are extremely light colored and were not extensively sampled. As expected, their organic-carbon contents are very low (Fig. 4). In contrast, samples from the off-reef facies (facies B) of the barrier complex and the platform carbonates (facies A—Keg River, Nahanni, and Lonely Bay Formations) have a wide range of carbon contents (Fig. 4). Skall (1975) considers that facies B represents deposition in a gently sloping distal fore-reef area between an organic-barrier facies to the south and a basinal facies to the north (Fig. 3). The wide range of organic contents reflects this transitional environment. The platform carbonates represent a similar environment in terms of wave energy (Skall, 1975). The presence of organic-carbon contents as high as 3% in both facies indicates that restricted conditions must have occurred locally with the development of euxinic conditions. South of the barrier complex in the Rainbow–Zama area, the Keg River Formation also is considered to be organic rich and bituminous (Barss, Copland, and Ritchie, 1970).

The basinal deposits (facies F—bituminous shale and limestone members) contain exceptionally high organic-carbon contents (Fig. 4), which are characteristic of this facies. Mineralogical analyses show that the F facies is richer in carbonate (>80%) than the equivalent bituminous limestone and shale unit, which lies north of the barrier. The remaining inorganic mineral is mainly quartz. Preservation of organic matter in facies F is due to the development of euxinic conditions in the area north of the barrier. The development of these conditions may have been aided by refluxing of brines from the evaporite basin through the barrier complex and along the floor of the open sea to the north (Skall, 1975; K. Williams, personal communication).

Table 1 — Location and analytical results for samples with indigenous organic matter.

Table 2

Sample No.	Formation/Facies	Association	Depth (m)	Extract (% of Bitumen)	Hydrocarbons (% of Extract)	Saturates/Aromatics	Atomic H/C	Biodegradation	Reflectance (% R_o)
Soluble Bitumens									
17	Watt Mtn/L	d	*	100.0	18.9	0.30	1.42		0.13
19	Pine Point Gp/B	d	60.0	90.0	47.3	0.30			
66	Pine Point Gp/B	d	135.3	100.0	36.1	0.17		B	
79	Pine Point Gp/B	d	37.5	100.0	35.1	0.29			0.18
113	Pine Point Gp/B	d	139.0	100.0	48.4	0.53		PB	
121	Watt Mtn/L	c	*	90.0	29.0	0.30		B	
122	Watt Mtn/L	c	*	100.0	25.4	0.20			
123	Watt Mtn/L	c	*	100.0	22.7	0.25		PB	
124	Watt Mtn/L	c	*	89.0	26.8	0.23		B	
125	Watt Mtn/L	c	*	100.0	25.2	0.23		B	0.17
127	Watt Mtn/L	c	*	56.0	35.5	0.01			0.35
131	Slave Pt/M,N	d	*	80.0	54.1	0.60		PB	
139	Pine Point Gp/K	d	*	100.0	57.6	0.32			
143	Pine Point Gp/H, I	d	*	100.0	30.9	0.60			0.13
144	nd	c	*	88.0	18.4	0.05	1.25	B	0.18
Insoluble Bitumens									
15	Watt Mtn/L	c	*	39.0	44.0	0.40			
16	Watt Mtn/L	c	*	10.0	29.0	0.39			
18	Watt Mtn/L	a	*	20.0	46.6	0.20	1.11		0.65
48	Pine Point Gp/K	a, b	127.4	0.4	53.2	0.32		PB	1.11
49	Pine Point Gp/K	a, b	131.4	9.0	21.7	0.43			
57	Pine Point Gp/B	a	*	11.0	33.6	0.40			0.52
62	Pine Point Gp/K	a	84.3	0.3	39.2	0.33	0.89	PB	1.04
71	Pine Point Gp/K	a	56.4	0.7	53.0	1.04	0.91		0.97
87	Pine Point Gp/J	a, b	101.5	1.5	42.0	1.20		PB	
90	Pine Point Gp/H, I	a	68.0	0.9	62.7	0.32	0.90		0.90
107	Slave Pt/N	a	36.9	29.8	47.2	0.69		B	
108A	Slave Pt/N	a	37.8	0.5	25.3	0.53	0.82	B	
108B	Slave Pt/N	a	37.8	0.6	47.7	0.15		B	
116	Pine Point/K	a	22.5	24.6	32.9	0.13		B	
117	Watt Mtn/L	d	*	42.6	42.4	1.33			0.85
Dispersed Bitumens—solubility not measured									
54	Pine Point Gp/K	a	40.2		31.1	0.15		B	
67	Pine Point Gp/B	d	160.3		37.6	0.21			
132	Slave Pt/M, N	b	*		53.0	0.39			
138	Pine Point Gp/K	a	*		12.8	0.18		PB	
153	Pine Point Gp/J	c	*		36.2	0.52		B	

a = associated with coarsely crystalline dolomite.
b = associated with native sulfur.
c = associated with Pb/Zn ore.
d = no particular association.

B = biodegraded.
PB = partly biodegraded.
* = from open pit.

N.B.: White coarsely crystalline dolomites are identified as facies K only where their original facies are in doubt.

Table 2 — Location and analytical results of bitumens.

Selected samples have been examined under the microscope to determine the form and distribution of the organic matter. Reflected-light microscopy under both white light and reflected fluorescent light shows that the organic matter is extremely finely divided and exhibits only faint fluorescence. The organic matter consists of a delicate mesh arranged parallel to the bedding. Large algal structures were recognized in certain samples of facies F, but in general the organic matter was too finely divided to allow recognition of discrete organic components. In certain samples under white light it was possible to see blebs of uniform, low-reflecting material, which were interpreted as bitumen or pyrobitumen. This material appears to be largely insoluble in solvents, since samples containing it generally had a lower yield of extract relative to organic carbon, and generally it was absent from those samples showing high extractability. Reflectance data obtained on these bitumens ranged from 0.33% R_o to 0.7% R_o but showed no relationship with depth. A possible mode of origin for this material is discussed subsequently.

Characteristics of Extracts and Kerogens

Samples <200 m (650 ft)

The extracts from shallow depths (<200 m or 650 ft) show some unusual and contradictory characteristics. Many of the extracts have a low proportion of hydrocarbons (<30%), a high ratio of pristane and phytane relative to n-alkanes, and low saturate-to-aromatic ratios (Table 1; Fig. 5). Such features are typical of samples from the immature or very early stages of hydrocarbon generation in conventional marine source rocks (Tissot and Welte, 1978). At the same time, however, the extract and hydrocarbon yields expressed in milligrams per gram of organic carbon (mg/g C_{org}) show a large variation. Over half the samples have yields which equal or exceed the values normally encountered in source rocks at the peak stage of hydrocarbon generation (extract yield 250–350 mg/g C_{org}; hydrocarbon yield 50–120 mg/g C_{org}) (Tissot and Welte, 1978). Values of extract yield greater than 600 mg/g C_{org} indicate that more than half the organic matter in the rock is in soluble form.

High hydrocarbon yields (> 160 mg/g C_{org}) are usually encountered where there is staining brought about by petroleum migration or contamination (Powell, 1978; Snowdon, 1980). Under these circumstances the organic-carbon values are generally low, and the proportion of hydrocarbons in the extracts is high. In Figure 6 the extract yield is plotted as a function of the percentage of hydrocarbons in the extract and saturate to aromatic ratios for those samples occurring at less than 200 m (650 ft) depth. While there is a general relationship between the extract yield and the percentage of hydrocarbons in the extract, there is no apparent relationship to organic carbon. Some of those samples with the highest extract yields also have high organic-carbon contents. Three samples are clearly stained. In one case the high yield (1686.3 mg/g C_{org}, sample 13) can be attributed to oil migrating from a mature source. This sample has a high proportion of hydrocarbons in the extract and a high saturate-to-aromatic ratio. A gas chromatogram of the saturate fraction shows that the extract has been biodegraded. The extremely high extract yield of 1686.3 mg/g C_{org} is theoretically impossible and indicates loss of hydrocarbons and thus carbon from the sample during acid digestion prior to carbon analysis. Samples 8 and 12 appear to be contaminated with pipe grease, based on the diagnostic hump of naphthenic alkanes.

It is difficult to attribute the remaining high yields to staining by mature hydrocarbons because of the overall similarity in composition of the low- and high-yield extracts and their immature aspect. Rather, the finely laminated nature of the organic matter, in these rocks suggests

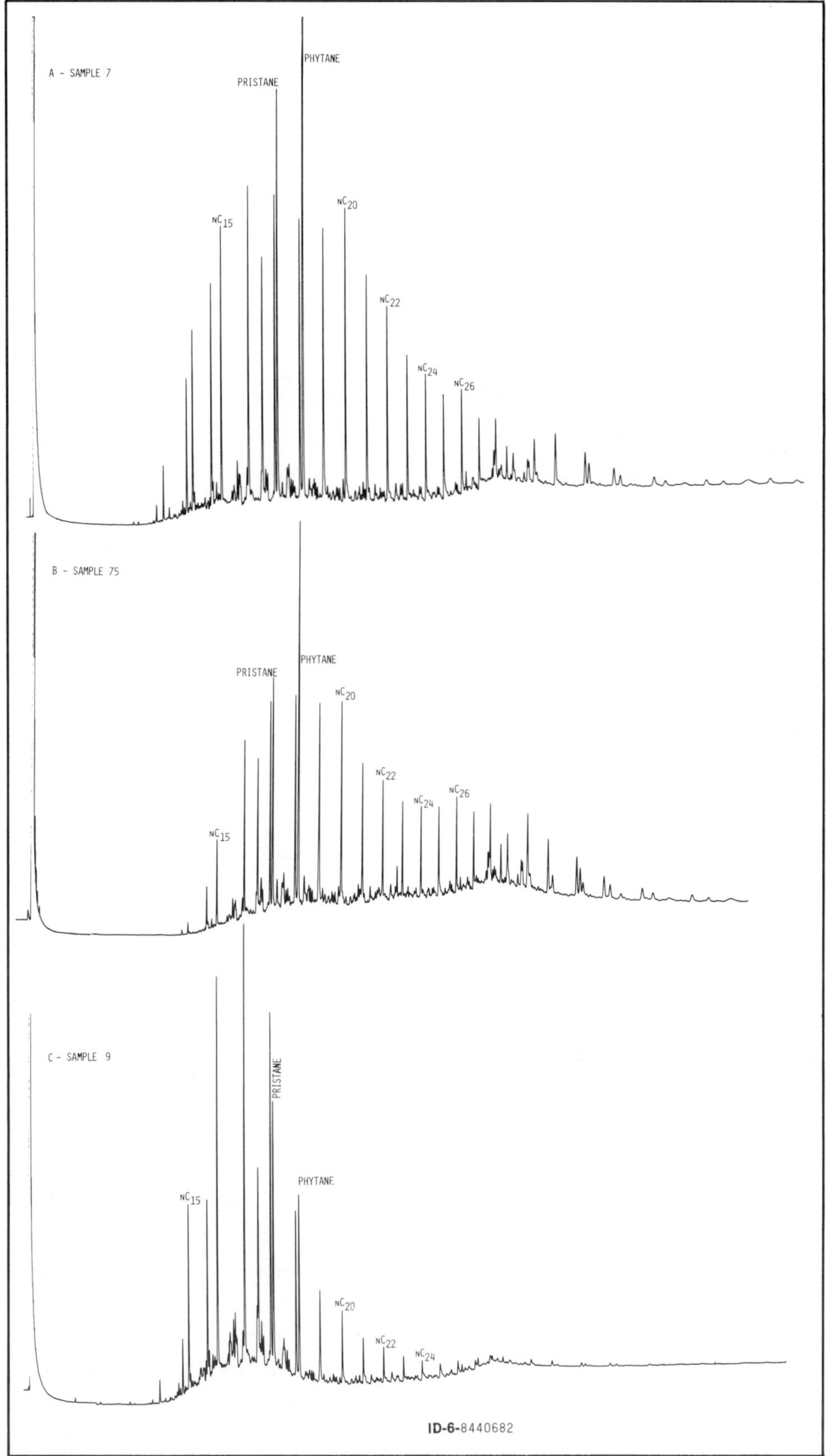

Figure 5 — Representative gas chromatograms (capillary column) of saturated hydrocarbons from shallow samples (< 200 m or 650 ft depth).

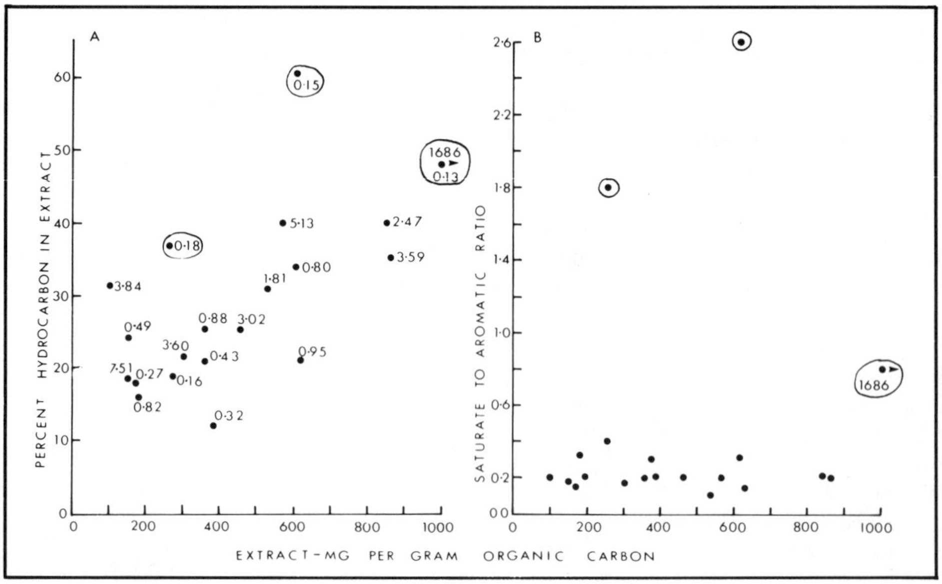

Figure 6 — Variation in hydrocarbon content of extracts and saturate-to-aromatic ratio with extract yield for samples less than 200 m (650 ft) depth. Figures adjacent to points in A are organic-carbon values. Circled samples are those definitely thought to be stained or contaminated: samples 8, 12, 13 (Table 1).

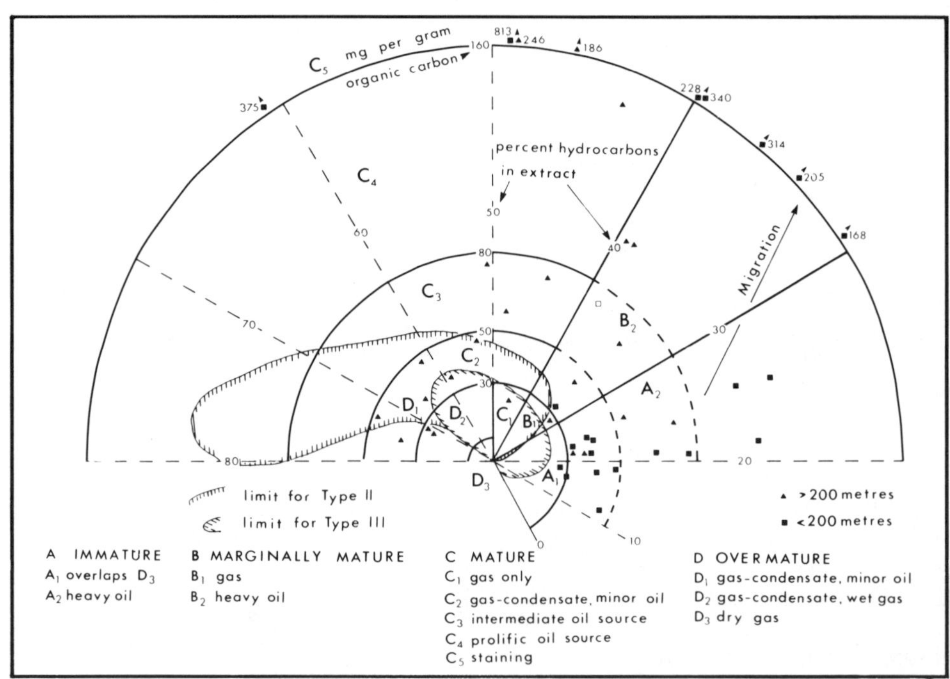

Figure 7 — Source-rock potential of samples based on extract data. Diagram is a plot on polar coordinates (Powell, 1978): Γ = yield of hydrocarbons in mg per gram organic carbon; Θ = percentage of hydrocarbons in extract. Fields for type II and type III organic matter are derived from Powell (1978).

that the soluble organic matter is indigenous and that local accumulations arise by migration within the parent rock. The effects of migration and the unique composition of these extracts are illustrated in Figure 7. Here the hydrocarbon yields are plotted against percentage of hydrocarbons in the extract, using polar coordinates. This diagram has been used for interpretation of extract data in terms of state of maturation, quality of organic matter, and hydrocarbon product (Powell, 1978). The diagram is based on the observation that extracts obtained by a solvent system comprising methanol and chloroform change in a regular fashion with maturation (e.g., Tissot and Welte, 1978). In Figure 7 the radial distance from the origin represents the hydrocarbon yield in mg/g C_{org} and the angle θ is related to the percentage of hydrocarbons in the extract. The horizontal axes have been arbitrarily set at 20 and 80%. In conventional source rocks at the immature stage (zone A, Fig. 7) the hydrocarbon yields and percentage of hydrocarbons in the extract are low (< 10 mg/g C_{org} and 25%, respectively). With increasing maturation (zone B-C, Fig. 7) the hydrocarbon yields and percentage of hydrocarbons in the extract increase, and the compositional field curves out from the origin in a counterclockwise direction. Increasing distance from the origin indicates increasing hydrocarbon-source potential (zone C_1-C_3, Fig. 7). As the overmature zone is reached, the hydrocarbon yields decrease owing to cracking, and the trend lines curve back to the origin (zones D_1 to D_4). The data from the shallow samples at Pine Point fall outside the range of extract yield and composition in

Table 3

Sample	%C	%H	%N	Ash (%)	Atomic H/C
Ore-Field Area (shallow depth <200 m)					
4	75.66	7.88	2.21	3.84	1.25
6	71.54	7.51	1.84	17.88	1.26
7	69.44	7.00	1.79	15.94	1.21
Petroleum-Exploration Wells (intermediate depth 400–600 m)					
26	53.24	3.62	1.13	37.62	1.12
37	81.15	7.80	2.93	3.20	1.15
28	82.01	8.55	1.54	1.35	1.24

Table 3 — Kerogen data.

conventional source rocks (Fig. 7). It is unusual to encounter hydrocarbon yields as high as 120-140 mg per gram of organic carbon and representing less than 30% of the extract, since this latter feature is usually associated with immature samples. Samples with hydrocarbon yields greater than 160 mg/g C_{org} show an enrichment in hydrocarbons over the lower yield samples owing to local migration within the source.

There is some variation in the character of the gas chromatograms of saturates from the shallow samples. Generally the chromatograms show a high ratio of the isoprenoid alkanes, pristane and phytane, to the normal alkanes (Fig. 5A, B); low pristane-to-phytane ratios (<1); an n-alkane distribution that reaches a maximum of n-C_{16}–n-C_{18}; and often an even–odd predominance in the n-alkane distribution. This type of distribution is characteristic of immature or marginally mature samples containing organic matter of microbial or algal origin deposited under extremely reducing conditions (Tissot and Welte, 1978). Within the barrier complex of the ore-field area, this distribution is particularly characteristic of the basinal facies (facies F). In contrast, gas chromatograms of saturate hydrocarbons from samples of the platform carbonate (facies A) show no particular predominance of pristane and phytane over the n-alkanes; they do show a smooth distribution of n-alkanes, reaching a maximum of n-C_{16}–n-C_{18}, and a pristane-to-phytane ratio above 1 (Fig. 5C). These samples were deposited in less severely reducing conditions. Samples from the fore-reef facies (facies B) and from the bituminous limestone and shale unit (facies F–equivalent away from the barrier) can show either set of characteristics, although in the latter case the even-to-odd predominance is never apparent, presumably because of the slightly greater thermal alteration from greater depth of burial.

Extremely high yields of extractable organic matter relative to total organic matter have been observed previously from immature samples of carbonates, phosphorites, and phosphatic shales (Powell, Cook, and McKirdy, 1975; Claypool, Love, and Maughan, 1978). This result contrasts with normal shales, from which the yields of extract at the immature stage of thermal catagenesis are typically low (<100 mg/g C_{org}). Palmer and Zumberge (1981) observed that the proportion of extractable organic matter in extracts from Mediterranean evaporites is also extremely high. This feature appears, therefore, to be characteristic of extremely reducing conditions in which the extent of the polymerization is retarded so that the ratio of extract, particularly NSO and asphaltenes, to insoluble organic matter is abnormally high.

The behavior of organic matter of this type during catagenesis is somewhat uncertain. With increasing burial depth (Fig. 8) there is an overall decrease in extract yield but a corresponding increase in the proportion of hydrocarbons in the extract. This observation suggests that NSO and asphaltenes are being removed from the system. Similar effects have been observed during catagenesis of phosphorites and a Tertiary carbonate sequence from northwest Australia (Powell, 1975; Powell, Cook, and McKirdy, 1975). In both cases there appeared to be loss of asphaltic components from the extract. The most likely mechanism is polymerization of the NSO–asphaltene fraction to form insoluble kerogen/pyrobitumen. Such a mechanism would explain the occurrence of bitumens or pyrobitumens with varying degrees of reflectivity in many of the samples. Direct evidence of this process was obtained from the discrete bitumens, which are discussed below. In this respect, the atomic hydrogen-to-carbon ratios of the total organic matter from selected samples are of interest (Table 3). The organic matter was isolated by acid digestion without prior extraction. Samples 4 and 7 from the ore-field area have high extract yields, and the atomic hydrogen-to-carbon ratios of the organic-matter concentrates are similar (1.25 and 1.21). Furthermore, they are like sample 6 (1.26), which has a low extract yield, suggesting that the overall composition of the kerogen is similar to the extract at this level of maturity. Three samples also were analyzed from the bituminous limestone and shale unit at slightly greater depth. In this case the two samples with somewhat lower extract yields had lower atomic hydrogen-to-carbon values (1.12 and 1.15), whereas the sample with a higher extract yield also had a higher atomic hydrogen-to-carbon ratio (1.24). These results suggest that in these samples the kerogen is more depleted in hydrogen than the extract, a condition which may be attributed to catagenesis. The atomic hydrogen-to-carbon ratios indicate that the organic matter is probably type II of Tissot et al (1974).

Samples >200 m (650 ft)

With increasing burial depth (>200 m or 650 ft), the extract and hydrocarbon yields relative to organic carbon are generally lower than were obtained from the shallow samples (Figs. 7, 8). The deeper samples tend to have a higher proportion of hydrocarbons in the extracts, although there is much variation (Fig. 8) amongst the shallower samples. The deeper samples have higher saturate-to-aromatic ratios, reflecting their greater degree of catagenesis (Table 1; Fig. 9).

Gas chromatograms of the saturated hydrocarbons of the deeper samples show smooth distributions of n-alkanes and a smooth naphthene hump, reflecting their higher degree of alteration (Fig. 10). Pristane and phytane occur in less concentration than the adjacent n-alkanes, and the pristane-to-phytane ratios are approximately unity. Gas chromatograms of the saturated fraction of the

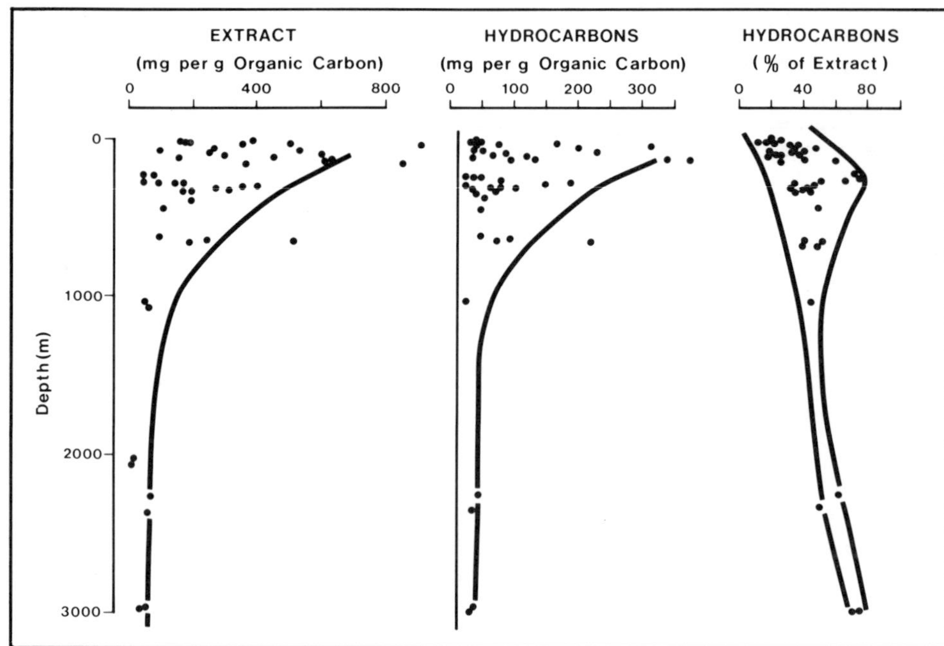

Figure 8 — Extract and hydrocarbon yields as a function of depth.

deepest samples (below 2,000 m or 6,500 ft) indicate that the extracts have undergone extensive catagenesis because few hydrocarbons are recorded above n-C_{20} (Fig. 10C). A number of the deep samples show enhanced concentrations of the normal C_{20} to C_{21} hydrocarbons. The presence of these hydrocarbons is attributed to contamination from core-box wax (Snowdon and Powell, 1978).

The dramatic decline in hydrocarbon yields is a combination of a change in facies type and depth of burial. The following tentative limits have been placed on the zones of catagenesis: immature to marginally mature, <400 m (1,300 ft); mature, 400 to 2,000 m (1,300–6,500 ft); overmature, $>2,000$ m (6,500 ft). These depths are based on extract yields and character of the gas chromatograms.

Relation of Bitumens and Heavy Oil to Host Rocks

Bitumen Occurrence

Conventional use of the terms bitumen and kerogen with respect to petroleum source rocks describes those materials that are soluble and insoluble, respectively, in organic solvents (e.g., Tissot and Welte, 1978). On the other hand, accumulations of black, tarry organic material or brittle, obsidian-like organic material in the vugs of carbonate rocks are commonly termed bitumens (e.g., Rogers, McAlary, and Bailey, 1974). The latter usage is the usage of the term bitumen herein.

The bitumen and heavy oil in the ore-field area occurs in vugs in the following forms: (1) as liquid "heavy oil" oozing from freshly broken surfaces within most of the carbonate units of the barrier complex; (2) as soft, dark-brown, tarry, asphalt-like material; and (3) as moderately hard to hard, black, glassy, commonly brittle, obsidian-like material, especially in dolomites of facies K of the barrier complex but also in other carbonate units. Some of this latter material is friable and commonly is intimately mixed with large calcite crystals and, locally, with free sulfur. Large accumulations of bitumen weighing several hundred grams are not uncommon.

The bitumens are divided into two forms on the basis of solubility. One form is largely soluble, and the other, mostly insoluble (Table 2). The degree of solubility of the bitumens is not always related to their physical appearance in the field. Thus some glassy and brittle bitumens are soluble in organic solvents, whereas others are not. The soluble bitumens include all the liquid "heavy oil" and all the soft, tarry, asphalt-like material noted above, and some of the hard, black, glassy material. Apparently, soluble bitumen can occur within any of the stratigraphic units developed within the ore-field area, but it appears to be most abundant within facies B (off-reef) of the Pine Point Group, facies L of the Watt Mountain Formation, and facies N of the Slave Point Formation (Table 2). Soluble bitumen is present in drill cores and in samples collected from ore-body open pits (Table 2). In contrast, the insoluble bitumens are intimately associated with white, coarsely crystalline dolomite (Table 2).

Biodegradation

Many of the soluble bitumens, and some of the insoluble bitumens, are biodegraded, as indicated by the gas-chromatographic character of their saturated, hydrocarbon fractions (Fig. 11). Biodegradation usually occurs where hydrocarbons are in contact with circulating ground water, which brings oxygen and bacteria in contact with the hydrocarbons (Winters and Williams, 1969). The bacteria selectively metabolize the n-alkanes, followed by the isoprenoid alkanes, pristane and phytane (Bailey, Jobson, and Rogers, 1973). Gas chromatograms of biodegraded, partly biodegraded, and nonbiodegraded bitumens are illustrated in Figure 11. Fully biodegraded bitumens are those that show complete removal of n-alkanes and isoprenoid alkanes; partly biodegraded samples are those that show removal of some or all n-alkanes and some isoprenoid alkanes. Biodegraded and partially biodegraded samples are indicated in

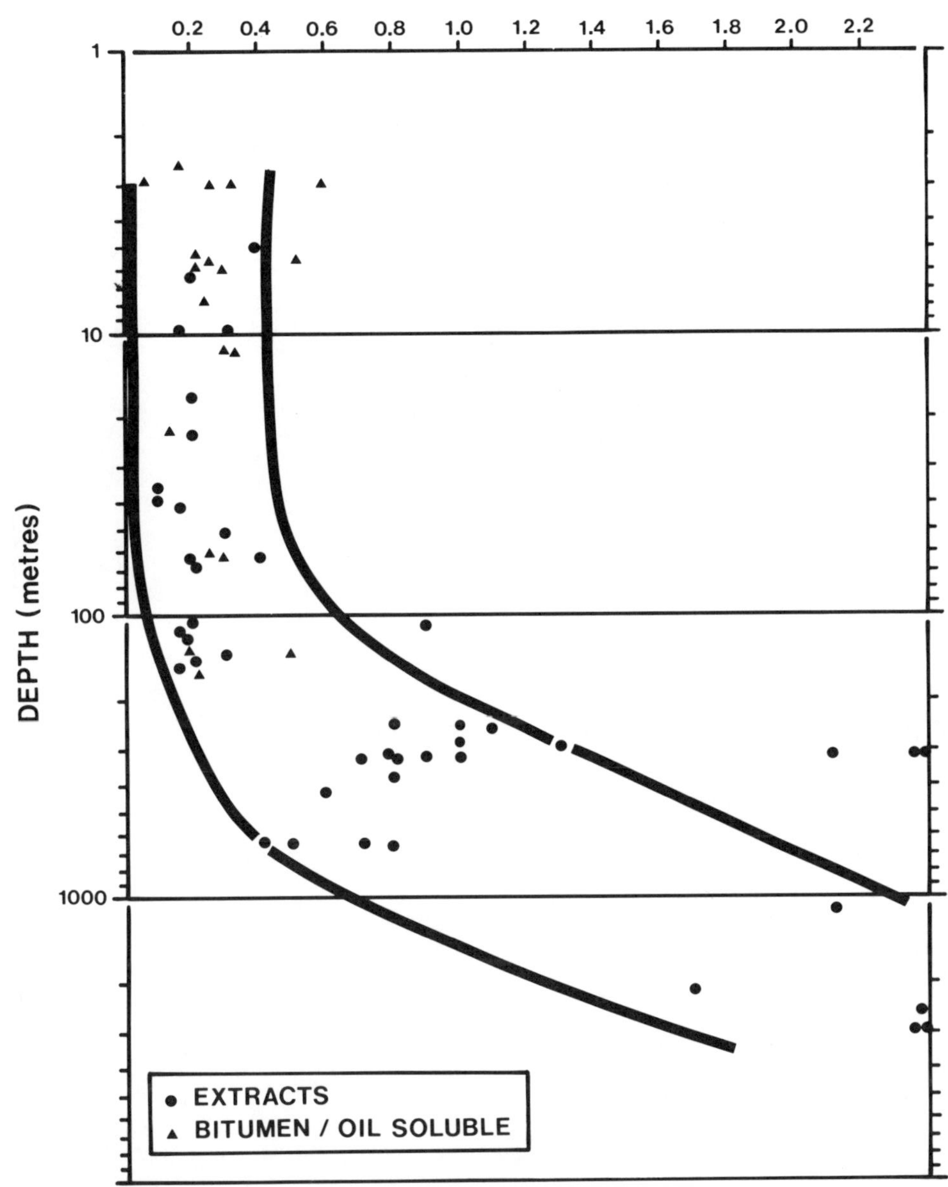

Figure 9 — Variation in saturate-to-aromatic ratio with depth in noncontaminated samples. Note that depth scale is logarithmic.

Table 2. There is no particular pattern to biodegradation, although surface and near-surface samples may show more evidence of biodegradation. Biodegraded samples are confined to samples taken from the ore-field area.

Soluble Bitumens

It is difficult to determine whether those soluble bitumens in which there is a predominance of pristane and phytane over the adjacent n-alkanes are thermally immature or partly biodegraded (e.g., Fig. 11A). The former explanation is preferred, because similar gas-chromatographic patterns are observed in most of the rock extracts from the area (Fig. 5). These extracts were derived from indigenous hydrocarbons that could not have been expected to have undergone biodegradation. This conclusion is supported in that the saturated fractions from certain of the non-biodegraded bitumens show low pristane-to-phytane ratios and an even-to-odd predominance, as do the thermally mature source rocks described above (Figs. 5, 11). Further, the saturate-to-aromatic ratios of the bitumens are similar to those of extracts from the immediately surrounding rocks and do not have the higher values (> 1) that would be expected if the bitumens had migrated from the basin to the west (Fig. 9). Local migration of the bitumen from the thermally immature rocks of the Pine Point barrier appears to be a consequence of the high hydrocarbon source potential of these immature rocks (see below).

Insoluble Bitumens

Many of the bitumens are essentially insoluble in organic solvents. In particular, those bitumens intimately associated with white, coarsely crystalline dolomites are largely insoluble (Table 2). The extractability of the essentially insoluble bitumens ranges from as low as 0.3% to as high as 30% (Table 2). The proportions of hydrocarbons in the small amounts of extracts obtained from these bitumens, and the saturate-to-aromatic ratios, cover the range of values observed in the soluble bitumens (Table 2). Generally, the extracts from the samples of low extractability show a mature distribution of hydrocarbons with a smooth naphthenic hump (Fig. 12). Sample 18 is an exception and has a distribution more like the soluble samples (Fig. 12).

The question arises, "What is the relationship between the soluble bitumens and the insoluble bitumens?" The only reasonable explanation is that the insoluble bitumens were once similar to the soft, soluble bitumens but that they have undergone local heating, possibly in the presence of sulfur, which has resulted in polymerization of the NSO/asphaltene fraction to an insoluble polymer. This hypothesis of local thermal alteration is supported by the fact that the reflectance of many of these insoluble bitumens is relatively high, at 0.85% R_o, which classifies them as pyrobitumens of the epi-impsonite type and contrasts with the reflectance of the soluble bitumens at $<0.35\%$ R_o (Table 2). In polished sections under the microscope, the insoluble bitumens also show evidence of flow structure, which is typical of bitumens that have been melted. The atomic H/C ratio of the insoluble bitumens is lower than that of the soluble bitumens, which also supports a thermal-alteration concept (Table 2). Sample 18 may represent an intermediate case between soluble bitumens and insoluble bitumens. Its extractability is relatively high (20.0%), and the atomic hydrogen-to-carbon ratio of the unextracted bitumen at 1.11 is in between the ratios for the two types of bitumens. The ratio for the extract residue (0.95) is similar to the other insoluble bitumens. Furthermore, the gas-chromatographic character of the saturated hydrocarbons of sample 18 is like the soluble samples (compare Figs. 11a and Fig. 12a), and its reflectivity at 0.65% R_o is lower than the other insoluble bitumens at $>0.85\%$ R_o. The association of these bitumens with white, coarsely crystalline dolomite (hydrothermal dolomites of Fritz and Jackson, 1972), and uncommonly with free sulfur, supports the interpretation that these insoluble bitumens are thermally altered. On the basis of available fluid-inclusion data and carbon-isotopic data, a temperature of

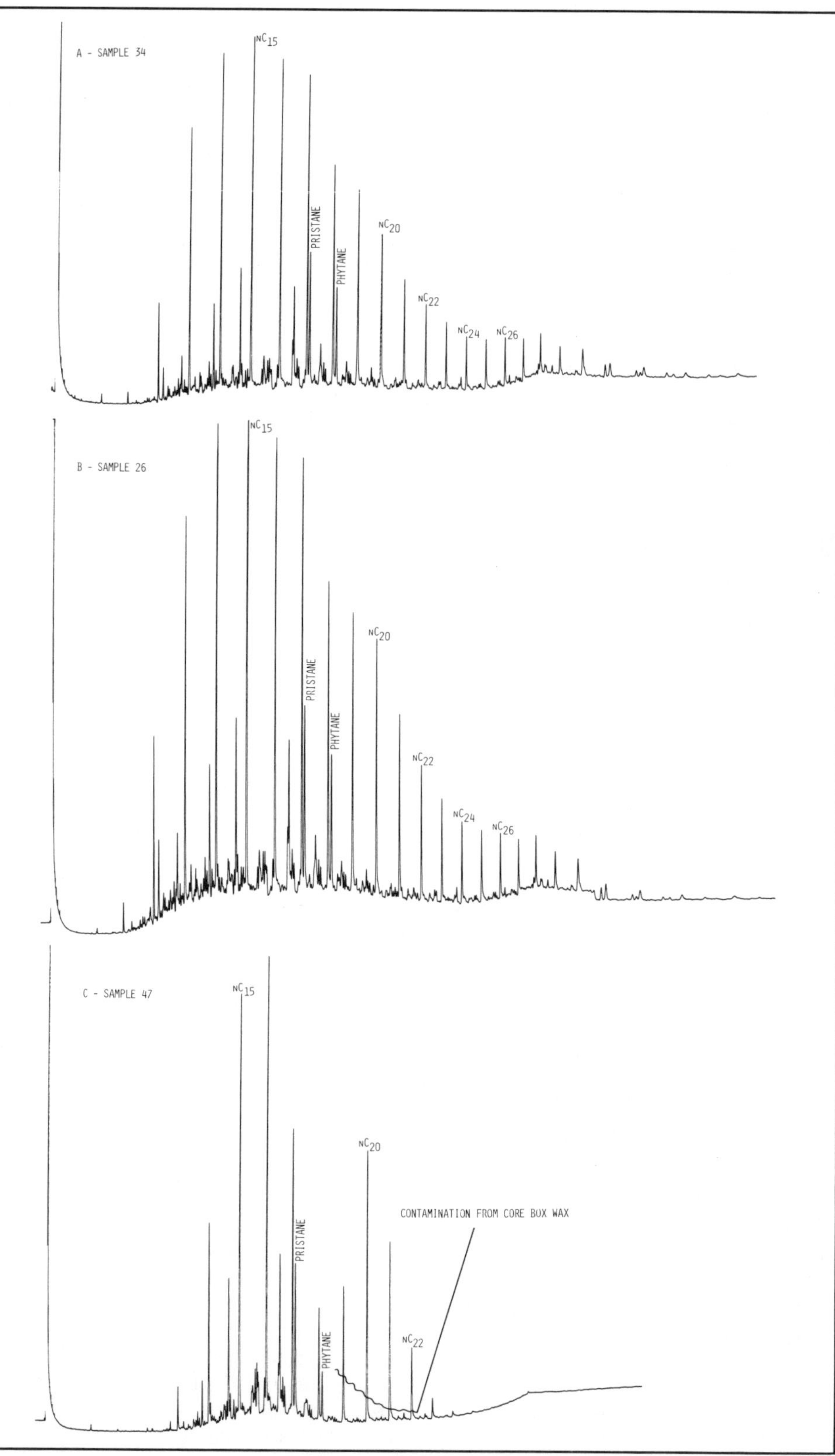

Figure 10 — Representative gas chromatograms (capillary column) of saturated hydrocarbons from deep samples (>200 m or 650 ft depth).

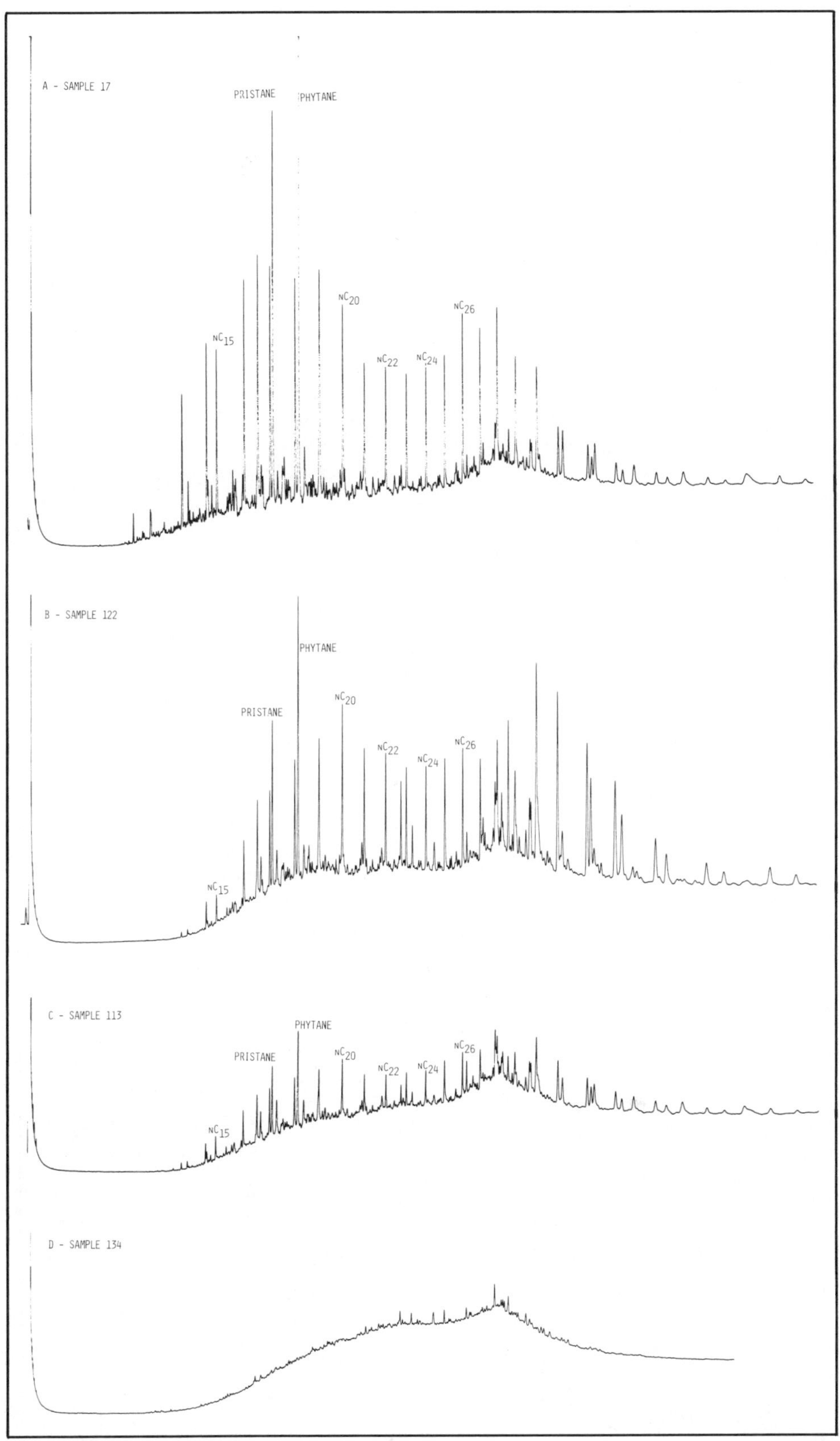

Figure 11 — Gas chromatograms of saturated hydrocarbons of soluble bitumens, showing various degrees of biodegradation: A = non-biodegraded; B = non-biodegraded; C = partially biodegraded; D = biodegraded.

approximately 100°C (212°F) is postulated for the dolomitization event (Skall, 1975).

Source-Rock Potential

The only truly applicable definition of a source rock is "a rock that has demonstrably been the source for migrated oil and gas." This definition infers that a correlation has been made between the hydrocarbon in the reservoir and its source. Potential source rocks are considered to have all the attributes of source rocks in that they contain a suitable organic-matter type at the correct level of maturation. It is also possible to recognize "immature potential source rock." These rocks have all the attributes of source rock except maturity. One might postulate that these immature potential sources could become potential source rocks, given sufficient burial. These considerations are important in evaluating a frontier region (Powell, 1978), but the distinctions become blurred in this case, as discussed below.

Source-rock quality is a function of two variables, the quality of the organic matter and the amount of the organic matter. The quality of the organic matter is best expressed in terms of hydrocarbon yield in milligrams per gram of organic carbon (mg/g C_{org}). Thus a hydrocarbon concentration of 700 ppm in a rock containing 1% organic carbon is far more significant in terms of source potential than the same hydrocarbon concentration in a rock with 4% organic carbon. The degree of saturation of the insoluble organic matter is considerably higher in the former than in the latter and will more readily yield hydrocarbon for migration. On this basis, Powell (1978) considered that hydrocarbon yields greater than 80 mg/g C_{org} indicate excellent source-rock potential for oil in respect to organic-matter type; yields of 50 to 80 mg/g C_{org} indicate good source potential for oil; yields of 30 to 50 mg/g C_{org} indicate marginal source potential for oil, and yields less than 30 mg/g C_{org} indicate no potential for oil. According to these criteria, many of the samples in this study have good to excellent source potential as far as quality of organic matter is concerned (Table 1; Fig. 7). This includes many of those shallow samples that show immature characteristics. In fact, it might be construed, according to the definitions outlined above, that some of these shallow samples are actually source rocks, because primary heavy oil has been shown to have migrated from them. In contrast, other samples from shallow depth have low hydrocarbon yields, and the atomic hydrogen-to-carbon ratio of the kerogen indicates that with further maturity they would generate hydrocarbons. As such, they could be classed as immature potential source rocks.

Powell (1978) derived a diagram for interpreting source-rock potential from extract and hydrocarbon yields. This diagram shows that many of the shallow samples are immature or have heavy-oil potential (Fig. 7). The real question is whether the so-called bitumens represent early heavy oil that has undergone some inspissation in the shallow regime or whether they represent migratory bitumens such as gilsonite. The former appears likely, since immature heavy oils are known to occur in similar settings in the eastern Mediterranean (De Groot, 1980). Certainly at slightly higher levels of catagenesis, and with the removal of the NSO and asphaltene components by polymerization, these rocks would continue to be good to excellent potential source rocks. This is illustrated by the more deeply buried samples from the petroleum-exploration wells (Fig. 7; Table 1).

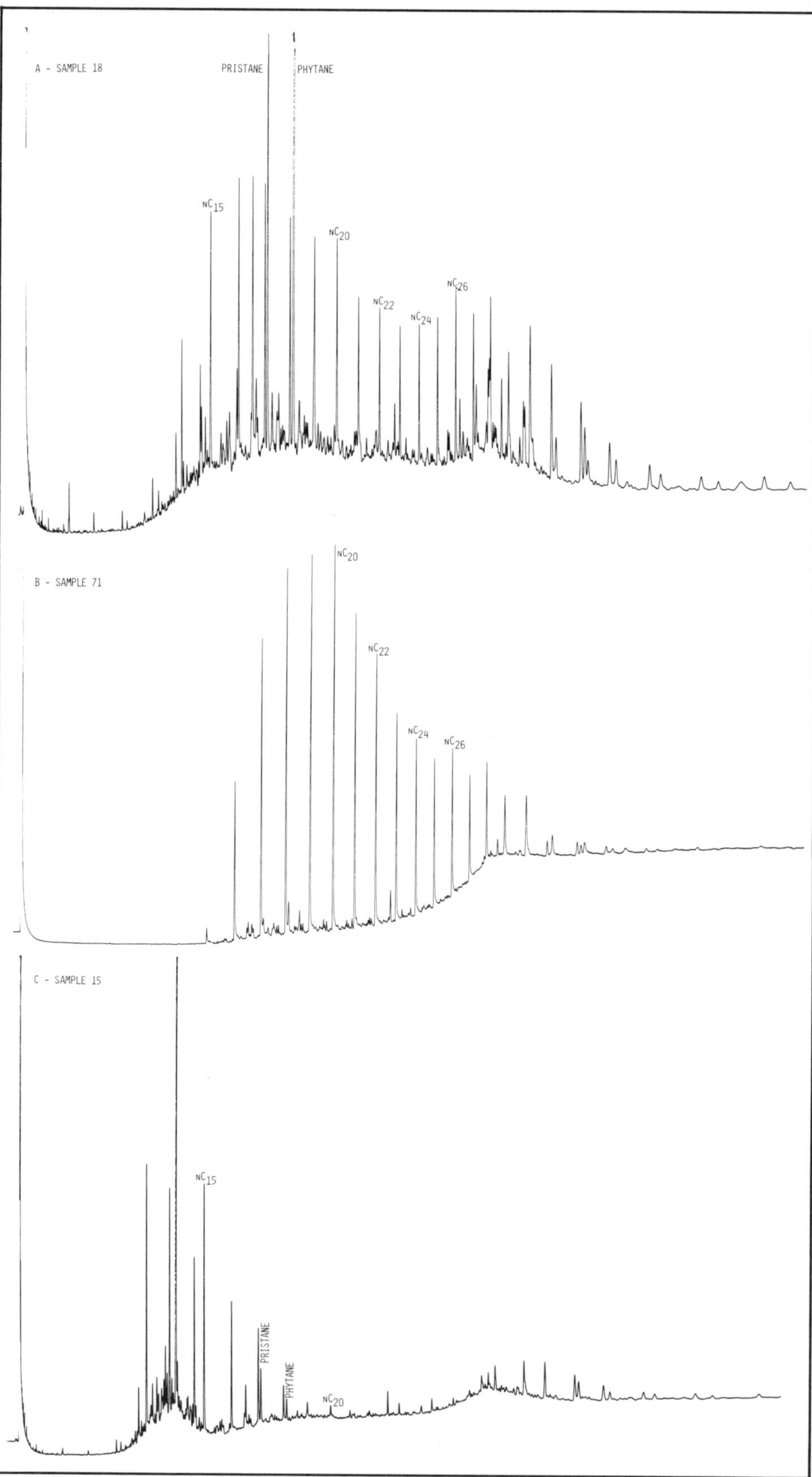

Figure 12 — Representative gas chromatograms of saturated hydrocarbons of partly soluble (A) and largely insoluble (B, C) bitumens.

Characteristics of Oil Samples

Of the 13 oil samples examined, 11 were from the Rainbow field, 1 was from the Zama field, and 1 was from the Slave Point Formation west of the ore field. The oil sample from the Slave Point Formation was recovered from a shallow diamond-drill hole (about 600 m or 1,970 ft) whereas the Zama and Rainbow samples came from depths of 1,414 and 1,734 m (4,639 and 5,690 ft). These latter samples were recovered from the Keg River pinnacle reefs.

The gas chromatogram of saturated hydrocarbons of the Zama sample (Fig. 13) shows many of the characteristics seen in chromatograms of the extracts from the barrier reef in the ore-field area, although the oil is obviously somewhat more mature. These characteristics include a pristane-to-phytane ratio of less than 1, a high phytane-to-n-C_{18} ratio, and a high content of steranes and triterpanes. The saturate-to-aromatic ratio of the Zama samples is 0.8, reflecting the higher degree of maturation of this oil when compared to the bitumens and extracts from the Pine Point area. The

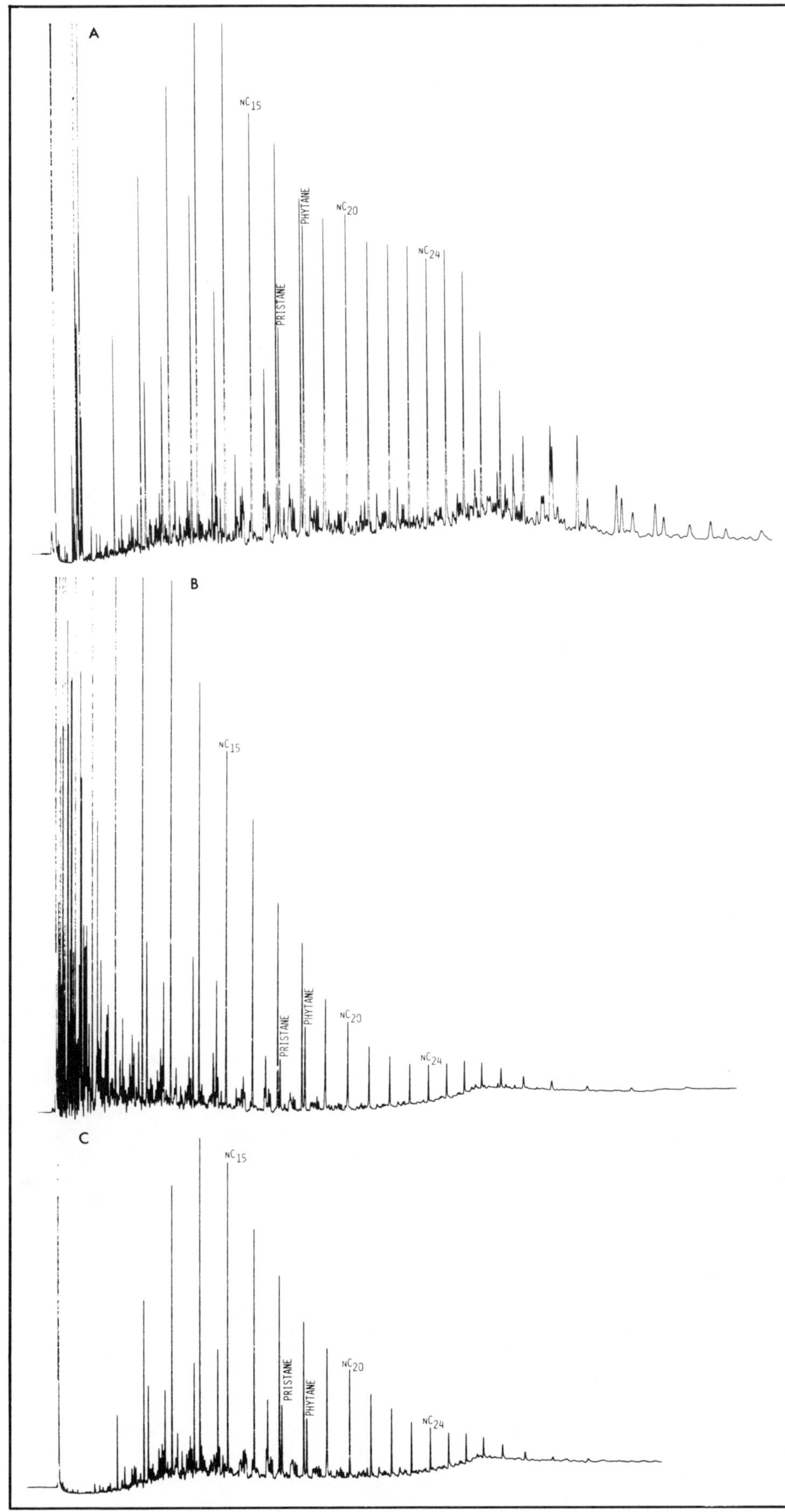

Figure 13 — Gas chromatograms of saturated hydrocarbons from conventional oil samples: A = Zama oil; B = Rainbow oil; C = oil from Slave Point Formation.

more deeply buried Rainbow samples are even more mature and have lower sterane and triterpane contents, lower ratios of phytane to n-C_{18} (Fig. 13), and higher saturate to aromatic ratios (>2). The difference in maturity between the Rainbow and Zama oils also is reflected in the composition of the gasoline-range fraction. Paraffin Index I[2] (Thompson, 1979) has a value of 1.01 in the Zama sample, whereas the Rainbow samples fall in the range 1.97–2.70.

The sample from the Slave Point Formation west of the ore-field area resembles the samples from a less euxinic-source facies at a slightly higher level of maturation than was encountered in the Pine Point mining area. The pristane-to-phytane ratio of this Slave Point oil is greater than 1, and the n-C_{17}-to-pristane ratio also is substantially greater than 1 (Fig. 13). The chromatogram resembles those from samples 26 and 34 (Fig. 10), which are from depths of 640.9 and 320 m (2,103 and 1,050 ft), respectively. This Slave Point oil comes from below 400 m (1,300 ft) and probably has not migrated extensively.

The Rainbow and Zama oils are derived from a source rock deposited under highly reducing conditions similar to those under which facies F sediments were deposited. The likely source is either a restricted facies of the Keg River platform or the bituminous laminated evaporites that immediately overlie the platform carbonate between the reefs in the evaporite basin behind the barrier-reef complex (Davies and Ludlum, 1973).

In addition to the oil, partially soluble bitumens occur extensively in the Rainbow and Zama pinnacle reefs (Rogers, McAlary, and Bailey, 1974). Their percentage of extractability ranges from 0.4 to 30%, and the atomic hydrogen-to-carbon ratio of the extracted residues is 0.53 to 0.74. Rogers, McAlary, and Bailey (1974) consider that gas-deasphalting is the primary cause of the formation of these reservoir bitumens, but they also

[2]Paraffin Index I = (2 − methylhexane + 3-methylhexane)/(dimethylcyclopentane isomers: 1c3-, 1t3-, 1t2-).

acknowledge the role of thermal processes in converting the asphalt precipitates into soluble bitumens. The data from the Pine Point region in this study suggest that mild thermal events are a viable mechanism for converting NSO and asphaltenes into pyrobitumens. The lower atomic hydrogen-to-carbon ratios of the Rainbow and Zama bitumens reflect a greater level of catagenesis than that of the bitumens in the ore-field area.

SUMMARY AND CONCLUSIONS

The barrier-reef complex developed on a widespread carbonate platform and affected Middle Devonian sedimentation throughout a wide area of western Canada. The preservation of organic matter in the various facies of the barrier complex can be related directly to their environments of deposition. The platform carbonate developed in relatively deep water, and euxinic conditions prevailed from time to time, allowing the preservation of organic matter. With the development of the barrier complex, euxinic conditions developed as a result of stratification in front of the barrier. The fore-reef facies (facies B) occupied a transitional position between moderately euxinic conditions and the extremely euxinic conditions in which the basinal facies (facies F) was formed. The extremely reducing environment resulted in anomalous conditions of preservation that prevented extensive polymerization of the organic materials into kerogen. As a result, extracts from relatively immature rock in the Pine Point area may represent up to half the organic matter, consisting mostly of NSO and asphaltenes. The extremely reducing conditions are indicated by the low pristane-to-phytane ratios (< 1) and the even-to-odd predominance in the n-alkane distributions. Where reducing conditions were less severe—e.g., in platform carbonates or the bituminous shale and limestone unit north of the barrier and in parts of the fore-reef facies (facies B)—the pristane-to-phytane ratios are higher (≥ 1) and no even/odd predominance exists in the n-alkanes.

Accumulations of bitumen in the ore-field area are attributed to local migration from the immediately surrounding host rocks and appear to be a consequence of the extremely high extract-to-organic-carbon ratios observed in the unstained rocks. These bitumens are similar in composition to the extracts from surrounding rocks in having a high content of NSO and asphaltenes, a low saturate-to-aromatic ratio, a high ratio of pristane and phytane to n-alkanes, a low pristane-to-phytane ratio, and an even-to-odd predominance in the n-alkanes. These migrated bitumens have been converted to insoluble bitumens during the dolomitization processes that led to the formation of the coarsely crystalline dolomite. This dolomitization occurred at a somewhat higher temperature than was achieved by normal burial diagenesis.

ACKNOWLEDGMENTS

This study would not have been possible without the cooperation of Cominco, Ltd., which allowed access to diamond-drill cores and open pits at Pine Point, Northwest Territories of Canada. The author also is indebted to R. W. Macqueen, who cooperated with the author on the aspects of the study concerned with lead–zinc mineralization and who collected the samples from Pine Point. K. Williams provided invaluable advice concerning aspects of the regional geology and collected samples from the petroleum-exploration wells. Esso Resources (Canada), Ltd., and Aquitaine (Canada), Ltd., kindly provided samples of oils from the Rainbow and Zama fields. W. Kalkreuth conducted the microscopic work. Finally, the author is indebted to M. Northcott, S. D'Entremont, R. Fanjoy, and R. Davidson for technical assistance.

NOTE ADDED IN PRESS

Subsequent work by Powell and Macqueen (1984) has shown that insoluble bitumens are formed from soluble bitumens by reaction with elemental sulfur during dolomitization and mineralization.

REFERENCES CITED

Bailey, N.J.L., A.M. Jobson, and M.A. Rogers, 1973, Bacterial degradation of crude oil: Comparison of field and experimental data: Chemical Geology, v. 11, p. 203–221.

Barss, D.L., A.B. Copland, and W.D. Ritchie, 1970, Geology of Middle Devonian reefs, Rainbow area, Alberta, Canada, *in* M.T. Halbouty, ed., Geology of giant petroleum fields: AAPG Memoir 14, p. 19–49.

Basset, H.G., and J.G. Stout, 1967, Devonian of Western Canada, *in* International Symposium on the Devonian System, Calgary, 1967: Calgary, Alberta, Alberta Society of Petroleum Geologists Proceedings, v. 1, p. 717–752.

Claypool, G.E., A.H. Love, and E.K. Maughan, 1978, Organic geochemistry, incipient metamorphism and oil generation in black shale members of Phosphoria Formation, Western Interior United States: AAPG Bulletin, v. 62, p. 98–120.

Creaney, S., 1980, The organic petrology of the Upper Cretaceous Boundary Creek Formation, Beaufort–Mackenzie Basin: Bulletin of Canadian Petroleum Geology, v. 28, p. 112–129.

Davies, G.R., and S.D. Ludlum, 1973, Origin of laminated and graded sediments, Middle Devonian of western Canada: Geological Society of America Bulletin, v. 84, p. 3527–3546.

De Groot, K., 1980, Origin, migration and accumulation of hydrocarbons, *in* Panel discussion: Heyden, London, 10th World Petroleum Congress Proceedings, v. 2, p. 50.

Fritz, P., and S.A. Jackson, 1972, Chemical and isotopic characteristics of Middle Devonian dolomites from Pine Point, northern Canada: International Geological Congress, 24th, Montreal, sec. 6, p. 230–243.

Hriskevich, M.E., 1970, Middle Devonian reef production, Rainbow area, Alberta, Canada: AAPG Bulletin, v. 54, p. 2260–2281.

Hunt, J.M., 1972, Distribution of carbon in crust of earth: AAPG Bulletin, v. 56, p. 2273–2277.

Kyle, J.R., 1980, Controls of lead–zinc mineralization, Pine Point district, Northwest Territories, Canada: Mining Engineering, v. 32, p. 1617–1626.

Macqueen, R.W., and T.G. Powell, 1983, Organic geochemistry of the Pine Point lead–zinc ore field and region, Northwest Territories, Canada: Economic Geology, v. 78, p. 1–25.

Macqueen, R.W., et al, 1975, Devonian metalliferous shales, Pine Point region, District of Mackenzie: Geological Survey of Canada Paper 75–1, pt. A, p. 553–556.

Palmer, S.E., and J.E. Zumberge, 1981,

Organic geochemistry of U. Miocene evaporite deposits in Sicilian Basin, Sicily, *in* J. Brooks, ed., Maturation studies and fossil fuel exploration: New York, Academic Press, p. 393–426.

Powell, T.G., 1975, Geochemical studies related to the occurrence of oil and gas in the Dampier sub-basin, Western Australia: Journal of Geochemical Exploration, v. 4, p. 441–466.

——, 1978, An assessment of the hydrocarbon source rock potential of the Canadian Arctic Islands: Geological Survey of Canada Paper 78–12, 82 p.

——, P.J. Cook, and D.M. McKirdy, 1975, Organic geochemistry of phosphorites—relevance to petroleum genesis: AAPG Bulletin, v. 59, p. 618–632.

Powell, T.G., and R.W. Macqueen, 1984, Precipitation of sulfide ores and organic matter: sulfate reactions at Pine Point, Canada: Science, v. 224, p. 63–66.

Rogers, M.A., J.D. McAlary, and N.J.L. Bailey, 1974, Significance of reservoir bitumens to thermal maturation studies, Western Canada Basin: AAPG Bulletin, v. 58, p. 1806–1824.

Skall, H., 1975, The paleoenvironment of the Pine Point lead–zinc district: Economic Geology, v. 70, p. 22–45.

Snowdon, L.R., 1980, Resinite—a potential petroleum source in the Upper Cretaceous/Tertiary of the Beaufort Mackenzie Basin, *in* A. Miall, ed., Facts and principles of world oil occurrence: Canadian Society of Petroleum Geologists Memoir 6, p. 509–522.

——, and E. Peake, 1978, Gas chromatography of high molecular weight hydrocarbons with an inorganic salt eutectic column: Analytical Chemistry, v. 50, p. 379–381.

Snowdon, L.R., and T.G. Powell, 1978, N-alkane contamination of Western Canadian drill cores: Bulletin of Canadian Petroleum Geology, v. 26, p. 156–158.

Thompson, K.F.M., 1979, Light hydrocarbons in subsurface sediments: Geochimica et Cosmochimica Acta, v. 43, p. 657–672.

Tissot, B., and D.H. Welte, 1978, Petroleum formation and occurrence: Berlin, Springer–Verlag, 538 p.

Tissot, B., et al, 1974, Influence of nature and diagenesis of organic matter in formation of petroleum: AAPG Bulletin, v. 58, p. 499–506.

Winters, J.A., and J.C. Williams, 1969, Microbiological alteration of petroleum in the reservoir: Paper presented at Symposium on Petroleum Transformation in Geologic Environments, American Chemical Society, Division of Petroleum Chemistry, September 7–12, New York City, Paper PETR 86: E22–E31.

Carbonate Source Rocks in the Jurassic Smackover Trend of Mississippi, Alabama, and Florida

John H. Oehler
Conoco, Incorporated, Exploration Research, Ponca City, Oklahoma[1]

The Smackover trend is a belt of carbonate, evaporite, and clastic rocks of Late Jurassic age that rims the Gulf Coast of the United States from Texas to the Florida panhandle. This study focuses on the eastern end of the trend, where recent exploration has resulted in numerous oil and gas discoveries. Stratigraphic and geochemical data indicate that the oil and gas were generated from algal-rich lime mudstones (argillaceous content typically less than 6%) of the lower Smackover Formation. The hydrocarbons are present in carbonate grainstones of the overlying upper Smackover Formation and/or sandstones of the underlying Norphlet Formation.

Results of this study and of similar studies in other areas of the world indicate that carbonate rocks with good source potential are typically light-brown to black, fine-grained, laminated limestones. Such rocks were laid down in a rather restricted range of depositional environments, the most important of which (very shallow settings with restricted circulation and isolated basins on broad platforms) occupy specific, predictable positions on carbonate shelves. Recognition of these facts can aid explorationists in locating reservoirs associated with carbonate source rocks.

INTRODUCTION

Whether carbonate rocks can be source beds for petroleum is an issue that has been debated for many years. Arguments on the affirmative side of the controversy tend to be megascopic in scale and stratigraphic in nature. An example is the following statement by Hunt (1967, p. 225): "The fact that 40 percent of the petroleum in major oil fields is in carbonate reservoirs, many of which are completely surrounded by carbonate rocks, indicates that carbonates can be oil source beds." In contrast, arguments on the opposing side tend to be microscopic in scale and geochemical or petrographic in nature. Such arguments might be summarized as follows: "Carbonate rocks are generally lean in organic matter, and, because of early cementation, they are too tight to be effective source beds; they may generate some hydrocarbons, but the hydrocarbons cannot get out."

It appears that this conflict exists to greater or lesser degrees within most major petroleum companies, and as a consequence it can lead to incoherent exploration strategies.

In this paper, I shall briefly describe evidence supporting a carbonate source for oil and gas in the eastern part of the Smackover trend (where the case is fairly clear-cut and no other reasonable source seems possible), then address the broader subject of the nature of carbonate source rocks in general.

EXPLANATION OF TERMS

Before proceeding, it is necessary to define some terms. In this report, unless otherwise specified, the term carbonate refers to limestones. However, many of the geochemical characteristics described here for limestones may apply equally well to primary dolomites. Because this report deals with source rocks, the terms carbonate and limestone as used here refer to fine-grained rocks. In this context, they are synonymous with lime mudstone and micrite, the calcareous equivalents of shales. Finally, because it is natural to compare the source character of limestones with that of shales, and because the two lithologies grade into one another, the terms limestone, shale, and calcareous shale should be defined with respect to relative argillaceous and calcareous contents. Generally accepted definitions are illustrated in boldface type in Figure 1.

As this figure shows, considerable latitude in the argillaceous content of limestones is permissible. This is partly due to the fact that there is no precise definition of argillaceous limestone. To add more balance to this system of terminology, I have defined "argillaceous limestone" and "limestone" as indicated in quotation marks in Figure 1. It should be remembered, however, that many carbonate petrologists define limestones as those rocks containing more than 50% calcareous minerals.

It should also be borne in mind that these definitions do not address the practical question of scale. For example, if argillaceous material were concentrated in recognizable layers, how thick would the layers have to be before they were called shale stringers? Clearly, there is ample room for the splitting of nomenclatural hairs.

Suffice it to say that all of the Smackover micrites described in this report contain less than 15% argillaceous material, and most contain less than 6%.

SMACKOVER SOURCE ROCKS

The Smackover trend in the United States is a belt of carbonate, evaporite, and clastic rocks of Late Jurassic age that rims the Gulf Coast from the Florida panhandle through Texas (Fig. 2; Mexican equivalents are not shown). It has been an important petroleum fairway for decades and continues to yield new discoveries each year. One area of recent intensive activity within this trend encompasses southern Mississippi, southern Alabama, and extreme western Florida (Fig. 3). This is the principal study area of the present report.

The Smackover trend in this area contains numerous oil and gas fields ranging

[1]Present address: (U.K.), Ltd., London, England.

in size from one-well discoveries like Barnett field to major accumulations like the Jay complex (estimated ultimate recovery of 430 million stock-tank barrels of crude oil and condensate plus 676 billion ft^3 of gas; Sigsby, 1976).

These fields produce from upper Smackover carbonates and/or Norphlet sandstones (Fig. 4) within a sequence of rocks that is sealed above and below by anhydrites and salt, respectively, and that contains organic-rich micrites (in the lower Smackover) but no significant shales. At some localities outside the study area, the Norphlet is shaly and may be a source rock, but within the study area it is a fairly clean sandstone with no significant source potential. Thus, stratigraphic relationships alone suggest that the reservoirs must have been charged from micrites of the lower Smackover, the only fine-grained, organic-rich rocks in the sequence.

It is the stratigraphic relationships noted above and illustrated in Figure 4 that provide the impetus for further study of lower Smackover micrites as examples of carbonate source rocks. We will consider first the geochemical character of the micrites, then the geochemistry of the produced oils and gases, and finally some oil–source correlation data.

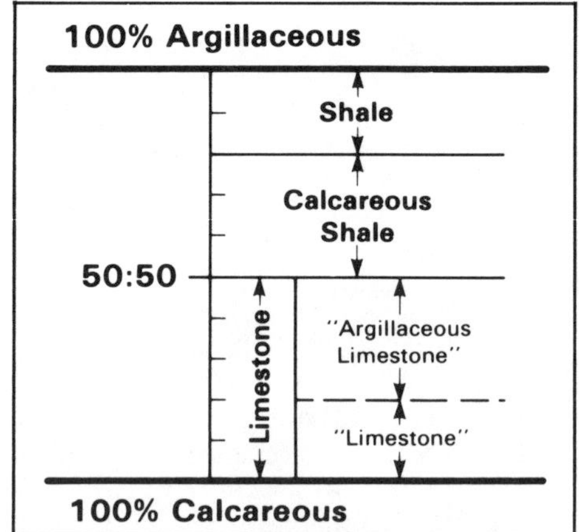

Figure 1 — Nomenclature of calcareous and argillaceous rocks, based on component concentrations. In this system, marls contain 35–65% clay and 35–65% carbonate. (See Bates and Jackson, 1980.)

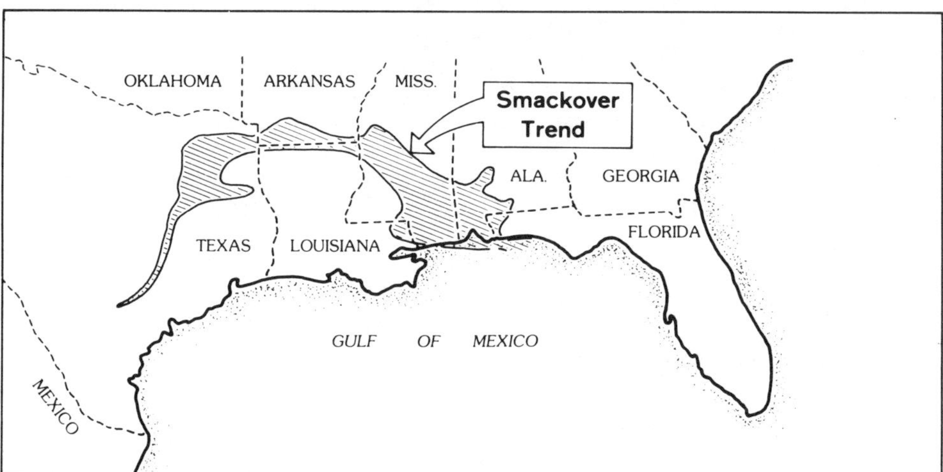

Figure 2. Index map showing location of the Smackover trend within the United States.

ORGANIC GEOCHEMISTRY

Representative organic geochemical data for lower Smackover micrites in the study area are shown in Table 1. Total organic carbon (TOC) concentrations typically are in the range of 0.1 to 1.0%, with an average of about 0.5%. Kerogen in the micrites is dominated by a mixture of amorphous and herbaceous types; the amorphous material is mainly algal in origin. Hydrocarbon contents vary widely depending on organic-carbon content and thermal maturity. Mature samples (TAI 3- to 3) from the Jay field area and from localities updip from Jay have C_1–C_4 hydrocarbon contents up to 270,000 ppm (by volume), C_5–C_7 hydrocarbon contents up to 28,000 ppm (by volume), and C_{15+} hydrocarbon contents up to 300 ppm (by weight). These values indicate good to very good source richness for liquids and associated gas. Overmature samples (TAI $>3+$) from deep (18,000–21,000 ft; about 5,500–6,400 m) tests are lean in C_{15+} hydrocarbons (as would be expected) but have C_1–C_4 (mostly C_1)

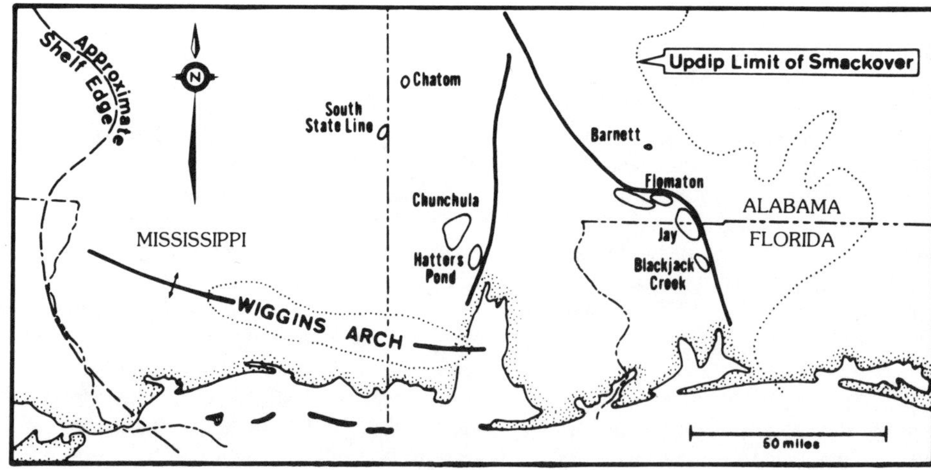

Figure 3. Location map showing Smackover and Norphlet fields, from which petroleum samples were collected for this study, plus some major structural elements in the area.

hydrocarbon contents up to 560,000 ppm, which indicate very good source richness for dry gas.

The richest rocks in the section are algal-laminated micrites of the basal Smackover (Fig. 5). According to Sigsby (1976), this basal laminated sequence (which may be up to 200 ft or 60 m thick)

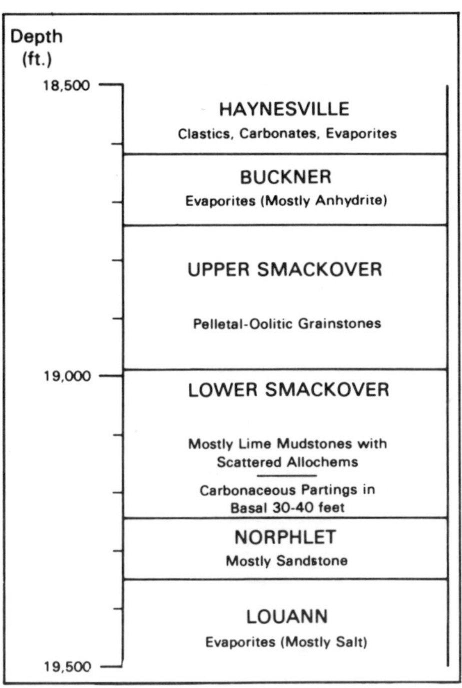

Figure 4 — Typical stratigraphic section through the Smackover and associated formations in the area shown in Figure 3 (from a well in Baldwin County, Alabama).

Table 1

Parameter	Range of Values	Average	Number of Samples	Number of Wells
Mature Samples				
TOC (%)	0.05–2.52	0.48	22	8
EOM (ppm)	40–727	259	22	8
EOM/TOC (%)	1.9–15.2	7.0	22	8
C_{15} + HC (ppm)	31–326	134	21	8
HC/TOC (%)	0.7–8.5	3.6	21	8
C_1–C_4 HC (ppm)	175,000–270,000	223,000	2	2
C_5–C_7 HC (ppm)	15,700–28,000	21,900	2	2
Overmature Samples				
C_1–C_4 hydrocarbons (ppm)	600–563,000	42,100	48	4
C_5–C_7 hydrocarbons (ppm)	0–4,300	144	48	4

TOC = total organic carbon; EOM = extractable organic matter; HC = hydrocarbons.

Table 1 — Organic contents of typical lower Smackover micrites.

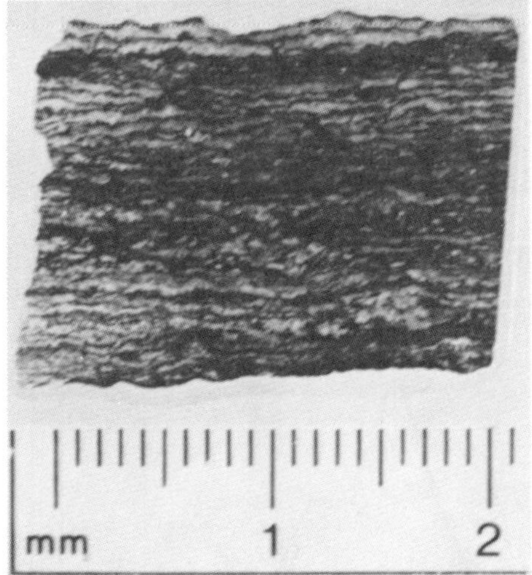

Figure 5 — Thin section of algal-laminated micrite from the lower Smackover. (Sample is from 15,700 feet in a well near Jay field.)

was deposited in a hypersaline, supratidal to shallow intertidal environment of low relief that received little clastic input from the arid land. The hypersalinity of such settings restricts faunal diversity and permits the widespread growth of algal-mat communities, a phenomenon that is reflected in the fabric of the rocks (Fig. 5).

Geochemical correlation data (Fig. 6; Table 2) for production samples from seven fields in the study area suggest that the oils and gases belong to one major family and were generated from the same or chemically similar source rocks (samples from the eighth field, South State Line, were too contaminated with corrosion inhibitors to permit useful analysis). Differences in the data primarily reflect differences in the maturities of the samples; for example, the liquids range from black oil with an API gravity of 48° to gas condensate with an API gravity of 68°.

Finally, three types of oil–source correlation data comparing Smackover and Norphlet petroleum liquids (oils and condensates) with extracts from lower Smackover micrites (Fig. 7) provide evidence that the micrites are the source of the produced hydrocarbons.

Thus, organic geochemical data support stratigraphic relationships in indicating that micrites of the lower Smackover are the source of oil and gas in the overlying upper Smackover and the underlying Norphlet reservoirs. The micrites are algal laminated and typically contain less than 15% argillaceous material. Their organic carbon, bitumen, and C_{15+} hydrocarbon contents are about average (or perhaps somewhat lower than average) for lime mudstones generally.

CHARACTERISTICS OF POTENTIAL CARBONATE SOURCE ROCKS

Based on descriptions given in this volume and elsewhere (e.g., Burwood, 1980; Klopp, 1975), we can outline a number of features that appear to be characteristic of good carbonate source rocks in general. This, in turn, should assist geologists in recognizing such rocks in outcrop and in cuttings and core samples.

To summarize in a single statement, we can say that carbonate rocks with good source potential are typically *light-brown to black, fine-grained, laminated limestones.*

Color—Brown to black colors usually reflect relatively high organic-carbon contents, although, in some cases, a black color may be due to the presence of finely disseminated pyrite. If the colors are due to organic material, as is usually the case, then brown often signifies lower

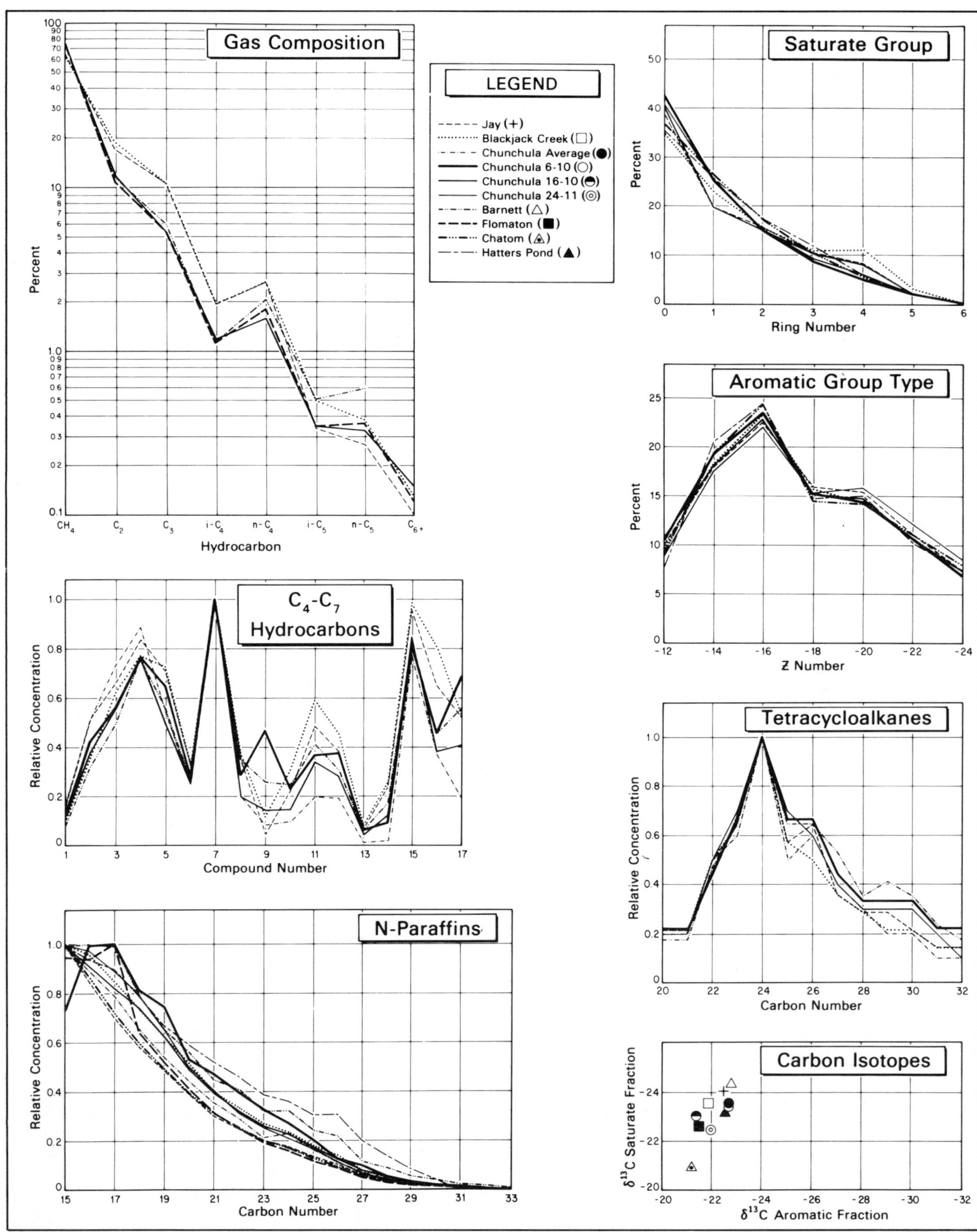
Gas Composition
Percent
Hydrocarbon
CH4
C2
C3
i-C4
n-C4
i-C5
n-C5
C6+
LEGEND
Jay (+)
Blackjack Creek (□)
Chunchula Average (●)
Chunchula 6-10 (○)
Chunchula 16-10 (◓)
Chunchula 24-11 (◎)
Barnett (△)
Flomaton (■)
Chatom (◬)
Hatters Pond (▲)
Saturate Group
Percent
Ring Number
Aromatic Group Type
Percent
Z Number
C4-C7 Hydrocarbons
Relative Concentration
Compound Number
Tetracycloalkanes
Relative Concentration
Carbon Number
N-Paraffins
Relative Concentration
Carbon Number
Carbon Isotopes
δ13C Saturate Fraction
δ13C Aromatic Fraction

Figure 6 — Geochemical correlation data for Smackover and Norphlet petroleum samples. (See Table 2 for identification of numbered compounds in the C_4–C_7 hydrocarbon fraction.)

organic maturity than does black.

Grain Size—In carbonates, as in clastics, fine-grained rocks tend to have higher organic contents than coarse-grained rocks.

Lamination—Crenulate to wavy laminations are particularly promising because they commonly represent microbial (algal and bacterial) mats, and microbial organic matter is the best generator of oil. Flat laminations may be due to argillaceous material. In either case, laminated micrites tend to be more organic-rich than unlaminated ones. Further, laminations of both types appear to facilitate primary migration of bitumens by serving as pathways for fluid movement (for example, Table 3 shows that well-laminated samples of lower Smackover micrite have lower EOM/TOC and HC/TOC ratios than their poorly laminated counterparts, apparently reflecting more efficient release of bitumens from the well-laminated rocks).

Mineralogy—Aside from some primary or syndepositional dolomites, such as those of the Observatory Hill Beds in the Officer basin of Australia (McKirdy and Kantsler, 1980; McKirdy et al, this volume), most dolomites are lean in organic matter and consequently have negligible source potential.

Figure 8 shows an example of a potential carbonate source rock with the characteristics just described. It is a light- to medium-brown, algal-laminated micrite from the Ordovician of Wisconsin.

Of course, possession of these characteristics is no guarantee that a particular carbonate rock is a potential hydrocarbon source (just as blackness and well-developed fissility are no guarantee that a particular shale is a potential source). However, these characteristics are easily recognizable in the field, should alert the geologist to the possibility of carbonate source rocks in the area, and are sufficient grounds for submitting such rocks for geochemical analysis.

Table 2

Compound Number	Compound Name(s)
1	isobutane
2	n-butane
3	isopentane
4	n-pentane
5	cyclopentane; 2,3-dimethyl butane; 2-methyl pentane
6	3-methyl pentane
7	n-hexane
8	methyl cyclopentane; 2,2-dimethyl pentane; 2,4-dimethyl pentane
9	benzene
10	cyclohexane
11	2-methyl hexane; 2,3-dimethyl pentane
12	3-methyl hexane
13	1,cis-3-dimethyl cyclopentane
14	3-ethyl pentane; 1,trans-2-dimethyl cyclopentane
15	n-heptane
16	methyl cyclohexane
17	toluene

Table 2 — Identification of numbered compounds in plot of C_4–C_7 hydrocarbon fraction in Figure 6.

Table 3

Depth (ft)	Degree of Lamination	TOC (%)	EOM (ppm)	EOM/TOC (%)	HC (ppm)	HC/TOC (%)
15,699	Well laminated	0.32	173	5.4	113	3.5
15,699	Poorly laminated	0.18	180	10.0	106	5.9
18,505	Well laminated	0.70	630	9.0	282	4.0
18,505	Poorly laminated	0.42	570	13.6	297	7.1

Table 3 — Comparison of organic contents in well-laminated and poorly laminated micrites from lower Smackover in vicinity of Jay field, Florida.

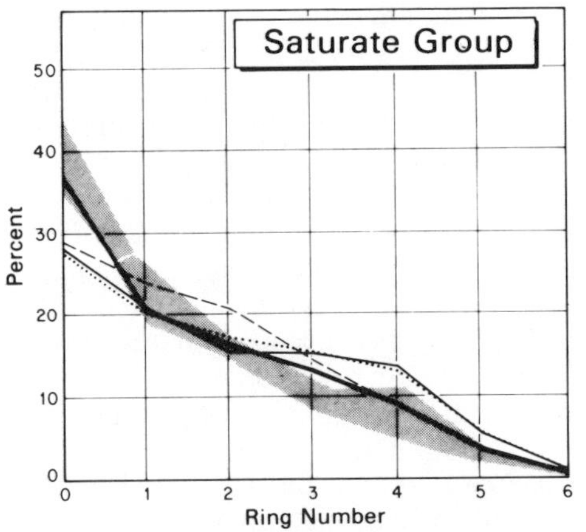

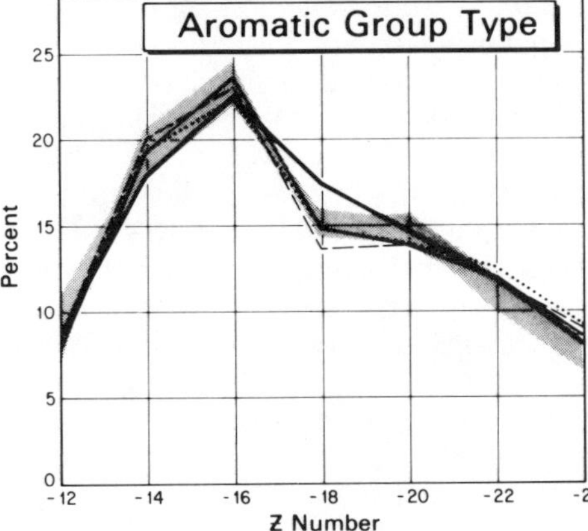

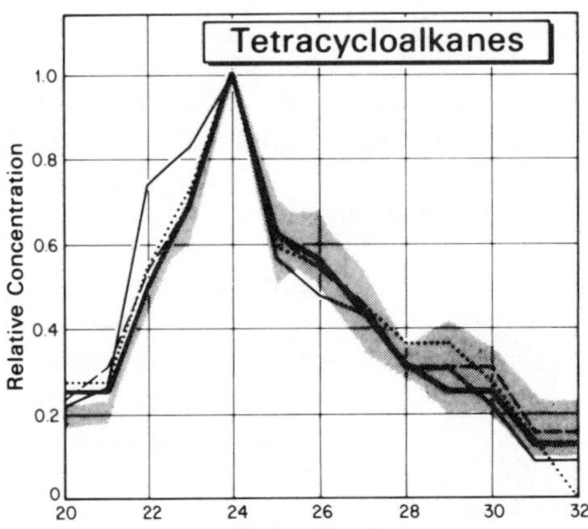

Figure 7 — Geochemical correlation data for Smackover and Norphlet petroleum liquids and lower Smackover rock extracts. The shaded bands enclose data for the liquids (see Fig. 6), and the lines represent data for typical source-rock extracts.

Figure 8 — Potential carbonate source rock from the Ordovician of Wisconsin, illustrating characteristics described in the text. (Sample courtesy of R. K. McLimans.)

DEPOSITIONAL ENVIRONMENTS THAT FAVOR FORMATION OF POTENTIAL CARBONATE SOURCE ROCKS

In the preceding section, I outlined criteria that can be used for recognizing possible carbonate source rocks in the field and at the well site. A significant fact for exploration on a larger scale is that these kinds of rocks are laid down in a rather restricted range of depositional environments, the most important of which occupy specific, predictable positions on carbonate shelves.

Before considering particular depositional settings individually, it will be helpful to understand what general environmental conditions must be met in order to form carbonate rocks with good petroleum-source potential. There are four basic requirements, as listed below:

1. Low influx of materials from the land, including terrestrial organic matter, terrigenous sediment, and fresh water. For nearshore depositional settings, this normally translates into (1) an arid climate and (2) a bordering terrain and drainage area of low topography.

2. High input of autochthonous, microbial organic matter (both algal and bacterial), because this kind of organic matter is the best source of oil. This requires that the depositional surface (or, at least, the waters overlying the depositional surface) be well illuminated and rich in inorganic nutrients, such as phosphate and nitrate.

3. Exclusion of predators and burrowers so that there is minimal consumption and recycling of microbial organic matter through the food chain. This means that conditions at the sediment–water interface (and preferably in the overlying water column, as well) should be inhospitable for organisms such as molluscs, worms, and fish. Such stressful conditions are best afforded by hypersalinity and low levels of dissolved oxygen.

4. Maintenance of euxinic bottom waters. This is necessary for preservation of organic matter. Euxinic conditions are achieved through restricted water circulation either from topographic barriers or from a stratified water column with little or no vertical mixing.

One set of depositional environments that satisfy these criteria consists of quiet, shallow subtidal to supratidal, hypersaline, shelf settings of low relief that receive little runoff from the land. These include coastal sabkhas, intertidal flats, evaporative embayments, and restricted lagoons. In these environments, mesosaline to hypersaline conditions restrict faunal diversity so that most predatory and burrowing organisms are excluded. In their absence, phytoplankton and algal-mat communities flourish. There is little to no dilution by organic matter, sediment, or water from the land. The high salinity reduces the equilibrium concentration of dissolved oxygen, and stratification of the brines inhibits mixing so that anoxic bottom conditions are maintained and relatively large proportions of the produced organic matter are preserved. Although algal mats in the sabkha and intertidal environments may be exposed to air intermittently, bacterial activity continually maintains reducing conditions below a few millimeters under the mat surface.

Some examples of such rocks are those of the lower Smackover, lower Sunniland (Palacas, Anders, and King, this volume; Pontigo et al, 1979), and Salina A-1 units (Gardner and Bray, this volume), as well as the Observatory Hill Beds in the southeastern part of the Officer basin (McKirdy et al, this volume) and the "infra-Cambrian" sub-salt carbonates of Oman (unpublished data).

Another type of depositional environment that satisfies the four criteria listed above is that of large, evaporative, alkaline, playa lakes that receive little runoff from the surrounding terrain. In such lakes, elevated salinity and alkalinity restrict faunal diversity, allowing algal populations to thrive. Terrigenous input is low and seasonal. However, there may be extensive precipitation of authigenic minerals other than calcium carbonate (e.g., hydrated and unhydrated carbonates, bicarbonates, and silicates such as shortite, gaylussite, nahcolite, trona, magadiite, kenyaite, and so on). Stratification of the waters inhibits mixing and results in anoxic bottom conditions that promote preservation of the microbial organic matter.

Examples of carbonate source rocks deposited in this kind of environment include the Observatory Hill Beds in the northeastern part of the Officer basin (McKirdy et al, this volume) and "oil shales" in the Parachute Creek Member of the Green River Formation.

A third favorable environment is isolated basins and subbasins on very broad, low-relief carbonate shelves. Topographic restriction of water movement into and out of these basins results in stagnant conditions that reduce faunal diversity and enhance preservation of microbial organic matter. The shelf itself acts as a trap against influx of terrigenous materials.

A number of prolific carbonate source beds in the Middle East (Murris, 1980) fall into this category, including those of the Hanifa and Khatiyah formations.

All of the depositional environments discussed above are in relatively shallow water: on carbonate shelves or in lacustrine settings. "Deeper water" lime mudstones (Wilson, 1969) constitute a fourth possible carbonate source-rock facies, although I know of no undisputed case in which a significant petroleum accumulation has been traced to a source rock of this type.

According to Wilson (1969), such sediments tend to accumulate at the bases of carbonate shelves, as fringing deposits on the peripheries of deep basins. In this environment, influx of terrigenous material is very low, although occasional turbidity flows of sand size or larger grained carbonate clastics from the shelf edge may dilute the lime muds. Organic input is mainly from phytoplankton and zooplankton in overlying, near-surface waters and occurs largely in the form of fecal pellets from higher organisms. Low levels of dissolved oxygen in the bottom waters restrict predators and burrowers and enhance the potential for organic preservation in the sediments. Periodic plankton blooms in near-surface waters may result in periodically high rates of organic deposition, leading in turn to the formation of an organically laminated sediment. Such sediments also may contain an abundance of calcareous and siliceous microorganism tests; high concentrations of the latter may give rise to chert layers and nodules within the limestone.

As examples of this kind of deposit,

Wilson (1969) cited, among others, the Permian Lamar, Radar, and Bone Springs limestones of west Texas and New Mexico; the Devonian Duvernay limestones of the Leduc area in Alberta; and the Jurassic–Cretaceous Tamaulipas limestones of the Golden Lane in Mexico. Some of these may be petroleum source rocks (e.g., the Tamaulipas limestones in the Golden Lane and their equivalents in the Reforma–Campeche area; Meyerhoff, 1980). However, as was noted above, convincing evidence of this has not been reported. Obtaining convincing evidence (if it exists) may be difficult, because these kinds of limestones commonly are associated laterally with shales that also could be petroleum source rocks.

Knowing the particular types of depositional environments in which potential carbonate source rocks are most likely to form, and knowing where these environments are situated on carbonate shelves, should enable the explorationist to recognize packages of such rocks on seismic sections and evaluate their relationships to migration paths and possible reservoirs.

CONCLUSIONS

Evidence presented here and elsewhere in this volume leaves little doubt that fine-grained, organic-rich carbonates are important sources of crude oil and natural gas. Indeed, such rocks have been credited with having charged some of the largest petroleum accumulations known. In the Smackover trend, they have charged numerous reservoirs, including those of the Jay complex, which is one of the largest oil reserves discovered in the United States since Prudhoe Bay field in Alaska.

In general, carbonate source rocks tend to be oil prone, to occur in close proximity to good reservoir facies (reef complexes, oolite buildups, sandstones), and to be associated with evaporite units that form excellent reservoir seals.

Potential carbonate source rocks share certain visually discernible characteristics (color, grain size, fabric, and so on) that can assist field and well-site geologists in selecting samples for geochemical analysis. Moreover, depositional environments of potential carbonate source rocks are reasonably well understood, and this knowledge can assist explorationists in mapping source–reservoir relationships. This is not to say that exploration for reservoirs charged from carbonate sources is a simple matter. It is certainly no easier than finding reservoirs charged from shales. However, merely acknowledging the fact that carbonate source rocks exist may, for some explorationists, open new avenues of thought that can lead to the discovery of reserves that otherwise might have remained untapped.

REFERENCES CITED

Bates, R.L., and J.A. Jackson, eds., 1980, Glossary of geology, second edition: Falls Church, Virginia, American Geological Institute, 751 p.

Burwood, R., 1980, Carbonate source rocks for six million barrels of oil per day, Zagros Foldbelt, S.W. Iran (abs.): Geological Society of America Abstracts with Programs, v. 12, p. 397.

Gardner, W.C., and E.E. Bray, 1984, Oil and source rocks of Niagaran reefs in the Michigan Basin, *in* J.G. Palacas, ed., Petroleum geochemistry and source-rock potential of carbonate rocks: AAPG Studies in Geology 18, this volume.

Hunt, J.M., 1967, The origin of petroleum in carbonate rocks, *in* G.V. Chilingar, H.J. Bissel, and R.W Fairbridge, eds., Carbonate rocks: New York, Elsevier, p. 225–251.

Klopp, H.C., 1975, Petrographic analysis of the Sunniland Formation, an oil-producing formation in S. Florida: Provo, Utah, Brigham Young University Geology Studies, v. 22, p. 3–27.

McKirdy, D.M., and A.J. Kantsler, 1980, Oil geochemistry and potential source rocks of the Officer Basin, South Australia: Australian Petroleum Exploration Association Journal, v. 20, no. 1, p. 68–86.

———, et al, 1984, Hydrocarbon genesis and organic facies in Cambrian carbonates of the eastern Officer basin, South Australia, *in* J.G. Palacas, ed., Petroleum geochemistry and source-rock potential of carbonate rocks: AAPG Studies in Geology 18, this volume.

Meyerhoff, A.A., 1980, Geology of Reforma–Campeche shelf: Oil and Gas Journal, v. 78, no. 16, p. 121–124.

Murris, R.J., 1980, Middle East: stratigraphic evolution and oil habitat: AAPG Bulletin, v. 64, p. 597–618.

Palacas, J.G., D.E. Anders, and J.D. King, 1984, South Florida basin, a prime example of carbonate source rocks of petroleum, *in* J.G. Palacas, ed., Petroleum geochemistry and source-rock potential of carbonate rocks: AAPG Studies in Geology 18, this volume.

Pontigo, F.A., Jr., et al, 1979, South Florida's Sunniland oil potential: Oil and Gas Journal, v. 77, p. 226–232.

Sigsby, R.J., 1976, Paleoenvironmental analysis of the Big Escambia Creek–Jay–Blackjack Creek Field area: Gulf Coast Association of Geological Societies Transactions, v. 26, p. 258–278.

Wilson, J.L., 1969, Microfacies and sedimentary structures in 'deeper water' lime mudstones, *in* G.M. Friedman, ed., Depositional environments in carbonate rocks: SEPM Special Publication 14, p. 4–19.

South Florida Basin—A Prime Example of Carbonate Source Rocks of Petroleum

James G. Palacas, Donald E. Anders, and J. David King
U.S. Geological Survey, Denver Federal Center, Denver, Colorado

Stratigraphic, lithologic, and geochemical data provide ample evidence that carbonate rocks and not shales are the source beds of commercial oil in carbonate grainstone reservoirs of the Lower Cretaceous Sunniland Limestone of the South Florida basin.

Detailed crude-oil–source-rock correlations, including gas-chromatographic–mass-spectrometric analyses of steranes, and of tricyclic and pentacyclic terpanes, indicate that the algal-sapropelic, organic-rich argillaceous limestones in the lower Sunniland Limestone, particularly the downdip, basinal facies, are the probable major source of upper Sunniland oil.

Conventional geochemical maturity parameters, plus biomarker isomerization ratios, indicate that the Sunniland oils and source rocks are marginally mature. The sterane ratios, commonly in conjunction with the extended hopane isomerization ratios, proved invaluable in deciphering the thermal maturity of carbonate rocks in south Florida and in pinpointing the upper threshold of the conventional oil-generation zone.

INTRODUCTION

The South Florida basin is an excellent sedimentary setting in which to study and understand the evolution of petroleum and source-rock potential in a carbonate–evaporite sequence. Crucial properties of the basin are as follows: (1) it contains a nearly continuous, thick sequence of carbonate–evaporite rocks (more than 15,000 ft or 4,570 m); (2) it is characterized by a simple burial history uncomplicated by any major tectonism; and (3) it produces commercial oil from a single stratigraphic unit.

In the past, because most carbonate rocks were presumed to have been deposited in shallow-water, well-oxygenated environments—environments not conducive to the accumulation and preservation of substantial amounts of organic matter—the geologic community generally doubted the effectiveness of carbonate rocks as major sources of petroleum. Black bituminous shales deposited under anoxic conditions were usually regarded as the primary source beds for oil.

The prime purpose of this paper, in conjunction with other papers in this volume, is to present sufficient geologic and geochemical evidence to show that carbonate rocks of the South Florida basin are indeed sources of petroleum. Besides examining the stratigraphic framework and lithology and the quality, quantity, and thermal maturity of the organic matter of the rocks themselves, data are given on the crude oils. A second and perhaps more difficult objective is an attempt to identify the specific or probable source facies by means of source-rock–crude-oil correlations. In addition to conventional correlation techniques, such as gas-chromatographic patterns of saturated hydrocarbons, pristane/phytane ratios, and saturated/aromatic ratios, gas-chromatographic–mass-spectrometric analyses of biological marker compounds (steranes, tricyclic and pentacyclic terpanes) were extensively used. The biomarker patterns proved to be the most diagnostic factors in delineating the probable source facies.

GEOLOGIC SETTING

Figure 1 shows the approximate boundaries of the South Florida carbonate–evaporite basin and the general structural configuration of the Lower Cretaceous section of the basin. Figure 2 shows the generalized stratigraphic column formally established in south Florida for the section of rock reported on in this study. Approximate equivalent units (some with informal names) used by Exxon Co., U.S.A., for the past several decades and commonly referred to in the oil industry as well as by one of the authors in previous related papers (Palacas, 1978a, 1978b, 1984), are also given for comparison.

In the South Florida basin, which is part of the larger South Florida–Bahamas Platform province (Winston, 1971a), more than 15,000 ft (4,570 m) and perhaps as much as 20,000 to 30,000 ft (6,100 to 9,140 m) (Applegate, personal communication) of predominantly limestones, dolomites, and evaporites (mostly anhydrites) and minor amounts of shale were deposited. Major faults and extensive vertical fractures are unknown, especially in the shelf area of the onshore part of the basin. The only significant clastic sections in the basin are the Holocene–Pleistocene sands and sandstones at or near the surface and the basal transgressive clastic facies (average 300 ft or 90 m) that immediately overlie igneous basement (Fietz, 1976).

The basin is characterized by a low geothermal gradient that generally is about 1°F/100 ft (1.8°C/100 m) (Griffin et al, 1969), but locally, particularly in the vicinity of some of the oil fields, the gradient may be as much as 1.5°F/100 ft (2.7°C/100 m) (Reel and Griffin, 1971). Because of the generally low geothermal gradient and the less than mature nature of the oils and carbonate rocks at the Sunniland level (discussed later), the basin is considered marginally mature.

The Sunniland Limestone, the only producing formation in south Florida, averages about 280 ft (85 m) in thickness and lies at an average depth of about 11,500 ft (3,535 m). It is composed principally of interbedded porous to impervious, organically lean to organically rich limestones and dolomites; shales are scarce to absent. The richer and more

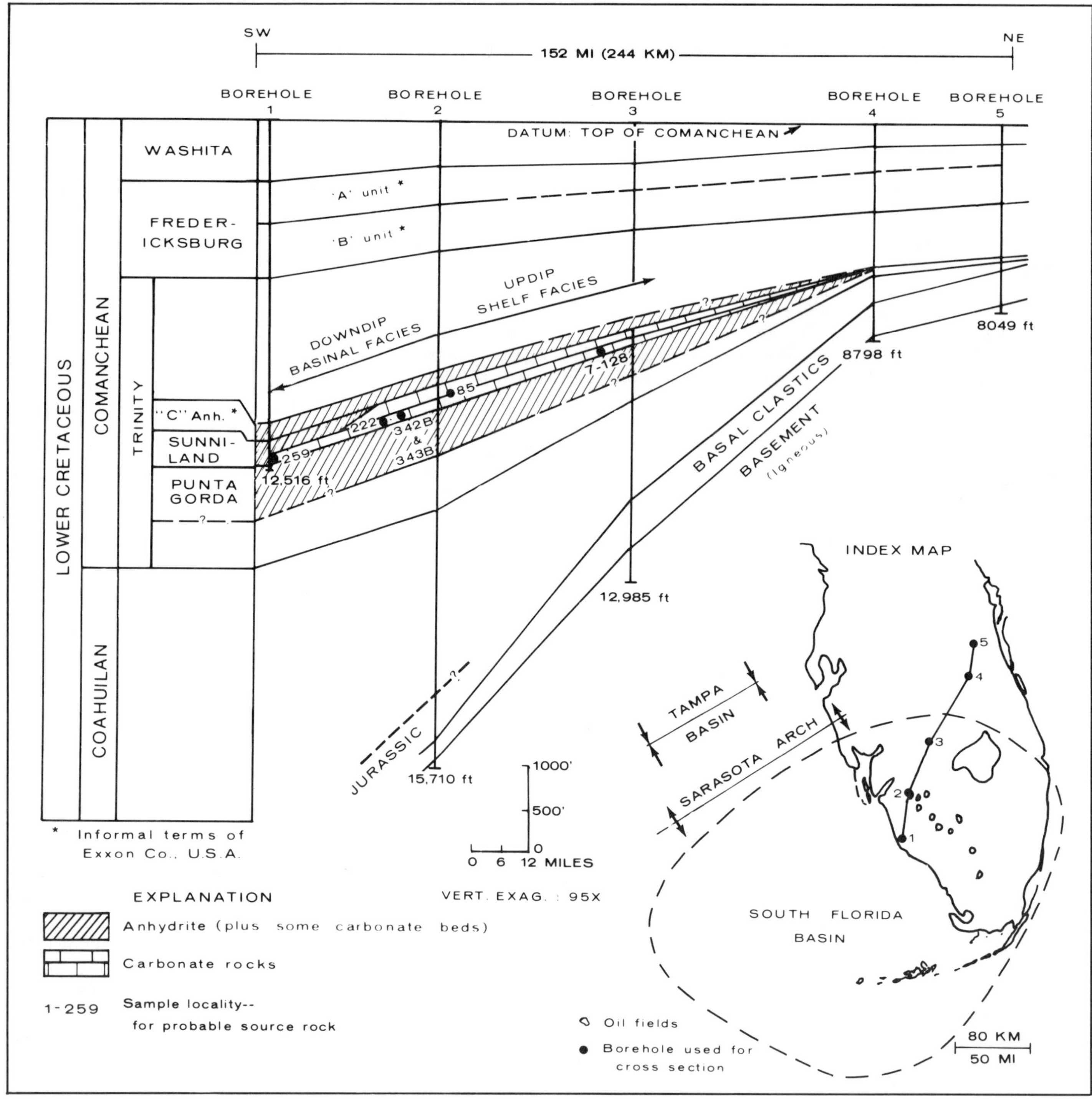

Figure 1 — Generalized northeast–southwest cross section of Lower Cretaceous rocks of south Florida, with index map showing line of section, outline of South Florida basin, and adjoining structural features and oil fields. Sample localities of probable source rocks are shown at approximate borehole positions, except for sample 259, which is in borehole 1. See Table 1 for sample depths. Cross section is modified from Fietz (1976); borehole names in Fietz (1976); outline of South Florida basin modified from Applegate (in press).

argillaceous carbonate units are predominantly in the lower Sunniland Limestone. Downdip, the upper Sunniland carbonate rocks are replaced by anhydrites (Fig. 1).

One of the key features of the Sunniland Limestone, as illustrated in Figure 1, is that it is essentially sealed by the overlying impervious anhydrites of the Lake Trafford Formation and the underlying impervious anhydrites of the Punta Gorda Anhydrite. This fact, coupled with the absence of known faults or extensive fractures, should preclude the migration of oil into or out of the formation. Hence, geologic evidence alone indicates that the

SOUTH FLORIDA BASIN

System	Series/Stage	Group	Published stratigraphic terms	Terminology used by Exxon Co., U. S. A.; Gulfian units after Palacas (1978b)
TERTIARY	PALEOCENE		CEDAR KEYS FM	CEDAR KEYS FM
UPPER CRETACEOUS	GULFIAN		PINE KEY FM 3,000'	GULFIAN 'A' GULFIAN 'B' GULFIAN 'C'
LOWER CRETACEOUS	COMANCHEAN	NAPLES BAY GROUP	CORKSCREW SWAMP FM 450'	WASHITA
			ROOKERY BAY FM 500'	FREDERICKSBURG 'A'
			PANTHER CAMP FM 350'	
		BIG CYPRESS GROUP	DOLLAR BAY FM 475'	FREDERICKSBURG 'B'
			GORDON PASS FM 475'	TRINITY 'A'
			MARCO JUNCTION FM 350'	
		OCEAN REEF GROUP	RATTLESNAKE HAMMOCK FM 600'	TRINITY 'B'
			LAKE TRAFFORD FM 150'	TRINITY 'C' ANHYDRITE
			SUNNILAND LS 280'	U. SUNNILAND LS L. SUNNILAND LS
		GLADES GROUP	PUNTA GORDA ANHYDRITE 600'	PUNTA GORDA ANHYDRITE
		LEHIGH ACRES FM	ABLE MBR 290'	TRINITY 'F'
			TWELVE MILE MBR 320'	
			WEST FELDA SH MBR 40'	TRINITY 'G'
	COAHUILAN		PUMPKIN BAY FORMATION 1050'	COAHUILAN 'A'
			BONE ISLAND FM 1300'	COAHUILAN 'B' ★
? JURASSIC	?	?	? WOOD RIVER FM 1700' ?	
TRIASSIC			IGNEOUS BASEMENT	BASEMENT

Figure 2 — Generalized stratigraphic chart of lower Paleocene to Upper Jurassic(?) rocks in South Florida basin. Equivalent units used by Exxon Co., U.S.A., are given for comparison. Published stratigraphic terms are reported as follows: Gulfian: Meyerhoff and Hatten (1974); Comanchean: Applin and Applin (1965); Oglesby (1965); Winston (1971a, 1971b, 1976); pre–Punta Gorda rocks: Applegate et al (1981). Rocks spanning the interval from the base of Coahuilan 'A' to at least the top of the basal clastic unit, which is the basal part of the Wood River Formation (left column), are considered by Exxon as Early Cretaceous in age.

Sunniland carbonate rocks are the probable major source of Sunniland oil. Detailed geochemical data presented later in this paper support this conclusion.

Most of the oil (260 million bbl in place; C. H. Tootle, Florida Bureau of Geology, personal communication) has accumulated in several carbonate grainstone reservoir units in the upper part of the upper Sunniland Limestone. The reservoir facies formed as northwest–southeast–trending isolated patch reefs, bioclastic mounds, banks, or pods in shoal-type environments on the shelf portion of the basin (Fig. 3). Minor amounts of oil also have accumulated in a highly fractured micritic limestone in the lowermost part of the lower Sunniland Limestone. Only 134,000 bbl of this lower Sunniland oil (Fietz, 1976) have been recovered from the one well at Lake Trafford field (shut in) in Collier County. Reservoir temperatures in upper Sunniland fields generally range from 195°–200°F (91°–93°C).

SAMPLES AND METHODS

Total organic carbon (TOC), reported in weight percent (Table 1), was measured using a wet oxidation method modified from Bush (1970). Carbonate carbon was determined by measuring the CO_2 released by acid treatment prior to the oxidation step. Carbonate carbon content is multiplied by 8.33 and reported as weight percent $CaCO_3$.

Out of about 120 Sunniland core samples that were analyzed, total-organic-carbon and other data are given only for 32 rock samples deemed to be possible source rocks (Table 1). Figure 3 shows the location of all boreholes from which samples were obtained. Eighty-eight other Sunniland core samples were analyzed but were omitted because either they were reservoir-type or oil-stained rocks, or they contained less than 0.4% TOC, a value considered to be the lower limit for source-rock potential in south Florida (Palacas, 1978b). Two values less than 0.4% TOC also are listed in Table 1, but their inclusion is discussed later in the text. Actually, more than 700 core and drill-cuttings samples (including the Sunniland samples) were analyzed for organic carbon in order to obtain a rough approximation of the petroleum potential for the entire stratigraphic section studied (Paleocene to Upper Jurassic) in south Florida. The distribution and source evaluation of most of these organic-carbon data are described in the reports by Palacas (1978a, 1978b); Palacas, Daws, and Applegate (1981); Applegate, Winston, and Palacas (1981); and Palacas (1984).

All analyses referred to in this report, unless stated otherwise, were made solely on carefully prepared, thoroughly cleaned core samples. In this way the problems of contamination from drilling fluids and caving inherent in drill-cuttings samples were either voided or greatly minimized.

Extractable organic matter of powdered rock samples was determined by Soxhlet extraction with chloroform. Elemental sulfur was removed during the extraction procedure with activated copper strips. Initially, the organic extract (without prior deasphalting) was separated by silica gel–liquid chromatography into three fractions: saturated hydrocarbons (or alkanes), aromatic hydrocarbons, and asphaltics, each fraction eluted successively with n-heptane, benzene, and 1:1 benzene–methanol. The amount of each fraction was determined gravimetrically. Most of the data in Table 1 reflect this method, but later in the investigation the chloroform extracts were refined further by deasphalting in n-heptane after solvent removal. The quantity of n-heptane insoluble material or asphaltene fraction also was determined gravimetrically. The heptane-soluble material was then subjected to liquid chromatography, as just described, to yield three similar fractions. However, in the modified deasphalting method, the benzene–methanol eluate is referred to as the resin fraction.

By deasphalting the organic extract, a "purer" aromatic fraction was commonly achieved (i.e., a fraction free of dark-colored substances, presumably asphal-

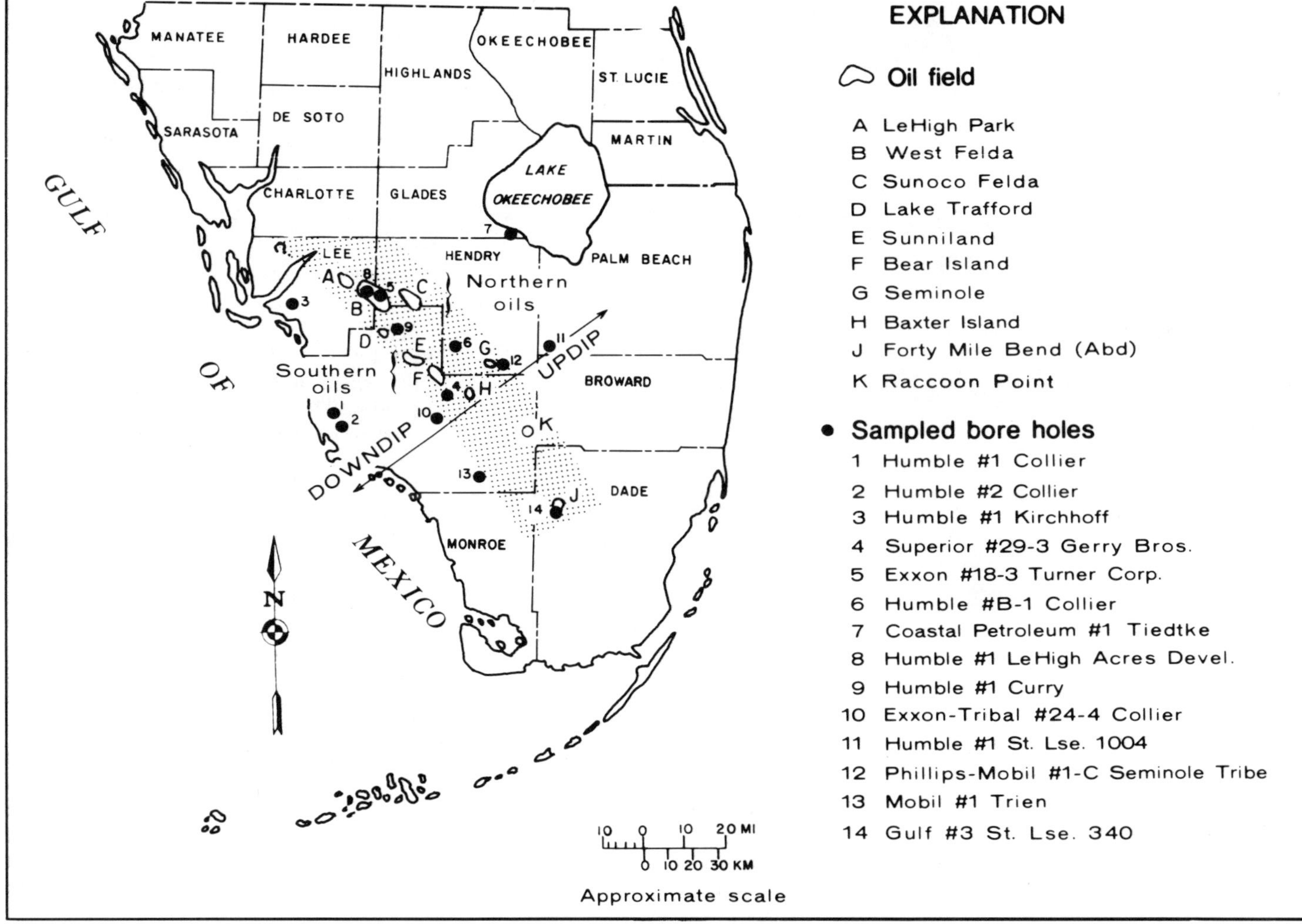

Figure 3 — Map showing locations of sampled boreholes, oil fields, and approximate boundary of present Sunniland oil-producing trend (stippled).

tenes and/or resins). By this procedure, the aromatic fraction usually had lower concentrations, resulting in higher saturated/aromatic (sat/arom) ratios. Hence, in Table 1 the sat/arom ratios for the most part should be considered as minimum values.

Eighteen upper Sunniland crude-oil samples and 11 extracts from oil-saturated reservoir units (which for convenience are herein also referred to as oil samples) were treated in the same manner as the rock extracts.

Gas-chromatographic (GC) analyses of the C_{15+} saturated hydrocarbons were performed with a Hewlett-Packard Model 5710A instrument equipped with a 30-m × 0.32-mm fused-silica capillary column. Two different coatings were used during the course of the analyses. Initially, an SE-30 coating was used, followed by an SE-54 coating. Column conditions were as follows: helium-carrier-gas flow rate was 25 cm/sec, splitless injection at 325°C (617°F), and temperature-program rate was 50° to 300°C (122°–572°F) at 4°C (7°F)/min (or 6°C [11°F]/min).

Gas chromatography–mass spectrometry was accomplished using a Hewlett-Packard 5880 GC directly coupled to a Kratos MS-30 mass spectrometer. The 5880 GC was equipped with a 50-m × 0.32-mm bonded phase SE-54 fused-silica capillary column. GC parameters were: 25 cm/sec helium carrier flow, injector temperature of 340°C (644°F), column temperature programmed from 80° to 160°C (176°–320°F) at 10°C/min (18°F/min), then from 160° to 340°C (320°–644°F) at 3°C/min (5°F/min). The column external to the GC oven was heated isothermally at 340°C (644°F). The mass-spectrometer ion source was operated at 70eV with a pressure of 5 × 10^{-6} torr at a temperature of 200°C (392°F).

Structural assignments of C_{29} steranes were determined by relative retention time and mass spectra obtained by magnetic exponential downscan at 1000 resolution and 4 KV accelerating potential. Relative intensities of peaks at m/z 217 and m/z 218 were used to determine the isomerization of the 14- and 17-carbon position (14β,17β steranes having peak intensities of m/z 218 ≥ m/z 217). The relative intensities of peaks at m/z 149 and m/z 151 yielded information on the position of the hydrogen atom at the 5-carbon position (5α[H], producing a more intense m/z 149 peak). Structural assignments of the pentacyclic triterpanes were determined in a similar fash-

Table 1

Sample Number[1]	Depth[2] (ft)	TOC (wt%)	$CaCO_3$ (wt%)	EOM (ppm)	HC/EOM (%)	non-HC/EOM (%)	SAT/AROM Ratio	EOM/TOC (%)	HC/TOC (%)	Atomic H/C Ratio	S_1 mgHC/g rock	S_2 mgHC/g rock	HI mgHC/g TOC	OI $mgCO_2$/g TOC	PI $S_1/(S_1 + S_2)$	T_{max} (°C)
							Upper Sunniland Limestone									
12-324	11,401	0.91	77	1,750	51	49	1.1	19	9.8	1.2	0.53	3.55	390	60	0.13	431
12-328	11,422	1.41	85	1,290	71	29	1.1	9.1	6.5	1.1	.53	5.44	386	46	.09	421
5-98	11,503	.50	86	950	79	21	1.0	19	15	1.2	.37	3.81	761	38	.09	428
8-155	11,525	.49	97	810	45	55	.8	17	7.4	1.0	.17	1.54	314	105	.01	423
8-165	11,600	.37	74	510	69	31	1.2	14	9.4	–	.10	.77	209	127	.11	434
9-208	11,682	2.95	–	–	–	–	–	–	–	–	.63	20.6	696	42	.02	429
4-82	11,684	.55	89	760	57	43	.6	14	8.0	1.1	.20	1.71	312	36	.10	429
4-83	11,689	.56	90	950	55	45	.8	17	9.3	1.2	.24	1.55	276	181	.14	427
10-217	11,720	1.36	79	2,070	76	24	1.3	15	12	–	.85	6.24	459	51	.12	430
1-254	12,192	2.51	75	3,340	62	38	1.3	13	8.2	–	.77	15	628	39	.05	430
1-255	12,231	1.14	–	–	–	–	–	–	–	–	.77	6.61	580	49	.10	432
1-256	12,247	.54	75	850	65	35	1.5	16	10	–	.14	1.53	283	146	.08	431
1-257	12,300	1.05	69	1,640	64	36	1.5	16	9.9	1.2	.28	3.70	357	104	.07	430
							Lower Sunniland Limestone									
14-702	11,475	7.04	–	–	–	–	–	–	–	–	2.23	45.7	649	30	.05	424
5-105	11,662	.59	87	520	84	16	1.6	9.0	7.4	1.1	.28	1.41	238	44	.17	431
5-107[3]	11,682	.26	86	600	82	18	.9	23	19	–	.25	.42	160	107	.37	433
4-85	11,718	1.62	93	1,590	54	46	.3	9.8	5.3	1.2	.40	8.82	544	26	.04	429
13-335[3,4]	11,729	1.77	–	–	–	–	–	–	–	–	3.72	6.24	352	32	.37	425
13-336[4]	11,735	.56	–	–	–	–	–	–	–	–	.23	2.31	412	51	.09	431
13-341[4]	11,755	.66	–	–	–	–	–	–	–	–	.21	1.38	209	115	.13	431
13-341A[4]	11,759	3.64	–	–	–	–	–	–	–	–	2.45	22.0	604	31	.10	432
13-342[4]	11,761	1.65	71	–	–	–	–	–	–	–	2.06	6.87	416	43	.23	430
13-342A	11,765	2.33	84	–	–	–	–	–	–	–	1.35	9.26	397	30	.12	436
13-342B	11,767	12.3	56	7,500	62	38	.8	6.1	3.8	–	5.03	104	845	15	.04	431
13-343	11,769	3.14	64	–	–	–	–	–	–	–	2.46	18.9	601	19	.11	436
13-343A	11,772	11.5	64	–	–	–	–	–	–	–	4.20	72.5	630	28	.05	431
13-343B	11,773	7.41	79	8,780	45	55	.5	12	5.4	1.3	2.22	45.1	608	12	.04	434
13-345[3]	11,815	.78	–	–	–	–	–	–	–	–	1.45	2.39	306	68	.37	427
10-221[4]	11,834	.75	74	870	65	35	1.2	12	7.4	–	.24	1.69	226	80	.12	431
10-222[4]	11,843	1.14	82	1,620	62	38	1.3	14	8.8	–	.46	4.02	352	50	.10	430
1-259	12,380	2.30	83	4,190	70	30	.3	18	13	–	.94	4.86	211	29	.16	441
1-260	12,387	1.36	82	–	–	–	–	–	–	–	.99	2.33	171	50	.30	436

[1]Prefix in sample number = borehole number; see Figure 3 for location of boreholes.
[2]Depth = depth to top of core sample.
[3]These samples may be oil stained.
[4]These samples actually occur in lowermost part of upper Sunniland unit; see text for explanation.

Table 1 — Geochemical and Rock-Eval pyrolysis data on Sunniland possible source rocks.
TOC = total organic carbon; EOM = extractable organic matter; HC = hydrocarbons; Sat = saturated hydrocarbons; Arom = aromatic hydrocarbons; S_1 = volatile HC; S_2 = pyrolytic HC; HI = hydrogen index, S_2/TOC; OI = oxygen index, S_3/TOC; PI = production index or transformation ratio; – = not determined. TOC by Rinehart Laboratories, Wheatridge, Colorado; total carbonate content by Rinehart Laboratories and Analytical Laboratories, U.S. Geological Survey, Denver, Colorado; atomic hydrogen-to-carbon ratio (H/C) of kerogens by Core Laboratories, Dallas, Texas; Rock-Eval analyses by Ted Daws, U.S. Geological Survey, Denver, Colorado.

ion. Isomerization at the 17- and 21-carbon positions was determined by comparison of the relative intensities of m/z 191 and m/z 149 + n14, n being equal to the number of carbon atoms in the side chain at the 21-carbon position ($17\alpha,21\beta$ was assumed to yield a relatively more intense m/z 191 than m/z 149 + n14 peak).

Relative distributions of steranes and terpanes were determined by multiple ion detection; this was accomplished by switching the accelerating potential to the specified mass at a constant magnetic field and subsequently scanning over the mass peak ±250 ppm for 100 milliseconds for each mass monitored. Calculation of relative peak areas yielded results within approximately ±10% of results calculated by peak height. In cases of higher signal-to-noise ratio with little interference, peak areas were used as calculated by a Kratos DS-55 mass-spectrometry data system.

$^{13}C/^{12}C$ ratios of the saturated- and aromatic-hydrocarbon fractions were determined on a Finnigan-MAT 251 mass spectrometer. The hydrocarbon fractions were converted to CO_2 by combustion under oxygen with subsequent purifica-

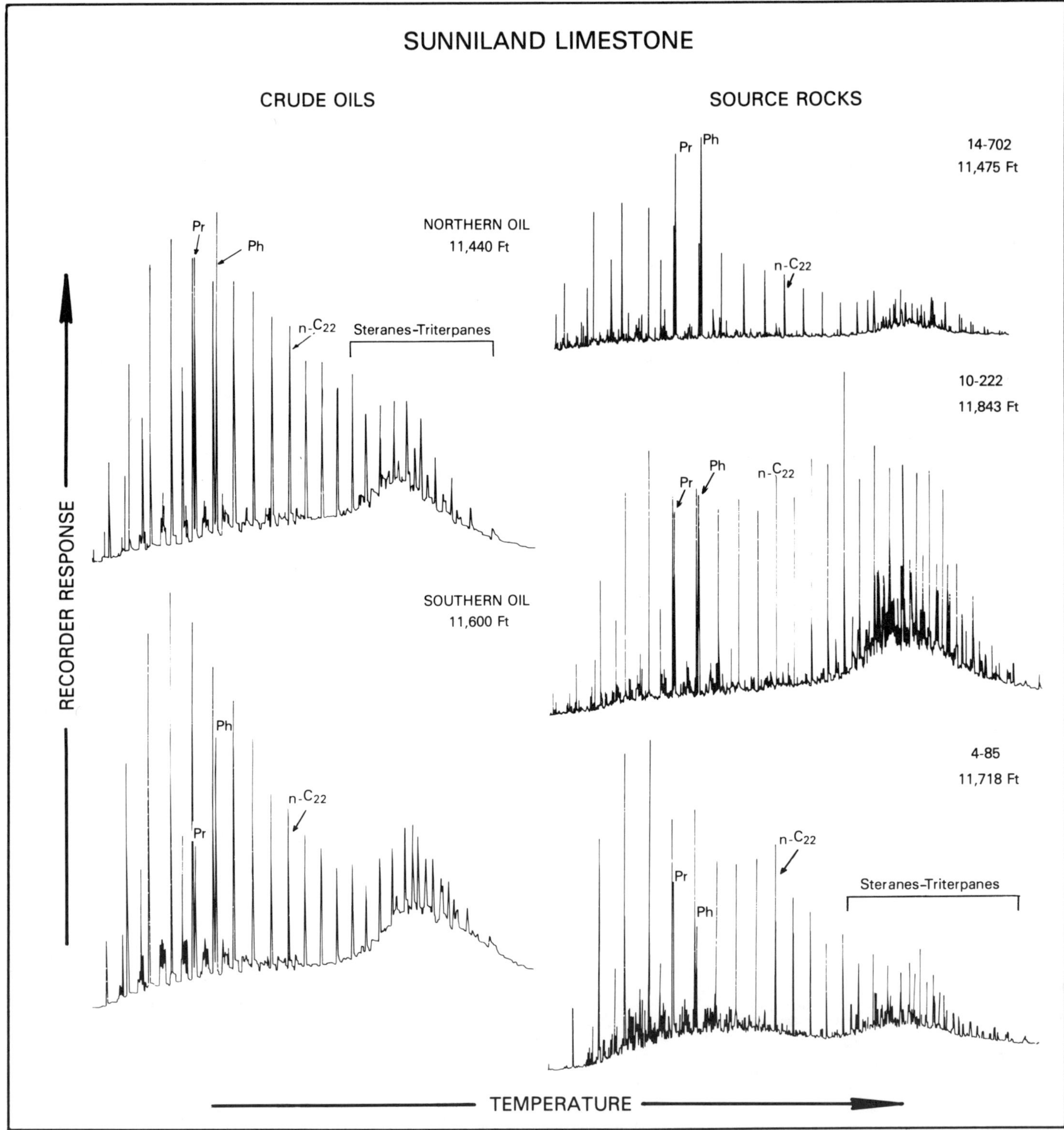

Figure 4 — Capillary gas chromatograms of C_{15+} alkane distributions of typical Sunniland oils and possible source rocks. Compositional variations in GC traces are due to slight differences in organic matter inherited from original depositional environment. Some dissimilarities in shape of GC traces are due to differences in GC column conditions. An SE-30 column was used for rock extracts, an SE-54 column for oils. Pr = pristane; Ph = phytane.

Table 2

Oil Group[1]	Depth (ft)	S (wt%)	GOR	°API	C_{15+} Fraction (wt%) HC	NonHC	V/Ni	$\delta^{13}C$ PDB (o/oo) Sat	Arom	Pr/n-C_{17}	Ph/n-C_{18}	Pr/Ph
Northern[2]	11,400	3.2	93	25	68	32	0.8	−27.8	−26.3	0.7	1.2	0.7
Southern[3]	11,580	3.1	100	25	66	34	2.1	−26.1	−26.1	.4	.7	.5

[1]See Figure 3 for location of oil groups.
[2]Average values based on analyses up to seven samples.
[3]Average values based on analyses up to eight samples.

Table 2 — Geochemical analyses (average values) of upper Sunniland oils, south Florida. (GOR = gas–oil ratio (cu ft/bbl); HC = hydrocarbons; V = vanadium; Ni = nickel; Pr = pristane; Ph = phytane.

tion in a high-vacuum system. $^{13}C/^{12}C$ ratios are expressed in the usual δ-notation: $\delta^{13}C$, per mil = $\{[(^{13}C/^{12}C)\ \text{sample}/(^{13}C^{12}C)_{PDB}] - 1\} \times 10^3$ relative to the PDB marine-carbonate standard.

EXPLANATION OF TERMS

In this paper, *carbonate rocks* refer chiefly to those sedimentary rocks that contain 50% or more of carbonate minerals but may include marlstones that contain less than 50% carbonate content.

A *possible source rock* is a rock that has a sufficient amount of oil-generating organic matter but lacks specific oil-source data that confirm whether it indeed is an effective source rock. A *probable* or *effective source rock* is one in which a definitive oil–source-rock correlation has shown that the rock has yielded hydrocarbons comparable to those of associated oils in reservoirs.

Organic facies, as proposed by Rogers (1980), refers to a rock body characterized by a specific amount, type, and source (e.g., marine vs. terrestrial) of organic matter and type of depositional environment, with special emphasis on type of organic matter.

CHARACTERISTICS OF SUNNILAND OILS

The gross composition of Sunniland crude oils is given in Table 2. Although the oils that were sampled belong to the same broad family of oils, the oils can be divided into two groups based on differences in certain gross or conventional properties. In bulk-compound composition, the Sunniland oils are between the aromatic–intermediate oils (e.g., Middle East) and the paraffinic–naphthenic oils (e.g., Paris basin and Aquitaine basin) (Tissot and Welte, 1978).

The oils have high sulfur contents (2.6–4.5%), low API gravities (19–27°), unusually small amounts of gas in solution (gas–oil ratio about 100:1), and chemical compositions (based on liquid chromatographic fractions) similar to other worldwide oils believed to be derived from carbonate sources (Tissot and Welte, 1978).

The distribution of C_{15+} alkanes is illustrated by capillary gas chromatograms (GC) in Figure 4. The GC traces reveal (1) a wide range of n-alkanes with carbon numbers of n-C_{38} or higher, (2) moderate to high C_{15}–C_{20} isoprenoid peaks at which pristane and phytane are dominant, (3) a distinct hump of unresolved branched-cyclic compounds in the C_{28}–C_{32} region, and (4) appreciable amounts of steranes–terpanes relative to adjacent linear alkanes. The GC traces also showed no apparent evidence of biologic degradation; i.e., no systematic or random removal of n-alkanes in relation to the entire alkane distribution was observed. The absence of biodegradation is supported by the fact that there are no signs of intrusion or mixing of oxygenated waters, a necessary condition if biodegradation had occurred. On the contrary, the formation waters are exceedingly saline and generally have total dissolved salt contents of about 230,000 ppm and as much as 260,000 ppm (Babcock, 1962). The high salinity of these brine waters is attributed at least in part to the original saline or evaporitic conditions of the depositional environment.

One of the unique properties of these oils is the even preference of n-alkanes in the region C_{22}–C_{30}. Also characteristic of these oils is the relatively high ratio of phytane to n-C_{18} (often > 1) and the low Pr/Ph ratio, ranging from 0.4 to 0.9 and averaging about 0.6. These molecular properties are commonly found also in other oils derived from carbonate–evaporite sequences (Tissot and Welte, 1978; Connan, 1981; Palacas, 1984). These properties are attributed to marine-derived source material deposited under highly reducing conditions (Welte and Waples, 1973; Tissot and Welte, 1978; Connan, 1981). In contrast, immature to marginally mature oils derived from marine deltaic shale–sandstone sequences commonly show odd preference or smooth distributions of n-alkanes, Pr/Ph > 1, and Ph/n-C_{18} < 1. Predominantly land-derived source material deposited under more oxidizing conditions probably accounts for these differences.

The molecular compositions of 19 upper Sunniland oil samples were characterized in detail by analysis of three series of polycyclic alkanes using computerized GC-MS (Ensminger et al, 1974; Seifert and Moldowan, 1978; Zumberge, 1983). Figure 5 shows typical mass fragmentograms of the tricyclic diterpanes (m/z 191), pentacyclic triterpanes (m/z 191), and steranes (m/z 217) in Sunniland oils. The distribution patterns are remarkably similar for all 19 oil samples analyzed despite the fact that the samples were collected from several different zones at 14 widely scattered localities covering about 1,000 mi^2 (2,590 km^2).

The nearly identical biomarker compositions strongly suggest that the oils were derived from the same organic facies. However, differences in conventional geochemical properties (e.g., the last six items in Table 2) indicate that the oils were derived from at least two distinct but closely related facies. For example, assuming no changes in ratios owing to

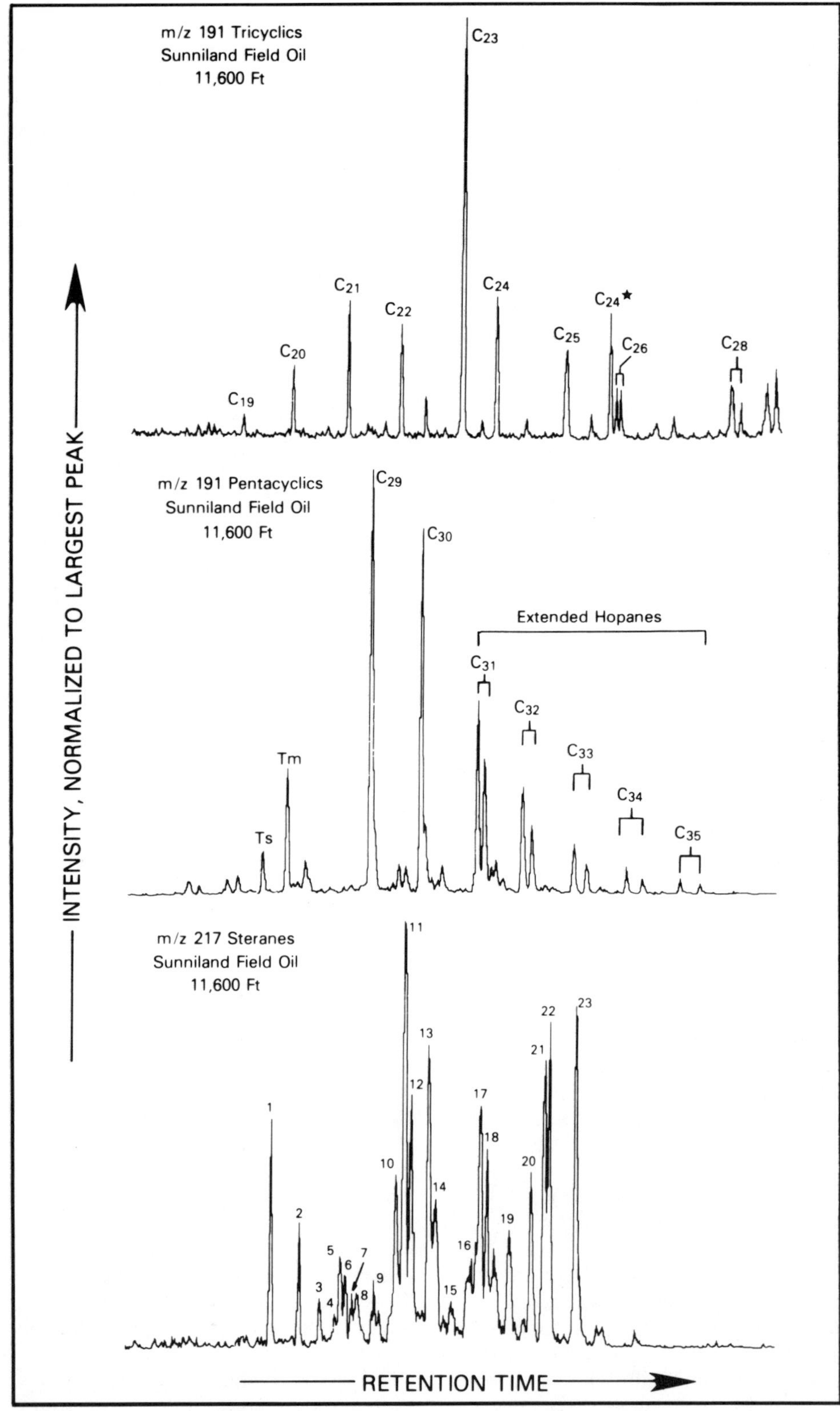

Figure 5 — Mass fragmentograms of tricyclic terpanes (m/z 191), pentacyclic terpanes (m/z 191), and steranes (m/z 217) in a typical upper Sunniland oil. Common names for labeled peaks in pentacyclic terpane distribution: Ts = 18 (H)-trisnorneohopane; Tm = 17 (H)-trisnorhopane; C_{29} = norhopane; C_{30} = hopane. Identified steranes are listed in Table 3.

migration effects, most of the northern oils are distinguished by Ph/n-C_{18} ratios >1 and must have been derived from source facies characterized by Ph/n-C_{18} >1 (Table 2; Fig. 4). In like manner, most of the southern oils (Ph/n-C_{18} <1) must have been derived from organic facies with Ph/n-C_{18} ratios <1. These consistent differences in elemental, isotopic, and molecular geochemical properties of the Sunniland oil subgroups illustrate that distributions of polycyclic alkanes alone may not always be conclusive in differentiating specific sources. This example also underscores the fact

Table 3

	Compound	Elemental Composition
1	13β,17α-diacholestane (20S)	$C_{27}H_{48}$
2	13β,17α-diacholestane (20R)	$C_{27}H_{48}$
3	13α,17β-diacholestane	$C_{27}H_{48}$
4	13α,17β-diacholestane	$C_{27}H_{48}$
5	24-methyl-13β,17α-diacholestane	$C_{28}H_{50}$
6	24-methyl-13β,17α-diacholestane	$C_{28}H_{50}$
7	14β,17β-cholestane	$C_{27}H_{48}$
8	24-methyl 13α,17β-diacholestane	$C_{28}H_{50}$
9	24-methyl 13β,17α-diacholestane	$C_{28}H_{50}$
10	14α,17α-cholestane (20S)	$C_{27}H_{48}$
11	14β,17β-cholestane (20R)	$C_{27}H_{48}$
	+24-ethyl-13β,17α diacholestane	$C_{29}H_{52}$
12	14β,17β-cholestane (20S)	$C_{27}H_{48}$
13	14α,17α-cholestane (20R)	$C_{27}H_{48}$
14	24-ethyl-13β,17α-diacholestane	$C_{29}H_{52}$
15	24-ethyl-13β,17α-diacholestane	$C_{29}H_{52}$
16	14α,17α-ergostane (20S)	$C_{28}H_{50}$
17	14β,17β-ergostane (20R)	$C_{28}H_{50}$
18	14β,17β-ergostane (20S)	$C_{28}H_{50}$
19	14α,17α-ergostane (20R)	$C_{28}H_{50}$
20	14α,17α-stigmastane (20S)	$C_{29}H_{52}$
21	14β,17β-stigmastane (20R)	$C_{29}H_{52}$
	+5β,14α,17α-stigmastane*	
22	14β,17β-stigmastane (20S)	$C_{29}H_{52}$
23	14α,17α-stigmastane (20R)	$C_{29}H_{52}$

*Detected only in immature samples.

Table 3 — List of identified steranes.

that, whenever feasible, a balanced analytical approach should be used in which both gross or conventional and detailed molecular analyses are used for oil–oil and oil–source correlations.

On the other hand, as will be shown later, the detailed molecular biomarker techniques were found to be indispensable for distinguishing between possible and effective source rocks for these oils.

CHARACTERISTICS OF SUNNILAND SOURCE ROCKS

Lithology

Those Sunniland rocks judged to be possible source rocks by virtue of their inherent geochemical properties have the following lithologic characteristics. The source rocks are predominantly brown to medium- to dark-gray, slightly fossiliferous, commonly pelletal, micritic limestones. Foraminiferal tests and microshell fragments are common, but megafossils are rare. Total carbonate content (Table 1) varies from as low as 50–60% (typical of rocks classified as marlstones) to as much as 90–98% (typical of true limestones). The remaining noncarbonate content consists primarily of argillaceous matter. Of all the Sunniland core samples analyzed, none seems to fit the "pure" shale classification (i.e., much less than 30% carbonate content), strongly suggesting that bona fide shales in the Sunniland are absent to scarce. Hence, from the standpoint of lithology alone, shales can be dismissed readily as possible sources of Sunniland oil.

Laminated and stylolitic structures characterize many, but not all, of the possible source rocks. Commonly associated with the stylolites are oil stains and/or solid carbonaceous residues. Also common to the Sunniland Limestone are algal structures that occur both in fine-grained source-type rocks (Sears, 1972; Klopp, 1975) and in reservoir-type rocks (Means, 1977). In the fine-grained micritic rocks, algal material can be of many forms, among which are pellets (Klopp, 1975) and laminae or bands that commonly are wavy and irregular in appearance, probably reflecting stromatolitic structures (Sears, 1972; M. Malek–Aslani, written communication).

In evaluating all the possible source rocks that were analyzed, one important observation should be emphasized—that carbonate source beds are not limited only to the laminated organic facies. Some black, organic-rich beds are not laminated, showing that some burrowing or bioturbation took place. A case in point is sample 1-259 (Fig. 1; Table 1), which is a dark-gray to black, argillaceous, dense, organic-rich micrite (TOC, 2.3%) that has no apparent laminations. In thin section, the sample appears to be a wackestone with abundant pelloids and microdebris (M. Malek–Aslani, written communication). On the basis of its organic-carbon and extractable-organic-matter content, HC/TOC ratio, alkane GC fingerprint, and biomarker contents, the micrite certainly has all the earmarks of a probable source rock.

Total Organic Carbon

The total-organic-carbon content of the Sunniland possible source rocks generally varies between 0.4 and 3.0%, may be as high as 12.3% or more (Table 1), and averages about 1.8%. This average organic-carbon value, in general, does not include values for rocks that have <0.4% TOC. Such lean rocks are abundant, especially in the upper half of the Sunniland Limestone and in the updip shelf areas. Rather, the average total-organic-carbon value is primarily based on those possible source rocks in the lower Sunniland and in the lower part of the upper Sunniland (e.g., boreholes 4, 10, 13, and 1; Table 1), particularly in downdip basinal areas.

Although 0.4% has been selected as a working lower limit of organic-carbon content for petroleum source rocks in south Florida (Palacas, 1978b), a few samples that contain less than 0.4% TOC (e.g., 8-165, 5-107, Table 1) were also included because of other seemingly favorable characteristics. Some of these characteristics are the presence of bituminous laminae or bands, high ratios of extractable organic matter to total organic carbon, or a combination of these and other factors.

The average total-organic-carbon content (1.8%) for Sunniland source rocks is about 2.5 times higher than the average value (0.67%) of 118 carbonate source rocks analyzed from 18 basins worldwide (Tissot and Welte, 1978, p. 96) and more than 5 times greater than the average content (0.33%) for carbonate rocks in general (Ronov and Yaroshevsky, 1969; Hunt, 1972). Therefore, from the standpoint of total-organic-carbon content alone, the Sunniland carbonate source rocks seem to contain more than suffi-

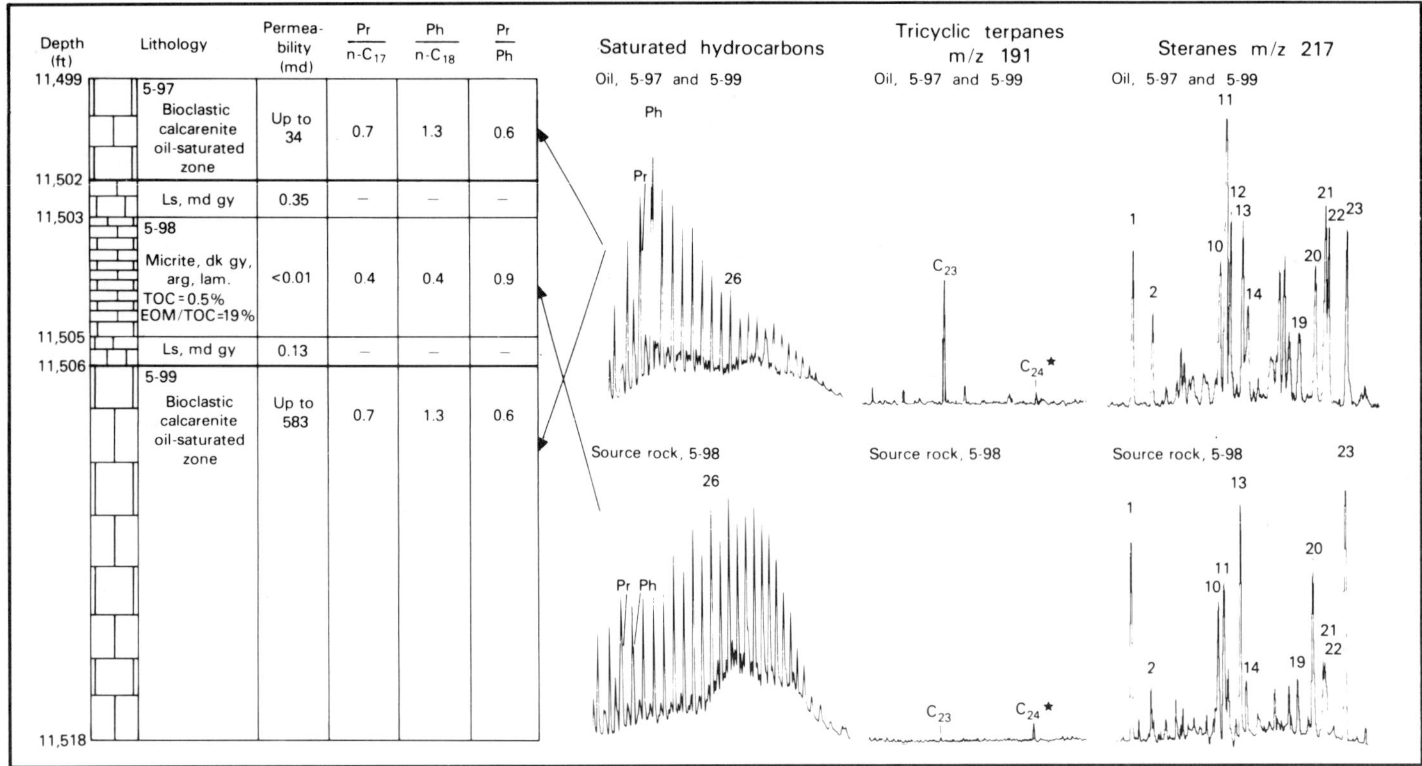

Figure 6 — Comparison of saturated hydrocarbon, sterane (m/z 217), and tricyclic terpane (m/z 191) distributions from two permeable oil-saturated reservoir units (samples 5-97 and 5-99) and an interbedded impermeable, laminated micrite (5-98). Chromatograms for only one oil sample are given, since both oils (5-97 and 5-99) are essentially identical. Note the lack of correlation between the micrite source rock and the oils, strongly indicating that no diffusion of oil into the micrite has taken place. GC's of saturated hydrocarbons were run on packed columns coated with SE-30. Identified steranes are listed in Table 3. Pr = pristane; Ph = phytane.

cient amounts of organic matter to generate commercial quantities of petroleum.

Extractable Organic Matter

Possible Sunniland source rocks are rich in chloroform-extractable organic matter, as indicated by the absolute yields ranging from 509 to more than 8,700 ppm (of the dry weight of rock) (Table 1). The relative high proportion of soluble substances, however, is best demonstrated when related to total organic carbon. Accordingly, the extractable-organic-matter content as a percentage of total organic carbon (EOM/TOC) averages about 14.7%, but values may be as high as 23% (Table 1). The Sunniland average is significantly greater than that for worldwide carbonate source rocks (11.6%), and even much greater than that for worldwide shale source rocks (8.1%) (Tissot and Welte, 1978, p. 96).

Compositionally, the chloroform extracts of Sunniland source rocks on a gross level are characterized by greater proportions of hydrocarbons than nonhydrocarbons (resins and asphaltenes) (about 64 and 36%, respectively). The ratio of saturated hydrocarbon to aromatic hydrocarbons (SAT/AROM) varies widely from 0.3 to 1.6 and averages roughly 1.0. The wide variation of these ratios is probably due in large part to source differences, but other factors may also play a role, such as unequal maturation effects or variable adsorption behavior of organic and inorganic constituents that cause differences in composition of expelled hydrocarbons. Part of the variability, however, may be due also to the different analytical techniques used (deasphalting vs. nondeasphalting).

The hydrocarbon richness of Sunniland source rocks is further manifested in the amount of extractable hydrocarbons (saturate plus aromatic) as measured against the percentage of total organic carbon (HC/TOC, in %). The HC/TOC value ranges widely from 3.8 to 19.0% and averages 9.4% (Table 1). This relatively high average HC/TOC ratio, especially when compared to the average for worldwide carbonate source rocks (5.0%) and shale source rocks (4.3%) (Tissot and Welte, 1978, p. 96), attests to the hydrogen-rich, oil-generating character of the Sunniland organic matter.

The wide range of HC/TOC values for the most part reflects the wide variation in organic facies and their different response to thermal stresses. It may be argued that some of the higher hydrocarbon yields are due to oil staining despite the fact that all precautions were taken to preclude obviously stained samples. For those samples that contain seemingly anomalous amounts of hydrocarbons (e.g., HC/TOC $>12\%$), available evidence indicates that the hydrocarbons are largely, if not entirely, indigenous to the host rock.

A case in point is core sample 5-98, which is representative of a 2-ft (0.6-m) impermeable (<0.01 md), laminated micrite that is sandwiched between two permeable (as much as 583 md) oil-saturated producing reservoir units (Fig. 6). Compelling geochemical evidence indicates that the relatively high hydrocarbon content of the laminated micrite

(HC/TOC = 15.0%) is not a result of migrated oil from without but rather a result of *in-situ* generation from an oil-prone type of organic matter. The rock extract, in terms of overall saturated-hydrocarbon distribution, isoprenoid ratios, and biomarker distributions, showed essentially no resemblance to the oil in the encompassing oil-saturated rservoir units. Thus, there is no evidence of diffusion from the reservoir units even over long periods of geologic time.

Microscopic examination of a thin section of sample 5-107 (lowermost lower Sunniland) also indicated that the high proportion of hydrocarbons (HC/TOC = 19%) was probably due to the concentration of bituminous organic matter along stylolites in an otherwise organically lean, medium-gray micritic limestone. Compositionally, the hydrocarbons are similar to those of Sunniland oils. The source of the relatively high concentration of organic matter may be directly related to the process of stylolitization wherein dispersed organic matter and other insoluble residues are concentrated along irregular seams during pressure dissolution of the host carbonate rock.

If this conclusion is correct, then this may be the first geochemically documented example of a mechanism for hydrocarbon expulsion in carbonate rocks. This hypothesis is not new but was proposed decades ago by Ramsden (1952) and Dunnington (1954). There is, of course, an alternative explanation. the high HC/TOC ratio may simply be due to staining by Sunniland oil migrating along the stylolites. If the latter hypothesis is true, we propose that stylolites may well be one of the main avenues of primary migration in carbonate rocks. Stylolites are abundant in the Sunniland Limestone.

Gas-Chromatographic (GC) Analyses

The extractable organic matter is further characterized, on a molecular scale, by capillary gas-chromatographic analyses of the C_{15+} alkanes (Fig. 4).

In general, most of the Sunniland possible source rocks have geochemical properties and overall GC patterns not unlike those of the oils. However, rarely is a single source-rock GC pattern (or for that matter, a biomarker pattern) exactly identical to that of the oils. There are three possible explanations for this slight dissimilarity when detailed molecular compositions of crude oils and possible source-rock extracts are compared. First, the expulsion and migration process may create differences in the alkane distributions; second, the oils represent a mixture of contributions from several organic facies within the carbonate sequence; and third, the actual source-rock facies may not have been sampled. We prefer the second explanation.

In summary, the favorable lithology, ideal stratigraphic position, and comparable gross and molecular features (as determined by gas chromatography) all argue convincingly that the rocks designated as possible source beds are the probable sources of the Sunniland oils. However, polycyclic alkane biomarker analyses indicate significant differences in the source materials, suggesting that only certain organic facies are the major probable source rocks.

Gas-Chromatographic–Mass-Spectrometric Analyses

Tricyclic Terpanes

Of the three series of biomarker compounds that were analyzed, the tricyclic distributions appeared to be the simplest and most useful series of biomarkers in differentiating effective source rocks from possible source rocks. The tricyclic as well as the pentacyclic terpane distributions were determined by examining the m/z 191 ion by GC-MS. For the purposes of this report, only the tricyclic molecules from C_{19} to C_{28} are discussed herein (Fig. 7), although the tricyclic compounds may extend in carbon number as high as C_{35} (Moldowan, Seifert, and Gallegos, 1983). Furthermore, within this tricyclic series there may be present, in varying amounts, tetracyclic terpanes, such as the C_{24} and C_{26} components, which either coelute with or elute near the C_{26} and C_{28} tricyclic components, respectively.

In evaluating the various tricyclic distributions in the Sunniland carbonate rocks as well as in the older and more deeply buried rocks that were analyzed, the tricyclics apparently can be grouped conveniently into two end-member types. One end-member type is characterized by a pattern with a maximum C_{23} tricyclic terpane peak (Fig. 7A) and the other with a maximum C_{24} tetracyclic terpane peak (Fig. 7B). Most of the other tricyclic patterns fit between the two end-member types. Interestingly, of all the patterns examined, only the end-member type with the dominant C_{23} tricyclic peak closely resembles the distinctive high C_{23} tricyclic type found in *all* Sunniland crude oils (Fig. 7C). Consequently, we propose that those facies with the maximum C_{23} tricyclic pattern are the major source rocks for the Sunniland oil, assuming, of course, that all other geologic and geochemical criteria also are met and that no significant changes occur in the tricyclic distributions owing to migration from source to reservoir.

Those source facies that have tricyclic patterns similar to the other end-member pattern (i.e., the high C_{24} tetracyclic peak pattern) are not considered major contributors of hydrocarbons to Sunniland oils, though the possibility that they make some minor contributions cannot be excluded. However, certain organic facies characterized by the high C_{24} tetracyclic pattern, either in the Lake Trafford Formation or in the upper Sunniland, may be sources for a Lake Trafford oil that also has a similar pattern (Fig. 7D). The Lake Trafford oil is a noncommercial oil deposit about 100 ft (30 m) or more stratigraphically above the top of the Sunniland Limestone. Good oil shows in the Lake Trafford Formation have been observed in the Exxon 34-4 Collier well, sec. 34, T49S, R32E, Collier County, and in the Exxon 36-1 Oleum well, sec. 36, T50S, R31E.

Pentacyclic Terpanes

The pentacyclic terpanes belong mostly to the $17\alpha(H),21\beta(H)$ hopane series, with molecules ranging from C_{27} to C_{35} or more (Fig. 8). Hopane (C_{30}) is the most abundant component, followed usually by norhopane (C_{29}). The component $17\alpha(H)$-trisnorhopane (Tm) commonly exceeds its counterpart $18\alpha(H)$-trisnorneohopane (Ts). The $17\beta(H),21\alpha(H)$ series of molecules (moretanes) and the compound gammacerane also are present but in low concentrations, and the $17\beta(H),21\beta(H)$ hopanes either are absent or are present in trace amounts. The extended hopanes, C_{31}–C_{35}, which occur as doublets or stereoisomeric pairs, commonly show either a staircase distribution pattern (i.e., uniformly lower amounts with increasing molecular weight) (Fig. 8A) or an irregular pattern in which the even-predominant molecules C_{32} and/or C_{34} are more abundant than the adjoining C_{31}

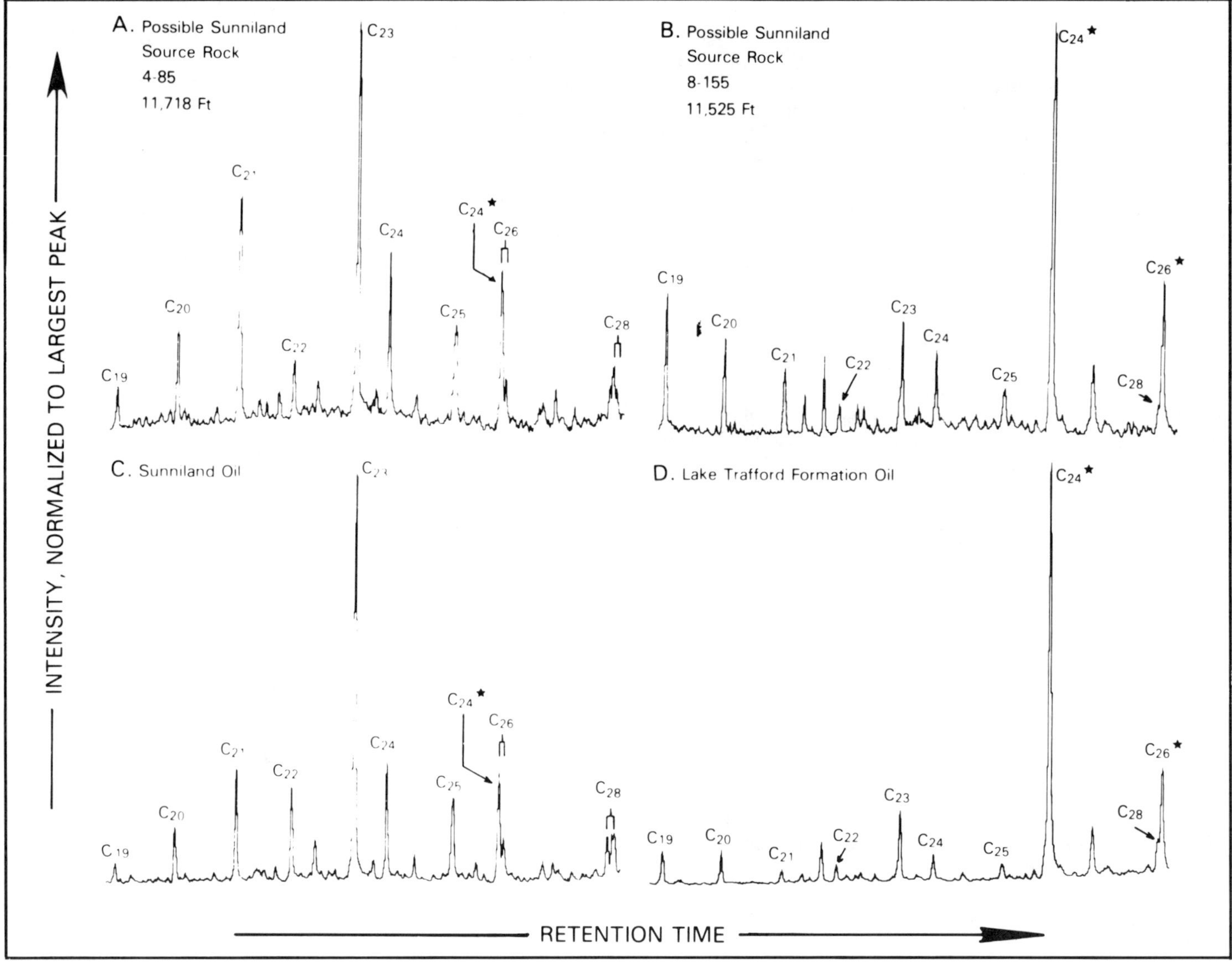

Figure 7 — Comparison of end-member tricyclic terpane patterns (m/z 191) of Sunniland possible source rocks and associated crude oils. A. a source-rock pattern with a maximum C_{23} tricyclic peak; B. a source-rock pattern with a maximum C_{24} tetracyclic peak; C. typical pattern of a Sunniland crude oil; D. typical pattern of a Lake Trafford Formation crude oil. Asterisk indicates tetracyclic terpane compounds.

and C_{33} molecules, respectively (Fig. 8B). The Sunniland carbonate rocks that have been judged as probable source rocks mostly exhibit the staircase pattern, which is the diagnostic pattern for all the upper Sunniland oils.

One of the characteristic features of the pentacyclic distributions in both Sunniland rocks and oils is their predominance over the tricyclic and sterane distributions (Fig. 9). The pentacyclics, however, from a worldwide perspective, do not always constitute the major polycyclic biomarker group in rock extracts or crude oils. For example, in the Toarcian shales of the Paris basin and in the Cretaceous source rocks of the Persian Gulf the steranes are more abundant than the triterpenoids (Tissot and Welte, 1978, p. 117). In most of the crude-oil accumulations in the Lyons Sandstone of Permian age in northern Colorado, the hopane (pentacyclic) molecules are entirely absent, whereas the tricyclics and steranes are intact and in normal abundance (J. L. Clayton, written communication).

Steranes

Figure 10 shows m/z 217 mass fragmentogram patterns representative of sterane distributions in various Sunniland possible source rocks. Sterane identifications are listed in Table 3.

In their studies of modern sediments, Huang and Meinschein (1976, 1979) have shown that the C_{27}, C_{28}, and C_{29} sterols, and presumably their reduced equivalents (the steranes), can be used as indicators of depositional environments. An abundance of C_{29} over C_{27} steranes would suggest a terrestrial source, whereas C_{27} steranes predominating over C_{29} steranes would indicate mainly a marine or aquatic source.

The application of this relationship in south Florida is not so clear. All the available geologic and geochemical evidence indicates that most of the organic matter in the Sunniland carbonate rocks is of marine origin. Yet biomarker analyses

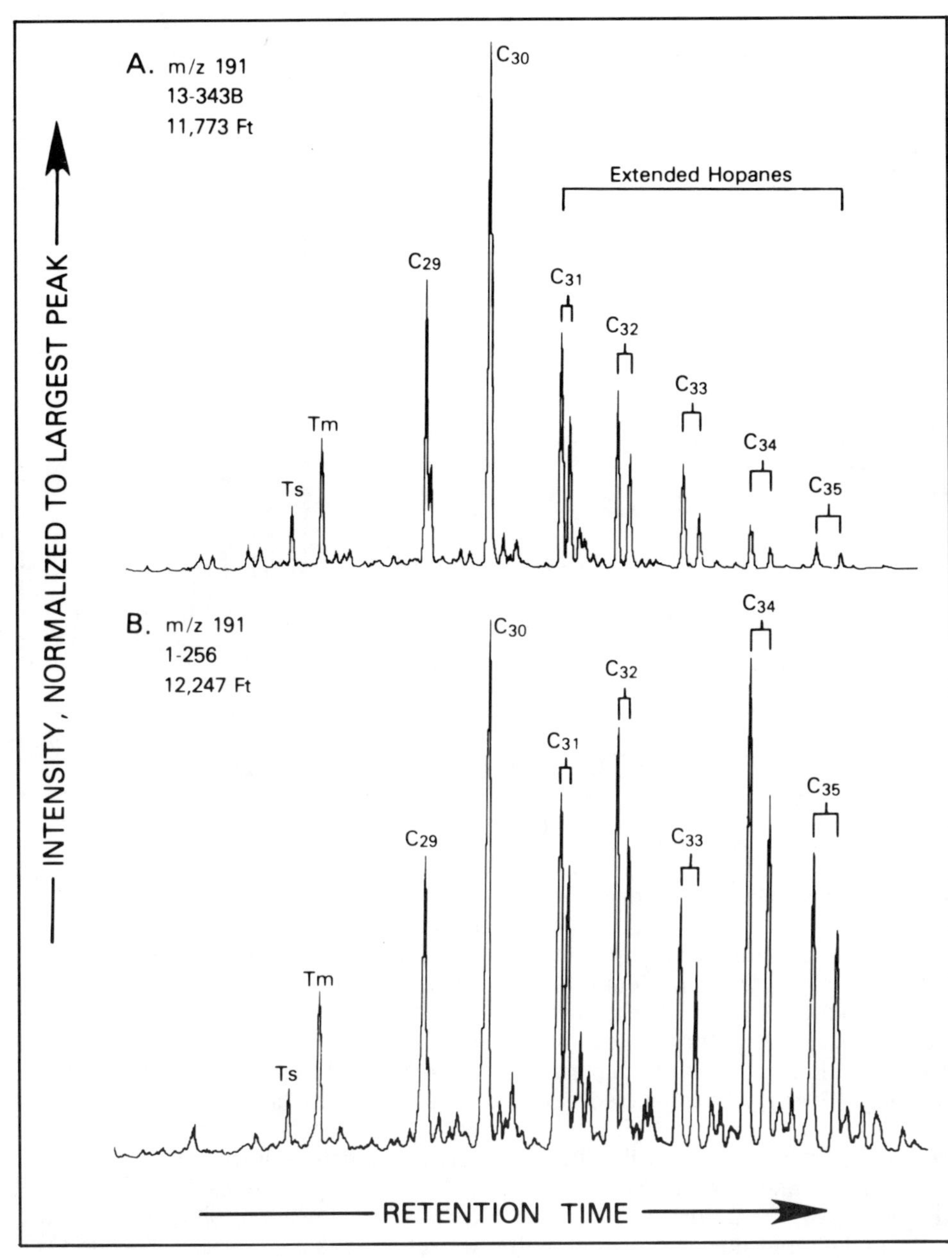

Figure 8 — Comparison of end-member patterns of pentacyclic terpanes (m/z 191) in Sunniland possible source rocks. A. This end member has a regular or staircase extended hopane pattern. Such patterns are found in both upper and lower Sunniland source rocks. B. The other end member is characterized by an irregular extended hopane pattern with C_{32} and/or C_{34} hopanes exceeding the C_{31} and C_{33} hopanes, respectively. This pattern is more common in the upper Sunniland source rocks.

indicate that the C_{29} sterane concentrations exceed or at best are equivalent to the C_{27} sterane concentrations. Possibly, in a carbonate–evaporite environment, conditions such as intense microbiologic activity and highly preservative saline conditions could lead to a greater production of the C_{29} steranes. In glancing at the m/z 217 fragmentograms (Fig. 10), the presence of a large peak at the $\beta\beta$20R-C_{27} position (peak 11) gives the appearance that the C_{27} steranes as a whole are in greater abundance relative to the C_{29} steranes. This is not true. Peak 11 is significantly enhanced owing to the coelution of the 24 ethyl-13β, 17α diacholestane component.

The rearranged steranes (diasteranes) (peaks 1 to 9, but in particular 1 and 2) occur in relatively abundant quantities in nearly all the Sunniland carbonate rocks (Figs. 10A and B) except for certain facies in the lower part of the lower Sunniland Limestone. At or near the base of the section, the diasterane distributions of two rock extracts and five widely scattered oils or oil shows are dramatically different—the diasteranes (peaks 1 and 2) are either absent or are present in minor or trace amounts (e.g., Figs. 10C and 11).

Based on laboratory experiments, researchers such as Rubenstein, Sieskind, and Albrecht (1975) and Sieskind, Joly, and Albrecht (1979) have shown that diasteranes can form from precursor sterol substances through the acid-catalytic activity of clays. However, in the submature carbonate–evaporite environment of south Florida, no correlation has been observed between clay content and abundance of diasteranes. In fact, irrespective of (1) geologic age (Paleocene to Early Cretaceous), (2) depth of burial (maturity) (<5,000 ft [1,524 m] to about 15,000 ft [4,570 m]), and (3) abundance of organic matter, some carbonate rock samples that contain little or *no* clay, have relatively large concentra-

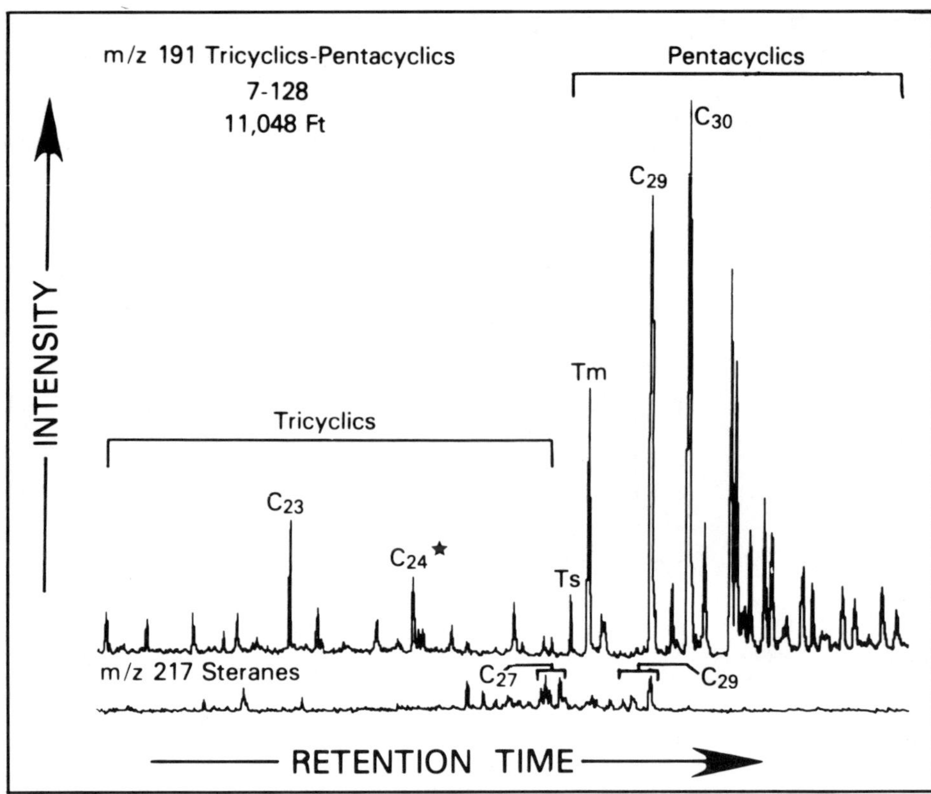

Figure 9 — Traces of m/z 191 and m/z 217 fragmentograms superimposed at the same relative intensity show that the sum of the relative concentrations of cyclic biomarkers are: pentacyclics > tricyclics > steranes.

tions of diasteranes. On the other hand, some carbonate samples that contain as much as 20 to 40% argillaceous material (presumably consisting mostly of clays) contain relatively minor amounts of diasteranes.

Biodegradation also has been advanced as a possible means for explaining increased quantities of rearranged steranes relative to regular steranes (Seifert and Moldowan, 1979; Curiale, Harrison, and Smith, 1983). However, this process is applicable only to migratable hydrocarbons in contact with oxygen-containing waters and bacteria either in subsurface reservoir-type settings or in surface or near-surface structures. It does not apply to the south Florida rock extracts, which occur originally as dispersed organic matter in tight, nonreservoir rocks at depths as great as 12,750 ft (3,886 m), well beyond any conceivable intrusion of meteoric waters.

In short, the likely explanation is that the diasteranes in the south Florida carbonate rocks were generated sometime during the early stages of diagenesis ($R_o \leq 0.3$; TAI ≤ 1.5) probably under mild thermal stress (<55°C [130°F]), and with or without the catalytic support of clays. Consideration, however, should be given also to the possibility that some if not all the diasterane precursors might have been a product of microbial activity in a highly reducing, hypersaline environment. Although little evidence is available to test such a hypothesis, the recent documentation by Simoneit, Halpern, and Didyk (1980) gives some credence to this idea. In their studies of lipids in a fresh-water, alpine lake in Chile, Simoneit, Halpern, and Didyk found C_{26}–C_{29} diasterenes in a 1-cm-thick algal mat and in the underlying anoxic black mud. With slightly more burial and continued anoxia, it is not unreasonable to suggest that the unstable diasterenes could be easily reduced to their diasterane counterparts. Interestingly, traces of tricyclic diterpanes (C_{20}–C_{25}) and the 17β(H),21β(H) hopanes also were discovered in the algal mat and underlying mud.

One important application of biomarker distributions with possible far-reaching implications is the use of the nonrearranged C_{29} sterane isomeric ratios as thermal-maturity indicators. The excellent foundational studies of Mackenzie et al (1980, 1984) and Seifert and Moldowan (1981) have demonstrated in several different siliciclastic sedimentary basins that the ratios of $\alpha\alpha$20S to $\alpha\alpha$20R (peaks 20 and 23, Fig. 10) and the ratio of the $\beta\beta$(20R + 20S) to $\alpha\alpha$(20S + 20R) (peaks 21 + 22 and 20 + 23, respectively) systematically change with maturation until isomerization-equilibrium values are reached. The study of these same sterane ratios in the south Florida carbonate–evaporite basin confirms and reinforces their findings.

The integrity of the results, however, hinges upon the full resolution of all four C_{29} sterane peaks (20, 21, 22, 23, Fig. 11). For example, if the interface between the gas chromatograph and the mass spectrometer is not heated adequately, the successively higher molecular-weight components are not as fully resolved as the prereceding ones. This inadequacy was particularly evident in the early stages of our study, when the GC-MS interface was heated to only 280°C (536°F). We noticed a significant loss of resolution at the 20R peak (23), resulting in a deceptively "more mature" signature (Fig. 11A). Increasing the heating element at the interface to 335°–340°C (635°–644°F) results in a decisively better resolution of the peaks and a more reliable maturity index (Fig. 11B).

Other workers seem to have had the

Figure 10 — Traces of m/z fragmentograms, illustrating typical sterane distributions in Sunniland possible source rocks. Note the relatively high amounts of diasteranes (peaks 1 and 2) in samples 8-155 (A) and 13-343B (B) but minor amounts in sample 1-259 (C), which is from the lower Sunniland Limestone. Identification of steranes is shown in Table 3.

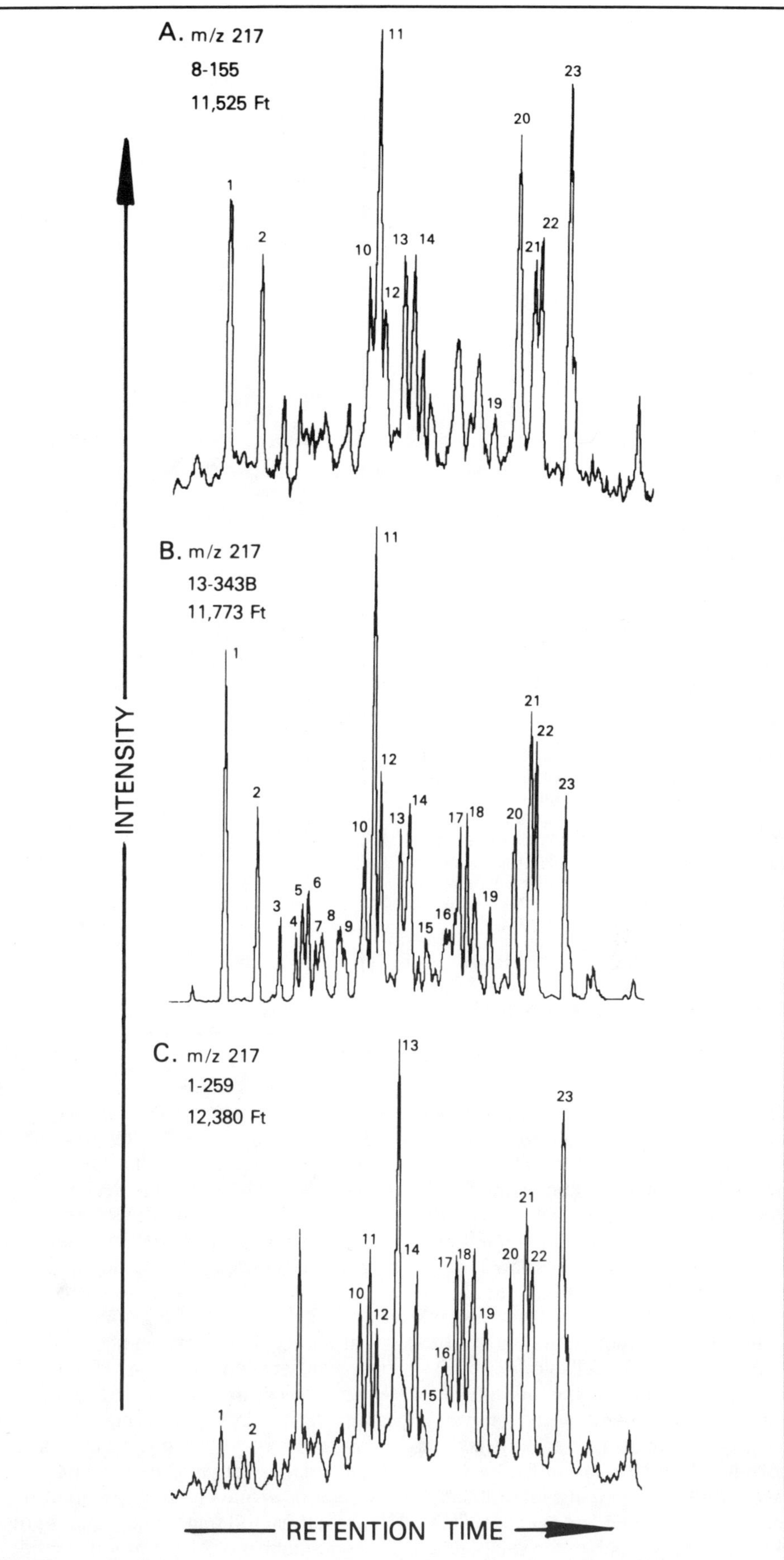

same problem with this inadequate sterane resolution. Therefore, caution should be exercised in interpreting the sterane maturity-index results unless appropriate instrumental conditions have been achieved. Higher molecular-weight components such as the extended hopanes also showed less than optimum resolution under the lower heating conditions, although the stereoisomeric ratios were not affected as seriously as were the C_{29} sterane ratios.

Kerogen

The oil-proneness of Sunniland source rocks, as indicated earlier by the characterization of the soluble organic matter (e.g., average HC/TOC ratio = 9.4%), is substantiated by the analysis of the dispersed insoluble organic matter called kerogen. Microscopic examination of 37 Sunniland and older carbonate-rock samples showed that the organic matter consists principally of algal-amorphous particles (commonly >80%), smaller amounts of exinite or herbaceous material (generally <20% but as much as 45%), and still less of vitrinite, inertinite and other particles (generally <10% and often trace amounts).

Elemental analysis of kerogen concentrates of 16 Sunniland samples (10 of which are listed in Table 1) and 3 older carbonate source rocks indicate that the atomic hydrogen-to-carbon (H/C) ratio is relatively high, commonly ranging from 1.0 to 1.3 and averaging about 1.15. These H/C ratios reflect both the hydrogen richness and the relative low degree of thermal maturity of the organic matter.

These analyses strongly suggest that most of the solid organic matter in the Sunniland and associated carbonate rocks belongs to the type I and type II kerogen classes (Tissot and Welte, 1978), derived chiefly from algal, planktonic, and microbial residues that accumulated

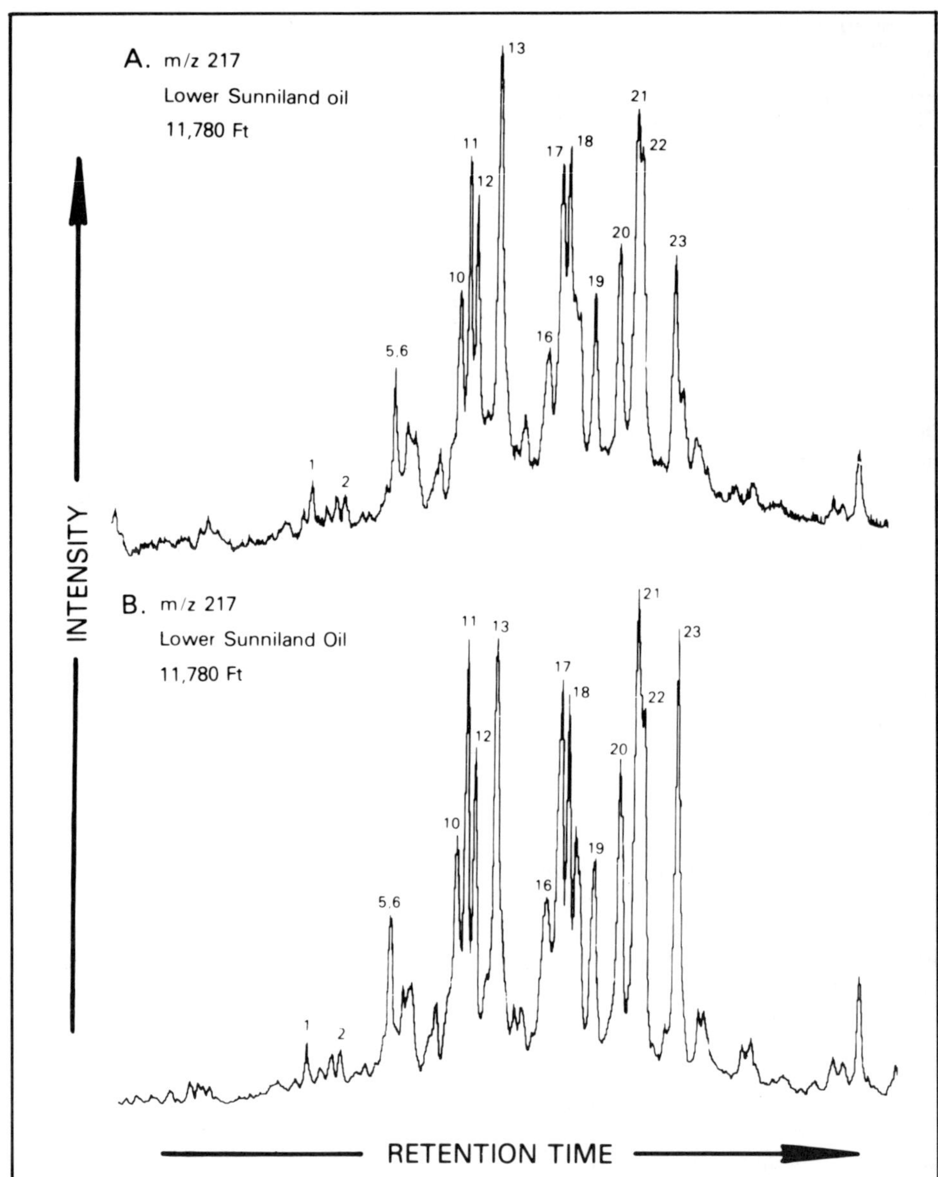

Figure 11 — Comparison of m/z 217 fragmentograms of the same sample (a lower Sunniland oil), illustrating variations in C_{29} sterane ratios owing to different instrumental conditions. A. Inadequate heating (280°C [536°F]) at GC-MS interface produces unreliable "mature-looking" C_{29} sterane ratios. Note that 20S/(20S + 20R) ratio, peaks 20/(20 + 23) is about 0.52, a maturity level within isomerization equilibrium values (0.50–0.55) proposed by Mackenzie et al (1984). B. Adequate heating at elevated temperatures (335°–340°C [635°–644°F]) results in a more reliable sterane-maturity index. The ratio 20S/(20S + 20R) is now about 0.39.

in saline to hypersaline, moderately to highly reducing marine environments.

Rock-Eval pyrolysis data (Table 1) add further support to the above interpretation. For example, for those samples that contain about 1.0% or more TOC the hydrogen-index (HI) values (S_2/TOC), which are comparable to atomic H/C ratios (Espitalié et al, 1977), are also relatively high, commonly ranging from 350 to 845 mgHC/gTOC (Table 1; Fig. 12).

Another pyrolysis-type measure of the hydrocarbon-generating capability of Sunniland organic matter is the genetic potential (sum of $S_1 + S_2$; Table 1). According to the classification scheme suggested by Tissot and Welte (1978, p. 447), nearly one-half of the samples analyzed from the Sunniland, especially those samples from the downdip region that have genetic potentials >6 mgHC/g rock, are rated as good to perhaps excellent source rocks. Another one-fifth or more of the samples that have genetic potentials ranging from 2 to 6 mgHC/g rock are rated as moderate source rocks. Although the remaining one-third or so of the source rocks have values predominantly between 1 and 2 mgHC/g rock, which according to the scheme of Tissot and Welte (1978, p. 477) are rated as non-oil-source rocks, we suggest that the sum total of these organically leaner carbonate rocks, by virtue of other geochemical properties, still might contribute significant amounts of hydrocarbons to the overall petroleum accumulation.

MATURITY OF OILS AND SOURCE ROCKS

Conventional Parameters

On the basis of many properties that commonly are ascribed to thermal immaturity, Sunniland crude oils are considered to be marginally mature. Some of these properties are (1) low API gravities (19–27°), (2) high sulfur contents (2.6–4.5%), (3) low gas–oil ratios (100:1), (4) a high proportion of C_{15+} hydrocarbon molecules (80–90%) relative to those below C_{15}, (5) relatively high amounts of isoprenoids with respect to the adjoining

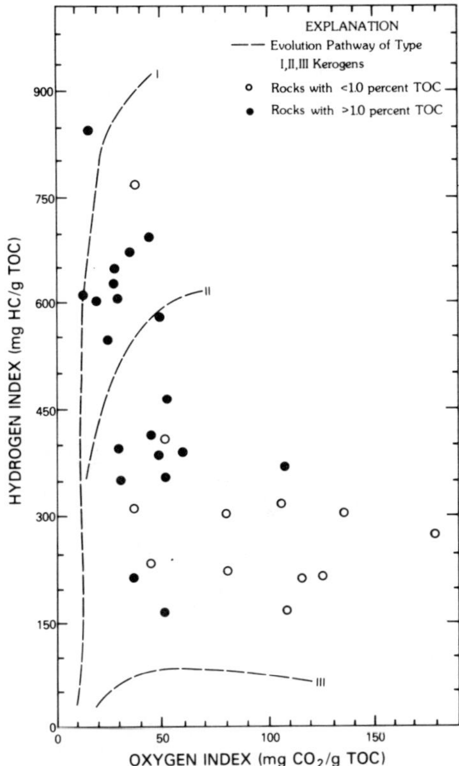

Figure 12 — Plot of Rock-Eval hydrogen indices (HI) versus oxygen indices (OI) of Sunniland possible source rocks. Samples with relatively high oxygen indices are generally restricted to rocks that contain less than 1.0% TOC. The high OI values are attributed in part to the indigenous organic matter deposited under more oxic or less reducing conditions. The high values also may be due in part to instrumental causes. Palacas, Daws, and Applegate (1981) showed that for organically lean carbonates some or most of the pyrolytic CO_2 was derived from the dissociation of carbonate minerals during pyrolysis.

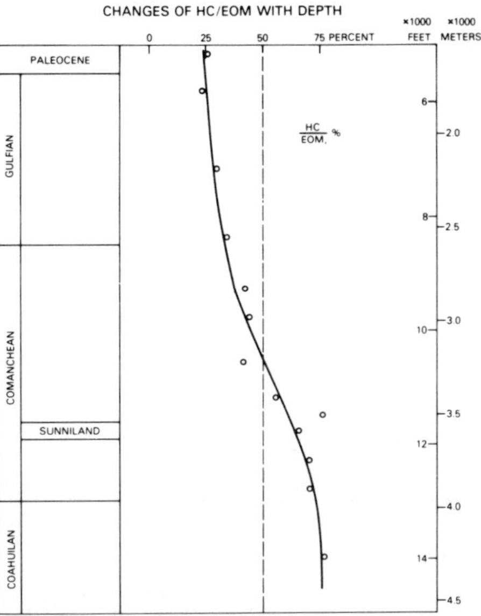

Figure 13 — Plot of the ratio of hydrocarbons to extractable organic matter (HC/EOM, in percent) with depth of burial, illustrating continuous diagenetic–catagenetic effects with increasing thermal maturity. Each datum point is an average of several to as many as 15 values for each successive stratigraphic unit. Note that under shallow, immature conditions the hydrocarbon content is much less than 50%, whereas at the Sunniland level and below it is much higher than 50%.

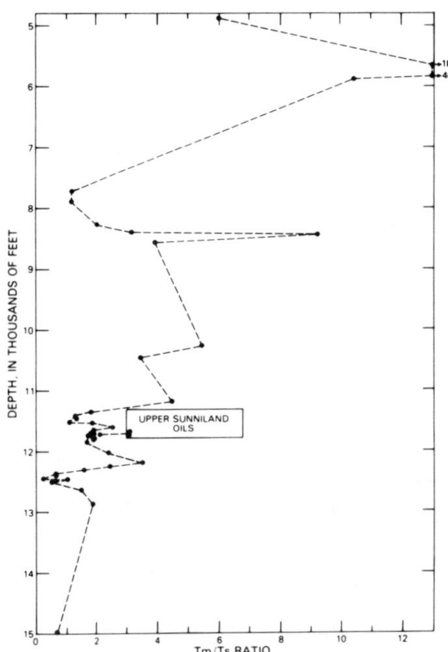

Figure 14 — Plot of Tm/Ts ratio (17α[H]-trisnorhopane/18α[H]-trisnorneohopane) for individual carbonate-rock samples with depth. Although there is an overall decrease in this ratio with depth, there are also significant fluctuations at several stratigraphic levels owing to organic-facies changes. For comparison, the ratios of upper Sunniland oils (15 samples) are indicated within the boxed area.

n-alkanes (this includes a phytane/n-C_{18} ratio commonly greater than 1) (Fig. 4), (6) a low proportion of linear (n-)alkanes (20-25%) relative to branched-cyclic compounds, (7) Pr/Ph ratios <1, (8) distinct even predominance in the C_{22}-C_{30} n-alkane distribution, and (9) a distinct hump in the sterane-triterpane region. It follows that such properties demand that the source of the oil should also be from rocks that were immature to marginally mature at the time of expulsion. Possible carbonate source rocks within the Sunniland Limestone seem the most likely to meet these criteria.

In addition to the properties (4 through 9) just listed, other standard, immature thermal indicators characterize the Sunniland source rocks. For example, thermal-alteration indices (TAI) of kerogen concentrates commonly range from 1+ to 2, consistent with an immature to marginally mature level of thermal alteration (Staplin, 1969). Although less reliable because of the scarcity of vitrinite particles (sometimes a few particles or less per sample), vitrinite-reflectance measurements (R_o) generally range from 0.4 to 0.6%. These low maturity values agree with the TAI values. The relatively high atomic H/C ratios (generally 1.1 to 1.3) of kerogen concentrates also reflect a low thermal history (Table 1).

In plotting Rock-Eval pyrolysis data on a modified van Krevelen diagram (which depicts the thermal evolutionary path of three specific types of kerogen), the HI values for most of the richer source rocks (>1.0% TOC) plot in the upper or immature region of the type I and type II kerogens (Fig. 12).

Another pyrolysis parameter commonly used to monitor the level of maturity is T_{max}, or the temperature corresponding to the maximum generation of pyrolytic hydrocarbons (S_2). T_{max} measurements normally increase systematically with increasing depth and temperature, particularly when one predominant type of organic matter is being monitored. Although credibility of T_{max} measurements of south Florida carbonate rocks was sometimes questioned because differences of as much as 10°C (18°F), or more were noted for some samples rerun at different times, the bulk of the data nevertheless indicates that T_{max} values generally vary between 421° (790°) and 441°C (826°F), with most values near 430° to 435°C (Table 1). If T_{max} at 430° to 435°C (806°–815°F) can be considered indicative of the onset of the oil-generation zone (Tissot and Welte, 1978, p. 454), then these T_{max} values seem to be in harmony with the other parameters and thus reflect a marginal maturity for most of the samples.

Of all the bulk geochemical parameters considered, the ratio of hydrocarbons to extractable organic matter (HC/EOM) seems to illustrate best the systematic maturity changes with increasing depth of burial in the carbonate–evaporite rocks of south Florida (Fig. 13). At the Sunniland level, the average HC/EOM ratio is about 64%, more than twice as great as the average ratio in the immature Paleocene–Gulfian 'A' (Fig. 2) level (HC/

Table 4

Number of Samples	Hopanes: Tm/Ts	Hopanes: C_{29}/C_{30}	Extended Hopanes: C_{31} 22S/22R	Extended Hopanes: C_{32} 22S/22R	Extended Hopanes: C_{33} 22S/22R	C_{29} Steranes: 20S/20R	C_{29} Steranes: $\beta\beta/\alpha\alpha$	C_{29} Steranes: 20S/(20S + 20R)	C_{29} Steranes: $\Sigma\beta\beta/(\Sigma\beta\beta + \Sigma\alpha\alpha)$
Upper Sunniland Oils									
15*	4.1 (2.4–6.8)	1.0 (0.8–1.3)	1.3 (1.2–1.4)	1.4 (1.3–1.5)	1.5 (1.3–1.7)	0.7 (.5–.8)	1.0 (0.7–1.2)	0.4 (0.3–0.4)	0.53 (.49–.56)
Upper Sunniland Possible Source Rocks									
10	2.1 (1.1–3.5)	0.6 (0.4–0.7)	1.4 (1.3–1.4)	1.4 (1.2–1.6)	1.5 (1.4–1.7)	0.7 (.6–.9)	0.5 (0.3–0.8)	0.4 (0.4–0.5)	.35 (.24–.44)
Lower Sunniland Possible Source Rocks									
8	1.9 (0.6–3.2)	0.8 (0.5–1.4)	1.4 (1.3–1.6)	1.4 (1.3–1.6)	1.5 (1.4–1.6)	0.7 (.5–.9)	0.8 (0.3–1.3)	0.4 (0.3–0.5)	.45 (.24–.58)

*Sterane ratios based on 11 samples.

Table 4 — Comparison of average biomarker ratios of upper Sunniland oils and Sunniland possible source rocks.
Ranges in parentheses; Tm = 17α(H)trisnorhopane; Ts = 18α(H)trisnorneohopane; C_{29}/C_{30} hopanes = norhopane/hopane; C_{29} steranes: 20S = 5α(H),14α(H),17α(H)20S; 20R = 5α(H),14α(H),17α(H)20R; ββ = 5α(H),14β(H),17β(H)20R; αα = 5α(H),14α(H),17α(H)20R; Σββ = 5α(H),14β(H),17β(H)(20R + 20S); Σαα = 5α(H),14α(H),17α(H)(20S + 20R).

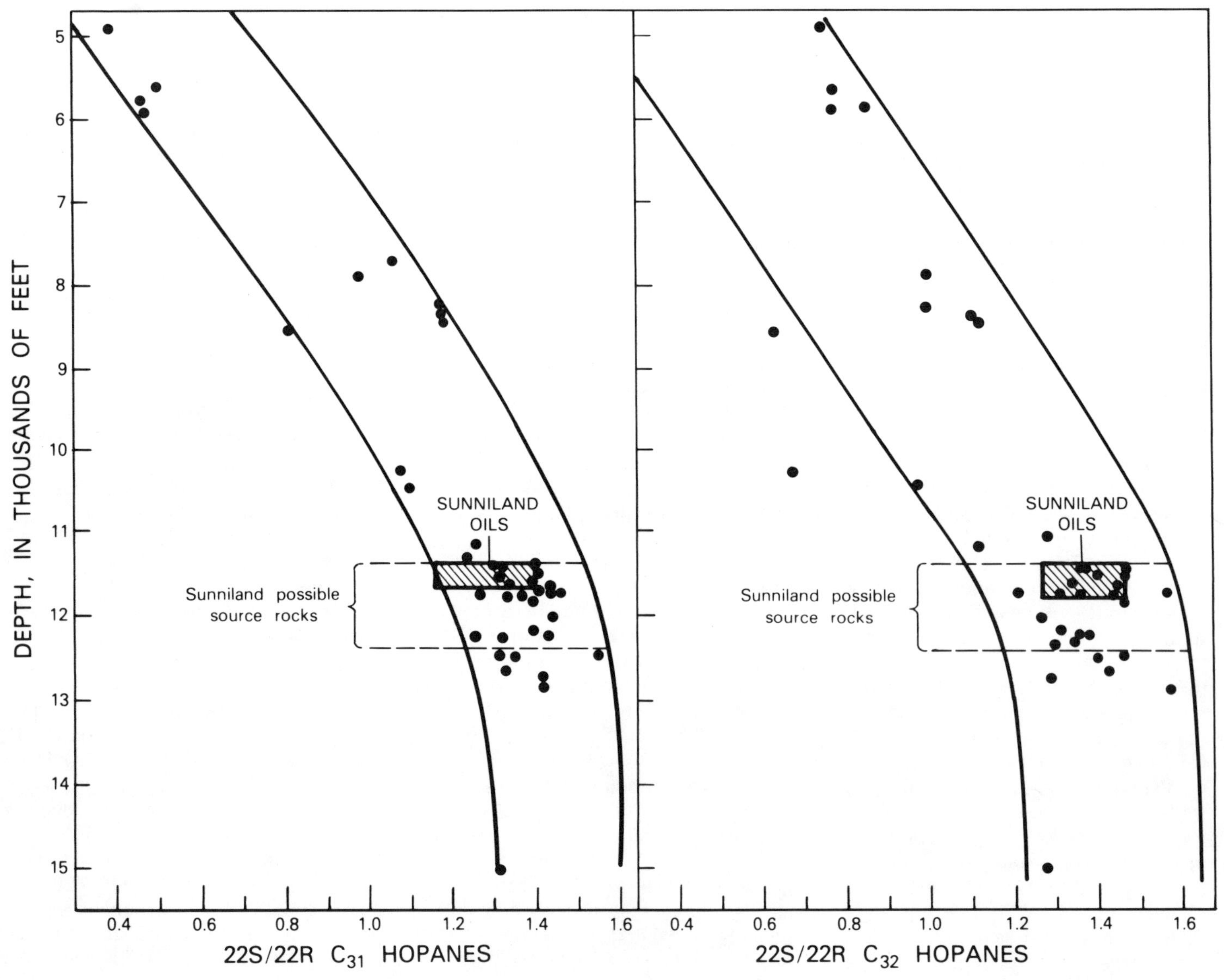

EOM ~ 25%), strongly suggesting that new hydrocarbons have been generated from the solid kerogens and/or from the soluble nonhydrocarbons.

Biomarker Parameters

In considering isomeric ratios of biomarkers as maturity indices, most of the data indicate that most of the Sunniland possible source rocks are marginally mature—a maturity level consistent with the above conventional indicators. Because of the potential usefulness of the isomerization ratios in monitoring levels of thermal stress in rocks and oils of various sedimentary basins of the world, it seems appropriate to discuss in greater detail their significance and applicability in marginally mature carbonate rocks such as in south Florida.

Seifert and Moldowan (1978) proposed that the ratio of 17α(H)-trisnorhopane/18α(H)-trisnorneohopane (Tm/Ts) can be used as an indicator of thermal maturity. They showed that with increasing thermal stress the less stable Tm decreases in relative concentration, whereas the more stable Ts increases. On the basis of this finding, the upper Sunniland oils, which have ratios as high as 6.8 and average 4.2 (Table 4; Fig. 14) would be considered immature or at best marginally mature. In fact, as Figure 14 indicates, the upper Sunniland oils appear to be less mature than all the Sunniland and older rock extracts down to depths of 15,000 ft (4,570 m). If Tm/Ts is strictly temperature dependent, then one inference that may be drawn is that the oils were generated and expelled at an earlier, less mature stage and that the residual organic matter in the rocks subsequently sustained a greater rate of maturation than the oils in the reservoirs.

However, because of several major fluctuations (Fig. 14), we believe that the Tm/Ts ratio also can be strongly influenced by source differences. Furthermore, because of the significant disparity in average Tm/Ts ratios between the oils and the bulk of the source rocks, the Tm/Ts ratio of the oils also may in part reflect migration effects. Therefore, the relative role of the various factors—thermal stress, source differences, or chromatographic effects owing to migration—in the controlling of the Tm/Ts ratio in the Sunniland oils is unresolved.

Nevertheless, from the perspective of the wide range of burial history (4,800–15,000 ft [1,460–4,570 m]), our data confirm Seifert and Moldowan's (1978) findings that Tm/Ts does in general decrease with depth and hence is temperature dependent. The variation of the Tm/Ts ratio from such high values as 46.0 in the immature, shallow Paleocene–upper Gulfian rocks (4,800–5,800 ft [1,460–1,760 m]) to values as low as 0.66 in the mature, deeply buried lowermost Cretaceous rocks (~15,000 ft [4,570 m]) bears this out (Fig. 14). However, equally important is the fact that significant departures from the expected Tm/Ts value can occur at depth owing to significant differences in organic facies and, with respect to oils, possibly also to migration effects, all of which suggests that some caution should be exercised in interpreting these kinds of data.

The stereoisomeric ratios, 22S/22R, of the extended hopanes (C_{31}–C_{35}) also have been used as maturity indicators (Ensminger et al, 1974; Seifert and Moldowan, 1978, 1980; Mackenzie et al, 1980). In general, 22S/22R ratios less than about 1.2 are considered immature, whereas ratios of about 1.3 or greater indicate that the rocks are near or within the oil-generation window. Using these values as guidelines, the Sunniland possible source rocks, characterized by average ratios of 1.4, 1.4, and 1.5 for the C_{31}, C_{32}, and C_{33} hopanes, respectively (Table 4), probably would be placed somewhere within the oil-generation window.

Monitoring the hopane stereoisomeric ratios in carbonate rocks in the South Florida basin with depth of burial provides some clues as to the approximate maturity position of the Sunniland source rocks. Plots of the C_{31} and C_{32} 22S/22R ratios in Figure 15 show that the major break in slope (i.e., the depth at which the relatively mature values, commonly 1.3–1.5, begin to appear) is approximately at the Sunniland level, or at depths generally greater than 11,400 ft (3,475 m). This relationship suggests that the Sunniland Limestone, at least in the region of the present producing trend and the region immediately to the west of this trend (Fig. 3), is at the onset of the oil-generation zone. The 22S/22R values for older and deeper rocks (to depths of 15,000 ft or 4,570 m) are more or less comparable to those at the Sunniland level.

The C_{31} and C_{32} 22S/22R ratios for the Sunniland oils plot in the midst of the Sunniland source rocks (Fig. 15), indicating comparable maturity and probable genetic linkup.

The use of the C_{29} sterane ratio $\alpha\alpha$20S/($\alpha\alpha$20S + $\alpha\alpha$20R) or simply the 20S/(20S + 20R) ratio seems to be definitive among molecular indices in monitoring maturity changes with depths of burial. The plot of this ratio with depth for carbonate rocks in south Florida (Fig. 16A) reinforces the interpretations made on the C_{31} and C_{32} hopane plots. The principal change in slope also is at about the Sunniland level, once again indicating the possible onset of the oil-generation zone. Above the Sunniland level, rock extracts are characterized by values of less than 0.30, and below this level, to depths of about 13,000 ft (3,960 m) or more, values range from 0.35 to nearly 0.50.

Because thermally mature 20S/(20S + 20R) ratios exist at the proposed isomerization equilibrium values of 0.50–0.55 (Mackenzie et al, 1984), the Sunniland possible source rocks, which have a significantly lower average value, 0.4 (Table 4), are considered marginally mature. The same maturity designation holds for the Sunniland oils that have almost the same average value (0.4) (Table 4). The marginally mature level of the Sunniland oils is depicted in Figure 16A, in which they are compared to the North sea oils, which are considered thermally mature (Mackenzie et al, 1984).

The plot of the other C_{29} sterane ratio, $\Sigma\beta\beta/(\Sigma\beta\beta + \Sigma\alpha\alpha)$, with depth of burial in carbonate rocks and oils of south Florida is not so clear. For example, in Figure 16B there is no clear-cut trend in the sterane ratios with depth as there is in the preceding 20S/(20S + 20R) sterane plot. One explanation for this, based on mass spectral analyses, is the presence of the 5β(H),14α(H),17α(H) C_{29} sterane, which coelutes with the 5α(H),14β(H),17β(H)20R counterpart in the less mature samples (generally less than 11,000 ft [3,350 m]). Trace amounts to no 5β(H),14α(H),17α(H) C_{29} steranes were detected in the more mature sam-

Figure 15 — Changes of stereochemical ratios of extended hopane compounds in carbonate rocks of south Florida with increasing burial depth (maturity). Left, isomerization ratios for the 22S/22R C_{31} hopanes. Right, isomerization ratios for the 22S/22R C_{32} hopanes. For comparison, the respective ratios of the upper Sunniland oils are plotted in the hachured area.

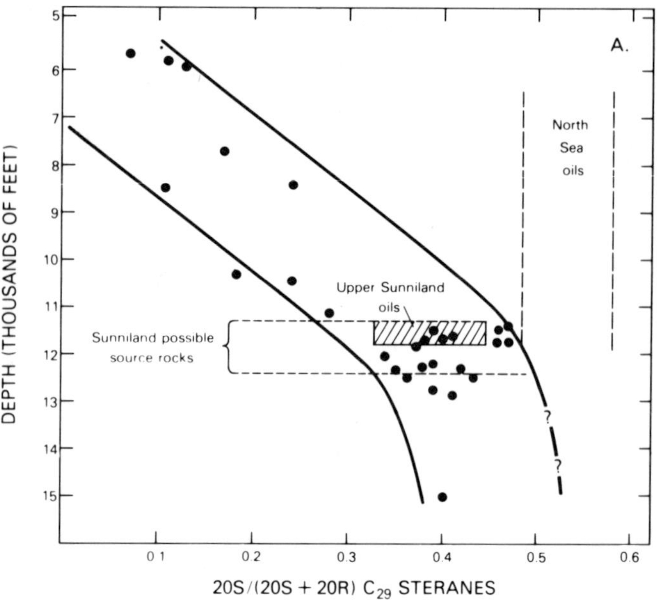

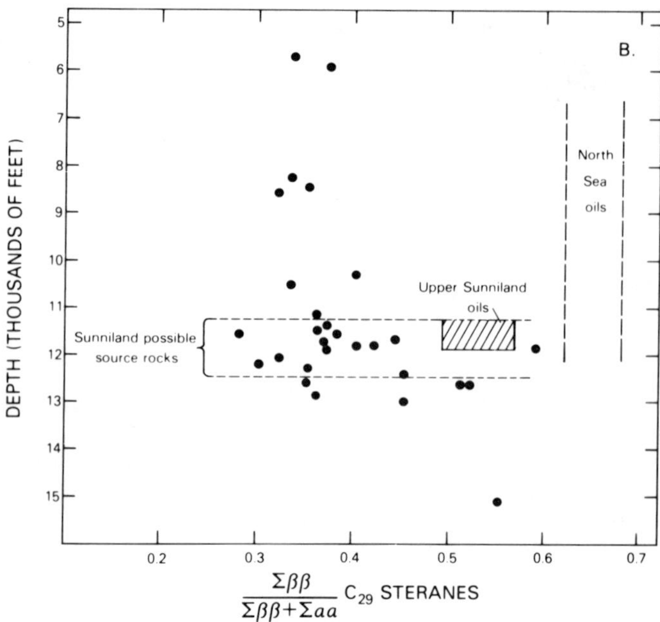

Figure 16 — Changes of isomerization ratios of C_{29} steranes in carbonate rocks of south Florida with increasing burial depth (maturity). A. Isomerization ratios 20S/(20S + 20R) for the C_{29} steranes. B. Isomerization ratios $\Sigma\beta\beta/(\Sigma\beta\beta + \Sigma\alpha\alpha)$ for the C_{29} steranes. Ratios for Sunniland oils are plotted in hachured areas. For comparison, the approximate ratios of the thermally mature North Sea oils are indicated within the dashed vertical bands. (After Mackenzie et al, 1984.)

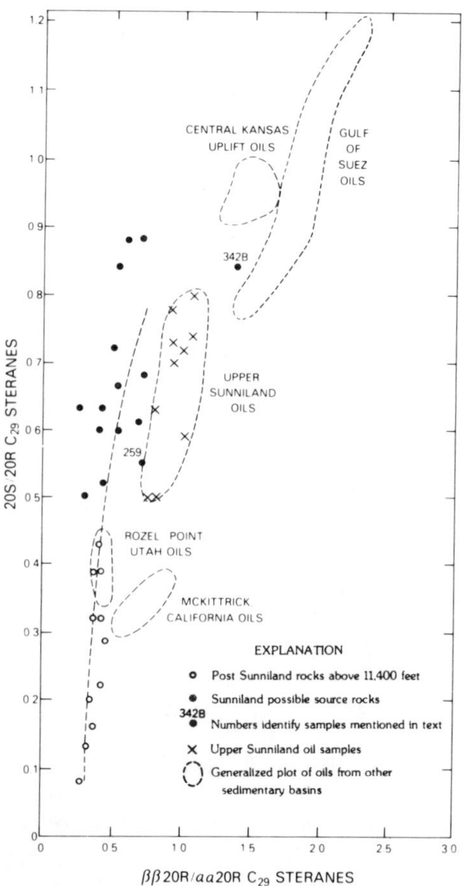

Figure 17 — Maturity-migration cross plot of C_{29} steranes in carbonate rocks of south Florida (after Seifert and Moldowan, 1981). Open-circle symbols indicate ratios for mostly post-Sunniland through lower Paleocene carbonate rocks above 11,400 ft (3,470 m). C_{29} sterane ratios of crude oils from different sedimentary basins in the world are given to provide some basis for estimating relative maturities. Sterane components identified in Table 4.

ples. Consequently, in the shallower, immature samples the ratios should be taken as maximum values. Perhaps if lower values were plotted for the shallower samples, a better vertical trend would emerge.

The applicability of the maturity-migration plot of Seifert and Moldowan (1981) was also tested by a cross plot of two C_{29} sterane ratios, the 20S/20R and the $\beta\beta/\alpha\alpha$ ratios (Fig. 17). The dashed curve drawn through the biomarker data represents an approximate trend of the ratios with increasing maturity for *unstained* carbonate rocks within and above the Sunniland Limestone in the South Florida basin. According to Seifert and Moldowan (1981), during oil migration the $\beta\beta$20R component is enriched relative to the $\alpha\alpha$20R component of the C_{29} steranes. Therefore, any significant deviation to the right of the trend would indicate migrated hydrocarbons.

The data in Figure 17 show that the upper Sunniland oils plot to the right of the curve, indicating that, at face value, migration effects are apparent. However, because of uncertainties associated with the C_{29} $\beta\beta$ to $\alpha\alpha$ sterane ratios, such as scatter of data, and because of inadequate sampling of the more mature probable source facies such as samples 1-259 and 13-342B, it would seem unwarranted at this time to attribute the apparent maturation discrepancy between source rocks and oils solely to migration effects.

In spite of the few uncertainties, the sterane cross plot seems to have great merit in comparing the relative maturities of rocks and oils not only in south Florida but also in other sedimentary basins of the world. For example, carbonate-derived oils of the Gulf of Suez (Rohrback, 1983) plot in a region indicating greater maturity than the Sunniland oils. The most thermally mature, higher API-gravity oils (as much as 44.6°) are in the upper part of the Suez oil plot, whereas the relatively less mature, lower gravity oils (13–27°API), as expected, are in the

Table 5

Sample Number	Facies	Pr/n-C_{17}	Ph/n-C_{18}	Pr/Ph	Hopanes		Extended Hopanes			C_{29} Steranes			
					Tm/Ts	C_{29}/C_{30}	C_{31} 22S/22R	C_{32} 22S/22R	C_{33} 22S/22R	20S/20R	$\beta\beta$/$\alpha\alpha$	20S/20S+20R	$\Sigma\beta\beta$/$\Sigma\beta\beta+\Sigma\alpha\alpha$
					Upper Sunniland								
12-324	Shelf	0.4	0.4	1.4	1.8	0.7	1.4	1.6	1.5	0.71	0.47	0.41	0.38
12-328	Shelf	.1	.2	≥ .9	1.3	.7	1.3	1.4	1.5	.88	.61	.47	.36
5-98	Shelf	.4	.4	.9	1.1	.6	1.4	1.4	1.4	.63	.27	.39	.24
8-155	Shelf	.4	.3	1.5	1.9	.7	1.3	1.5	1.7	.84	.54	.46	.38
8-165	Shelf	.3	.2	1.4	2.5	.6	1.4	1.3	1.4	.73	.76	.46	.44
10-217	Basinal	.9	1.2	.8	3.1	.5	1.3	1.2	1.4	.61	.69	.38	.40
1-254	Basinal	.7	1.1	.8	3.5	.6	1.4	1.3	1.4	.65	.51	.39	.33
1-255	Basinal				1.9	.4	1.4	1.4	1.7	.66	.48	.40	.35
1-256	Basinal	.5	.9	.6	2.4	.6	1.3	1.4	1.4	.60	.41	.38	.30
1-257	Basinal	.7	1.2	.7	1.6	.5	1.3	1.4	1.5	.72	.51	.42	.35
					Lower Sunniland								
14-702	Shelf	1.6	2.1	.9	3.2	1.4	1.6	1.5	1.5	.50	.28	.33	.24
5-105	Shelf	.8	.6	1.5	2.0	.6	1.4	1.4	1.4	.66	.53	.40	.39
5-107	Shelf	.2	.4	.6	2.0	.7	1.3	1.3	1.4	.79	1.2	.44	.56
4-85	Shelf	.9	.7	1.6	2.1	.8	1.5	1.6	1.6	.88	.71	.47	.42
13-342B	Basinal	1.4	2.5	.6	1.9	.5	1.4	1.4	1.5	.84*	1.3*	.46*	.58*
13-343B	Basinal	1.1	1.9	.7	1.9	.5	1.4	1.4	1.5	.84	1.3	.46	.58
10-221	Basinal	.6	.8	.8	–	–	–	–	–	–	–	–	–
10-222	Basinal	1.0	1.2	.9	1.7	.7	1.4	1.5	1.5	.63	.43	.39	.33
1-259	Basinal	.2	.3	.8	.6	1.2	1.1	1.3	1.5	.55	.72	.35	.45

*These are estimated values comparable to sample 13-343B. Analyses under previous lower temperature GC–MS interface conditions indicated that sterane ratios were essentially identical.

Table 5 — Biomarker ratios of upper and lower Sunniland possible source rocks.
Pr = pristane; Ph = phytane; all other ratios explained in Table 4.

lower part of the plot. The less mature Suez oils approximate the marginal maturity of the Sunniland oils, which have nearly similar API gravities (19–27°).

The plot of two oils from Rozel Point, Utah (one oil analyzed by the Denver USGS laboratory and the other by Cities Service Oil Co.; J. E. Zumberge, personal communication), indicate that the oils were derived from immature to marginally mature rocks. Rozel Point oil is interesting in that it may be slightly biodegraded and may be derived from a Tertiary lacustrine source rock. Barring any significant changes in sterane ratios by possible bacterial alteration, this sort of plot gives some idea of the thermal history of oil generation.

Oils from the Central Kansas uplift area, presumably derived from mature black shales with vitrinite reflectance of probably >0.7 to 0.8 (J. R. Hatch, unpublished data), also plot, as predicted, in a more mature position than the Sunniland oils. The comparison of these two oil groups is considered good because both were analyzed under the same conditions and using the same instruments.

The McKittrick oils of California, based on data extrapolated from two data points in Figure 10 of Seifert and Moldowan (1981), plot at a relatively lower maturity level than the Sunniland oils. The position of these McKittrick oils relative to the adjacent immature carbonate rocks of south Florida indicates that the oils also were generated under comparable immature conditions, with vitrinite reflectance, R_o, <0.5.

IDENTIFICATION OF PROBABLE SOURCE ROCKS

In answer to the first and foremost question addressed in this study, we believe that organic matter in shales (fine-grained siliceous rocks) can be eliminated as the source of the commercial oil accumulations in Sunniland reservoir rocks, because these shales are not present in the South Florida basin.

The second question addressed, concerning the identification of the specific source facies, is more difficult to answer, primarily because of the limited number of samples available from stratigraphic sections believed to contain the major oil source beds. Furthermore, determination of the probable source facies, using standard methodologies—i.e., geologic criteria such as favorable lithologies and stratigraphic position, and conventional geochemical properties such as TOC, ratios of EOM/TOC, vitrinite reflectance, Rock-Eval production indices, T_{max}, alkane GC fingerprints, and carbon-isotope ratio—proved inconclusive. On the other hand, the combined use of these bulk properties with computerized GC-MS analysis of polycyclic biomarker distributions enabled us to rule out certain possible source rocks and to focus on the likely or probable major source-rock facies.

Possible carbonate source beds in the overlying Lake Trafford Formation were eliminated as sources of Sunniland oil for several reasons. First, on the grounds of

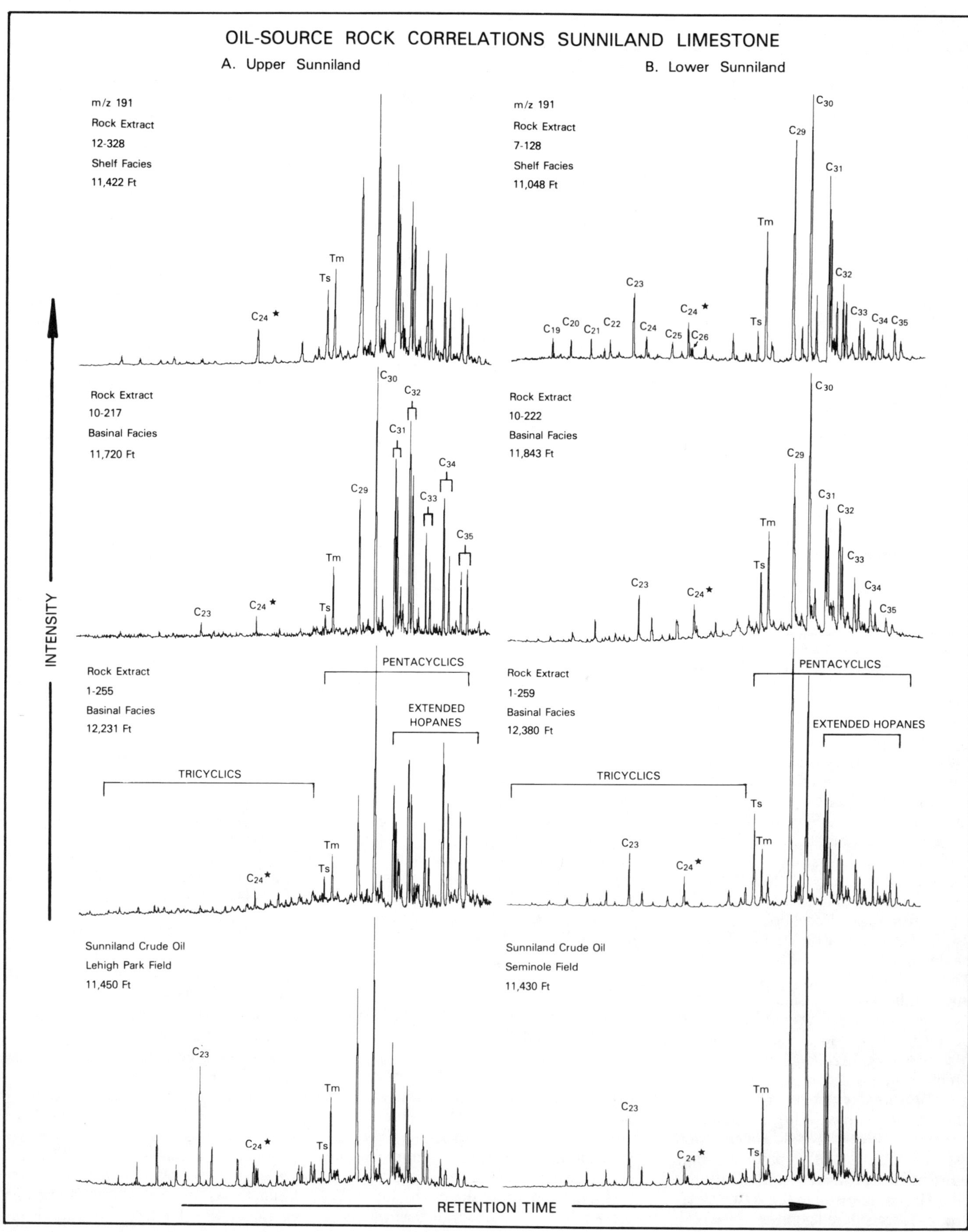
OIL-SOURCE ROCK CORRELATIONS SUNNILAND LIMESTONE
A. Upper Sunniland
B. Lower Sunniland
m/z 191
Rock Extract
12-328
Shelf Facies
11,422 Ft
m/z 191
Rock Extract
7-128
Shelf Facies
11,048 Ft
Rock Extract
10-217
Basinal Facies
11,720 Ft
Rock Extract
10-222
Basinal Facies
11,843 Ft
Rock Extract
1-255
Basinal Facies
12,231 Ft
Rock Extract
1-259
Basinal Facies
12,380 Ft
Sunniland Crude Oil
Lehigh Park Field
11,450 Ft
Sunniland Crude Oil
Seminole Field
11,430 Ft
PENTACYCLICS
EXTENDED HOPANES
TRICYCLICS
INTENSITY
RETENTION TIME

geologic reasoning, the presence of thick, impervious anhydrite beds overlying the Sunniland Limestone and the lack of any known major faulting would tend to prohibit oil from migrating downward into the Sunniland reservoirs. Second, the few possible source rocks that were analyzed show biomarker patterns unlike those of the Sunniland oils. Third, if some Lake Trafford carbonate rocks did yield Sunniland-type oil, certainly the interbedded Lake Trafford reservoir rocks should bear record of that Sunniland oil. On the contrary, the Lake Trafford Formation oils (obtained from two widely separated boreholes) bear no resemblance whatsoever to the Sunniland oils in all three biomarker distributions and in essentially all bulk geochemical and isotopic properties. Hence, combined evidence indicates that the Lake Trafford rocks are not the source beds of Sunniland oil, but rather certain Lake Trafford facies are the likely source of the Lake Trafford Formation oils.

Strata extending from the base of the Sunniland Limestone to the igneous basement also were investigated to determine possible deep sources for Sunniland oil. Total-organic-carbon and Rock-Eval pyrolysis analyses of nearly 300 core and cuttings samples showed that most of the rocks, especially those of the Pumpkin Bay, Bone Island, and Wood River Formations (Fig. 2), are too lean in organic matter (TOC $<0.4\%$) to have generated commercial amounts of oil (Palacas, 1978a; Applegate, Winston, and Palacas, 1981; Palacas, Daws, and Applegate, 1981). Also, if source-rock facies are present in untested areas that did generate commercial oil, the probability of the oil's migrating vertically upward, in the absence of known major faults and/or extensive fractures, through numerous thin and thick impervious anhydrite beds, particularly through the Punta Gorda Anhydrite, which contains an average of 500 ft (150 m) of anhydrite, is extremely unlikely.

Figure 18 — Source-rock–crude-oil correlations using combined tricyclic–pentacyclic distributions based on m/z 191 fragmentograms. A. Upper Sunniland possible source rocks reflecting variable burial depths and source facies. B. Lower Sunniland possible source rocks representing variable burial depths and source facies. See Figure 5 for explanation of symbols.

Further detailed analyses (GC and GC/MS) of 18 selected core samples of the Punta Gorda and Lehigh Acres Formations and the upper part of the Pumpkin Bay Formation indicated that none of the rock extracts correlated with the Sunniland crude oils. Admittedly these samples, taken from six different drill holes, are limited in their representation of such a large possible source-rock drainage area. Nevertheless, the available geochemical evidence, coupled with the geologic constraints, certainly suggests that the primary source of upper Sunniland oil is not from rocks below the Sunniland Limestone.

Anhydrites also were investigated as to their source potential. Upon visual inspection of many anhydrite core samples within and below the Sunniland Limestone, most appeared too lean in organic matter to generate commercial quantities of hydrocarbons. Total-organic-carbon contents of five representative samples, ranging from 0.05 to 0.11% TOC, attest to their leanness. In addition, GC-MS analysis of one of these lean samples showed that the pentacyclic and tricyclic fingerprints mismatched those of the Sunniland oils, reinforcing the judgment that these lean anhydrites are not major sources of Sunniland oil.

Interestingly, however, in the downdip part of the South Florida basin, where upper Sunniland shelf carbonates have been replaced by basinal anhydrites, certain relatively organic-rich facies were formed. Such facies should not be overlooked in any source-rock evaluation. A case in point is sample 1-255, which represents an anhydrite bed as thick as 16 ft (4.9 m), characterized by an appreciable amount of oil-prone organic matter (TOC = 1.14%, HI = 580 mgHC/gTOC, Table 1). However, biomarker distributions indicate that these upper Sunniland anhydrite facies, as represented by sample 1-255, are not probable sources of Sunniland oil.

Finally, possible carbonate source beds within the Sunniland Limestone itself are assessed as to their likelihood of being the probable or effective sources of Sunniland oil.

Various carbonate beds in the upper Sunniland Limestone were initially thought to be probable sources of Sunniland oil on the basis of reasonable oil-source correlations of bulk geochemical properties and favorable stratigraphic position. However, tricyclic and pentacyclic distributions show significant mismatches between the rocks and oils, indicating that, with one exception, the upper Sunniland possible source beds are not the major sources of Sunniland oil. The one exception is the basal part of the upper Sunniland Limestone, which in certain areas seems to be more akin, in source-rock character, to the rocks in the lower Sunniland Limestone. For this reason and for ease of categorization, samples from the basal upper Sunniland—4-85, 13-335, 13-336, 13-341, 13-341A, 13-342, 10-221, and 10-222 (Table 1)—are incorporated in this report into the unit designated as lower Sunniland.

In evaluating the sterane distributions of the upper Sunniland possible source rocks, they appear to be similar in general to those of the oils, except for the $\beta\beta$ to $\alpha\alpha$ ratios. As discussed previously, the rock extracts have a significantly lower average $\Sigma\beta\beta/(\Sigma\beta\beta + \Sigma\alpha\alpha)$ ratio (0.35) to that of the oils (average 0.53, Table 4). If these isomerization ratios are indeed strictly maturity dependent, as inferred from studies of Mackenzie et al (1980, 1984), then the relatively more mature oils could not have been derived from the closely associated, less mature source rocks. Hence, this disparity in $\beta\beta$ to $\alpha\alpha$ ratios would be another key factor in disqualifying these upper Sunniland rocks as probable sources of Sunniland oil. Alternatively, if the difference in $\beta\beta$ to $\alpha\alpha$ ratios is due largely to migration effects, as Seifert and Moldowan (1981) suggested, it still would not annul the previous assessment of noncorrelation between possible source rocks and oils of the upper Sunniland, primarily because of the significant mismatches in tricyclic and pentacyclic distributions. The migration effect, however, might make a difference in evaluating effective source rocks in the lower Sunniland unit.

Of all the stratigraphic units in south Florida that were studied, the bulk of the evidence points to certain organic facies in the lower Sunniland Limestone as the major source of upper Sunniland oil, especially the downdip basinal facies. This sequence of rocks (which also includes, in some areas, the lowermost part of the upper Sunniland unit) is characterized by (1) a preponderance of algal-sapropelic organic matter; (2) some of the highest organic-carbon contents observed in the Cretaceous rocks of south Florida, with total organic carbon as high as 12.3%; and (3) bulk chemical properties, thermal maturities, and biomarker distributions comparable to those of the oils.

The correlation parameter that proved to be the most diagnostic in pinpointing the lower Sunniland rocks as the prime source is the tricyclic terpane distribution. Of 12 lower Sunniland samples from 7 different boreholes that were analyzed, at least 9 and possibly 10 samples have tricyclic patterns comparable to those of the oils—i.e., patterns that have the C_{23} molecule as the dominant peak. In contrast, of 4 samples (2 rock extracts and 2 oils) from the Lake Trafford Formation, 11 samples from the upper Sunniland, and 13 samples from pre-Sunniland rocks, not one has a tricyclic pattern that resembles that of the Sunniland oils. Furthermore, 9 of 12 lower Sunniland rocks also have extended hopane patterns that are nearly identical to those of the oils, reinforcing the positive oil-source correlation observed in the tricyclic pattern. Figure 18B illustrates the good correlation of the combined tricyclic–pentacyclic distributions of three lower Sunniland rock extracts with that of a typical upper Sunniland oil.

The sterane distributions of possible source rocks in the lower Sunniland, like those in the upper Sunniland, are similar in most respects to those of the upper Sunniland oils, except again for the differences in $\beta\beta$ to $\alpha\alpha$ ratios (Table 4). Yet, in terms of overall average values, the lower Sunniland rocks have values closer to those of the upper Sunniland oils, once more suggesting a closer genetic link to the oils. Should the differences in values between most of the lower Sunniland source rocks and the upper Sunniland oils be due to migration effects, as discussed previously, nearly all the lower Sunniland possible source rocks would be considered as probable sources. On the other hand, if the differences are due strictly to maturation, the probable sources would be limited to the deeper basinal facies such as those represented by samples 13-342B, 13-343B, and 1-259, which have ratios compatible with the oils.

Obviously, the choice between these alternative explanations depends upon a better understanding of the possible effect of migration in biomarker distributions. Further laboratory migration-type experiments and more field studies involving crude-oil–source-rock correlations are needed in order to resolve these questions. Because sterane-isomerization ratios appear to be powerful tools not only in assessing maturity but also in source-rock–crude-oil correlations, they deserve special attention.

In comparing two other biomarker parameters, the Tm/Ts and the C_{29}/C_{30} hopane ratios, the relationship between the oils and the source rocks is not readily apparent. As discussed earlier and as Table 4 indicates, the Tm/Ts ratios of both upper and lower Sunniland source rocks are, in general, significantly less than those of the oils. Although migration effects and differing maturity histories may explain some of the disparity between oils and source rocks, the most plausible explanation focuses on differences in organic facies. For example, lower Sunniland samples 14-702 (Table 5) and 7-128 (Fig. 18B) have Tm/Ts ratios (3.2 and 4.3, respectively) that fit within the range of ratios for Sunniland oils (2.4–6.8) (Table 4). These samples may be representative of other, more plentiful organic facies in the downdip region that have not yet been examined because of limited sample control. (Sample 7-128 is not listed in Tables 1 and 5 because of its immaturity and unfavorable geographic position (next to Lake Okeechobee, Fig. 3); yet, based on conventional geochemical and biomarker parameters, it is nearly identical to Sunniland oil.)

In a similar fashion, the C_{29}/C_{30} hopane ratios for the Sunniland source rocks are also generally lower than the oils (Table 4), but, as a whole, the lower Sunniland rocks, with an average ratio of 0.8, are closer to the oils (avg. 1.0) than are the upper Sunniland rocks (avg. 0.6).

Conventional correlation parameters, such as Pr/Ph ratios, are in accord with the correlation interpretations provided by the polycyclic alkane biomarker parameters. Table 5 shows that the basinal facies are characterized predominantly by ratios <0.9, similar to those of the oils (Table 2), while the shelf facies, especially in the upper Sunniland, generally have ratios much greater than 0.9.

In summary, all the combined available evidence strongly indicates that the major source of upper Sunniland oil is the lower Sunniland Limestone, predominantly from the basinal facies, with probable minor contributions from other parts of the Sunniland Limestone. Finally, because of rapid horizontal and vertical changes in organic facies in this shallow-water carbonate environment wherein minor variations in molecular ratios would be readily expected, we believe that the upper Sunniland oil is derived from a blending of the individual probable source-rock units.

CONCLUSIONS

Geologic and geochemical evidence has shown that carbonate rocks and not shales are the source of Sunniland crude oil. Identification of the probable source-rock facies of the upper Sunniland oils, using geologic criteria (e.g., stratigraphic position) and conventional geochemical properties (e.g., TOC content, hydrogen richness of organic matter, GC alkane patterns, and isoprenoid ratios), proved inconclusive. On the other hand, the use of gas-chromatographic–mass-spectrometric analysis of biological marker compounds (steranes, tricyclic and pentacyclic terpanes) in conjunction with the bulk parameters was effective in delineating the specific source facies. The bulk of the evidence indicates that the organic-rich, argillaceous limestones in the lower Sunniland Limestone, particularly the basinal facies, are the primary source of Sunniland oil.

In the absence of an adequate number of vitrinite particles for confident vitrinite-reflectance measurements, and sometimes variable Rock-Eval T_{max} values—methods widely used for determining thermal-maturity levels—the sterane and pentacyclic isomerization ratios proved most helpful in deciphering the thermal history of the Sunniland carbonate rocks and in defining an upper threshold of the oil-generation zone.

ACKNOWLEDGMENTS

The authors are deeply indebted to Exxon Co., U.S.A., New Orleans, Louisiana, and Esso Production Research Co., Houston, Texas, for supplying most of the core and oil samples and general geologic information. For additional samples and geologic information, appreciation is also extended to: Mobil Oil Co., Diamond Shamrock Corp., Phillips Petroleum Co., Sun Oil Co., Shell Oil Co., Coastal Petroleum Co., Ashland Oil, Inc., Analytical Logging, Inc., the Florida Bureau of Geology, and the University of Tampa. We acknowledge the U.S. Geological Survey laboratory and office support of Sister Carlos, Ted Daws, Jeff Baysinger, Peter Gerrild (now deceased), Charles Threlkeld, and Dorothy Malone. Special gratitude is extended to George Claypool for enlightening discussions, helpful comments, and ongoing support.

REFERENCES CITED

Applegate, A.V., in press, The Brown Dolomite of the Lehigh Acres Formation in South Florida: Florida Bureau of Geology Report of Investigations 95.

——, G.O. Winston, and J.G. Palacas, 1981, Subdivisions and regional stratigraphy of the pre-Punta Gorda rocks (lowermost Cretaceous–Jurassic?) in South Florida: Gulf Coast Association of Geological Societies Transactions, v. 31, Supplement 2, p. 447–453.

Applin, P.L., and E.R. Applin, 1965, The Comanche Series and associated rocks in the subsurface in central and south Florida: U.S. Geological Survey Professional Paper 447, 84 p.

Babcock, C., 1962, Florida petroleum exploration and prospects: Florida Geological Survey Special Paper 9, 79 p.

Bush, P.R., 1970, A rapid method for determination of carbonate carbon and organic carbon: Chemical Geology, v. 6, p. 59–62.

Connan, J., 1981, Biological markers in crude oils, *in* J.F. Mason, ed., Petroleum geology in China: Tulsa, Oklahoma, PennWell Books, p. 48–70.

Curiale, J.A., W.E. Harrison, and G. Smith, 1983, Sterane distribution of solid bitumen pyrolyzates. Changes with biodegradation of crude oil in the Ouachita Mountains, Oklahoma: Geochimica et Cosmochimica Acta, v. 47, p. 517–523.

Dunnington, H.V., 1954, Stylolite development postdates rock induration: Journal of Sedimentary Petrology, v. 24, p. 27–49.

Ensminger A., et al, 1974, Pentacyclic triterpanes of the hopane type as ubiquitous geochemical markers: Origin and significance, *in* B. Tissot and F. Bienner, eds., Advances in organic geochemistry 1973: Paris, Éditions Technip, p. 245–260.

Espitalié, J., et al, 1977, Méthode rapide de caractérisation des roches mères, de leur potential pétrolier et de leur degré d'évolution: Revue de l'Institut Français du Pétrole, v. 32, p. 23–42.

Fietz, R.P., 1976, Recent developments in Sunniland exploration of South Florida: Gulf Coast Association of Geological Societies Transactions, v. 26, p. 74–78.

Griffin, G.M., et al, 1969, Geothermal gradients in Florida and southern Georgia: Gulf Coast Association of Geological Societies Transactions, v. 19, p. 189–193.

Huang, W.Y., and W.G. Meinschein, 1976, Sterols as source indicators of organic materials in sediments: Geochimica et Cosmochimica Acta, v. 40, p. 323–330.

—— and ——, 1979, Sterols as ecological indicators: Geochimica et Cosmochimica Acta, v. 43, p. 739–745.

Hunt, J.M., 1972, Distribution of carbon in crust of earth: AAPG Bulletin, v. 56, p. 2273–2277.

Klopp, H.C., 1975, Petrographic analysis of the Sunniland Formation, an oil-producing formation in S. Florida: Provo, Utah, Brigham Young University Geology Studies, v. 22, p. 3–27.

Mackenzie, A.S., et al, 1980, Molecular parameters of maturation in the Toarcian shales, Paris Basin, France: I. Changes in the configurations of acyclic isoprenoid alkanes, steranes and triterpanes: Geochimica et Cosmochimica Acta, v. 44, p. 1709–1721.

———, 1984, Biological marker and isotope studies of North Sea crude oils and sediments: London, 11th World Petroleum Congress Proceedings, 1983, v. 2, p. 45–56.

Means, J.A., 1977, Southern Florida needs another look: Oil and Gas Journal, v. 75, no. 5, p. 212–225.

Meyerhoff, A.A., and C.W. Hatten, 1974, Bahamas salient of North America: Tectonic framework, stratigraphy, and petroleum potential: AAPG Bulletin, v. 58, p. 1201–1239.

Moldowan, J.M., W.K. Seifert, and E.J. Gallegos, 1983, Identification of an extended series of tricyclic terpanes in petroleum: Geochimica et Cosmochimica Acta, v. 47, p. 1531–1534.

Oglesby, W.R., 1965, Folio of South Florida Basin, a preliminary study: Florida Geological Survey Map Series 19, 3 p., 10 maps.

Palacas, J.G., 1978a, Distribution of organic carbon and petroleum source rock potential of Cretaceous and Lower Tertiary carbonates, South Florida Basin, preliminary results: U.S. Geological Survey Open-File Report 78–140, 35 p.

——, 1978b, Preliminary assessment of organic carbon content and petroleum source rock potential of Cretaceous and Lower Tertiary carbonates, South Florida Basin: Gulf Coast Association of Geological Societies Transactions, v. 28, p. 357–381.

——, 1984, Carbonate rocks as sources of petroleum: Geological and chemical characteristics and oil–source correlations: London, 11th World Petroleum Congress Proceedings, 1983, v. 2, p. 31–43.

——, T.A. Daws, and A.V. Applegate, 1981, Preliminary petroleum source-rock assessment of pre–Punta Gorda rocks (lowermost Cretaceous–Jurassic?) in South Florida: Gulf Coast Association of Geological Societies Transactions, v. 31, p. 369–376.

Ramsden, R.M., 1952, Stylolites and oil migration: AAPG Bulletin, v. 36, p. 2185–2186.

Reel, D.A., and G.M. Griffin, 1971, Potentially petroliferous trends in Florida as defined by geothermal gradients: Gulf Coast Association of Geological Societies Transactions, v. 21, p. 31–36.

Rogers, M.A., 1980, Application of organic facies concepts to hydrocarbon source rock evaluation: Bucharest, 10th World Petroleum Congress Proceedings, 1979, v. 2, p. 23–30.

Rohrback, B.J., 1983, Crude oil geochemistry of the Gulf of Suez, *in* M. Bjoroy et al, eds., Advances in organic geochemistry 1981: Chichester, John Wiley, p. 39–48.

Ronov, A.B., and A.A. Yaroshevsky, 1969, Chemical composition of the earth's crust, *in* J.M. Hunt, ed., The earth's crust and upper mantle: American Geophysical Union Geophysical Monograph 13, p. 37–57.

Rubenstein, I., O. Sieskind, and P. Albrecht, 1975, Rearranged steranes in a shale: occurrence and simulated formation: Jour. Chem. Soc. Perkin Trans. I, p. 1833–1836.

Sears, S.O., 1972, Facies interpretations and diagenetic modifications of the Sunniland Limestone, South Florida—A thin section and x-ray diffraction analysis: Gainesville, Florida, University of Florida unpublished M.S. thesis, 80 p.

Seifert, W.K., and J.M. Moldowan, 1978, Applications of steranes, terpanes and monoaromatics to the maturation, migration and source of crude oils: Geochimica et Cosmochimica Acta, v. 42, p. 77–95.

—— and ——, 1979, The effect of biodegradation on steranes and terpanes in crude oils: Geochimica et Cosmochimica Acta, v. 43, p. 111–126.

—— and ——, 1980, The effect of thermal stress on source-rock quality as measured by hopane stereochemistry, *in* A.G. Douglas and J.R. Maxwell, eds., Advances in organic geochemistry 1979: Oxford, Pergamon Press, p. 229–237.

—— and ——, 1981, Paleoreconstruction by biological markers: Geochimica et Cosmochimica Acta, v. 45, p. 783–794.

Seifert, W.K., J.M. Moldowan, and R.W. Jones, 1980, Application of biological marker chemistry to petroleum exploration: Bucharest, 10th World Petroleum Congress Proceedings, 1979, v. 2, p. 425–440.

Sieskind, O., G. Joly, and P. Albrecht, 1979, Simulation of the geochemical transformations of sterols: superacid effect of clay minerals: Geochimica et Cosmochimica Acta, v. 43, p. 1675–1679.

Simoneit, B.R.T., H.I. Halpern, and B.M. Didyk, 1980, Lipid productivity of a high Andean lake, *in* P.A. Trudinger, M.R. Walter, and B.J. Ralph, eds., Biogeochemistry of ancient and modern environments: Canberra, Australian Academy of Sciences, and New York, Springer–Verlag, p. 201–210.

Staplin, F.L., 1969, Sedimentary organic matter, organic metamorphism, and oil and gas occurrence: Bulletin of Canadian Petroleum Geology, v. 17, p. 47–66.

Tissot, B., and D.H. Welte, 1978, Petroleum formation and occurrence: Berlin, Springer–Verlag, 538 p.

Welte, D.H., and D.W. Waples, 1973, Über die Bevorzugung geradzahliger n-Alkane in Sedimentgesteinen: Naturwissenschaften, v. 60, p. 516–517.

Winston, G.O., 1971a, Regional structure, stratigraphy, and oil possibilities of the South Florida basin: Gulf Coast Association of Geological Societies Transactions, v. 21, p. 15–29.

——, 1971b, The Dollar Bay Formation of Lower Cretaceous (Fredericksburg) age in South Florida, its stratigraphy and petroleum possibilities: Florida Bureau of Geology Special Publication 15, 99 p.

——, 1976, Six proposed formations in the undefined portion of the Lower Cretaceous section in South Florida: Gulf Coast Association of Geological Societies Transactions, v. 26, p. 69–72.

Zumberge, J.E., 1983, Tricyclic diterpane distributions in the correlation of Paleozoic crude oils from the Williston Basin, *in* M. Bjoroy et al, eds., Advances in organic geochemistry 1981: Chichester, John Wiley, p. 738–745.

Generation and Migration of Hydrocarbons in Upper Cretaceous Austin Chalk, South-Central Texas

George J. Grabowski, Jr.
Exxon Production Research Company, Houston, Texas

The Austin Chalk of south-central Texas was deposited on a carbonate ramp marginal to the Gulf of Mexico during the Late Cretaceous Epoch. Dark-colored, laminated, and sparsely burrowed chalks containing >1.5% total organic carbon (TOC) were deposited under disaerobic conditions and predominate in deep, basinal cores. Light-colored, thoroughly bioturbated chalks containing <1.5% TOC formed in oxygenated shelfal environments and predominate in shallow cores.

The kerogen in the Austin Chalk is uniformly composed primarily of amorphous, type I or II material irrespective of carbonate content or lithology. The kerogen occurs disseminated in the dark-colored chalks and concentrated in microstylolites and stylolites. Extractable organic matter (EOM) is most abundant in porous chalk between zones of pressure solution.

EOM in shallow cores (<5,000 ft or 1,525 m) contains 70% NSO (nitrogen, sulfur, and oxygen) compounds and 30% hydrocarbons. The saturated hydrocarbons are dominated by geochemical fossils in these shallow cores. With increasing depth of burial, the EOM becomes enriched in hydrocarbons, up to 60–80% in the deepest core (9,100 ft or 2,780 m). These hydrocarbons formed from alteration of kerogen and NSO compounds: the kerogen becomes progressively condensed and less aliphatic with increasing depth of burial. This alteration is independent of the carbonate content of the rocks. Peak oil generation occurs between 5,000 and 8,000 ft (1,525 and 2,440 m), with gaseous hydrocarbons becoming abundant below 8,000 ft (2,440 m).

EOM that was formed during catagenesis migrated in the Austin Chalk through micropores to zones of higher porosity. This migration caused enrichment of the EOM in both absolute amounts and relative proportions of hydrocarbons and resins in porous chalks. Larger molecules, such as asphaltenes, did not readily migrate in the chalk. Migration of EOM occurred at all depths in the Austin Chalk but was most prevalent in mature chalks below 5,000 ft (1,525 m). Changes in the composition of EOM owing to migration are independent of carbonate content or lithology and are dependent primarily on the maturity of the rock. The light (>39°API) crude oils produced from deep wells in the Austin Chalk are similar to and may be formed from migrated EOM.

INTRODUCTION

About 40% of the petroleum produced by non-Soviet countries comes from carbonate reservoirs (Hunt, 1967), of which a portion is believed to have originated in fine-grained carbonate rocks. Carbonate source rocks may form in a variety of depositional environments, including lakes (Green River Formation; Hunt, Stewart, and Dickey, 1954; Reed and Henderson, 1972; Tissot, Deroo, and Hood, 1978), evaporitic shallow-marine platforms (Mesozoic of the Middle East; Kirkland and Evans, 1981), and deeper marine basins (Niobrara Formation, Rice, this volume).

The specific characteristics of the organic matter in fine-grained carbonate rocks are believed to result from deposition and diagenesis of predominantly sapropelic organic matter under reducing, noncatalytic conditions (Hunt, 1967). However, the controls on the distribution and alteration of organic matter in fine-grained carbonate rocks are poorly understood. Models for the distribution of organic matter and the generation and migration of hydrocarbons in fine-grained carbonate rocks come principally from extrapolation of the results of studies of shales (McIver, 1967; Laplante, 1974; Tissot et al, 1974) and the effects of mineral matrix on the pyrolysis of kerogen (Espitalié, Madec, and Tissot, 1980; Horsfield and Douglas, 1980). Shales typically undergo extensive physical compaction during early diagenesis (Potter, Maynard, and Pryor, 1980), whereas carbonates may undergo physical and chemical (pressure-solution) compaction, dissolution, cementation, and mineral transformations during burial (Wilson, 1975). The differences between shale and carbonate diagenesis may influence the distribution of organic matter within the rocks. Further, the lower catalytic activity of carbonates in high-temperature, short-duration cracking reactions of organic molecules (Espitalié, Madec, and Tissot, 1980) may not accurately reflect the conditions of relatively low-temperature, long-duration catagenesis in nature.

The purpose of this paper is to describe and relate the abundance and composition of organic matter to the generation and migration of hydrocarbons in the Upper Cretaceous Austin Chalk of south-central Texas. These data may then be used for critical evaluation of the validity and constraints on the adoption of organic-geochemical models of shales to fine-grained organic-rich carbonate rocks.

BACKGROUND

Stratigraphy and Regional Geologic Setting

The Austin Chalk was deposited during the Coniacian and Santonian stages of

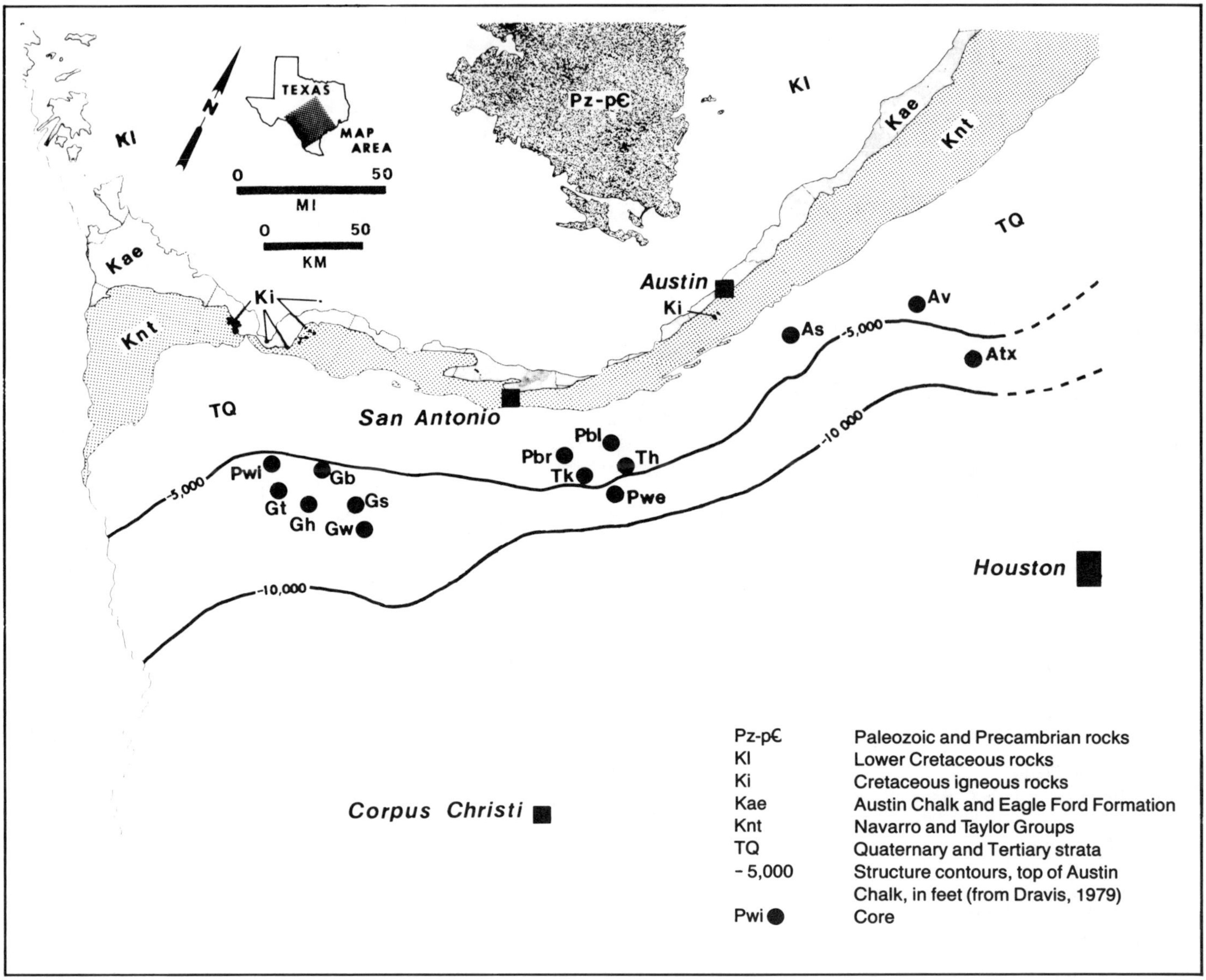

Pz-p€	Paleozoic and Precambrian rocks
Kl	Lower Cretaceous rocks
Ki	Cretaceous igneous rocks
Kae	Austin Chalk and Eagle Ford Formation
Knt	Navarro and Taylor Groups
TQ	Quaternary and Tertiary strata
- 5,000	Structure contours, top of Austin Chalk, in feet (from Dravis, 1979)
Pwi ●	Core

Figure 1 — Map of south-central Texas showing locations of cores and outcrop patterns of strata in the study area (modified from Dravis, 1979). Cores identified by abbreviations listed in Table 1.

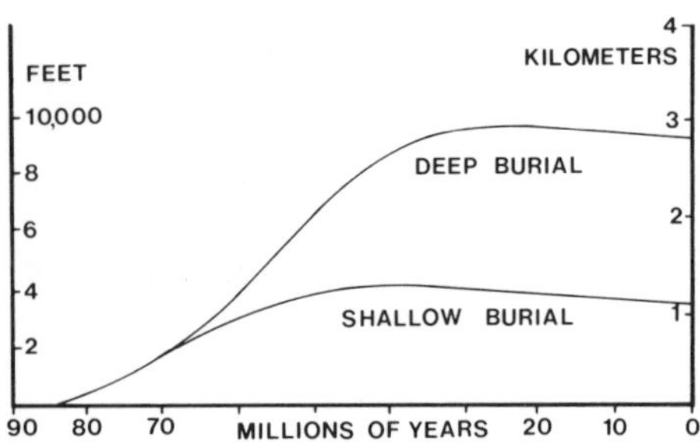

Figure 2 — Cumulative burial curves for deep and shallow Austin Chalk in south-central Texas (from Grabowski, 1981). Austin Chalk has been at near present-day depths of burial since 30 (deep cores) to 50 (shallow cores) m.y.b.p.

the Late Cretaceous Epoch (Dravis, 1979; Barrier, 1980), an interval of about 12 m.y. This was a time of a worldwide high stand of sea level (Vail, Mitchum, and Thompson, 1977), during which a large north–south epeiric sea existed in the western interior of the North American craton (Williams and Stelck, 1973). Chalks formed in the central parts of this seaway (Kauffman, 1979), including the Niobrara Formation (Rice, this volume), and in the southern opening into the Gulf of Mexico.

The Austin Chalk in south-central Texas formed on a gentle, southeasterly sloping ramp marginal to the Gulf of Mexico (Dravis, 1979). The chalk ranges in thickness from 100 to 650 ft (30 to 200 m) in outcrop (Fig. 1) but thickens basinward to about 1,000 ft (300 m) in the subsurface of south-central Texas. The Austin Chalk overlies the Eagle Ford Formation, a fine-grained, organic-rich carbonate similar to the Austin Chalk in the subsurface.

The sediments in the Gulf Coast region of south-central Texas have a relatively simple burial history. This area has been characterized by a southeasterly

Table 1

Core	County	Depth (ft)	Map Symbol
Anderson–Prichard 1 J.E. Standifer	Bastrop	2,302–2,329	As
Prairie Producing A-1 J. Blumberg	Guadalupe	3,169–3,336	Pbl
Prairie Producing 1 W. Brechtel	Wilson	3,337–3,369	Pbr
Tesoro 1 H.T. Hurts	Gonzales	4,462–4,565	Th
Tesoro 1 H.W. Kruger	Wilson	5,772–5,821	Tk
Getty 1 W.E. Beall	Frio	5,624–5,978	Gb
Amoco 1 L.&L. Varner–Wendler	Burleson	5,962–6,028	Av
Prairie Producing 1 J.M. Winterbotham	Zavala	6,265–6,368	Pwi
Prairie Producing 1 H.B. Weinert *	Wilson	6,734–6,900	Pwe
Getty 1 L.P. Samuels	Frio	6,980–7,130	Gs
Getty 1 L. Hurt*	LaSalle	6,767–7,304	Gh
Getty 1 G.E. Talbutt	LaSalle	7,225–7,266	Gt
Getty 1 J.T. Wilson**	LaSalle	8,052–8,136	Gw
Amoco 1-A Texas A&M*	Burleson	9,038–9,101	Atx

* = cores containing both Austin Chalk and Eagle Ford Formation.
** = core containing only Eagle Ford Formation.

Table 1 — List of cores, counties in which cores are located, sample intervals, and map symbols (corresponding to symbols in Fig. 1) of Austin Chalk.

(basinward) progradation of the shoreline since the Late Cretaceous, interrupted by several transgressions (Fisher, 1969). By late Eocene time, most of the study area was northwest and landward of the shoreline (Fisher et al, 1970), and sedimentation was diminishing. Sedimentation probably ceased during the Miocene, and only minor erosion occurred during the Pliocene and Quaternary. Generalized burial curves for the Austin Chalk in shallow (<5,000 ft or 1,525 m) and deep (>7,750 ft or 2,364 m) cores of this study (Fig. 2) indicate that the chalk has been at maximum burial since 30 to 50 m.y.b.p. (Grabowski, 1981).

Carbonate Sedimentology and Diagenesis

The Austin Chalk of south-central Texas is characterized by three size groups of sedimentary particles (Dravis, 1979): (1) mud composed mainly of whole and fragmented coccoliths and lesser amounts of clay minerals; (2) silt and fine sand composed of planktic foraminifers, calcispheres, and comminuted benthic skeletal material; and (3) medium to coarse sand and larger particles composed of whole and fragmented benthic skeletal material. The Austin Chalk contains a large proportion of benthic skeletal material, in contrast to chalks of the North Sea (Dravis, 1979).

Chalks in shallow (<5,000 ft or 1,525 m) cores and in outcrop are typically light colored, thoroughly bioturbated, and contain many coarse benthic skeletal grains. These chalks are composed principally of calcite (average 78% acid soluble), with lesser amounts of clay minerals, pyrite, glauconite, quartz, and organic matter (Dravis, 1979; Grabowski, 1981). Laminae and beds of calcareous shale and zones of wispy microstylolites (formed by pressure solution) occur in places in these chalks. These argillaceous layers contain an average of 53% acid-soluble material, mainly calcite (Table 2). The light-colored chalks were deposited in a shelfal environment in water tens of meters or less deep (Dravis, 1979).

Chalks in cores between 5,000 and 7,750 ft (1,525 and 2,364 m) are composed of variable amounts of light-colored chalks like those in the shallow cores, interspersed with dark-colored, laminated chalks containing solitary burrows and rare breccias (Dravis, 1979; Grabowski, 1981). The laminae are light-colored, generally composed of concentrations of silt- and sand-sized planktic foraminifers and calcispheres cemented with calcite. These dark-colored chalks contain greater than 75% acid-soluble material (calcite), with calcareous shales and zones of wispy microstylolites containing greater than 47% acid-soluble material (Table 2). The alternating light and dark chalks in these cores of intermediate depth record fluctuations in the oxygenation of the depositional environment, with the laminated and sparsely bioturbated dark-colored chalks forming in oxygen-poor settings. The rare breccias formed from basinward transport of carbonate sediments from the shallow shelf.

Chalks in deep (>7,750 ft or 2,364 m) cores are predominantly dark colored, laminated, and sparsely bioturbated. The purest chalks contain about 90% acid-soluble material, the remainder being clay minerals, pyrite, and organic matter (Grabowski, 1981). The argillaceous chalks and calcareous shales contain greater than 35% acid-soluble material. These dark-colored chalks in places contain continuous laminae of pyrite. They probably formed in nearly anoxic basinal settings below wave base (Dravis, 1979). The overall trend in the Austin Chalk is an increasing proportion of dark-colored chalks with increasing depth of burial.

Diagenesis of the Austin Chalk in south-central Texas involved the following processes (Dravis, 1979), in order of occurrence: (1) early physical compaction (argillaceous sediments were affected only by physical compaction); (2) stabilization of aragonite and high-magnesium calcite by solution and reprecipitation of low-magnesium calcite; (3) cementation by low-magnesium calcite (ferroan calcite more common in deeper cores); (4) pressure solution (wispy microstylolites in argillaceous chalks, stylolites in pure chalks); and (5) fracturing (hairline vertical fractures owing to burial compaction, restricted to nonargillaceous chalks). The large proportion of metastable carbonate constituents (most of the benthic skeletal grains) in the Austin Chalk accounts for the rapid lithification and porosity loss of the Austin Chalk (Dravis, 1979).

Reservoir Quality

The porosity of the Austin Chalk in the subsurface of south-central Texas is generally low and decreases more abruptly with depth compared to chalks from the North Sea (Fig. 3). The porosity is predominantly micro-interparticle between pieces of coccoliths and planktic foraminifers and calcispheres, with lesser amounts of intraparticle porosity (Dravis, 1979). The permeability of this chalk matrix is low (generally <0.01 md), which impedes the flow of petroleum through the rocks (Dravis, 1979). Fracturing enhances the permeability of the chalk but is relatively rare, even in some oil-producing intervals (Dravis, 1979). The fractures occur in organic-rich and organic-poor chalks, are commonly

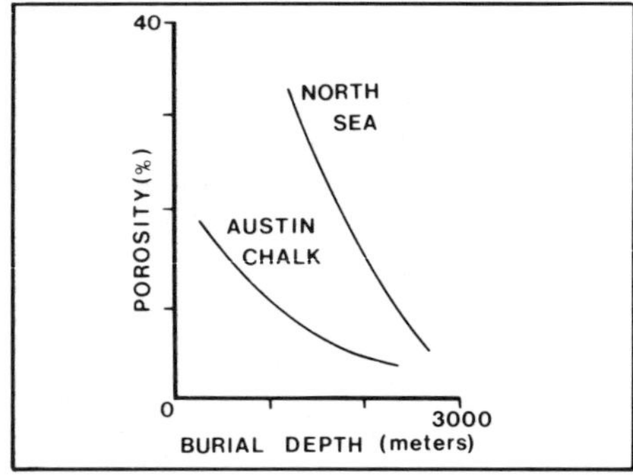

Figure 3 — Diagram comparing porosity reduction of Austin Chalk to North Sea chalks (from Dravis, 1979), showing the continuously lower porosity of the Austin Chalk with depth throughout the range of depths observed in this study.

occluded with calcite cement, but are absent from argillaceous zones, including calcareous shales and microstylolites. The reservoirs, therefore, are divided into many zones separated by less permeable argillaceous layers.

METHODS

Samples of the Austin Chalk and Eagle Ford Formation were collected from 15 cores (Table 1) ranging in depth from 2,300 to 9,100 ft (700 to 2,780 m). Details of the sampling and rock preparation are included in Grabowski (1981). Samples of each lithology were taken, so that the sampling was not random and some minor lithologies may be over-represented in the sample population. Six samples come from the Eagle Ford Formation.

Thin sections and polished chips were examined with transmitted-, reflected-, and fluorescent-light microscopy to describe the lithology and to determine the distribution of kerogen and extractable organic matter (EOM). Carbonate content was determined by weight loss during HCl-acid treatment of powdered rock. Organic-carbon contents were determined by subsequent combustion of the acid-insoluble material in a Leco carbon analyzer (Grabowski, 1981).

Powdered samples were analyzed on a Rock-Eval pyrolysis instrument (Espitalié et al, 1977) by GeoChem Laboratories, Inc. Light hydrocarbons (C_1–C_{13}) were determined by pyrolysis-gas chromatography, the results of which are included in Hunt (this volume).

Organic matter was extracted, fractionated and analyzed according to procedures outlined in Grabowski (1981). EOM was dissolved in purified chloroform during ultrasonic (30 min) and Soxhlet (24 hr) extractions of powdered rock. Copper metal was added to remove free sulfur from the extracts.

The extract was concentrated, and an aliquot was taken for determination of

Table 2

Core	Depth (ft)	Acid Soluble (%)	TOC (%)	EOM (ppm)	EOM/TOC (%)	% of EOM			
						Asph	Res	Arom	Sat
As	2,300.5	76.2	0.41						
	2,305	80.0	0.56						
	2,314.7	89.5	0.75						
	2,316	87.6	0.68						
	2,322	42.4	0.35						
Pbl	3,169	81.6	0.10						
&	3,274	73.3	1.63	270	1.7	9	70	9	12
Pbr	3,281.2	83.1	0.15	150	10.0				
	3,290	80.9	1.33	1,140	8.6	8	71	15	6
	3,293	80.1	5.54						
	3,327	55.2	5.20	9,080	17.5	4	75	19	2
	3,336	65.6	4.20						
	3,350	48.2	1.36	400	2.9	38	28	24	10
Th	4,462	89.0	0.75						
	4,486	52.6	0.30						
	4,520	80.9	1.22	150	1.2	15	58	16	10
	4,560	91.5	3.03	190	0.6				
	4,560.6	88.4	1.12						
	4,565.5	88.7	3.17	120	0.4				
Tk	5,772.5	36.8	0.60	250	4.2				
	5,792	70.4	1.08	280	2.6	4	49	31	16
	5,792.8	71.0	0.59	460	7.8	32	30	30	8
	5,795	82.8	0.61	165	2.7				
	5,821	48.6	0.79	550	7.0	5	46	31	18
Av	5,962	67.9	3.23						
	5,969.6	79.5	1.42	1,760	12.5	14	54	26	6
	6,011.8	65.9	3.27						
	6,015.7	76.3	0.95	800	8.4	17	30	23	30
	6,020.2	81.1	0.82	690	8.4	19	51	26	4
	6,023.4	84.0	1.02	1,060	10.4	14	53	28	5
Gb	5,647	76.9	0.13						
	5,668	88.0	0.11	510	46.4				
	5,672	69.5	5.26	4,970	9.5	4	37	26	23
	5,691	94.8	0.36	2,700	84.4	2	25	36	37
	5,705	64.3	0.24	300	12.5				
	5,749.8	46.8	0.25	274	11.0	15	38	30	17
	5,768	80.0	2.76	2,190	7.9	3	46	21	30
	5,770	75.3	5.07	5,915	11.7	9	51	25	15
	5,944	64.0	2.27	2,630	11.5	15	35	25	25
	5,946	72.9	2.22	2,970	12.6	9	47	25	19
	5,946	72.9	2.22	2,625	11.8	12	52	22	14
	5,973	67.4	2.01	13,000	64.7	7	49	18	26
	5,974	88.1	2.57	7,220	28.0				
	5,978	87.7	1.68						
Pwi	6,265	67.1	1.75	6,410	36.6	7	32	24	37
	6,275.3	59.8	1.98	7,280	36.8	4	33	30	33
	6,289	98.8	1.31	4,920	37.6	7	27	25	41
	6,290	79.9	1.18	4,350	36.9	5	21	28	46
	6,292	90.6	0.63	3,100	49.2	3	32	27	38
	6,315	67.8	1.30	6,180	47.5	1	25	24	50
	6,330	61.9	1.22	5,890	48.3	4	27	24	45
	6,334.5	84.1	1.03	5,630	54.7	5	24	29	42
	6,345	77.1	1.44	5,800	40.3	8	23	27	42
	6,355	58.9	1.79	7,820	43.7	3	22	24	51
	6,357	67.7	1.49	6,590	44.3	3	20	30	47
	6,365	58.6	2.30	10,000	43.5	2	35	15	48
	6,368.4	54.2	2.38	9,190	38.6	4	25	24	47

Table 2

Core	Depth (ft)	Acid Soluble (%)	TOC (%)	EOM (ppm)	EOM/TOC (%)	% of EOM Asph	Res	Arom	Sat
Pwe	6,765	87.2	0.10	250	25.0				
	6,770.7	85.8	0.23	290	12.6				
	6,859.5	88.5	1.00	5,280	52.8	31	37	26	6
	6,878	85.3	1.09	2,870	26.3	11	50	21	18
	6,883.2	87.9	0.92	3,510	38.2	34	33	24	9
	6,887.6	91.1	0.65	3,550	54.6	28	24	19	29
	6,894.3	85.0	1.49	4,080	27.4	22	30	20	28
	6,892	83.2	1.49	5,110	34.3				
	6,897.7*	65.3	2.65	7,630	28.8	13	44	24	19
	6,899.6*	73.7	2.19	12,700	58.0	6	57	20	17
Gs	6,987.1	79.8	0.66	53	0.8				
	7,005.4	91.0	0.75 }	6,720	96.0	1	23	32	44
	7,005.6	77.0	0.64 }						
	7,019.1	76.5	0.25	170	6.8				
	7,022.2	78.9	0.41	3,450	84.1	2	22	35	41
	7,027.2	81.2	0.05	160	32.0				
	7,034.2	87.0	0.05	37	7.4				
	7,114.5	77.4	2.63						
	7,115.1	85.5	2.04	5,580	27.4	1	47	20	32
	7,119.1	89.3	1.49	5,545	36.5	2	36	29	33
	7,122		2.23	6,230	27.9	7	41	26	26
	7,130	84.6	1.15	2,870	24.9	3	34	23	40
Gh	6,767	86.1	0.17	420	24.7	4	16	19	61
	6,858	72.0	0.12	280	23.3	4	33	12	51
	7,086.2	86.6	1.37	3,690	26.9	1	29	21	49
	7,121.1	90.9	0.87	2,090	27.1	1	29	24	46
	7,203.3	88.7	1.84	3,760	20.4	7	17	23	53
	7,204.2	90.9	1.30	3,750	28.8	4	27	23	46
	7,258.3	66.6	3.95	3,000	7.6	1	30	32	37
	7,304.2*	53.8	2.54	7,640	30.1	2	23	21	54
Gt	7,225	66.3	3.37	1,370	4.1	3	29	24	44
	7,230.1	85.9	4.30	950	2.2	4	32	29	35
	7,232.5	78.8	2.35	3,990	17.0	5	21	25	49
	7,253.4	64.7	1.54	2,490	16.2	7	25	27	41
	7,255.2	91.1	0.97	6,400	66.0	1	11	23	65
	7,255.2	91.1	0.97	5,570	57.4	1	17	24	58
	7,266.4	85.2	2.17	5,110	23.5	2	38	18	42
Gw	8,062*	49.0	2.82	8,250	29.3	2	31	22	45
	8,136*	58.9	5.67	15,730	27.7	17	19	34	30
Atx	9,038	90.4	5.15	1,050	2.0	3	19	30	48
	9,048	90.3	7.36						
	9,065	73.0	0.77	1,230	16.0	2	20	24	54
	9,069	62.7	1.41	2,100	14.9	2	18	26	54
	9,071	35.3	3.08						
	9,080.3	52.4	2.05	6,220	30.3	2	6	16	76
	9,090	58.6	2.42						
	9,097	76.9	1.92	3,340	17.4	4	10	21	65
	9,101*	56.1	1.47	2,810	19.1	7	11	18	64

Cores identified by map symbol (Table 1) and sample depth (in ft); sum of all fractions of EOM equals 100%; Asph = asphaltenes; Res = resins; Arom = aromatic hydrocarbons; Sat = saturated hydrocarbons; asterisks indicate samples of Eagle Ford Formation.

Table 2—Results of carbonate and organic-carbon analyses and extractions and fractionations of extractable organic matter (EOM).

EOM. Asphaltenes were precipitated from the extract in cold n-hexane, and the remaining filtrate was fractionated on an alumina–silica gel column into three parts: saturated hydrocarbons (eluted in n-hexane), aromatic hydrocarbons (eluted in benzene), and resins (eluted in benzene–methanol mixture).

Saturated hydrocarbons were analyzed on a Perkin–Elmer 990 gas–liquid chromatograph using compensating 18-ft (5.4-m) salt eutectic columns. Flame-ionization detectors were used with helium carrier gas with the following temperature program: isothermal at 100°C (212°F) for 2 min, increase at 24°C/min (43°F/min) to 300°C (572°F), isothermal at 300°C (572°F) for 15 min.

Kerogen was extracted according to standard procedures outlined in Grabowski (1981). Kerogen slides were examined under transmitted light for organic-matter type and thermal-alteration index (TAI). Some of the demineralized and concentrated kerogen was analyzed for wt% carbon, hydrogen, and nitrogen on Perkin–Elmer 240 elemental analyzers by GeoChem Laboratories, Inc., and Robertson Research (U.S.), Inc.

Stable-carbon isotopes were measured on selected samples of kerogen and EOM by Coastal Sciences Laboratories, Inc.

RESULTS

Organic-Carbon and Carbonate Content

Dark-colored chalks and calcareous shales contain the most organic carbon (generally >1.5 wt% TOC), irrespective of depth of burial or carbonate content. In addition, the maximum values of TOC are approximately the same (4–8%) at all depths. These dark-colored, organic-rich chalks and calcareous shales occur as thin beds (in. or cm scale) separated by thicker (ft or m scale) intervals of light-colored, organic-poor chalks in shallow (<5,000 ft or 1,525 m) cores (Dravis, 1979; Grabowski, 1981). Deeper cores contain a larger proportion and thicker intervals of dark-colored, organic-rich chalks, until at 9,000 ft (2,745 m) the core is composed almost entirely of dark-colored chalks. The general trend of increasing average TOC for each core with increasing depth can be seen in Table 2 by comparing the shallowest (<2,300 ft or 700 m) and deepest (>9,100 ft or 2,780 m) cores.

The acid-soluble (primarily carbonate) contents of the Austin Chalk range from 35 to 99 wt%, with most samples containing 60–90 wt% carbonate. The average carbonate content of the cores is largest in the shallowest cores and is lowest in samples from the Eagle Ford Formation. There is no statistical correlation between organic-carbon and carbonate content.

Visual Examination of Kerogen

Kerogen in the Austin Chalk and Eagle Ford Formation is predominantly type I

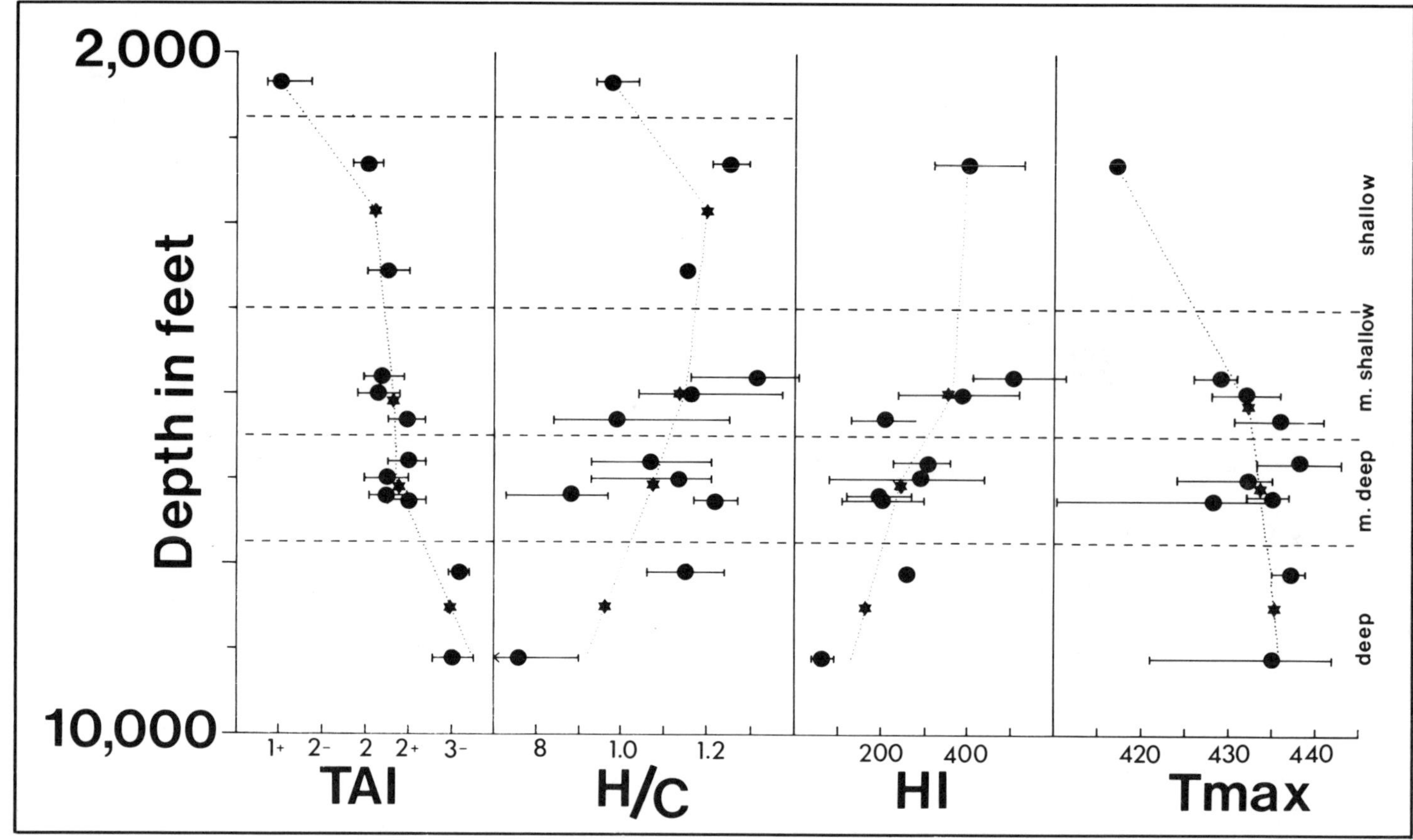

Figure 4 — Plots of thermal-alteration index (TAI), hydrogen/carbon atomic ratio (H/C), hydrogren index (HI), and maximum temperature (°C) of S_2 pyrolysis (T_{max}) versus depth of burial of Austin Chalk. Points represent mean values for each core; bars indicate range of values to each core. Dotted lines are placed through mean values (asterisks) for each depth interval indicated on right side of graph (shallow = <5,000 ft or 1,520 m; moderately, shallow = 5,000 to 6,500 ft or 1,520 to 2,000 m; moderately deep = 6,500 to 7,750 ft or 2,000 to 2,360 m; deep = >7,750 ft or 2,360 m).

or type II amorphous material (generally >95%), with minor amounts of structured plant material (pollen, spores, and woody and coaly material). Type III or structured kerogen is equally rare in high-carbonate chalk, argillaceous chalk, and calcareous shale.

The kerogen occurs disseminated in dark-colored chalks and in stylolites and microstylolites in both light- and dark-colored chalks. Kerogen is absent in solitary burrows and in cemented microfossil-rich laminae that occur within the dark-colored, organic-rich chalks.

The color and degree of alteration of the kerogen show some variability, depending on organic-matter type. The amorphous kerogen is commonly darker than the associated structured kerogen, especially in high-carbonate chalks in shallow cores (<5,000 ft or 1,525 m) that formed in shelfal environments. The color of the amorphous kerogen changes progressively with depth burial (Fig. 4) from light yellow (TAI = 1+, 2−) at 2,300 ft (700 m) to medium brown (TAI = 3−, 3) at 9,100 ft (2,780 m).

Elemental Analysis of Kerogen

The chemical composition of the kerogen, as determined by elemental analysis (Table 3), varies slightly between samples from a core but shows considerable variation between mean values for cores from the same depth interval (Fig. 4). For example, the hydrogen/carbon atomic ratio (H/C) for the three moderately shallow cores (Gb, Av, Pwi) ranges from 0.99 to 1.31 with a standard deviation of ±0.10. This variation is greater than the analytical error determined from replicate analyses and does not correlate with organic-carbon or carbonate contents of the samples.

The H/C atomic ratio is lowest in the deepest core (Atx, average 0.76) and highest in the shallower cores (especially Pbl, average 1.25, and Gb, average 1.31). Excluding the anomalously low values for the shallowest core (average 0.97), the H/C atomic ratio shows a general decrease with increasing depth of burial (Fig. 4).

The nitrogen (%N) and combined oxygen–sulfur (%O + S) contents of the kerogen are relatively uniform with depth of burial (Table 3). The nitrogen/carbon atomic ratio (100N/C) is relatively high and shows considerable variability within and between cores, with the lowest average values in the deepest core (Atx = 2.44, Gw = 2.28 average 100N/C) and the highest average value in a shallow core (Th = 4.41 average 100N/C).

Rock-Eval Pyrolysis

The pyrolysis data show more variability than the elemental-analysis data, both within and between cores (Table 4), above the level of experimental error determined by replicate analyses. The greatest variability within cores is shown by the "free hydrocarbon" index

(FI, mg volatile (S_1) hydrocarbons/g TOC) and the transformation index ($S_1/(S_1 + S_2)$, mg volatile hydrocarbons/mg total hydrocarbons). The largest differences occur in the Gs core, with FI varying from 28 to 543 mg HC/g TOC and $S_1/(S_1 + S_2)$ varying from 0.08 to 0.78. The variability in these parameters does not correlate with either organic-carbon or carbonate contents of the samples.

Variations between cores are apparent in most parameters determined from pyrolysis. The lowest mean values for FI (26 and 22 mg HC/g TOC; Table 4) and $S_1/(S_1 + S_2)$ (0.05 and 0.06) occur in two shallow cores (Pbl and Av, respectively). Higher mean values (up to 288 mg HC/g TOC and 0.63, respectively) for these parameters are found in cores below 6,000 ft (1,830 m).

The mean values for hydrogen index (HI, mg pyrolyzable (S_2) hydrocarbons/g TOC) for shallow, moderately shallow, moderately deep, and deep Austin Chalk (as shown in Fig. 4) decrease with increasing depth of burial. The average HI ranges from 503 mg HC/g TOC (Gb core, 5,900 ft or 1,800 m) to 65 mg HC/g TOC (Atx core, 9,100 ft or 2,780 m). The values for oxygen index (OI, mg CO_2/g TOC) are consistently low, ranging from a mean of 48 mg CO_2/g TOC (Av core, 6,000 ft or 1,830 m) to 13 mg CO_2/g TOC (Gw core, 8,100 ft or 2,470 m). The high HI and low OI values indicate either type I or II kerogen (Tissot and Welte, 1978).

The maximum temperature of S_2 pyrolysis (T_{max}, °C) is low in the shallowest core (Pbl, average 417°C or 783°F) but uniformly moderate in the other cores (between 428° and 438°C or 802° and 820°F).

Extractable Organic Matter

Extractable-organic-matter (EOM) contents vary greatly within and between cores (Table 2), above the level of experimental error determined by replicate analyses. EOM content ranges from 274 to 13,000 ppm in the Gb core, with the average EOM content ranging from 153 ppm (Th core) to 11,990 ppm (Gw core). The average EOM contents are lower in the shallow cores (<5,000 ft or 1,525 m). EOM content increases generally with increasing organic-carbon content of the sample. The EOM/TOC (%) values show similar variability within and between cores. The within-core variation of EOM/TOC is smaller in the shallow cores (0.4 to 17.5%) than in cores below about 5,000 ft (1,525 m) (0.8 to 96.0%). Mean EOM/TOC values increase from 0.7 to 8.1%.

The EOM/TOC percentage values show similar variability within and between cores. The within-core variation of EOM/TOC is smaller in the shallow cores (0.4 to 17.5%) than in cores below

Table 3

Core	Depth (ft)	% Ash	% C	% H	% N	% O+S	H/C	100N/C
As	2,300.5	31.43	58.19	4.60	2.21	35.0	0.95	3.25
	2,305	38.51	50.98	3.95	2.11	43.0	0.93	3.54
	2,314.7	37.40	47.63	4.13	1.81	46.4	1.04	3.26
	2,316	36.64	51.01	4.05	2.07	42.7	0.95	3.47
Pbl	3,293	37.72	49.65	5.01	2.06	43.3	1.21	3.55
	3,336	33.42	37.98	4.08	0.96	57.0	1.29	2.16
Th	4,560.6	37.84	49.01	4.64	2.91	43.4	1.14	5.10
	4,565.5	42.86	41.69	4.04	1.81	52.4	1.16	3.72
Gb	5,768	28.08	38.63	4.40	1.49	55.5	1.37	3.29
	5,770	14.14	59.90	7.03	2.32	30.7	1.41	3.33
	5,770	15.56	68.88	6.69	2.03	22.4	1.16	2.53
	5,946	27.26	44.67	4.78	2.19	48.4	1.28	4.19
	5,973	26.87	55.11	5.96	2.44	36.5	1.30	3.79
Av	5,962	21.30	57.61	6.57	2.82	33.0	1.37	4.19
	6,011.8	27.46	56.08	5.28	2.45	51.6	1.13	3.75
	6,015.7	34.26	52.00	4.74	2.21	41.1	1.09	3.62
	6,023.4	18.86	67.01	5.78	2.55	24.7	1.04	3.26
Pwi	6,265	39.54	60.54	4.50	1.75	33.2	0.89	2.48
	6,275.3	44.02	52.91	4.63	1.70	40.8	1.05	2.76
	6,289	48.96	47.62	4.05	1.50	46.8	1.02	2.70
	6,290	28.86	36.34	3.80	1.47	58.4	1.25	3.47
	6,292	23.84	51.64	3.93	1.78	42.4	0.91	2.95
	6,345	36.76	43.23	3.61	2.02	51.2	1.00	4.00
	6,365	42.10	56.10	3.92	1.29	38.7	0.84	1.97
Pwe	6,859.5	36.17	54.59	5.10	1.90	38.4	1.12	2.99
	6,859.5	37.16	58.78	4.84	1.16	35.2	0.99	1.69
	6,878	42.99	40.63	3.88	1.70	53.8	1.14	3.57
	6,892	33.65	34.86	4.46	1.41	59.3	1.21	3.47
	6,897.7*	37.12	60.98	4.74	1.24	33.0	0.93	1.75
Gs	7,005	39.86	57.00	4.43	1.42	37.2	0.93	2.13
	7,114.5	18.40	70.34	6.57	1.94	21.2	1.12	2.36
	7,115.1	18.09	61.18	6.15	2.45	30.2	1.21	3.43
	7,119.1	16.65	61.43	5.87	2.38	30.3	1.15	3.32
	7,130	18.55	57.69	5.72	2.10	34.5	1.19	3.12
Gh	6,767	45.36	51.15	3.11	1.47	44.3	0.73	2.46
	7,086.2	18.35	65.52	5.24	2.24	27.0	0.96	2.93
	7,203.3	17.70	61.77	4.98	2.89	31.2	0.97	2.89
	7,204.2	21.94	70.59	4.94	1.92	21.9	0.84	2.33
Gt	7,230	48.94	31.06	3.30	1.62	64.0	1.27	4.47
	7,232	28.66	51.23	5.27	1.82	41.7	1.23	3.04
	7,266.4	24.90	45.83	4.48	2.57	47.1	1.17	4.82
Gw	8,062*	40.93	46.16	4.75	1.32	47.8	1.24	2.44
	8,136*	25.37	62.96	5.56	1.55	29.9	1.06	2.11
Atx	9,038	23.77	54.39	3.56	1.84	39.8	0.79	2.89
	9,048	6.75	70.25	3.86	2.44	23.5	0.66	2.97
	9,048	11.22	80.65	4.21	2.19	12.9	0.63	2.32
	9,071	36.57	63.18	4.12	2.09	30.6	0.78	2.83
	9,080.3	44.03	52.84	3.94	1.27	41.9	0.90	2.06
	9,090	36.68	50.07	3.88	1.50	35.6	0.79	2.18
	9,097	15.75	68.76	3.68	1.57	26.0	0.64	1.95
	9,101*	41.59	55.21	3.06	1.49	40.3	0.66	2.31

Elemental contents given on an ash-free basis as weight percentages; percent oxygen plus sulfur (%O + S) calculated by difference. H/C = hydrogen/carbon atomic ratio; 100N/C = nitrogen/carbon atomic ratio, multiplied by 100. Asterisks indicate samples of Eagle Ford Formation.

Table 3—Results of elemental analysis of kerogen concentrates.

about 5,000 ft (1,525 m) depth (0.8 to 96.0%). Mean EOM/TOC values increase from 0.7 to 8.1% below 5,000 ft (1,525 m) up to a maximum of 42.9% (Pwi core) in deeper cores (Fig. 5). A pronounced decrease in the mean EOM/TOC value occurs in the deepest core (Atx), which is not apparent in the data on volatile hydrocarbons (FI) from pyrolysis. Excluding this deepest core, the general increasing trend of EOM/TOC matches that of FI (Fig. 5) with increasing depth of burial.

The bulk composition of the EOM changes with depth of burial (Fig. 6). EOM from shallow cores (< 5,000 ft or 1,525 m) contains an average of 70% NSO compounds (nitrogen, sulfur, and oxygen-containing compounds), 20% aromatic hydrocarbons, and 10% saturated hydrocarbons (Table 2). The ratio of resins to asphaltenes is less than 5.0, and the hydrocarbon-to-NSO-compound and saturated-to-aromatic-hydrocarbon ratios are both less than 1 (Fig. 8). The saturated-hydrocarbon fraction is characterized by C_{12-26} n-alkanes, an abundance of high-molecular-weight cycloalkanes (including steranes and triterpanes), and few isoalkanes (Fig. 7). C_{15-20} isoprenoids are recognizable, with pristane (C_{19}) and phytane (C_{20}) predominating.

With increasing depth of burial to between 5,000 and 6,500 ft (1,525 and 2,000 m), the content of NSO compounds decreases to about 50%, with the content of aromatic and saturated hydrocarbons increasing to 25% each (Fig. 6). The HC/NSO and Sat/Arom ratios (Fig. 8) both increase to about 1. The saturated-hydrocarbon fraction contains more n-alkanes in the C_{12-32} range and more low-molecular-weight cycloalkanes and isoalkanes, with a corresponding decrease in high-molecular-weight cycloalkanes and isoprenoids (Fig. 7). Some samples have a slight even-carbon-number n-alkane preference.

Below 6,500 ft (2,000 m), the EOM is composed of 20–40% NSO compounds,

Table 4

Core	Depth (ft)	S_1	S_2	S_3	FI	HI	$S_1 + S_2$ OI	$\frac{S_1}{S_1 + S_2}$	T_{max} (°C)
Pbl	3,290	0.12	4.19	0.93	9	315	70	0.03	416
	3,293	0.89	19.0	0.31	16	342	6	0.04	416
	3,336	2.25	22.1	0.56	54	526	13	0.09	419
Gb	5,672	1.30	28.7	0.60	25	546	11	0.04	427
	5,768	0.63	12.2	0.53	23	442	19	0.05	427
	5,770	1.49	27.3	0.56	29	538	11	0.05	426
	5,944	0.48	9.30	0.60	21	410	26	0.05	431
	5,945	0.62	9.60	0.69	28	432	31	0.06	431
	5,973	3.74	10.4	0.51	186	517	26	0.26	431
	5.974	1.63	12.7	0.41	63	492	16	0.11	428
	5,978	4.72	10.9	0.42	281	646	25	0.30	431
Av	5,962	0.87	16.7	0.69	27	514	21	0.05	431
	5.969.6	0.33	5.44	0.64	23	383	45	0.06	436
	6.011.8	1.01	14.9	0.84	31	455	26	0.06	428
	6,015.7	0.13	2.24	0.73	14	235	77	0.06	432
	6,015.7	0.13	1.,3	0.65	16	235	79	0.06	432
	6,023.4	0.20	3.48	0.44	20	341	43	0.05	434
Pwi	6,265	1.84	4.92	0.37	105	281	21	0.27	440
	6,275.3	3.65	3.92	0.37	184	198	19	0.48	436
	6,289	2.54	2.48	0.38	194	189	29	0.51	432
	6,290	1.42	2.81	0.38	120	238	32	0.34	438
	6,292	1.05	1.72	0.25	167	273	40	0.38	438
	6.315	3.17	1.86	0.34	244	143	26	0.63	438
	6,334.5	2.86	1.77	0.22	278	172	21	0.62	431
	6,345	3.30	2.44	0.33	229	169	23	0.57	441
	6.355	2.32	4.92	0.39	130	275	22	0.32	436
	6.357	3.40	1.92	0.26	228	129	17	0.64	436
	6.365	5.42	3.78	0.39	236	164	17	0.59	433
	6,368.4	3.41	5.54	0.24	343	236	10	0.38	436
Pwe	6,859.5	0.45	3.39	0.31	42	320	30	0.12	443
	6,878	1.01	3.75	0.21	93	344	19	0.21	438
	6,883.2	0.66	2.94	0.18	71	320	20	0.18	436
	6,894.3	0.95	4.17	0.24	64	280	16	0.19	436
	6,892	1.03	3.43	0.37	69	231	25	0.23	440
	6,897.7*	1.73	7.58	0.41	65	286	15	0.19	442
	6,899.6*	2.50	7.79	0.31	114	356	14	0.24	433

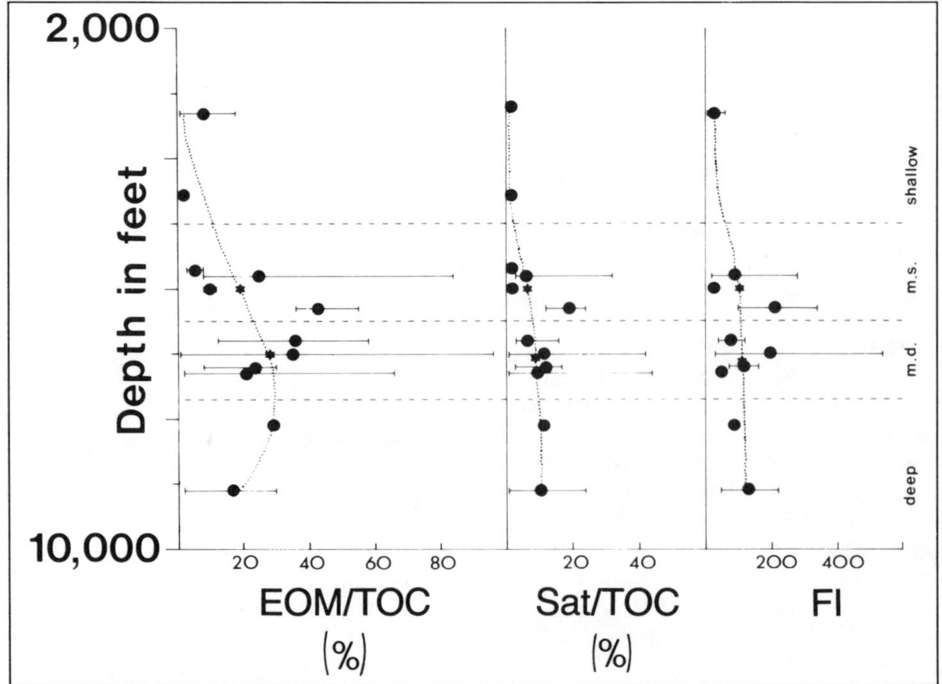

Figure 5 — Plots of percentage of extractable organic matter (EOM/TOC), percentage of extractable saturated hydrocarbons (SAT/TOC), and free-hydrocarbon index (FI, calculated as mg volatile hydrocarbons (S_1) divided by g TOC) versus depth of burial of Austin Chalk. Points represent mean values for each core; bars indicate range of values for each core. Dotted lines are placed through mean values (asterisks) for each depth indicated on right side of graph (as in Fig. 4).

Table 4

Core	Depth (ft)	S_1	S_2	S_3	FI	HI	$\frac{S_1 + S_2}{OI}$	$\frac{S_1}{S_1 + S_2}$	T_{max} (°C)
Gs	7,005.4	4.07	1.16	0.30	543	155	40	0.78	432
	7,005.6	0.18	0.53	0.70	28	83	109	0.25	424
	7,022.2	1.75	0.73	0.36	427	178	88	0.71	431
	7,114.5	1.04	11.8	0.50	39	446	19	0.08	434
	7,115.1	1.97	7.88	0.47	97	386	23	0.20	435
	7,119.1	2.28	5.95	0.48	153	399	32	0.28	433
	7,130	0.79	4.23	0.41	69	368	36	0.16	433
Gh	7,086.2	1.36	3.90	0.49	99	285	36	0.26	432
	7,203.3	1.21	3.00	0.41	66	163	22	0.29	437
	7,258.3	6.25	4.73	0.25	158	120	6	0.57	435
Gt	7,225	0.93	3.71	0.22	28	110	7	0.20	410
	7,230.1	2.21	7.42	0.41	51	173	10	0.23	422
	7,232.5	1.25	6.18	0.38	53	263	16	0.17	438
	7,253.4	0.71	2.74	0.44	46	178	29	0.21	435
	7,266.4	1.26	6.54	0.37	58	301	17	0.16	435
Gw	8,065*	2.34	6.99	0.41	83	248	14	0.25	437
	8,062*	2.39	6.76	0.47	85	240	17	0.26	436
	8,136*	4.60	15.7	0.48	81	277	8	0.23	439
	8,136*	4.98	14.6	0.64	91	266	12	0.25	435
Atx	9,038	5.31	2.85	0.32	103	56	6	0.65	438
	9,048	6.36	4.62	0.36	86	63	5	0.58	436
	9,065	0.71	0.30	1.01	92	40	40	0.70	436
	9,080.3	4.47	1.44	0.29	218	70	14	0.76	437
	9,090	3.98	2.28	0.34	164	94	14	0.64	421
	9,097	2.22	1.45	0.38	116	75	20	0.60	442
	9,101*	0.73	0.83	0.31	50	56	21	0.47	436

S_1 (volatile or "free" hydrocarbons) and S_2 (pyrolyzable hydrocarbons) given in mg hydrocarbons/g rock; S_3 (CO_2 from kerogen pyrolysis) given in mg CO_2/g rock; FI = "free hydrocarbon" index, S_1/TOC; HI = hydrogen index, S_2/TOC; OI = oxygen index, S_3/TOC; T_{max} = maximum temperature of S_2 pyrolysis, given in °C. Asterisks indicate samples of Eagle Ford Formation.

Table 4—Results of Rock-Eval pyrolysis of whole-rock samples.

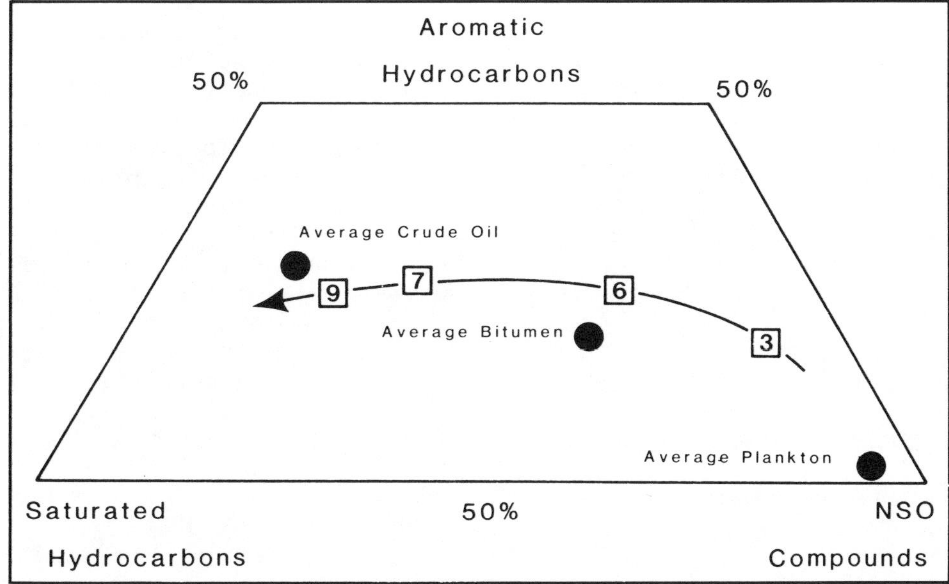

Figure 6 — Triangular plot of saturated and aromatic hydrocarbons and NSO compounds in Austin Chalk, indicating general trend in composition of extractable organic matter (EOM) with increasing depth of burial owing to catagenesis. Numbered boxes represent mean values for all cores from the following depth intervals: 3 = shallow (<5,000 ft or 1,520 m); 6 = moderately shallow (5,000 to 6,500 ft or 1,520 to 2,000 m); 7 = moderately deep (6,500 to 7,750 ft or 2,000 to 2,360 m); 9 = deep (>7,750 ft or 2,360 m). The compositions of average crude oil and bitumen from all lithologies and of plankton are taken from Tissot and Welte (1978).

25% aromatic hydrocarbons, and 35–55% saturated hydrocarbons (Fig. 6). The HC/NSO and Sat/Arom ratios (Fig. 8) are between 1 and 5 for these deeper samples of Austin Chalk. The saturated-hydrocarbon fraction is dominated by C_{12-32} n-alkanes, with little or no carbon-number preference, abundant low-molecular-weight cycloalkanes, isoalkanes, and isoprenoids (Fig. 9). Isoprenoids previously identified in samples from shallow cores are present in these deeper samples but are less prominent because of dilution by other saturated hydrocarbons in the EOM.

Major variations in the composition of EOM occur within each core. Samples with large values of EOM/TOC generally contain more hydrocarbons, especially saturated ones, and resins, and fewer asphaltenes than samples with small values of EOM/TOC (Fig. 10). The trends of these changes are consistent regardless of depth of burial (Fig. 11) and are independent of organic-carbon or carbonate content of the samples. The trends are most apparent in the cores below 5,000 ft (1,525 m).

The compositions of saturated hydrocarbons within each core do not vary greatly, despite differences in amounts and compositions of EOM. As is illustrated in Figure 9, the samples containing a higher ratio of EOM/TOC typically contain a smaller proportion of cycloalkanes spread more uniformly by carbon number.

A few samples contain an unusual distribution of saturated hydrocarbons unlike those of the adjacent samples. For example, sample Pwi 6290 shows a bimodal distribution of n-alkanes (Fig. 9), with dominant peaks at C_{17} and C_{26}. This sample, from a fractured chalk containing 1.18% TOC and 36.9% EOM/TOC, probably contains EOM from two different sources. Similarly, sample Gs 7022.2 shows a narrow distribution of saturated hydrocarbons like that found in the deeper Atx core (Fig. 9), except for the predominance of phytane over pristane. This sample, from a chalk containing only 0.41% TOC but 84.1% EOM/TOC, probably includes EOM from another source.

Stable-Carbon Isotopes

The stable-carbon-isotope data (Table 5) for the kerogen and fractions of EOM show little variation within or between cores. The isotopic ratio of the saturated-hydrocarbon fraction is generally the lightest (average $-27.9\ \delta^{13}C_{PDB}$), whereas the ratios of the kerogen (average $-26.5\delta^{13}C_{PDB}$) and nonhydrocarbons (average $-26.8\delta^{13}C_{PDB}$) are the heaviest.

Table 5

Core	Depth (ft)	‰ $\delta^{13}C_{PDB}$ Ker	Sat	Arom	Res	Asph
Gb	5,691	*	−28.0	−27.2	−27.0	
	5,768	−26.1	−28.1			
	5,770	−26.4	−28.3	−27.6	−26.8	
	5,946	−27.2	−29.6			
	5,973	−27.1	−27.8	−27.3	−27.1	−26.9
Av	6,015.7	−27.3	−29.1	−28.4	−27.7	
Pwi	6,290	−26.1	−27.9	−26.7	−26.4	−27.0
	6,292	−26.7	−27.8			
	6,345	−24.7	−26.9	−25.6	−25.1	
Pwe	6,878	−27.2	−27.3	−27.2	−26.9	−27.8
	6,894.3	−27.1	−27.4	−27.0	−26.9	−27.7
Gh	7,022.2		−28.3	−27.6	−27.1	
&	7,086.2	−26.2	−27.6	−26.5	−26.2	−26.2
Gs	7,119.1	−26.6	−28.1			
	7,130	−26.7	−28.4			
Gt	7,225	−25.7	−27.7			
	7,232.5	−26.0	−27.1	−26.5	−26.3	−26.2
	7,266.4	−26.3	−27.7			
Atx	9,038	−26.8	−28.2			
	9,065	−25.9	−27.8	−26.7	−26.4	
	9,097	−26.5	−27.7			

Ker = kerogen; Sat = saturated hydrocarbons; Arom = aromatic hydrocarbons; Res = resins; Asph = asphaltenes. Asterisk represents insufficient kerogen for analysis.

Table 5—Results of carbon-isotope analyses, in ppt $\delta^{13}C_{PDB}$.

INTERPRETATION AND DISCUSSION

Distribution of Organic Matter

Distribution of organic matter in fine-grained carbonate rocks is controlled by both depositional and diagenetic processes (Malek–Aslani, 1980). This contrasts with shales, in which the distribution of organic matter appears to be controlled largely by depositional processes (Schrayer and Zarella, 1963, 1966; Davis, 1970; Clayton and Swetland, 1980).

Organic-rich rocks (> 1.5 wt% TOC) are present in each core of Austin Chalk examined, regardless of depositional environment or lithology, and the maximum amount of organic carbon is approximately the same in each core. At least local conditions favoring deposition and preservation of fairly large amounts of organic matter existed in both shelfal and basinal settings. Most of the shallow cores from shelfal environments contain thoroughly bioturbated, light-colored, organic-poor chalks. Organic matter deposited in these chalks was largely destroyed, probably by the action of burrowing infauna.

Organic-rich, dark-colored chalks are increasingly abundant and thicker in the deeper, more basinal cores of Austin Chalk. This results in the general trend of increasing organic-carbon content with increasing depth of burial. Many of these dark-colored chalks contain solitary, oxidized burrows, suggesting disaerobic bottom conditions. Post-depositional alteration and destruction of organic matter in the subsurface Austin Chalk of south-central Texas are minor, perhaps causing the anomalously low H/C atomic ratios in the shallow As core. Organic matter is almost completely absent from outcrops of the chalk, however, suggesting destruction of the organic matter during Holocene weathering.

Kerogen occurs disseminated throughout the dark-colored, organic-rich chalks and in stylolites and microstylolites in both light-colored and dark-colored chalks. The kerogen is concentrated during pressure solution into stylolitic laminae, creating a heterogeneous distribution of extremely organic-rich zones alternating with less organic-rich chalks.

The distributions of EOM and of volatile hydrocarbons determined by pyrolysis are more complex than those of the kerogen. The amount of EOM generally increases with increasing organic carbon, suggesting generation and retention of EOM from the organic matter in the Austin Chalk. However, some chalks have very high levels of EOM/TOC (approaching 100%), which suggests that EOM has migrated into these chalks. In general, relatively porous, commonly fractured, organic-poor chalks have higher values of EOM/TOC than less porous, organic-rich chalks, suggesting that EOM has moved into available porosity.

Alteration of Organic Matter and Generation of Hydrocarbons

Organic matter in the Austin Chalk has been altered and hydrocarbons have been generated during the 6,800 ft (2,075 m) of burial represented between the deepest and shallowest cores. These processes follow the general scheme of catagenesis (the main zone of oil and wet-gas generation) determined from several rock sequences (McIver, 1967; Laplante, 1974; Tissot et al, 1974; Tissot and Welte, 1978) and from experiments (Robin, Rouxhet, and Durand, 1977).

The kerogen darkens (shown by the increase in TAI) as the hydrogen content of the kerogen (shown by the decrease in H/C and HI) decreases over the span of cores examined. The T_{max} increases over the initial 2,500 ft (760 m) of the interval but then remains relatively constant with depth. The oxygen index (OI) of the kerogen remains consistently low, and the nitrogen content (100N/C) remains consistently high during burial, as is the general case for amorphous, type I, or type II kerogen (Tissot and Welte, 1978). These data, in conjunction with limited infrared spectra of the asphaltene fraction of the EOM (Grabowski, 1981, p. 137–138), suggest that the kerogen and asphaltenes become progressively condensed, more aromatic, and less aliphatic with increasing levels of maturation.

Concurrent with the loss of hydrogen-rich functional groups from the kerogen and asphaltenes, the absolute

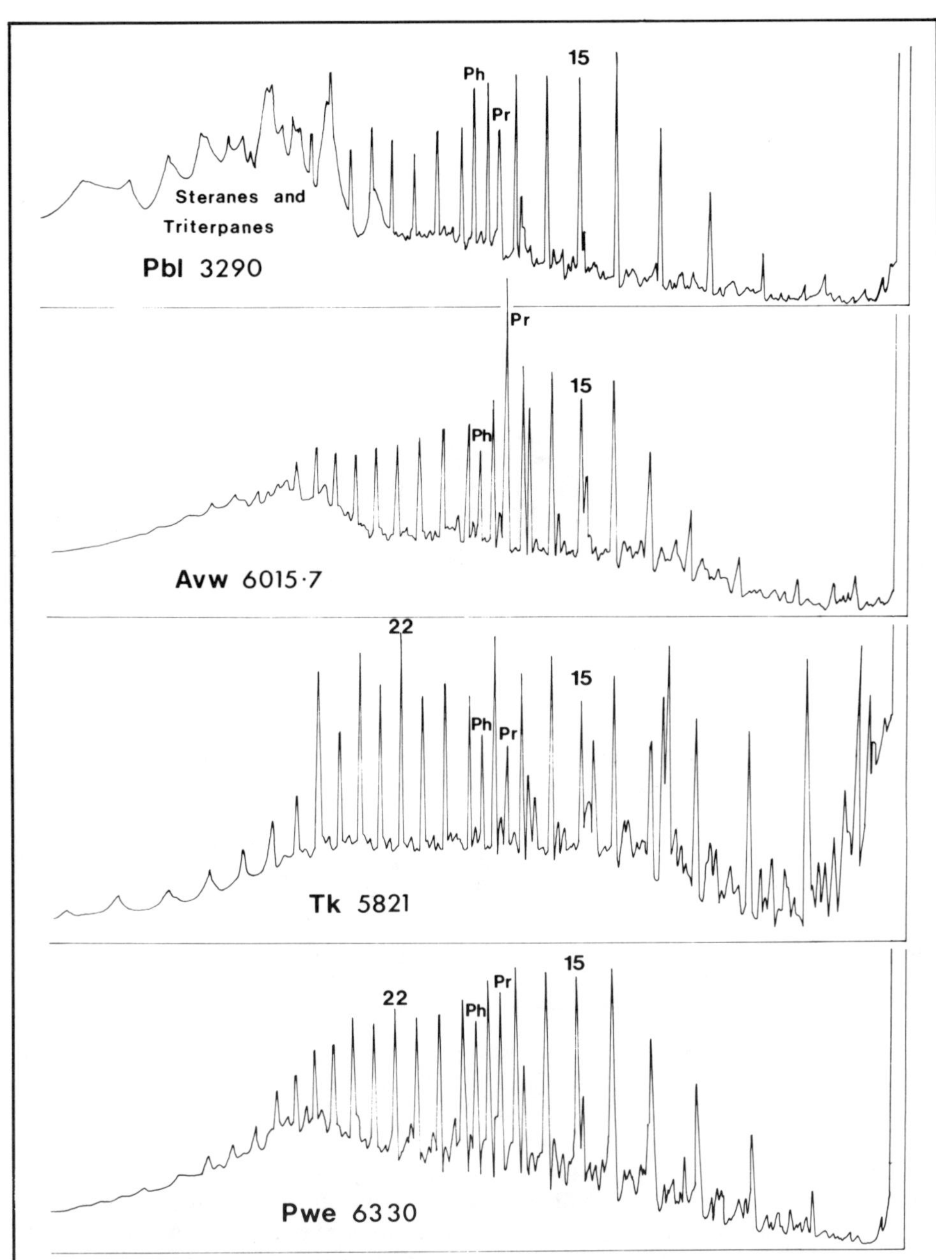

Figure 7 — Gas chromatograms of saturated-hydrocarbon fractions from shallow cores of Austin Chalk. Numbers refer to sample depths (in ft) in cores (abbreviations given in Table 1). Note predominance of steranes and triterpanes in the shallowest core, which becomes less pronounced in slightly deeper samples. Also note extreme even-carbon preference in sample Tk 5821 and large pristane (Pr) peak in sample AV 6015.7. Pr = pristane, Ph = phytane; numbers above peaks refer to carbon number of n-alkanes. Salt eutectic column, 18 ft (5.4 m); isothermal at 100°C (212°F) for 2 min, increase at 24°C/min (43°F/min) to 300°C (572°F), isothermal at 300°C (572°F) for 15 min.

amount of EOM, the ratio of EOM/TOC, and the FI increase with depth of burial. The increases in EOM/TOC and FI (Fig. 5) with increasing depth of burial are classical indications of oil generation that have been documented from many rock sequences (Tissot et al, 1971; Tissot et al, 1974; Tissot and Welte, 1978). The subsequent decreases in EOM/TOC and saturated hydrocarbons/TOC (Fig. 5) and increases in FI and light gases (Hunt, this volume) below 8,000 ft (2,440 m) most likely indicate cracking of hydrocarbons in the beginning of the principal zone of gas generation (Deroo et al, 1977; Tissot et al, 1974).

In the zone of peak hydrocarbon generation, saturated hydrocarbons of low molecular weight (C_{1-14}) are formed in the Austin Chalk (Hunt, this volume) as well as C_{15+} hydrocarbons. The C_{15+} hydrocarbons in deep (>6,500 ft or 2,000 m) samples show no carbon-number preference, suggesting that a uniform distribution of n-alkanes has been formed. The geochemical fossils present in the shallow samples (Fig. 7) exist in the EOM from deeper cores but have been diluted by newly formed saturated hydrocarbons. Saturated hydrocarbons form in larger quantities than aromatic hydrocarbons in cores below about 6,500 ft (2,000 m), suggesting some differences between "early" and "late" catagenesis of the organic matter in the Austin Chalk.

Some variations in the alteration of organic matter and generation of hydrocarbons exist between cores of Austin Chalk from the same depth of burial. The kerogen appears visually and chemically to be relatively uniform,

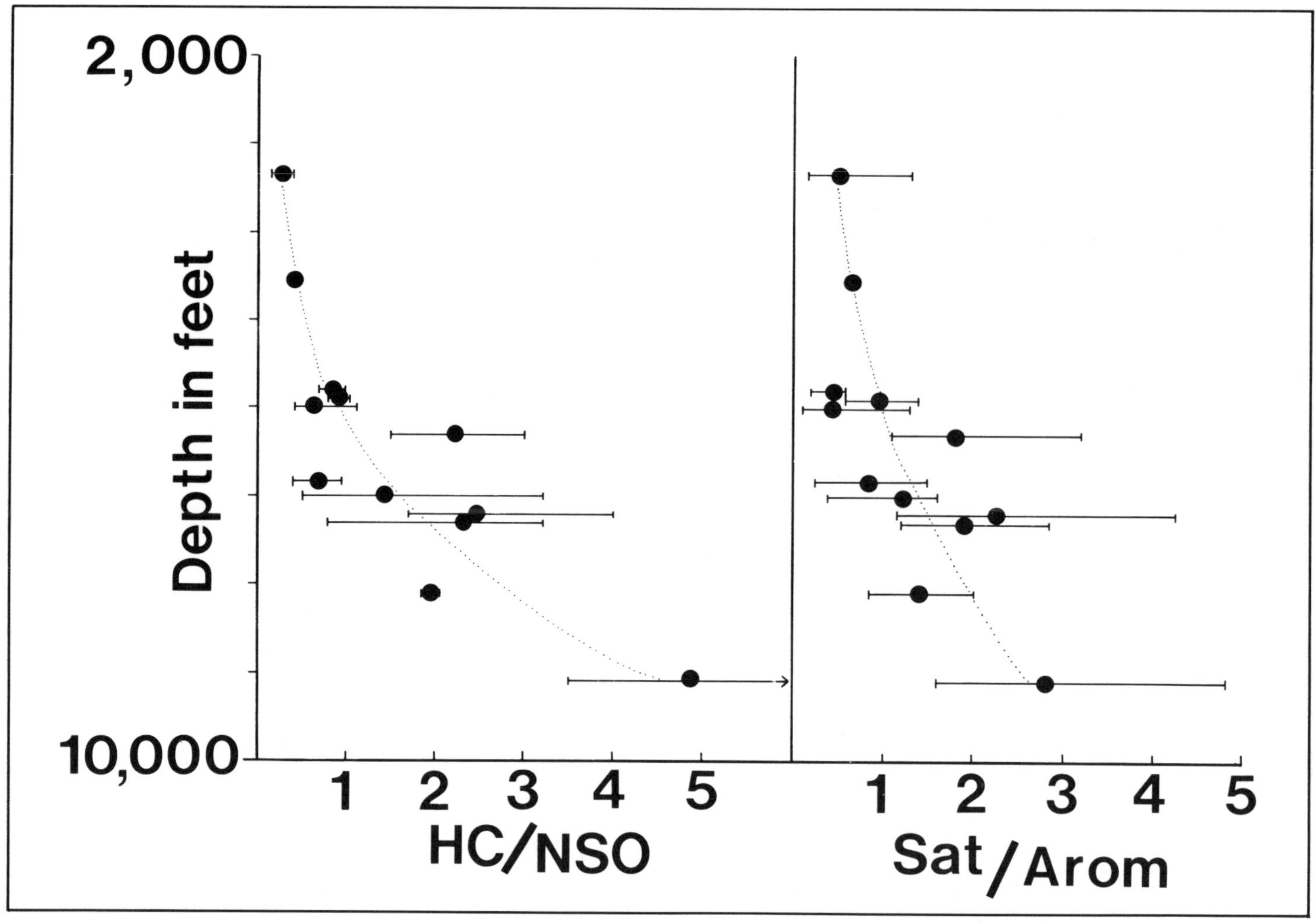

Figure 8 — Plots of hydrocarbon/NSO-compound and saturated/aromatic-hydrocarbon ratios of extractable organic matter (EOM) versus depth of burial of Austin Chalk. Points represent mean values for each core; bars indicate range of values for each core. Dotted line is placed through mean values.

suggesting that local differences in burial history and geothermal gradient may exist in south-central Texas.

The degrees of kerogen alteration and compositions of EOM and saturated hydrocarbons are generally similar within each core and are not related to the amount of acid-soluble (carbonate) material in the Austin Chalk. Samples with large values of EOM/TOC (>60%) occur in several cores (Table 2), generally in organic-poor (average 0.89% TOC) chalks (average 83% acid-soluble). The levels of kerogen alteration in these EOM-enriched samples are not higher than in the adjacent samples, suggesting that the high EOM/TOC values are due to addition of EOM from other rocks.

The absence of any correlations between kerogen alteration and hydrocarbon generation and lithology in the Austin Chalk suggests that the effects of specific minerals on organic matter during catagenesis were insignificant. Carbonate minerals, such as calcite, have been shown to have a lower catalytic activity toward cracking of hydrocarbons than clay minerals, especially smectites (Espitalié, Madec, and Tissot, 1980). If catalysis by clays were the predominant factor in the generation of hydrocarbons in carbonate rocks, then the degree of kerogen alteration and the composition of the EOM might be expected to be different in pure carbonates, argillaceous carbonates, and shales. In the Austin Chalk, kerogen alteration and EOM composition are the same in all lithologies.

An additional expectation of the hypothesis that clay catalysis is an important factor in petroleum generation in carbonate rocks is that petroleum should form at higher temperatures (greater depths of burial) in carbonates as compared with shales. The temperature profile determined from one well in the Austin Chalk (Fig. 12) suggests that the maximum oil generation in the cores studied occurs at about 85°C or 185°F (8,000 ft or 2,440 m). Using this value, the maximum oil generation in the Austin Chalk plots on the trend (Fig. 13) empirically determined by Connan (1974, Fig. 3) for time–temperature relationship of peak oil generation.

I conclude that, for the alteration of organic matter and generation of hydrocarbons in the Austin Chalk of south-central Texas: (1) the effects of minerals on the rate of alteration of kerogen and NSO compounds and on the composition of hydrocarbons generated during catagenesis are negligible; and (2) the composition of EOM in a source rock is determined primarily by the thermal

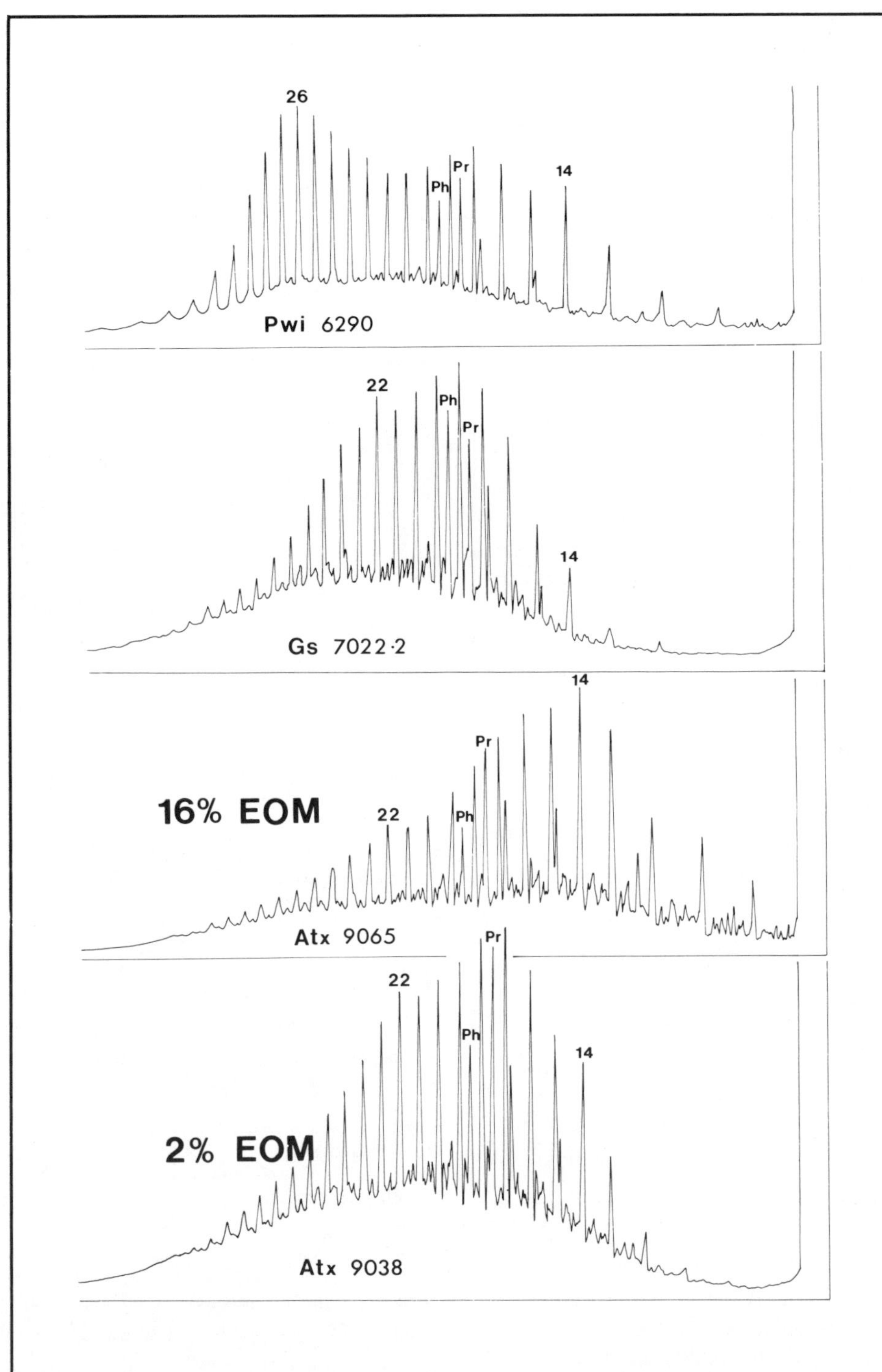

Figure 9 — Gas chromatograms of saturated-hydrocarbon fractions from deep cores of Austin Chalk. Numbers refer to sample depths (in ft) in cores (abbreviations given in Table 1). Note predominance of n-alkanes up to n-C_{32} in these samples compared to sample Pbl 3290 (shallow core) in Figure 7. Bimodal distribution of n-alkanes in Pwi 6290 probably is due to addition of C_{22-32} n-alkanes to this rock from another source. Sample Atx 9065 contains a higher proportion of extractable organic matter (EOM/TOC = 16%) and a more uniform distribution of saturated hydrocarbons than sample Atx 9038 (EOM/TOC = 2%). These differences are probably due to addition of migrated saturated hydrocarbons to sample Atx 9065 and expulsion from sample Atx 9038.

maturity of the rock, by the nature of the organic matter and its early diagenesis, and by the partitioning effects of migration. Similar conclusions were reached by Huc and Hunt (1980). Clay catalysis may be important in the generation of gas during later stages of catagenesis, but it appears to have had little influence on the formation of petroleum in the Austin Chalk.

Migration of Hydrocarbons

Migration of petroleum from carbonate source rocks has been considered problematical (Momper, 1980). One reason for this skepticism is the minor importance of physical compaction in carbonates during deep burial (Wilson, 1975). Physical compaction in shales is thought to provide a driving force for fluid migration (Tissot and Welte, 1978), and the absence of this driving force in carbonates limits primary migration by physical compaction to argillaceous sediments. However, two facts have commonly been ignored: (1) fluids probably are expelled from carbonates

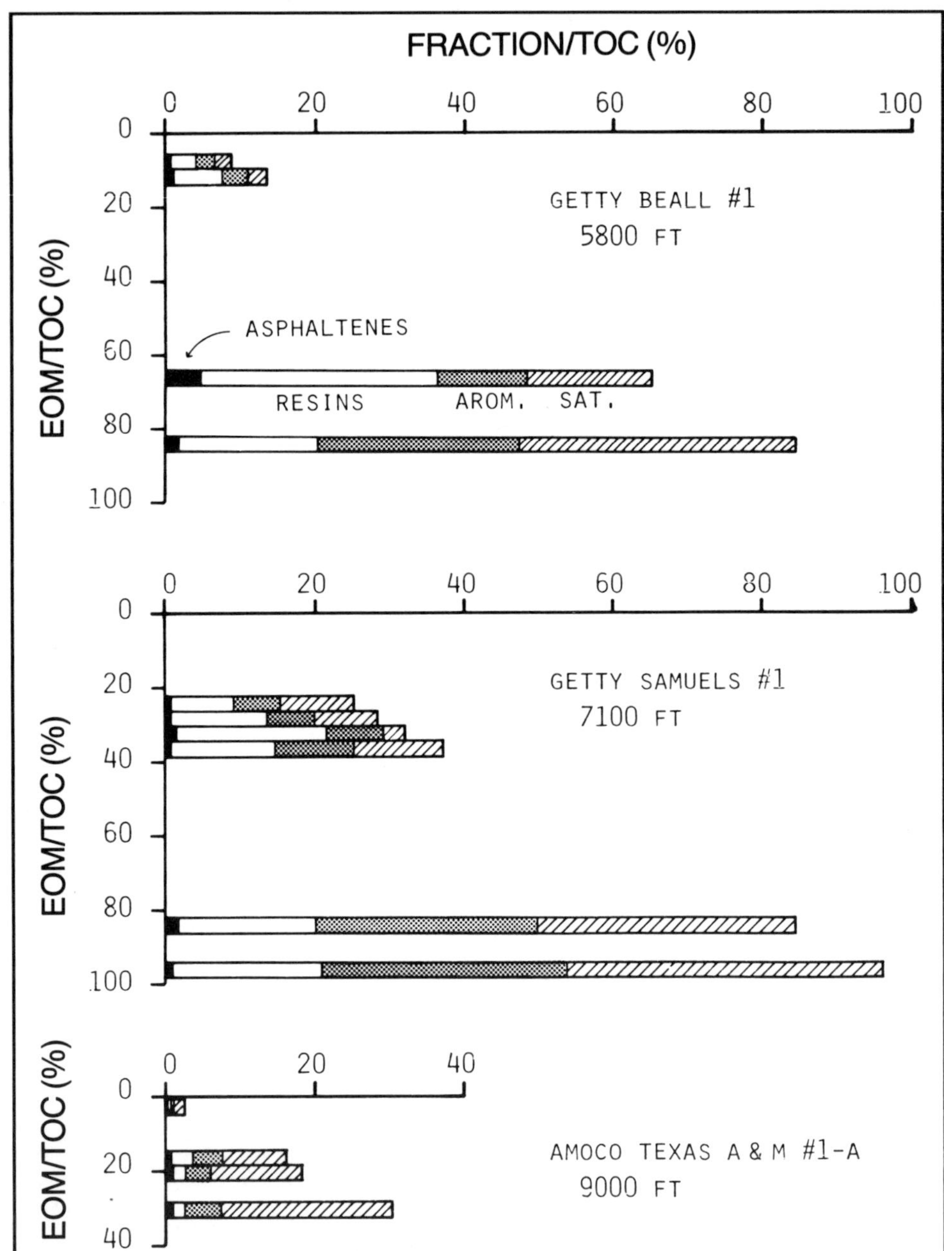

Figure 10 — Plots of ratio of each fraction of extractable organic matter (EOM) to organic carbon (in %) versus the ratio of EOM to TOC (in %) for three cores of Austin Chalk. In each core (average depth of core in ft) the absolute amount of asphaltenes (black portion, left part of bar) remains nearly constant, whereas the amounts of the other fractions (resins = white portion of bar; aromatic hydrocarbons = crosshatched portion of bar; saturated hydrocarbons = diagonally ruled portion of bar) increase with increasing proportion of EOM owing to addition of migrated EOM to the rock. The extent of this increase depends on the composition of EOM and maturity of the organic matter in each core, which are related in the Austin Chalk primarily with depth of burial.

during deep burial by pressure solution, a form of chemical compaction; and (2) fine-grained organic-rich carbonates generally contain micropores that can act as migration pathways and as reservoirs for petroleum. These two facts characterize the Austin Chalk in the subsurface of south-central Texas.

EOM/TOC (in %) and the bulk composition of EOM vary within each core studied, with the largest differences occurring in cores deeper than 5,000 ft (1,525 m). EOM/TOC (in %) is large (approaching 100%) in some organic-poor chalks, whereas it is small in some organic-rich, microstylolitic chalks. Since the composition and alteration of the organic matter are generally uniform within each core, differences in the abundance and composition of EOM are probably the result of migration of EOM.

The amounts of EOM that have migrated in each core appear to increase from the shallow, immature cores to the deeper (>5,000 ft or 1,525 m), more mature cores, as indicated by the larger range of values plotted in Figure 5. The extent of migration that occurred in the Austin Chalk appears to be related to the amount of mobile compounds generated, which increases with increasing depth of burial.

The relationship between extent of migration and availability of mobile compounds in the rocks is more clearly shown in Figure 10. In these three cases, the amount of each fraction of EOM per TOC (in %) is shown for each sample, arranged in order of increasing EOM/TOC. The amount of asphaltenes remains approximately the same in each case, whereas the amount of resins and hydrocarbons increases with increasing amounts of EOM/TOC. In other words,

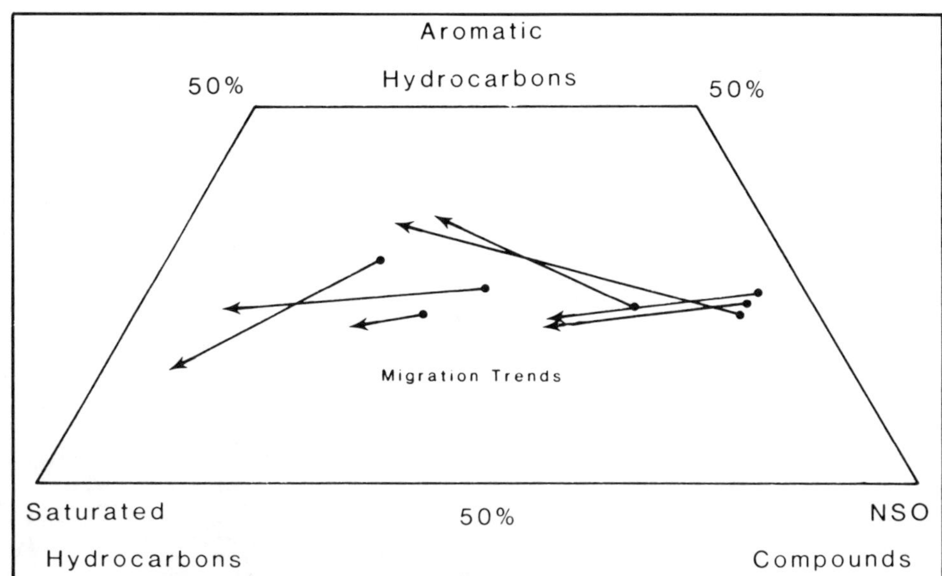

Figure 11 — Triangular plot of saturated and aromatic hydrocarbons and NSO compounds, showing trends owing to migration observed in each core of Austin Chalk. The dot is placed at the composition of the sample containing the lowest values of EOM/TOC; the tip of the arrow is placed at the composition of the sample containing the highest ratio of EOM/TOC.

the smaller, more mobile resins and hydrocarbons are responsible for the increase in the EOM/TOC ratio. The large asphaltene molecules do not appear to migrate readily in the Austin Chalk.

The changes in composition of EOM with increasing amounts of EOM/TOC (in %) are different in each of the three cores depicted in Figure 10. The Amoco 1-A Texas A&M core (9,000 ft or 2,745 m) is most enriched in saturated hydrocarbons compared with the other cores. Conversely, the Getty 1 Beall core (5,800 ft or 1,770 m) is most enriched in resins and asphaltenes. These changes correspond to the maturity and composition of the EOM in the cores: the more mature chalk in the deepest core contains EOM enriched in saturated hydrocarbons owing to preferential formation of these compounds, and therefore more saturated hydrocarbons are available for migration in this core. The shallower cores contain a less mature EOM with a larger proportion of NSO compounds and aromatic hydrocarbons available for migration. Hence, the compositional changes that occur during migration are strongly controlled by the mobile phases available for migration in the rock.

The compositions of saturated hydrocarbons do not change greatly with migration in the Austin Chalk. The gas chromatograms in Figure 9 show a "smoothing out" of the n-alkane distributions and relative depletions in cycloalkanes, similar to the observations of Vandenbroucke (1972).

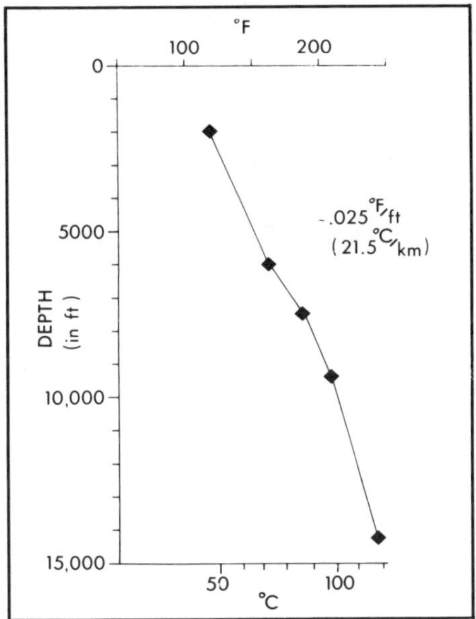

Figure 12 — Plot of core-hole temperature (as provided on logs) versus depth in the Amoco 1-A Texas A&M well. At the geothermal gradient indicated (about 21°C/km), maximum oil generation observed in this study occurs at about 85°C or 185°F (8,000 ft or 2,440 m).

Two lines of evidence suggest that most of the migration observed is local within the Austin Chalk. The carbon-isotopic ratios of individual fractions of EOM show essentially no change in samples with large values of EOM/TOC (Table 5, Gb 5691 and Gs 7022.22). The ratios of EOM and stable-carbon isotopes suggest a common source for the EOM in the Austin Chalk and Eagle Ford Formation. Furthermore, "reservoir" intervals (high EOM/TOC) occur between "source" intervals (low EOM/TOC) in the cores, within a distance of 2 to 30 ft (1–10 m). The presence of atypical distributions of saturated hydrocarbons in some samples suggests that some EOM also has come from outside the Austin Chalk and Eagle Ford Formation.

Similar changes in the composition of migrated crude oil have been attributed to adsorption of polar molecules to clay minerals and to the exclusion of large molecules from narrow pores along the migration pathway (Hunt and Jamieson, 1956; Bray and Evans, 1965; Tissot and Pelet, 1971; Vandenbroucke, 1972). However, migration occurred in the Austin Chalk and resulted in some changes in composition of EOM regardless of content of acid-soluble

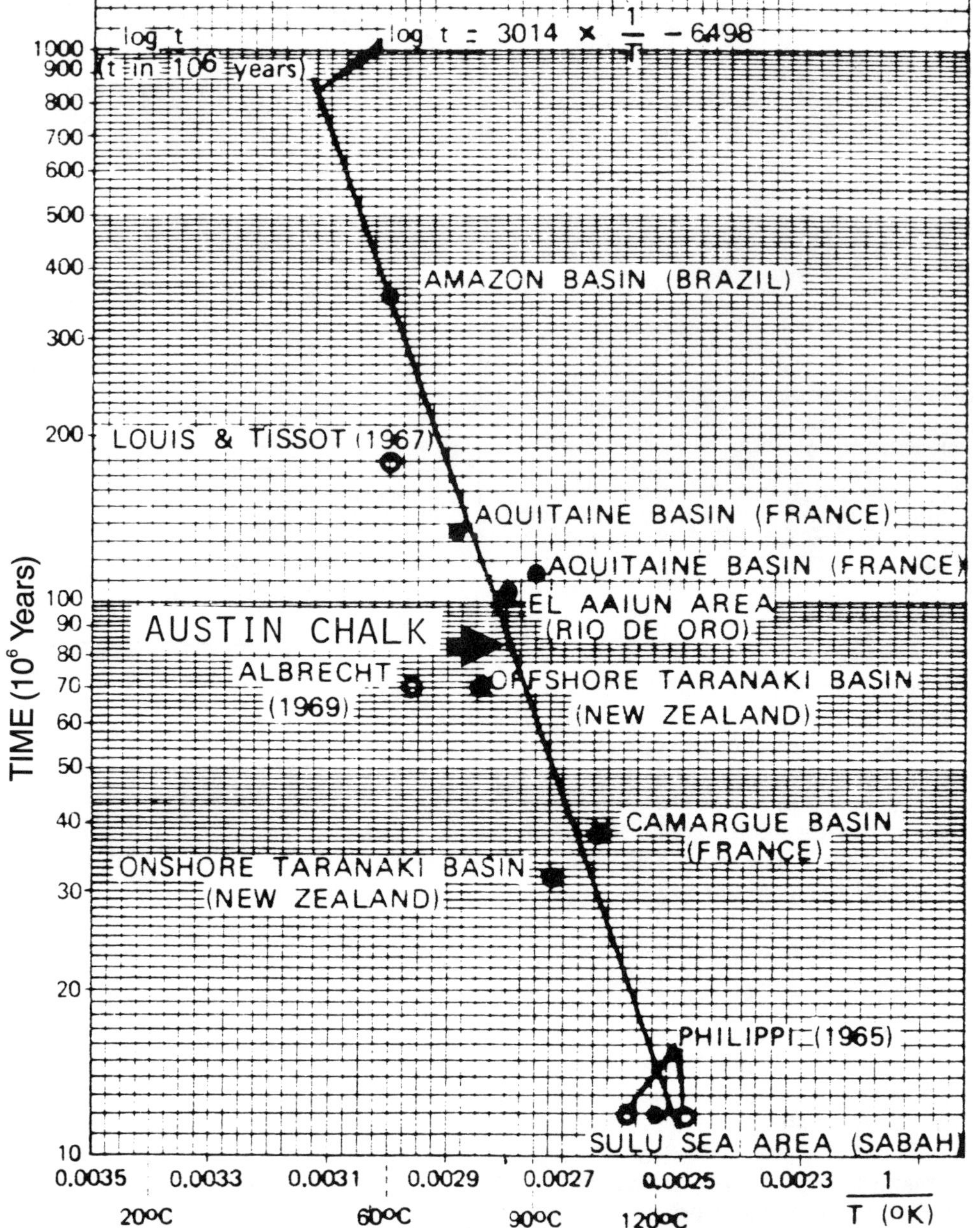

Figure 13 — Plot of time versus temperature of peak oil generation for many source rocks (from Connan, 1974, Fig. 3), on which position of Austin Chalk is plotted. Note that peak oil generation determined for Austin Chalk falls along line connecting points for other rock units, regardless of lithology.

(carbonate) material or polar minerals in the rock.

Most of the crude oils produced from the Austin Chalk have API gravities between 30° and 39°, as shown in Figure 14 (Scholle, 1977; Grabowski, 1981). These oils (called Type A in Figs. 14 and 15) have saturated/aromatic hydrocarbon ratios (Sat/Arom) between 1 and 4 and pristane/phytane ratios (Pr/Ph) less than 2, similar to most of the EOM analyzed in this study and probably derived from the Austin Chalk and Eagle Ford Formation. Lighter crude oils (Types B and C) are produced from many wells in the Austin Chalk, typically at greater depths of burial. They have larger Sat/Arom ratios (between 5 and 6) and generally higher Pr/Ph ratios (about 2) than the Type-A crude oil. The Type-B crude oils in particular are similar in composition to the "migrated" EOM in chalks with large values of EOM/TOC. These lighter crude oils may result from further maturation of the Austin Chalk and Eagle Ford Formation at greater depths of burial or from migration and enrichment of the rocks in EOM.

Some of the wells in the Austin Chalk produce a heavy crude oil (Type D), which is depleted in saturated hydrocarbons. These heavy oils are restricted to shallow depths of burial and either are extremely immature or biodegraded.

CONCLUSIONS

This study of the organic geochemistry of the Austin Chalk bears on three aspects of fine-grained organic-rich carbonate rocks: (1) the distribution of organic matter, (2) the effects of minerals on kerogen alteration and hydrocarbon generation, and (3) the controls on migration and its effects on composition of EOM.

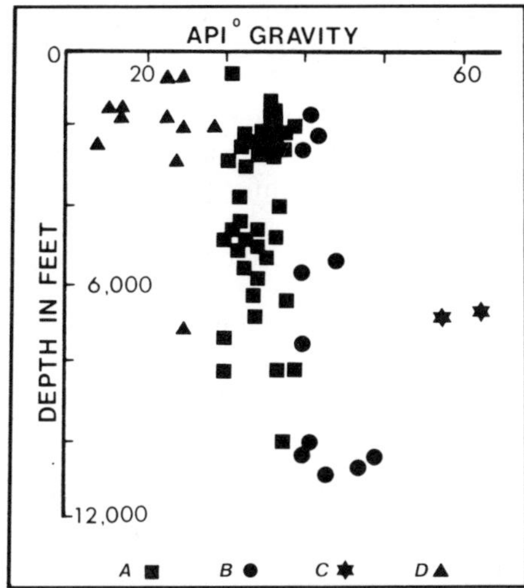

Figure 14 — Plot of API gravity of crude oils versus depth of producing intervals of Austin Chalk (data from Scholle, 1977). A = 30°–39° API; B = 40°–49° API; C = >50° API; D = <30° API.

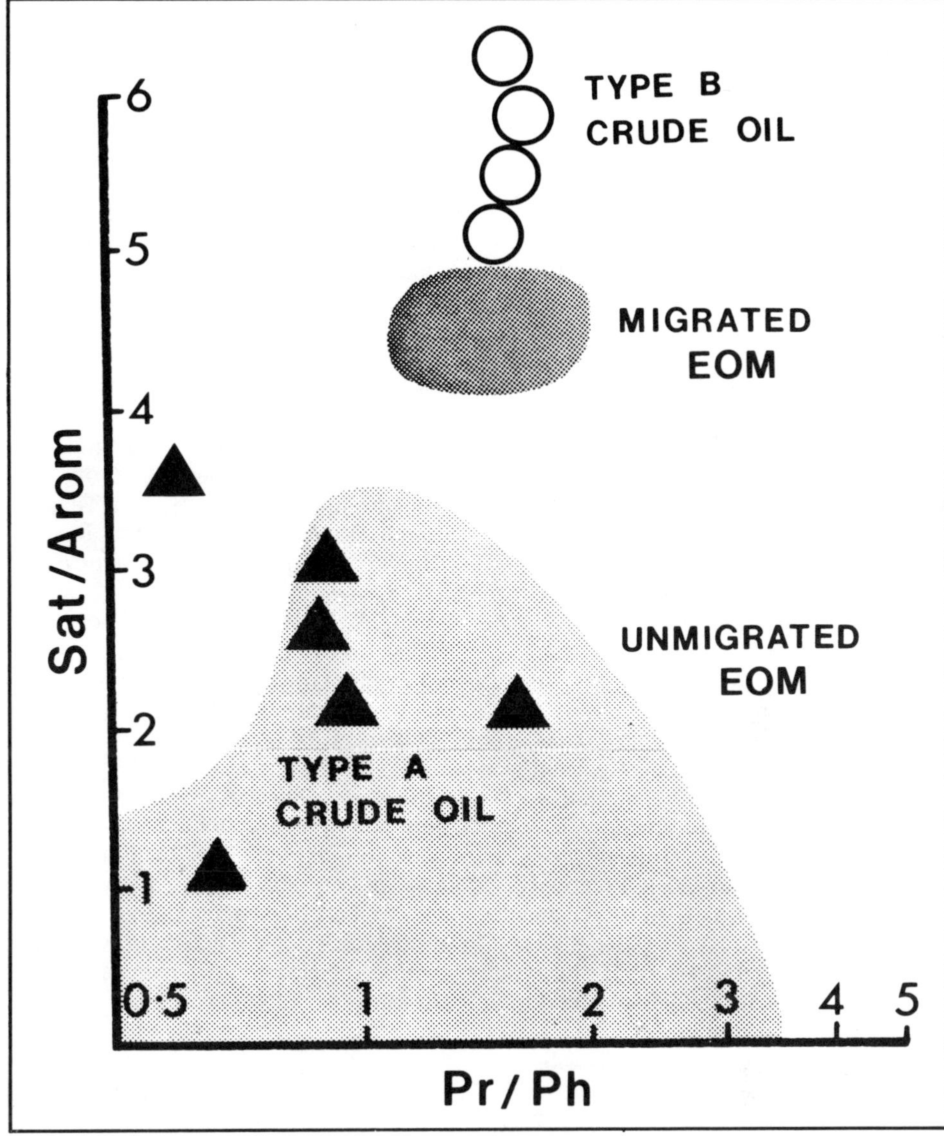

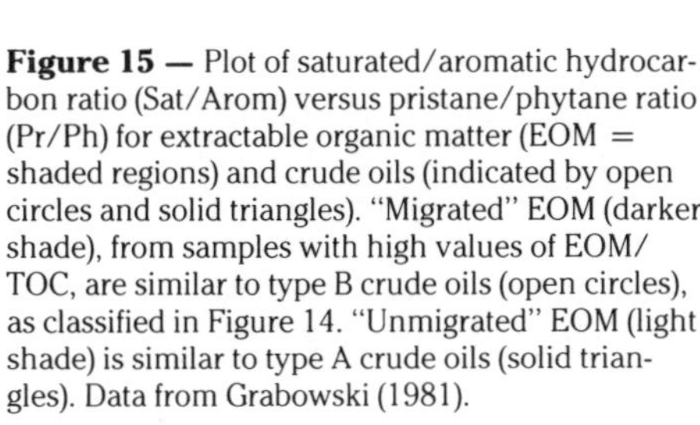
Figure 15 — Plot of saturated/aromatic hydrocarbon ratio (Sat/Arom) versus pristane/phytane ratio (Pr/Ph) for extractable organic matter (EOM = shaded regions) and crude oils (indicated by open circles and solid triangles). "Migrated" EOM (darker shade), from samples with high values of EOM/TOC, are similar to type B crude oils (open circles), as classified in Figure 14. "Unmigrated" EOM (light shade) is similar to type A crude oils (solid triangles). Data from Grabowski (1981).

Distribution of organic matter is a function of both depositional and diagenetic processes in the Austin Chalk. More organic matter is preserved in deeper, more basinal cores, which represent less aerobic depositional environments than the shallow, more shelfal cores. Kerogen occurs disseminated in the chalks and shales and concentrated in stylolites and microstylolites. These zones of pressure solution have less porosity than the surrounding chalks, and EOM has migrated into the porous chalks. A heterogeneous distribution of organic matter occurs in each core, owing to the irregular distribution of stylolitic and porous chalks.

Alteration of kerogen and NSO compounds and the generation of hydrocarbons in the Austin Chalk are similar to the processes reported for shales. Smaller compounds (hydrocarbons and resins) are formed from larger compounds (NSO compounds and kerogen) during catagenesis, with the larger compounds becoming depleted in hydrogen-rich functional groups. Large amounts of EOM are formed between 5,000 and 8,000 ft (1,525 and 2,440 m), whereas light hydrocarbons ($<C_{15}$) are more abundant at 9,000 ft (2,740 m). The extent of kerogen alteration and the amounts and compositions of EOM formed are unrelated to the amount of acid-soluble (carbonate) content in the Austin Chalk. Clay catalysis is not involved in the formation of petroleum in the Austin Chalk, which occurs at temperatures and times equivalent to other source rocks.

Large variations in the amounts of EOM in samples from the same cores indicate that EOM has migrated in the Austin Chalk. Some porous and fractured chalks contain higher EOM/TOC ratios than the surrounding organic-rich chalks. The EOM generally appears to have formed in the Austin–Eagle Ford interval and migrated into the porous reservoir lithologies. The EOM becomes enriched in hydrocarbons and resins during migration, with no relationship between the amount of acid-soluble material and the composition of EOM. The light crude oils (API >39°) produced from the Austin Chalk, at generally greater depths of burial, may be due to extreme maturation of organic matter or to migration of petroleum, including indigenous crude oils (API between 30° and 39°), within the chalk.

ACKNOWLEDGMENTS

This paper is based on a doctoral dissertation submitted to Rice University. I thank D. R. Baker and the other members of my examining committee for their guidance and support during the research. Suggestions and criticism were provided by J. G. Palacas, J. J. Dravis, J. P. Shannon, Jr., W. A. Young, E. E. Bray, W. R. Foster, and J. T. Smith during this work. A. B. Reaugh, M. Malek–Aslani, and J. M. Hunt provided assistance or analyses for part of this research. Duplicate analyses of organic carbon were provided by Exxon Production Research Co., and Conoco Research, Inc. Cores of the Austin Chalk and Eagle Ford Formation were provided by the Texas Bureau of Economic Geology, Getty Oil Co., Amoco Production Co., and Transocean Oil, Inc. Amoco Production Co. provided information on crude oil produced from the Austin Chalk. Preliminary versions of this paper were reviewed and greatly improved by the suggestions of J. J. Dravis, J. G. Palacas, V. E. Swanson, B. E. Torkelson, D. D. Rice, and one unidentified reviewer.

This research was supported by U.S. Department of Energy Grant no. DE-AS05-80ER10764, U.S. Geological Survey Grant no. 14-08-0001-G0649, Geological Society of America Grant no. 2454-79, and a grant from the Gulf Coast Association of Geological Societies. Amoco Production Co. and Chevron Oil Co. provided funds for this research, and Gulf Oil Foundation provided the author with a fellowship during the final year at Rice. The financial support of all of these organizations is gratefully acknowledged.

REFERENCES CITED

Barrier, J., 1980, A revision of the stratigraphic distribution of some Cretaceous coccoliths in Texas: Journal of Paleontology, v. 54, p. 289–308.

Bray, E.E., and E.D. Evans, 1965, Hydrocarbons in nonreservoir-rock source beds: pt. 1: AAPG Bulletin, v. 49, p. 248–257.

Clayton, J.L., and P.J. Swetland, 1980, Petroleum generation and migration in Denver Basin: AAPG Bulletin, v. 64, p. 1613–1633.

Connan, J., 1974, Time-temperature relation in oil genesis: AAPG Bulletin, v. 58, p. 2516–2521.

Davis, J.C., 1970, Petrology of Cretaceous Mowry shales of Wyoming: AAPG Bulletin, v. 54, p. 487–502.

Deroo, G., et al, 1977, The origin and migration of petroleum in the Western Canadian sedimentary basin, Alberta: Geological Survey of Canada Bulletin, v. 262, 136 p.

Dravis, J.J., 1979, Sedimentology and diagenesis of the Upper Cretaceous Austin Chalk Formation, south Texas and northern Mexico: Houston, Rice University unpublished Ph.D. thesis, 513 p.

Espitalié, J., M. Madec, and B. Tissot, 1980, Role of mineral matrix in kerogen pyrolysis: Influence on petroleum generation and migration: AAPG Bulletin, v. 64, p. 59–66.

Espitalié, J., et al, 1977, Méthode rapide de caractérisation des roches mères de leur potential pétrolier et de leur degré d'évolution: Revue de l'Institut Francais du Pétrole, v. 32, p. 23–42.

Fisher, W.L., 1969, Facies characterization of Gulf Coast Basin delta systems, with some Holocene analogues: Gulf Coast Association of Geological Societies Transactions, v. 19, p. 239–261.

———, et al, 1970, Depositional systems in the Jackson Group of Texas—their relationship to oil, gas and uranium: Gulf Coast Association of Geological Societies Transactions, v. 20, p. 234–261.

Grabowski, G.J., Jr., 1981, Origin, distribution and alteration of organic matter and generation and migration of hydrocarbons in Austin Chalk, Upper Cretaceous, Southeastern Texas: Houston, Rice University unpublished Ph.D. thesis, 262 p.

Horsfield, B., and A.G. Douglas, 1980, The influence of minerals on the pyrolysis of kerogens: Geochimica et Cosmochimica Acta, v. 44, p. 1119–1131.

Huc, A.Y., and J.M. Hunt, 1980, Generation and migration of hydrocarbons in offshore south Texas Gulf Coast sediments: Geochimica et Cosmochimica Acta, v. 44, p. 1081–1089.

Hunt, J.M., 1967, The origin of petroleum in carbonate rocks, *in* G.V. Chilingar, H.J. Bissell, and R.W. Fairbridge, eds., Carbonate rocks: New York, Elsevier, p. 225–251.

———, 1984, Carbonates as petroleum source rocks, *in* J.G. Palacas, ed., Petroleum geochemistry and source-rock potential of carbonate rocks: AAPG Studies in Geology 18, this volume.

Hunt, J.M., and G.W. Jamieson, 1956, Oil and organic matter in source rocks of petroleum: AAPG Bulletin, v. 40, p. 477–488.

———, F. Stewart, and P.A. Dickey, 1954, Origin of hydrocarbons of Uinta basin, Utah: AAPG Bulletin, v. 38, p. 1671-1698.

Kauffman, E.G., 1979, Cretaceous marine cycles of the western interior: Mountain Geologist, v. 6, p. 227–245.

Kirkland, D.W., and R. Evans, 1981, Source-rock potential of evaporitic environments: AAPG Bulletin, v. 65, p. 181–190.

Laplante, R.E., 1974, Hydrocarbon generation in Gulf Coast Tertiary sediments: AAPG Bulletin, v. 58, p. 1281–1289.

Malek–Aslani, M., 1980, Environmental and diagenetic control of carbonate source rocks (abs.): AAPG Bulletin, v. 64, p. 744–745.

McIver, R.D., 1967, Composition of kerogen—clue to its role in the origin of petroleum: Seventh World Petroleum Congress Proceedings, v. 2, p. 25–36.

Momper, J.A., 1980, Generation of abnormal pressures through organic matter transformation (abs.): AAPG Bulletin, v. 64, p. 753.

Oehler, J.H., 1984, Carbonate source rocks in the Jurassic Smackover trend of Mississippi, Alabama, and Florida, *in* J.G. Palacas, ed., Petroleum geochemistry and source-rock potential of carbonate rocks: AAPG Studies in Geology 18, this volume.

Potter, P.E., J.B. Maynard, and W.A. Pryor, 1980, Sedimentology of shale: New York, Springer–Verlag, 306 p.

Reed, W.E., and W. Henderson, 1972, Proposed stratigraphic controls on the composition of crude oils reservoired in the Green River Formation, Uinta Basin, Utah, *in* H.R.v. Gaertner and H. Wehner, eds., Advances in organic geochemistry 1971: Oxford–Braunschweig, Pergamon, p. 499–515.

Rice, D.D., 1984, Occurrence of indigenous biogenic gas in organic-rich, immature chalks of Late Cretaceous age, eastern Denver basin, *in* J.G. Palacas, ed., Petroleum geochemistry and source-rock potential of carbonate rocks: AAPG Studies in Geology 18, this volume.

Robin, P.L., P.G. Rouxhet, and B. Durand, 1977, Caractérisation des kérogénes et de leur évolution par spectroscopie infrarouge: Fonctions hydrocarbonées, *in* R. Campos and J. Goni, eds., Advances in organic geochemistry, 1975: Madrid, ENADIMSA, p. 693-716.

Scholle, P.A., 1977, Current oil and gas production from North American Upper Cretaceous chalks: U.S. Geological Survey Circular 767, 51 p.

Schrayer, G.J., and W.M. Zarella, 1963, Organic geochemistry of shales: I. Distribution of organic matter in siliceous Mowry Shale of Wyoming: Geochimica et Cosmochimica Acta, v. 27, p. 1033–1046.

——— and ———, 1966, Organic geochemistry of shales: II. Distribution of extractable organic matter in siliceous Mowry Shale of Wyoming: Geochimica et Cosmochimica Acta, v. 30, p. 415–434.

Tissot, B., and R. Pelet, 1971, Nouvelles donnés sur les méchanismes de genése et de migration du pétrole simulation mathematique et application a là prospection: Eighth World Petroleum Congress Proceedings, v. 2, p. 35–46.

Tissot, B., and D.H. Welte, 1978, Petroleum formation and occurrence: Berlin, Springer–Verlag, 538 p.

Tissot, B., G. Deroo, and A. Hood, 1978, Geochemical study of the Uinta Basin: formation of petroleum from the Green River formation: Geochimica et Cosmochimica Acta, v. 42, p. 1469–1485.

Tissot, B., et al, 1971, Origin and evolution of hydrocarbons in early Toarcian shales: AAPG Bulletin, v. 55, p. 2177–2193.

———, et al, 1974, Influence of the nature and diagenesis of organic matter in formation of petroleum: AAPG Bulletin, v. 58, p. 499–506.

Vail, P.R., R.M. Mitchum, Jr., and S. Thompson, III, 1977, Seismic stratigraphy and global changes of sea level, Part 4: Global cycles of relative changes of sea level, *in* C.E. Payton, ed., Seismic stratigraphy—applications to hydrocarbon exploration: AAPG Memoir 26, p. 83–97.

Vandenbroucke, M., 1972, Étude de la migration primaire: variation de composition des extraits de roche à un passage roche mère/réservoir, *in* H.R.v. Gaertner and H. Wehner, eds., Advances in organic geochemistry 1971: Oxford–Braunschweig, Pergamon, p. 547–565.

Williams, G.D., and C.R. Stelck, 1973, Speculations on the Cretaceous paleogeography of North America: Geological Association of Canada Special Paper, v. 13, p. 1–20.

Wilson, J.L., 1975, Carbonate facies in geologic history: New York, Springer–Verlag, 471 p.

The Cretaceous Austin Chalk of South Texas— A Petroleum Source Rock

John M. Hunt and Ann P. McNichol
Woods Hole Oceanographic Institution, Woods Hole, Massachusetts

Carbonate rocks commonly have been discounted as important source rocks because of their lower organic-carbon content and lower catalytic activity in comparison to shales. However, carbonate source rocks contain mostly sapropelic organic matter, which yields a higher percentage of oil earlier than the more humic organic matter of shales. Furthermore, carbonate source–reservoir sequences are at many places overlain by the perfect seal, evaporite, during the time of oil generation and accumulation. In contrast, many sandstone–shale sequences tend to leak petroleum during and after accumulation.

The richest source rocks in the world are the argillaceous and siliceous carbonates in formations such as the Green River of Utah, the La Luna of Venezuela, and the Nordegg of the Western Canada basin.

Carbonate rocks like the Cretaceous Austin Chalk of south Texas, which contains 60–90% $CaCO_3$, act as both source and reservoir rock. Light-hydrocarbon analyses and pyrolysis data both support the concept that most of the oil in the Austin Chalk is autochthonous.

INTRODUCTION

It has long been known that the organic matter of carbonate rocks yields a higher percentage of hydrocarbons on maturation than does that of shale sequences (Hunt, 1961; Baker, 1962). This difference in hydrocarbon yield is because the organic matter in carbonates is mostly of the sapropelic fluorescent amorphous-algal type, whereas that in shales contains a high proportion of humic material. The fresh-water or marine algae that are dominant in most carbonates yield more than twice as much oil as land-derived organic matter (Hunt, 1979, p. 274–278). The vitrinite-reflectance determination, which is a popular method for monitoring maturation, frequently gives erroneous data or no data in carbonates because of the lack of true vitrinite particles. The Eocene Green River Formation of Colorado and Wyoming and the Cretaceous La Luna Formation of Venezuela and Colombia contain large amounts of amorphous-algal kerogen.

The most prolific hydrocarbon sources are the carbonate rocks with varying amounts of either clay minerals or fine-grained quartz or both. About three-fourths of the source rocks of the Uinta basin hydrocarbons contain no clay minerals, but they do contain 20–60% fine-grained quartz (Hunt, Stewart, and Dickey, 1954). These impure carbonate rocks generally contain 10 or more times as much organic matter as pure limestone and dolomite. Because the organic matter is largely of the algal type, the hydrocarbon yield increases with organic-carbon content. For example, a sample of the argillaceous Duvernay Limestone of Alberta yielded 3,300 ppm hydrocarbon with about 8% kerogen, whereas the Ireton Limestone from the same area yielded 106 ppm hydrocarbon with 0.28% kerogen.

Some petroleum geochemists have believed that clay minerals or fine quartz must be present to catalyze the thermal breakdown of kerogen to form oil (Brooks, 1948; Goldstein, 1983). This has resulted in the identification of underlying or overlying shales as the source rocks of petroleum in carbonates. There is growing evidence, however, that carbonate rocks with little or no clay material, such as the Austin Chalk, do have a catalytic activity that does not differ greatly from shales. This has been attributed to oxides of metals, such as iron and aluminum, that are impurities in the carbonates (Hunt, 1979, p. 128).

It is generally accepted among petroleum geochemists that pore-water movement is involved in hydrocarbon migration, particularly for the lighter hydrocarbons (Magara, 1978, p. 277; Hunt, 1979, p. 208; Bonham, 1980; Bray and Foster, 1980; Hitchon, 1980). An argument used against carbonate source rocks is that they do not compact like shales and therefore may not provide the fluids needed for hydrocarbon migration. However, there are data that show a great deal of fluid movement through carbonates (Polak, 1954; McCrossan, 1961; Scholle, 1975). Chalk beds, for example, show compaction with increasing overburden. Carbonate rocks fracture more easily than shales, and some evidence shows that extensive microfracturing develops in carbonates, thereby providing both vertical and horizontal pathways for fluid movement. A stylolitic limestone may have lost 20–40% of its original rock volume during burial (Dunnington, 1967). Generally, organic matter becomes concentrated along stylolites, which represent avenues of migration.

Possibly the most favorable aspect of carbonates as source rocks is the common presence of evaporite seals. In sandstone–shale sequences, particularly stratigraphic traps, leakage of hydrocarbons is common, and the existence of a commercial accumulation of oil is largely determined by the relative rates of hydrocarbon leakage versus entrapment. It has been estimated that only 20–40% of migrated oil in sandstone–shale sequences is trapped (McDowell, 1975; J. A. Momper, personal communication, 1983). In some carbonate–evaporite sequences, there could be more than 50% of the migrating oil trapped. The huge petroleum reserves of the Middle East are partly a result of the multiplicity of evaporite layers in the Jurassic source–reservoir sequences.

The extractable organic matter of carbonate rocks, after burial to depths sufficient for oil generation, appears to have a lower ratio of nitrogen–sulfur–oxygen (NSO) compounds to aromatic hydrocar-

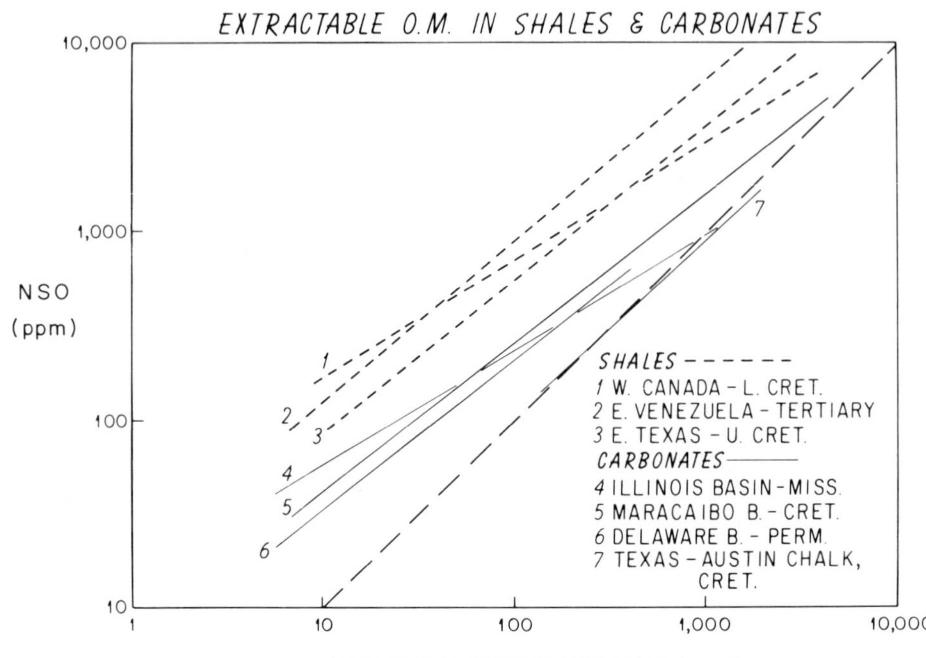

Figure 1 — Ratio of nitrogen, sulfur, and oxygen (NSO) compounds to aromatic hydrocarbons in the bitumen of shales and carbonates. Ratio equals 1 on heavy dashed line.

bons than does the extract of shales (Fig. 1). This may be partly because the marine organic matter of carbonates has a somewhat higher content of naphthenes than does the land-derived organic matter of shales. Partial conversion of the naphthenes to aromatics with increasing time and temperature could result in proportionally more aromatics in carbonate source rocks than in shales, as others have observed (Tissot and Pelet, 1971). The extractable organic matter of carbonates commonly contains about the same quantity of NSO's relative to aromatic hydrocarbons, whereas in shales the ratio may be as high as 10 (Fig. 1).

The extractable organic matter of carbonate rocks typically shows an even-carbon-number predominance in their straight-chain n-alkanes as compared to the odd-carbon-number predominance recognized in land-derived material. For example, Welte and Waples (1973) observed an even-carbon-number predominance in the n-C_{20} to n-C_{30} range in extracts of carbonate rocks. In a few cases, the even predominance has been observed in crude oils derived from carbonate rocks (Tissot et al, 1977).

Probably the earliest confirmation of a carbonate source rock by geochemical analyses was made by Hedberg (1931), who carried out a combined geological and geochemical study of the La Luna and Cogollo Limestones of Cretaceous age in northwestern Venezuela. The La Luna produces a medium-gravity oil from fractured limestone reservoirs that include the upper part of the underlying Cogollo Limestone. The La Luna is a facies that grades from calcareous shale to argillaceous limestone. Its carbonate content ranges from 50 to 95%, with the noncarbonate constituents consisting of organic matter, chert, and shale. The La Luna is of Turonian, or Late Cretaceous, age and is correlative with some of the black Cretaceous shales found in the eastern Atlantic Ocean by the Deep Sea Drilling Project (Tissot et al, 1980). The La Luna Limestone extends from Trinidad to Venezuela.

The La Luna was deposited as a *Globigerina* ooze under strongly reducing conditions; benthic fossils are completely absent. In contrast, much of the Cogollo Limestone, which contains sand and coarse clastics, was deposited in an oxidizing high-energy environment.

The organic-carbon content of the La Luna ranges up to 10%, and more than 10,000 ppm bitumen is extractable with carbon tetrachloride. Figure 2 shows the yield of bitumen versus organic carbon for the La Luna as compared to other carbonates. The average shale yields less than 10% of its organic carbon as extractable bitumen, whereas the average carbonate yields more than 20% (Hunt, 1979, p. 266). Data for all but one of the La Luna samples fall between shales and carbonates. If more than 50% of the organic carbon is extractable as bitumen, the presence of migrated hydrocarbons is indicated, which is the case for the one La Luna sample. Carbonate rocks commonly act as both source and reservoir, so it is not always simple to distinguish indigenous from migrated hydrocarbons.

Hedberg (1931) noted from microscopic examination of thin sections and slabs that the organic matter and the material extracted by chloroform in the La Luna was diffused throughout the calcareous sediment. In contrast, the Cogollo Limestone yielded bitumen only from fractures and microfractures; it was not uniformly diffused throughout the rock. Both the La Luna and Cogollo produce oil from fractured reservoirs, but oil in the Cogollo is limited to the upper part of the formation where it is in contact with the La Luna. Hedberg concluded that the La Luna Limestone was the major source of oil in the Cretaceous sediments of northwestern Venezuela.

A carbonate source rock more recently investigated is the Austin Chalk of south Texas. This chalk was deposited on a ramp marginal to the Gulf of Mexico during the Late Cretaceous Epoch. It is an organic-rich chalk containing up to 5% by weight of sapropelic, algal, marine organic matter. In the past few years, it

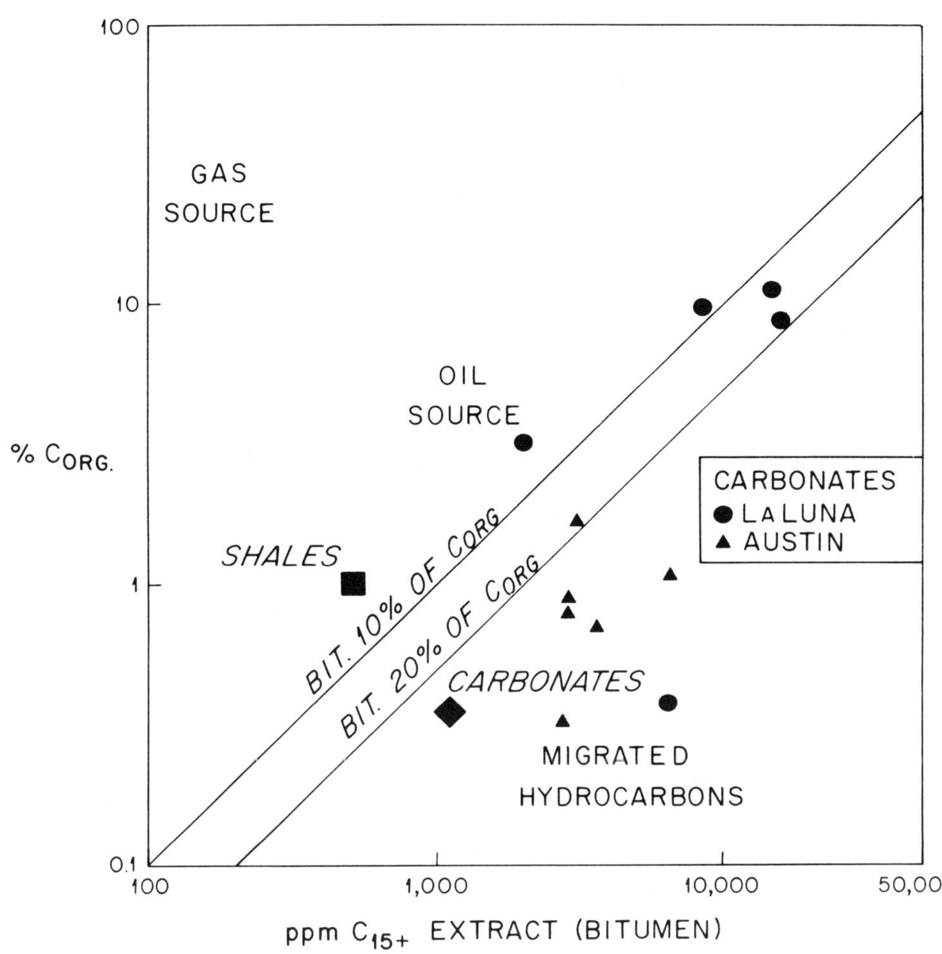

Figure 2 — Organic carbon versus extractable organic matter of shales and carbonates. Largest symbols are averages for about 800 shales and 300 carbonates. Data for La Luna samples from Hedberg (1931), for Austin Chalk, this report.

has been the target of exploration; a relatively light oil is produced from fractured reservoirs throughout the chalk.

A detailed study of the organic matter in this rock and its ability to generate hydrocarbons was made by Grabowski (1981). Cores of the Austin Chalk and the underlying Eagle Ford Formation were obtained from the University of Texas and several oil companies. These were analyzed by Grabowski for a series of organic and inorganic geochemical parameters. Tables of organic-carbon content, extraction data, kerogen analysis and Rock-Eval-pyrolysis data are in Grabowski's chapter in this volume. He also made some comparisons of the crude oils obtained from the Austin Chalk.

We obtained samples of this chalk from Grabowski for light-hydrocarbon analyses (C_1–C_{15}), using both a headspace and a thermal-distillation-pyrolysis technique. The objective of this study was to provide additional information concerning the source-rock potential of the Austin Chalk.

METHODS OF ANALYSIS

Hydrocarbons in the C_1 through C_8 range were determined by a headspace analysis as described by Whelan (1984). Basically, the technique involves placing 5 to 20 g of sediment in a stainless-steel vessel containing two stainless-steel ball bearings 4–8 mm in diameter. The vessel has a screw cap with a silicone rubber septum and gasket. Sediment and helium-bubbled distilled water are added to the vessel up to a calibration mark engraved on the inside, which leaves a 50-mL gas headspace after sealing. The vessel is placed in a glove bag; the bag then is evacuated and filled with helium so that the cap is screwed on in the helium atmosphere. The vessel is then removed from the bag and shaken vigorously for 5 minutes on a commercial paint shaker, followed by heating for 30 minutes in a 95°C (203°F) water bath. The headspace gas is then analyzed by gas chromatography (GC) in two stages.

The C_1–C_5 hydrocarbons are analyzed on a packed column, using OV-101, n-octane-Porasil C, or some other suitable GC substrate. Depending on the quantity of material detected in the packed column, a 1–15 mL sample of headspace gas is then analyzed by capillary GC for C_6–C_8 hydrocarbons, using a hexadecene/hexadecane/kel F (HHK) column as described by Schwartz and Brasseaux (1963). Compound identification is carried out by mass spectrometry.

The pyrolysis technique used was described by Whelan, Hunt, and Huc (1980). The analysis involves placing a weighed sample (10–30 mg) inside a 20-mm quartz tube with a 1-mm I.D. The sample is held in place by two plugs of quartz wool. The tube is then positioned in the platinum coil of a pyrolysis probe of a Reaction System 810 instrument (Chemical Data Systems). The probe containing the tube is placed in a cooled interface and heated rapidly in a helium stream to a temperature setting of 250°C

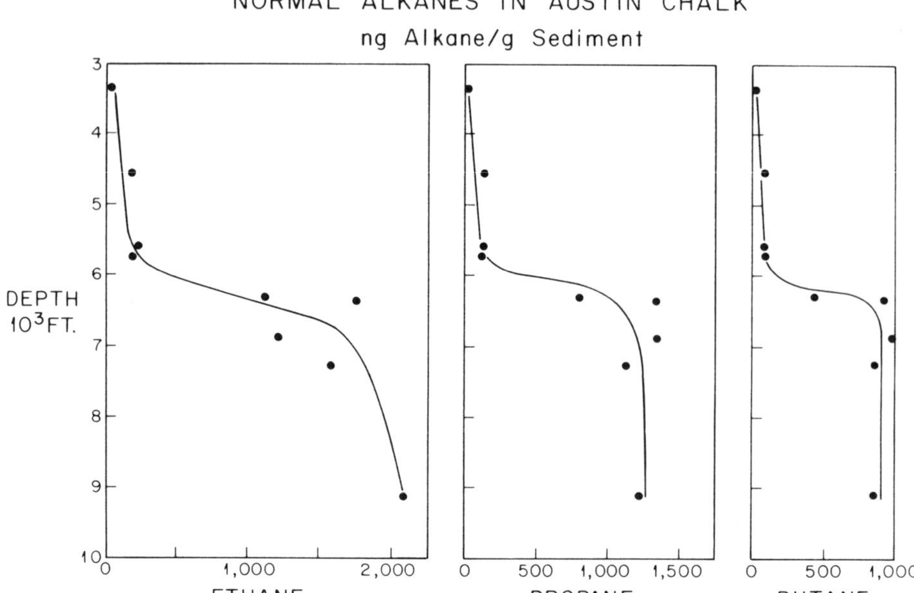

Figure 3 — Distribution of heavy gases in the Austin Chalk.

(482°F) for 5 minutes. An internal thermocouple reads the sample temperatures directly. After 5 minutes, the sample is heated from 250°C (482°F) to approximately 600°C (1,112°F) at a rate of 120°C/min (248°F/min). During both heating cycles the evolved hydrocarbons are swept out of the sample by the helium stream, which is split into two fractions. One fraction is sent to a flame-ionization detector. The resulting pyrogram is usually composed of two peaks, P_1, which contains the free hydrocarbons that can be thermally distilled ($t \leq 250°C$ [482°F]) out of the rock or clay, and the higher temperature P_2 peak, produced primarily from cracking products.

The second fraction of the split is collected in a stainless-steel trap packed with Tenax (60–80 mesh) at room temperature. The Tenax trap adsorbs hydrocarbons larger than C_7, while water is not retained. The trapped compounds are swept into loops cooled with liquid nitrogen in order to reduce sample volume before injection into a GC column. Peak areas of the chromatograms are measured with an electronic integrator. Individual hydrocarbons are initially identified by GC retention times and confirmed by combined gas chromatography–mass spectrometry (GC–MS).

Only cores of the Austin Chalk were analyzed. The samples were from 9 wells from depths of 3,351 to 9,095 ft (1,020 to 2,770 m). The wells extend over a distance of 180 mi (290 km), but both Grabowski's studies and ours indicate that the organic matter in the Austin Chalk is fairly uniform over this entire distance (for well locations, see Grabowski, 1981).

RESULTS AND DISCUSSION

The geothermal gradient in the study area is about 2.1°C/100 m (1°F/100 ft), and Grabowski (1981) estimated that the maximum burial of the chalk was about 9,840 ft (3,000 m). This indicates a maximum burial temperature of about 83°C (182°F), which would place this Cretaceous rock well within the oil window, based on the Lopatin time–temperature model (Hunt, 1979, p. 288). As will be shown later, the window extends from about 60° to 83° (140° to 182°F). The chalk is about 984 ft (300 m) thick in southeast Texas (Dravis, 1979). It apparently underwent solution compaction with periodic recrystallization, which caused degradation of some of the organic matter. This is one reason why conventional maturation techniques do not always yield definitive results. The organic matter in the Austin Chalk is distributed throughout the rock matrix, with some concentration along stylolites and microstylolites. For example, the color of the amorphous organic matter in the chalk changes from light to dark within a thickness of a meter or less. The sapropelic organic matter generally is several shades darker than the humic organic matter in the same sample, probably because of recrystallization (Grabowski, 1981). The samples had, at the most, only traces of vitrinite, so vitrinite reflectance could not be determined. By doing detailed studies on the less altered samples, Grabowski was able to observe a variation in the thermal-alteration index (TAI) from 1+ in the shallowest samples to 3 in the deepest. The TAI indicated that the Austin Chalk had passed through maturation stages equal to the coal ranks from lignite to low-volatile bituminous coal. This was confirmed by the atomic H/C ratio of the kerogen, which ranges from 1.4 in shallow samples to 0.6 in the deepest. The deepest samples have passed through the oil window and are in the beginning of the thermogenic-gas window. Additional confirmation was obtained by the temperature maximum of the second pyrolysis peak, which ranged from 399° to 422°C (750°–791°F). This T_{max} moves to higher temperatures as hydrocarbons are generated from the kerogen.

Grabowski's extraction data showed a large increase in hydrocarbons in the chalk at depths of 6,000–7,000 ft (1,800–2,100 m). Such a sharp increase with depth is typical of the beginning of an oil window (threshold of intense hydrocarbon generation). It would be expected that a more gradual change would have been observed if the hydrocarbons had been migrating from other source beds,

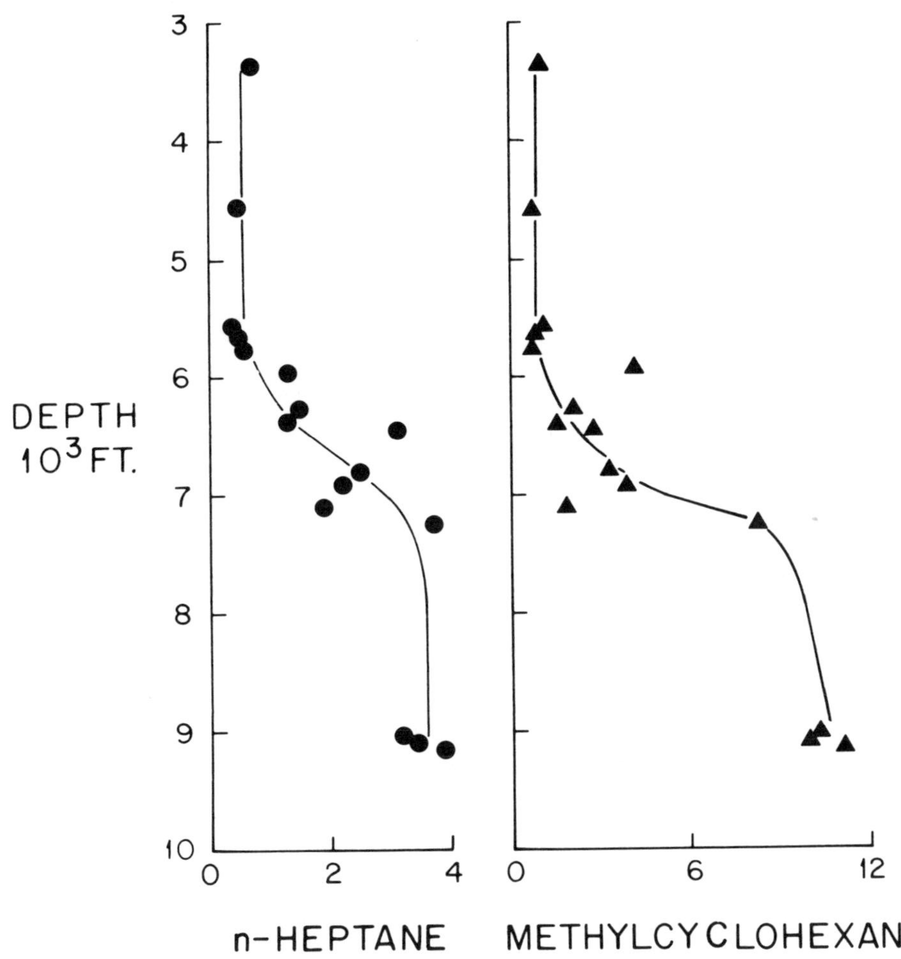

Figure 4 — Distribution of n-heptane and methylcyclohexane in the Austin Chalk.

such as the Eagle Ford Formation. Furthermore, Grabowski observed that the normal alkanes in some of the rock extracts had a strong even-carbon-number predominance in the C_{16} to C_{26} range. This would be characteristic of a carbonate source rock but not a shale (Welte and Waples, 1973). Grabowski concluded that much of the oil found in Austin Chalk reservoirs originated within the chalk, although he recognized that some oil in the lower part of the Austin Chalk possibly had migrated in from the underlying Eagle Ford. The results described below provide additional proof in the light-hydrocarbon distributions that the Austin Chalk is acting as a petroleum source rock.

Figure 3 shows the distribution of ethane, propane, and butane in Austin Chalk cores as determined by headspace analysis. The amount and distribution of these light hydrocarbons are particularly significant in distinguishing source versus migration processes, because these hydrocarbons are formed almost entirely by thermal processes and they migrate more readily than do the C_{15+} solvent-extractable (i.e., heavier) hydrocarbons. The distribution pattern in Figure 3 is not easily explained by diffusional migration from the Eagle Ford, in which case the distribution would tend to be gradational rather than sharp as previously mentioned. The large increase in yield of all three of these alkanes in the 6,000–7,000-ft (1,800–2,100-m) depth interval is characteristic of the beginning of an oil window.

This same increase was observed for many of the other hydrocarbons in the C_2 to C_8 range, two of which are shown in Figure 4. Hydrocarbon yields from the shallowest to the deepest samples increased by factors of 10 to 100.

Additional evidence that the chalk is a source rock was provided by the pyrolysis data. Figure 5 shows pyrograms of the Austin Chalk at two depths. The area of the first peak (P_1) is related to the free hydrocarbons in the rock, and the area of the second peak (P_2) is related to the amount of hydrocarbons formed by the thermal breakdown of the kerogen. This relationship applies primarily to hydrocarbons in the range of C_1 to about C_{22}, because the free, very high-molecular-weight hydrocarbons tend to come off in the second peak (Tarafa, Hunt, and Ericsson, 1983). Most of P_2 represents the amount of hydrocarbons that the kerogen is capable of generating if it is buried still more deeply. The curve for the Austin Chalk sample from 4,565 ft (1,390 m) shows that there is a small amount of free hydrocarbon in the chalk (P_1) but that the potential of the chalk to generate hydrocarbons is considerable (P_2). The curve for the sample from 9,093 ft (2,770 m)

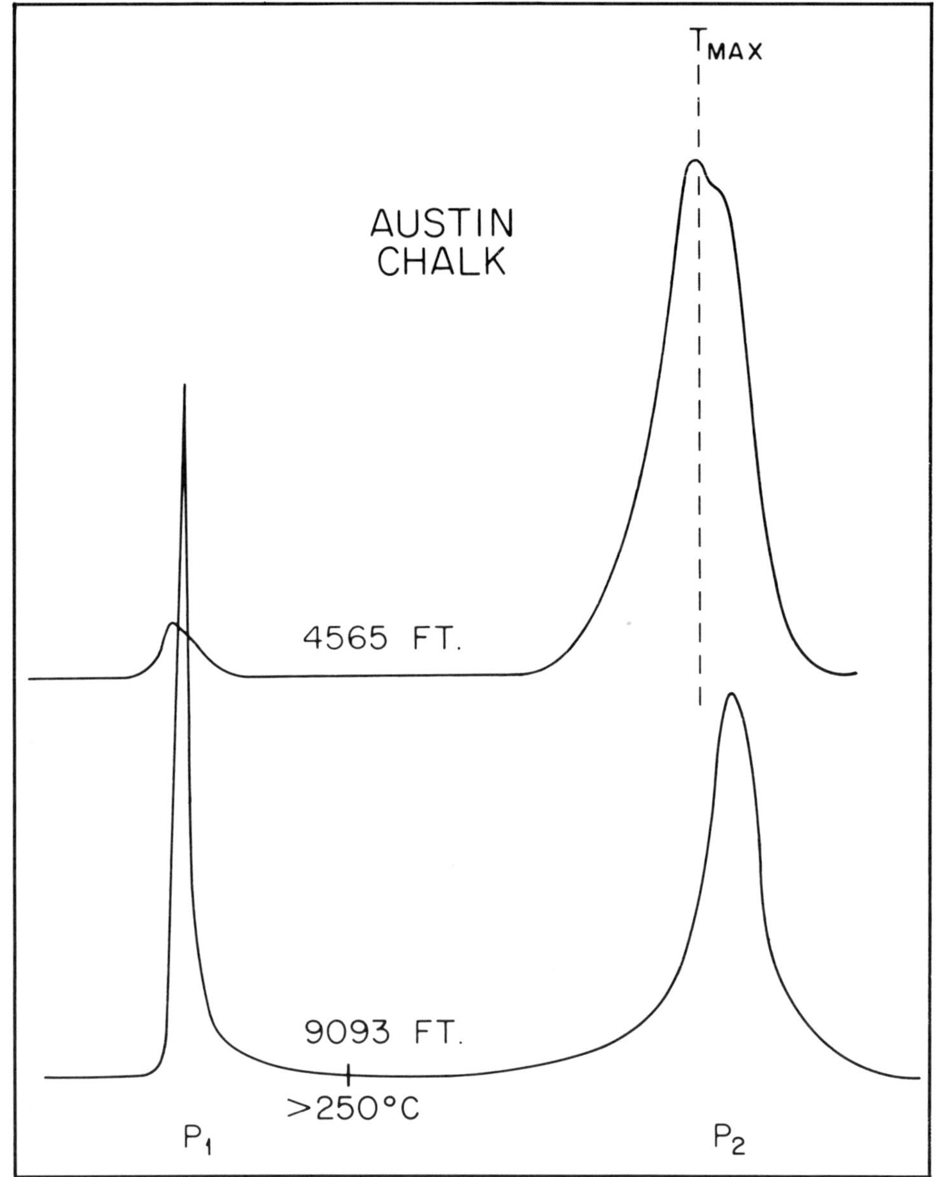

Figure 5 — Pyrograms of Austin Chalk, showing increase in P_1 with depth and shift of maximum temperature (T_{max}) of P_2.

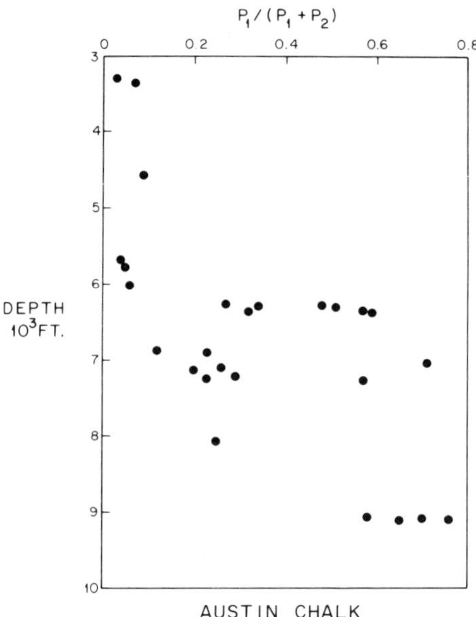

Figure 6 — Production index of Austin Chalk. Our data plus Grabowski's (1981).

shows a very large P_1, indicating that the chalk already has generated considerable quantities of hydrocarbons. P_2 is relatively small because an appreciable part of the kerogen has been converted to hydrocarbons. The degree of transformation of the kerogen into hydrocarbons is evaluated by the production index (PI), which is the ratio of the area of P_1 to the combined areas of P_1 and P_2. This index is shown in Figure 6. Our data and Grabowski's (1981) are combined, because the results are similar. The PI is low down to a depth of about 6,000 ft (1,800 m), at which point it increases abruptly to relatively high values. In the interval from 6,000 to 8,000 ft (1,800–2,400 m) considerable scatter indicates either differences in generating capability or loss of hydrocarbons caused by short-distance migration in the Austin Chalk. If the hydrocarbons move around as they are generated, P_1 will decrease, resulting in a corresponding decrease in the PI.

The consistent increases of P_1 and PI with depth are best explained by generation rather than by migrational changes. In the latter case, large and erratic changes in PI would be expected as observed—for example, by Barker (1974) and others.

The abrupt increase in PI at about 6,000 ft (1,800 m) is consistent with the headspace analyses in showing that this depth is the threshold of intense hydrocarbon generation within the Austin Chalk. The pattern is somewhat different than that observed in a sandstone–shale sequence penetrated by a South Padre Island well in the Texas Gulf Coast region (Huc and Hunt, 1980). In the latter case, there was a gradual increase in PI at the threshold of intense generation to maximum values of about 0.4. Also, there was less scatter of data within the oil window. The scatter may indicate differences in short-distance migration caused by the solution compaction and recrystallization phenomena that occur within carbonates, or differences in local concentrations of organic matter.

The total yield of hydrocarbons in the C_7 to C_{14} range relative to organic carbon for the Austin Chalk is compared to the shales in the South Padre Island wells in Figure 7. The hydrocarbon yields were determined from GC analyses of the P_1 fraction. The abrupt increase in yield at 6,000 ft (1,800 m) in the chalk contrasts with the more gradual increase at about 10,000 ft (3,000 m) in the shales. The time/temperature relationships for the two source rocks are about the same. The Cretaceous chalk is characterized by the oil window in the 60°–83°C (140°–182°F) range, the Miocene shales, by the oil window in the 110°–150°C (230°–302°F) range.

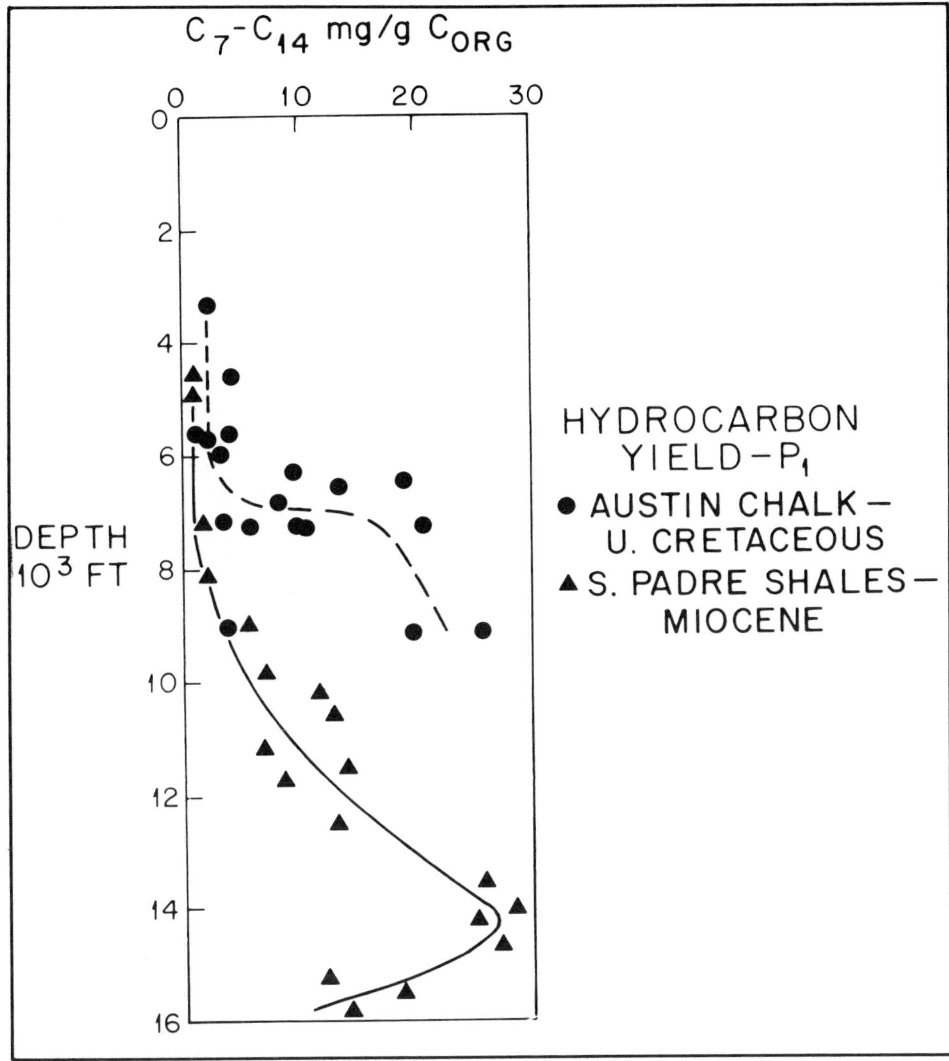

Figure 7 — Yield of hydrocarbons (C_7–C_{14}) relative to organic carbon, comparing Cretaceous chalk and Miocene shales.

Hydrocarbon distributions of the cracking-peak fraction (P_2, Fig. 8) for the Austin Chalk are all similar, indicating that the organic matter at different depths in the chalk is similar. Some of the P_2 hydrocarbon ratios undergo the same changes observed in the free hydrocarbon headspace data. Thus, the ratio of benzene to toluene cracked from the organic matter by pyrolysis follows the trend of decreasing with depth observed in the free hydrocarbons by ourselves and others. Thompson (1979) observed this decrease in free benzene and toluene in samples from wells in the Gulf of Mexico and California.

All three chromatograms in Figure 8 show the highest yields in the C_6 to C_{10} range and also show the double peaks characteristic of the alkenes and alkanes formed by cracking of sapropelic organic matter containing long-chain hydrocarbons. The alkene-to-alkane ratio decreases with depth, as has been observed in previous results from this laboratory (Huc and Hunt, 1980). The reason for this change (Fig. 8) in ratios is unknown but appears to be characteristic of the changes in the organic matter of a source rock that has generated petroleum with increasing depth and temperature.

Laboratory experiments generally have shown that the thermal cracking of large organic molecules requires activation energies considerably higher than those observed in nature (Greensfelder, Voge, and Good, 1949). This has caused some petroleum geochemists to conclude that natural processes involve some catalyst in order to lower the required activation energy. Studies by Goldstein (1983) showed that shales exert a catalytic activity in converting organic matter to petroleum. The procedure used involved measuring the catalytic activity of typical shales by a standard chemical reaction, which has been used to evaluate petroleum-cracking catalysts. This reaction is the acid-catalyzed decomposition of t-butylacetate to form isobutylene. In this work, Goldstein tested both the shales from the South Padre Island well and the Austin Chalk samples for catalytic activity; the results are shown in Table 1. The most surprising result of these data is that the Austin Chalk dem-

Table 1

South Padre Island Shales		Austin Chalk	
Depth (ft)	Activity*	Depth (ft)	Activity*
8,400	1.02	3,352	0.033
9,900	0.19	4,565	0.114
10,050	0.33	5,647	0.082
10,950	0.22	5,657	0.056
11,130	0.23	5,691	0.244
11,400	0.26	5,649	0.112
12,090	0.29	6,290	0.064
12,570	0.14	6,345	0.329
12,900	0.35	6,357	0.097
13,680	0.25	6,860	0.106
14,070	0.24	6,888	0.15
14,550	0.34	7,119	0.028
14,610	0.09	7,130	0.07
14,790	0.23	7,171	0.082
15,120	0.19	7,203	0.124
15,450	0.25	7,225	0.182
15,570	0.14	7,226	0.109
		9,065	0.182
		9,093	0.235
		9,095	0.203

*Activity, in cc, of isobutylene formed per minute by the decomposition of t-butylacetate by one gram of sediment sample (Goldstein, 1983).

Table 1 — Catalytic activity of well samples.

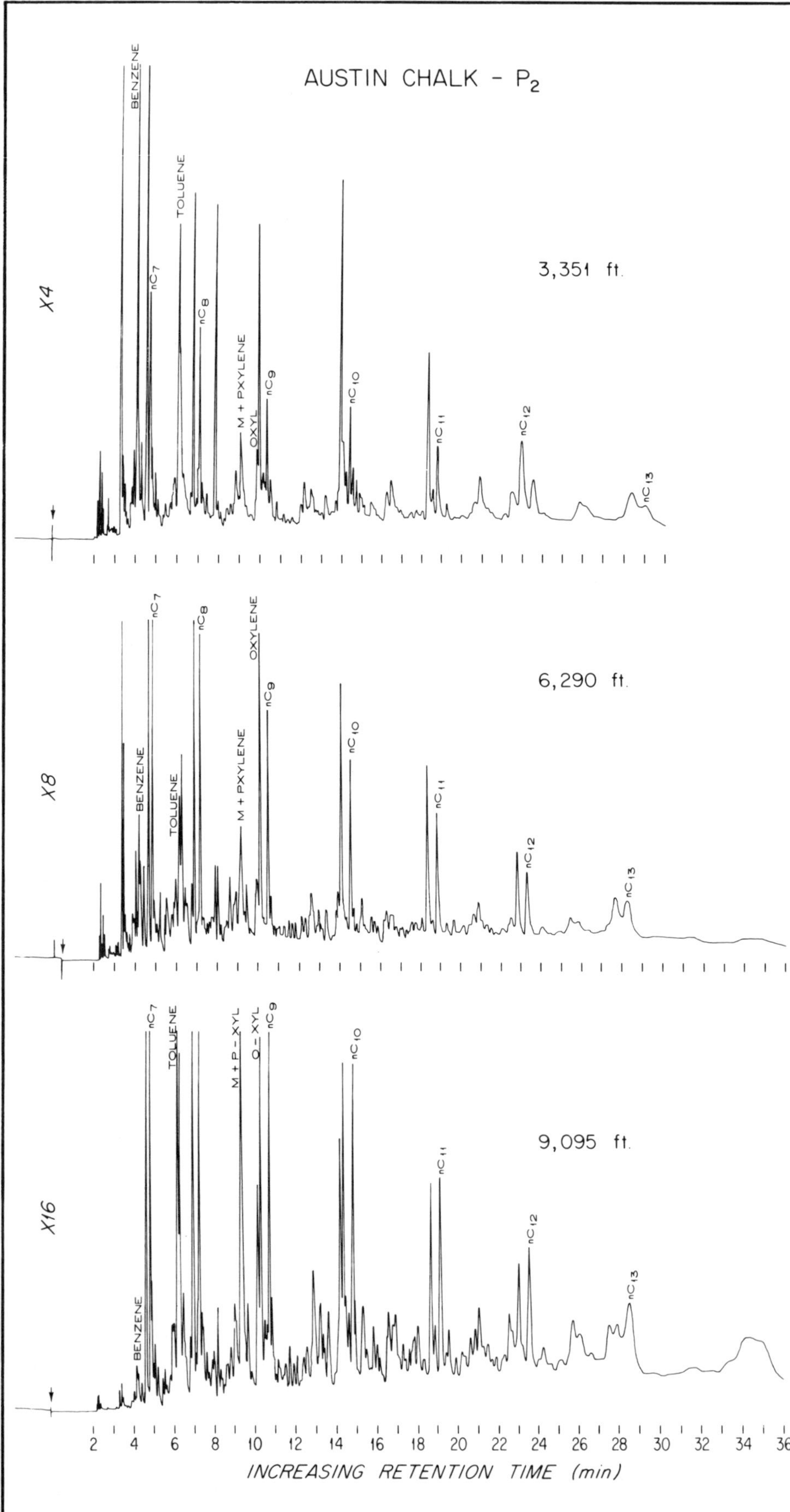

Figure 8 — Gas chromatograms of P_2 of Austin Chalk with increasing depth.

onstrates a catalytic activity that is almost half that of the South Padre Island shales (0.13 average for chalk compared to 0.28 for shales), and several chalk activities are comparable to the shale average. It is possible that oxides of metals, such as iron and aluminum, and other impurities in the chalk are more efficient catalysts than previously had been supposed. When pure calcium carbonate is tested by this technique, its activity is zero.

In summary, both the headspace and pyrolysis data support the concept that the Austin Chalk has acted as a petroleum source rock, with the threshold of intense generation having occurred at a depth of about 6,000 ft (1,800 m) or at a subsurface temperature of about 60°C (140°F). Some of the chalk, particularly in the lowest part of the formation, may contain migrated hydrocarbons, as indicated by Grabowski (1981), who determined that the ratio of extractable bitumen to organic carbon was more than 50% in some of these zones. Also, some of the reservoir oils in the chalk near the underlying Eagle Ford shales exhibited characteristics different from those within the bulk of the Austin Chalk. The overall conclusion made by Grabowski (1981) and further supported by our studies is that the Austin Chalk is the major source, and the Eagle Ford shales a minor source, for oils in reservoirs of the Austin Chalk.

ACKNOWLEDGMENTS

We thank Christine Burton, Martha Tarafa, and Jean Whelan for providing technical help; Nelson Frew for obtaining the mass spectra; and Dr. Ted Goldstein of Mobil Research Corp., Princeton, New Jersey, for carrying out the tests of acid catalytic activity on these samples. This work was supported by the Division of Basic Energy Science of the U.S. Department of Energy, contract no. EG-77-S-02-4392. This is WHOI Contribution no. 5138.

REFERENCES CITED

Baker, D.R., 1962, Organic geochemistry of Cherokee Group in southeastern Kansas and northeastern Oklahoma:

AAPG Bulletin, v. 46, p. 1621–1642.

Barker, C., 1974, Pyrolysis techniques for source-rock evaluation: AAPG Bulletin, v. 58, p. 2349–2361.

Bonham, L.C., 1980, Migration of hydrocarbons in compacting basins, *in* W.H. Roberts III and R.J. Cordell, eds., Problems of petroleum migration: AAPG Studies in Geology 10, p. 69–88.

Bray, E.E., and W.R. Foster, 1980, A process for primary migration of petroleum: AAPG Bulletin, v. 64, p. 107–114.

Brooks, B.T., 1948, Active-surface catalysts in formation of petroleum: AAPG Bulletin, v. 32, p. 2269–2286.

Dravis, J.J., 1979, Sedimentology and diagenesis of the Upper Cretaceous Austin Chalk Formation, south Texas and northern Mexico: Houston, Rice University unpublished Ph.D. thesis, 513 p.

Dunnington, H.V., 1967, Aspects of diagenesis and shape change in stylolitic limestone reservoirs: London, Elsevier, Seventh World Petroleum Congress Proceedings, v. 2, p. 339–352.

Goldstein, T.P., 1983, Geocatalytic reactions in formation and maturation of petroleum: AAPG Bulletin, v. 67, p. 152–159.

Grabowski, G.J., Jr., 1981, Origin, distribution and alteration of organic matter and generation and migration of hydrocarbons in Austin Chalk, Upper Cretaceous, Southeastern Texas: Houston, Rice University unpublished Ph.D. thesis, 262 p.

Greensfelder, B.S., H.H. Voge, and G.M. Good, 1949, Catalytic and thermal cracking of pure hydrocarbons: Industrial and Engineering Chemistry, v. 41, p. 2573–2584.

Hedberg, H.D., 1931, Cretaceous limestone as petroleum source rock in northwestern Venezuela: AAPG Bulletin v. 15, p. 229–246.

Hitchon, B., 1980, Some economic aspects of water–rock interaction, *in* W.H. Roberts III and R.J. Cordell, eds., Problems of petroleum migration: AAPG Studies in Geology 10, p. 109–120.

Huc, A.Y., and J.M. Hunt, 1980, Generation and migration of hydrocarbons in offshore south Texas Gulf Coast sediments: Geochimica et Cosmochimica Acta, v. 44, p. 1081–1089.

Hunt, J.M., 1961, Distribution of hydrocarbons in sedimentary rocks: Geochimica et Cosmochimica Acta, v. 22, p. 37–49.

———, 1979, Petroleum geochemistry and geology: San Francisco, W.H. Freeman, 617 p.

———, F. Stewart, and P.A. Dickey, 1954, Origin of hydrocarbons of Uinta Basin, Utah: AAPG Bulletin, v. 38, p. 1671–1698.

Magara, K., 1978, Compaction and fluid migration: practical petroleum geology: New York, Elsevier, 319 p.

McCrossan, R.G., 1961, Resistivity mapping and petrophysical study of Upper Devonian inter-reef calcareous shales of central Alberta, Canada: AAPG Bulletin, v. 45, p. 441–470.

McDowell, A.N., 1975, What are the problems in estimating the oil potential of a basin?: Oil and Gas Journal, v. 73, no. 23, p. 85–90.

Polak, L.S., 1954, Studies in the physical properties of rocks of the Ural–Emba region carried out during 1953: MNP SSSR Glavnefte–Geofizika, Kazakhstanskaya Geofizicheskaya Kontora.

Scholle, P.A., 1975, Applications of North Sea chalk diagenetic studies of petroleum exploration problems (abs): AAPG Annual Meeting Abstracts, v. 2, p. 93–94.

Schwartz, R.D., and D.J. Brasseaux, 1963, Resolution of complex hydrocarbon mixtures by capillary column gas liquid chromatography. Composition of the 28° to 114° portion of petroleum: Analytical Chemistry, v. 35, p. 1375.

Tarafa, M.E., J.M. Hunt, and I. Ericsson, 1983, Effect of hydrocarbon volatility and adsorption on source-rock pyrolysis: Journal of Geochemical Exploration, v. 18, p. 75–85.

Thompson, K.F.M., 1979, Light hydrocarbons in subsurface sediments: Geochimica et Cosmochimica Acta, v. 43, p. 657–672.

Tissot, B., and R. Pelet, 1971, Nouvelles données sur les mecanismes de genèse et de migration du petrole, Simulation mathematique et application a la prospection: Eighth World Petroleum Congress Proceedings, v. 2, p. 35–46.

Tissot, B., et al, 1977, Utilisation des alcanes comme fossiles geochimiques indicateurs des environnements géologiques, *in* R. Campos and J. Goni, eds., Advances in organic geochemistry 1975: Madrid, p. 117–154.

Tissot, B., et al, 1980, Paleoenvironment and petroleum potential of Middle Cretaceous black shales in the Atlantic basins: AAPG Bulletin, v. 64, p. 2051–2063.

Welte, D.H., and D. Waples, 1973, Über die Bevorzugung geradzahliger n-Alkane in Sedimentgesteinen: Naturwissenschaften, v. 60, p. 516–517.

Whelan, J.K., 1984, Volatile C_1–C_8 compounds in marine sediments, *in* G. Odham, L. Larsson, and P.A. Mardh, eds., Gas chromatography/mass spectrometry applications in microbiology: New York, Plenum Press, p. 381–414.

———, J.M. Hunt, and A.Y. Huc, 1980, Applications of thermal distillation-pyrolysis to petroleum source rock studies and marine pollution: Journal of Analytical and Applied Pyrolysis, v. 2, p. 79–96.

Source Rocks of the La Luna Formation (Upper Cretaceous) in the Middle Magdalena Valley, Colombia

John E. Zumberge
Cities Service Research, Tulsa, Oklahoma

Marlstone outcrops (~40–50% total carbonate content) of the Upper Cretaceous La Luna Formation in the Middle Magdalena Valley, Colombia, average 4.3 wt% total organic carbon and 2,500 ppm extractable hydrocarbons. Microscopic and elemental analyses of La Luna kerogen suggest that the organic matter in these La Luna Samples is of marine origin and at a maturity level within the oil-generation zone. Geochemical fossil distributions attest to a marine-planktonic and microorganism source, owing to abundant C_{27} steranes relative to C_{29} steranes, low quantities of C_{19} and C_{20} tricyclic diterpanes relative to extended (up to C_{30}) tricyclic terpanes, and the presence of ubiquitous hopane triterpanes derived from bacteria.

INTRODUCTION

The Cretaceous La Luna Formation has long been ascribed to as the source rock for much of the oil found in the Maracaibo basin of Venezuela (e.g., Hedberg, 1931; Young, Monaghan, and Schweisberger, 1977). Crude oils in the Middle Magdalena Valley, Colombia, which have accumulated in Tertiary fluvial-sand reservoirs, also have been generated, in part, from the Upper Cretaceous La Luna Formation, based on geochemical-correlation techniques such as stable-carbon-isotope ratios and geochemical fossil distributions (Zumberge, 1980). The objective of this present study is to describe both the gross organic geochemical properties (e.g., percentage of total organic carbon and quantities of extractable hydrocarbons) and the detailed geochemical character on a molecular level (e.g., sterane and terpane biomarker distributions) of this widespread and prolific calcareous source rock.

Outcrop samples of the La Luna Formation were collected along the eastern flank of the Nuevo Mundo syncline in the Middle Magdalena basin approximately 20 km (12.4 mi) west of Bucaramanga, Colombia (Fig. 1). Although the samples are from outcrops, care was taken to collect unweathered specimens; however, some of the observed variability of the data may be due to differential-weathering effects. All three members of the La Luna Formation (Galembo, Pujamana, and Salada) were sampled; the older Salada Member samples were collected ~5 km (3.1 mi) south of the Galembo and Pujamana Members sample-collection area (Quebrada La Sorda). In outcrop, the thickness of the formation ranges from 150 to 600 m (492–1,968 ft). In general, the La Luna Formation in this region can be described lithologically as consisting of dark-gray to black calcareous shales with varying amounts of interbedded limestones and some thin chert beds. In thin section, La Luna samples from the Pujamana and Salada Members contain abundant calcareous planktonic Foraminifera and other pelagic organisms, suggesting deposition in moderately deep water with restricted bottom circulation (M. W. Longman, personal communication). Photomicrographs of a Salada thin section (LL16) are shown in Figure 2; the textural relationship between organic matter (brown material in Fig. 2a) and inorganic minerals (especially the calcite-filled globigerinid Foraminifera) can be observed. Except for eolian-transported clays, little evidence of land-derived material, such as detrital quartz, was found. Foraminiferal tests that are not filled with calcite are commonly filled with oil, as was determined by fluorescence microscopy in reflected light of polished whole-rock specimens (S. E. Palmer, personal communication). These Colombia La Luna samples are similar with respect to lithology and organic-carbon content to La Luna rocks from northwestern Venezuela as described by Hedberg (1931) more than 50 years ago.

EXPERIMENTAL PROCEDURES

Rock samples were ground to < 100 mesh with subsequent carbonate dissolution with HCl and organic-carbon combustion using a Leco Carbon Analyzer in order to determine the approximate percentages of carbonate and total organic carbon (TOC). Powdered samples were Soxhlet extracted with chloroform/methanol (90:10, V:V) to remove extractable organic material followed by deasphalting in pentane after solvent removal. The pentane-soluble material was then separated by liquid chromatography into aliphatic hydrocarbon, aromatic hydrocarbon, and NSO (nitrogen–sulfur–oxygen–containing organic compounds) fractions on alumina and silica columns, and the quantity of each fraction was determined gravimetrically. The stable-carbon-isotopic compositions of both the aliphatic and aromatic hydrocarbon fractions were determined using the combustion method of Sofer (1980) with subsequent analysis on a V.G. Micromass 602 mass spectrometer. Capillary-gas-chromatographic separation of the aliphatic fraction was accomplished on a Hewlett–Packard 5880 gas chromatograph fitted with an SP-2100 fused silica column (12.5 m × 0.02 mm i.d.). Combined gas chromatography/mass spectrometry was performed on a Finnigin 4000 GC/MS/DS system by Global Geochemical Corp., Canoga Park, California, to determine sterane and terpane distributions. Microscopic and elemental analyses of kerogen concentrates (HCl and HF dissolution of inor-

Figure 1 — Location map showing the Middle Magdalena Valley, Colombia.

ganic matter followed by flotation in $ZnBr_2$) were made using a Zeiss Universal Microscope equipped for vitrinite reflectance and UV fluorescence measurements and Carlo Erba CHNO analyzers (Williams Brothers Co., Tulsa, Oklahoma).

RESULTS AND DISCUSSION

Total-organic-carbon values and approximate carbonate percentages for the three members of the La Luna Formation are presented in Table 1. The Galembo (youngest) Member samples (LL1–LL6) contain only small quantities of carbonate minerals (average = 2.4%) and average less than 0.9% TOC. The Pujamana and Salada Member marlstones, however, contain an average of 43.2 and 40.4% carbonate, respectively (excluding samples LL7 and LL8, which are from thin interbedded chert layers). These two relatively carbonate-rich lower members also contain almost five times more TOC (Pujamana average = 3.51%, and Salada average = 4.51%) than the carbonate-poor Galembo member. Sample LL11 (Pujamana Member) is a marlstone with fractures in-filled with asphaltite, a solid bitumen that is soluble in organic solvents (Hunt, 1979). An asphaltite vein that typically is up to 1 m thick also was sampled (LL20). This fracture-filling organic material is similar to the gilsonite common in the Uinta basin, Utah, and may represent an early, preserved stage of migration in which oil generated from La Luna kerogen has migrated short distances into associated fractures within the La Luna marlstones.

Microscopic examination of La Luna kerogen revealed predominantly fine-grained amorphous material with few well-defined vitrinite particles and pollen, suggesting little input of terrestrial organic matter. The amorphous kerogen fluoresces under UV light, indicating that it is prone to oil generation. Although few vitrinite-reflectance measurements were available for maturity estimates, an average value of 1.0 ± 0.2% R_o (N = 73) for all samples was obtained, which is well within the main phase of petroleum generation (Tissot and Welte, 1978). A spore-coloration index of 6.5 ± 0.8 (N = 216) and a T_{max} value of 446°C or 835°F (from Rock-Eval pyrolysis of a Salada marlstone) are consistent with this level of thermal maturity.

Table 2 lists the results of elemental analyses of five La Luna samples from the Pujamana and Salada Members; the atomic ratios of hydrogen to carbon and oxygen to carbon also are given. The H/C and O/C averages for the five kerogen samples are 0.85 and 0.05, respectively, which are characteristic of type II kerogen, placing them within the principal zone of oil generation on the van Krevelen diagram as defined by Tissot and Welte (1978). These results are consistent with the visual analysis of the La Luna kerogen.

Quantities of the various C_{15+} extractable fractions are presented in Table 1. As in the case of TOC, the marlstones of the Pujamana and Salada Members average more extractable indigenous organic matter than the shaly Galembo Member. Typically, the aromatic hydrocarbons are more abundant than the aliphatic (saturate) hydrocarbon fraction (averaging 1,638 ppm and 860 ppm, respectively) in the relatively carbonate-rich samples (excluding the asphaltite-containing LL11 sample). It is interesting to note, however, that the aliphatic hydrocarbons are more abundant than the aromatic hydro-

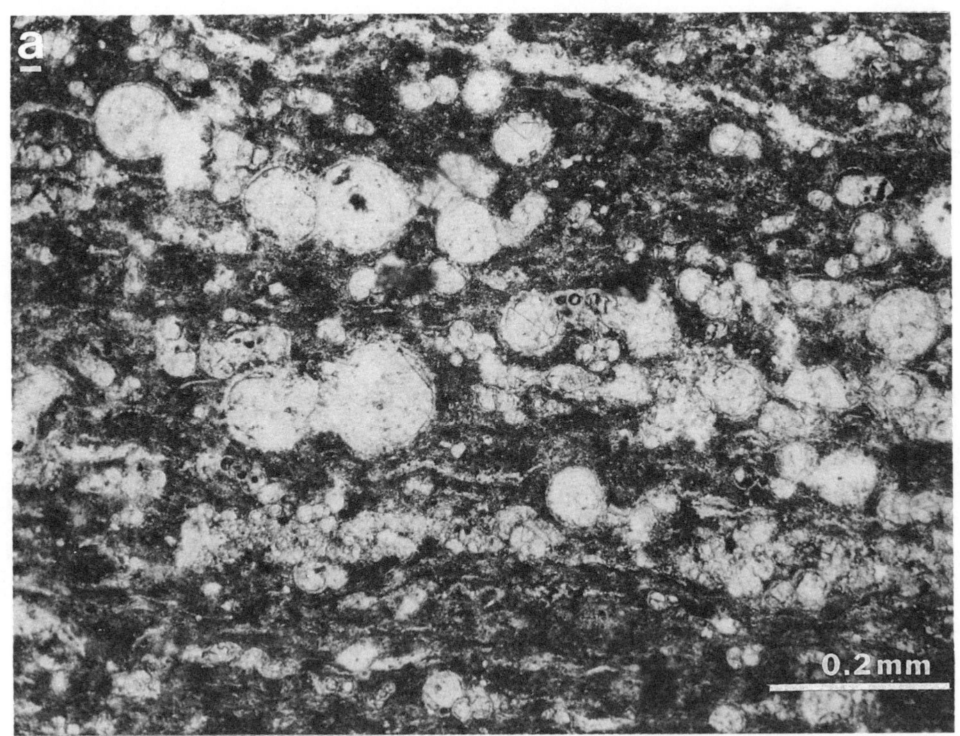

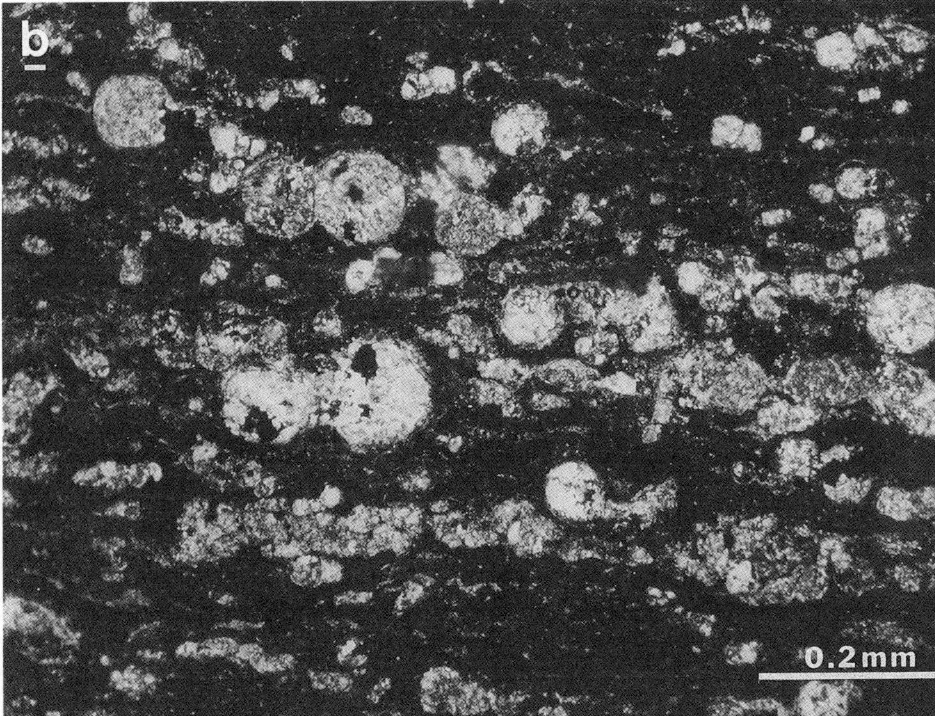

Figure 2 — Photomicrographs of a thin section from carbonate-rich La Luna sample LL16 (Salada Member); a, plane light; b, crossed nicols. The textural relationship between the brown organic matter and the calcite-filled globigerinid Foraminifera can be observed.

Table 1

Sample No. Member	Carbonates (wt%)	TOC (wt%)	Extractable C_{15+} Organic Material (ppm) Hydrocarbons Sat	Aro	Nonhydrocarbons NSO	Asph	Total (ppm)
LL1 Galembo	0.7	1.97	459	377	1,039	1,986	3,861
LL2 Galembo	1.9	0.57	605	542	935	523	2,605
LL3 Galembo	1.3	1.03	299	189	676	1,452	2,616
LL4 Galembo	1.5	0.79	294	249	580	385	1,508
LL5 Galembo	8.3	0.77	1,245	1,252	2,680	1,777	[b]6,954
LL6 Galembo	0.6	0.16	54	25	94	129	302
LL7 Pujamana	13.6	2.90	194	45	656	577	1,472
LL8 Pujamana	1.2	0.40	17	17	430	294	758
LL9 Pujamana	17.5	4.82	359	127	1,204	2,473	4,163
LL10 Pujamana	44.2	3.40	1,938	1,320	485	3,900	7,643
[a] LL11 Pujamana	45.0	3.37	1,000	5,626	4,975	28,075	39,676
LL12 Pujamana	51.9	4.41	303	1,145	834	3,302	5,584
LL13 Pujamana	47.0	4.78	455	1,166	663	3,105	5,389
LL14 Pujamana	53.5	3.99	438	1,224	729	3,679	6,070
LL15 Salada	46.5	3.77	627	1,180	1,499	3,170	6,476
LL16 Salada	38.2	4.52	1,223	2,806	1,749	4,134	9,912
LL17 Salada	37.8	4.60	918	1,966	1,579	2,543	7,006
LL18 Salada	41.5	4.70	1,081	1,979	2,111	4,725	9,896
LL19 Salada	37.8	4.94	1,253	3,033	1,651	4,722	10,659
[a] LL20 Pujamana	1.5	71.52	6,372	64,521	97,366	366,282	534,541

[a]Asphaltite material (solid bitumen) present, occurring as fracture fillings up to 1 m (3.3 ft) thick.
[b]This sample appears stained, owing to high quantities of total C_{15+} extractable organic material relative to the total organic carbon.

Table 1 — Organic geochemical data for La Luna source rocks.

Table 2

Sample	%C	%H	%O	%N	Atomic H/C	Atomic O/C
LL12	69.4	4.8	3.5	2.1	0.83	0.04
LL13	73.6	5.3	3.3	2.2	0.86	0.03
LL16	72.1	5.2	4.7	2.4	0.87	0.05
LL17	74.4	4.9	6.6	2.5	0.79	0.07
LL19	71.5	5.2	4.0	2.2	0.87	0.04

Table 2 — Elemental composition of La Luna kerogen.

carbons in the carbonate-poor Galembo samples. The nonhydrocarbon extractable fractions are yet more abundant than the hydrocarbon fractions. The asphaltite sample (LL20) contains primarily pentane-insoluble asphaltenes and NSO organic compounds; aromatic hydrocarbons are an order of magnitude greater than aliphatic hydrocarbons. Almost 6% of the TOC consists of extractable C_{15+} hydrocarbons in the relatively carbonate-rich La Luna samples, which is compatible with maturities estimated for La Luna kerogen and is indicative of good possible oil source rocks.

In Figure 3 a representative capillary gas chromatogram (sample LL15) of a La Luna C_{15+} aliphatic fraction is displayed. The isoprenoid phytane is typically more abundant than pristane (Pr/Ph = 0.70), and the n-alkane distribution indicates only a slight preference in odd- over even-numbered alkanes in the C_{28+} region (average OEP = 1.09). An even carbon preference is noticeable in the n-C_{20} to C_{28} range (average OEP = 0.956). Most of the aliphatic fraction consists of branched and cyclic alkanes, as evidenced by the large, unresolved "hump" under the n-alkane peaks. The appearance of this chromatogram is suggestive of organic matter derived from marine plankton and microorganisms deposited in a reducing environment.

The stable-carbon-isotopic compositions of the C_{15+} aliphatic- and aromatic-hydrocarbon extractable fractions

Table 3

Peak (Fig. 4A)	Sterane	Peak (Fig. 4C)	Triterpane
1	13β(H),17α(H)-diacholestane (20S)	A	18α(H)-22,24,30-trisnorhopane
2	13β(H),17α(H)-diacholestane (20R)	B	17α(H)-22,29,30-trisnorhopane
3	13α(H),17β(H)-diacholestane (20S)	C	17α(H),18α(H),21β(H)-28,30-bisnorhopane
4	13α(H),17β(H)-diacholestane (20R)	D	17α(H),21β(H)-30-norhopane
5	Rearranged C_{27} sterane	E	17β(H)21α(H)-normoretane
6	5α(H),14α(H),17α(H)-cholestane (20S)	F	17α(H),21β(H)-hopane
7	5α(H),14β(H),17β(H)-cholestane (20R)	G	17β(H),21α(H)-moretane
8	5α(H),14β(H),17β(H)-cholestane (20S)	H	17α(H),21β(H)-30-homohopane (22S)
9	5α(H),14α(H),17α(H)-cholestane (20R)	I	17α(H),21β(H)-30-homohopane (22R)
10	5α(H),14α(H),17α(H)-24-methylcholestane (20S)	J	gammacerane +
11	5α(H),14β(H),17β(H)-24-methylcholestane (20R)	K	17β(H),21α(H)-homomoretane
12	5α(H),14β(H),17β(H)-24-methylcholestane (20S)	L	17α(H),21β(H)-30,31-bishomohopane (22S)
13	5α(H),14α(H),17α(H)-24,-methylcholestane (20R)	M	17α(H),21β(H)-30,31-bishomohopane (22R)
14	5α(H),14α(H),17α(H)-24-ethylcholestane (20S)		
15	5α(H),14β(H),17β(H)-24-ethylcholestane (20R)		
16	5α(H),14β(H),17β(H)-24-ethylcholestane (20S)		
17	5α(H),14α(H),17α(H)-24-ethylcholestane (20R)		

Table 3 — Identified sterane and triterpane components.

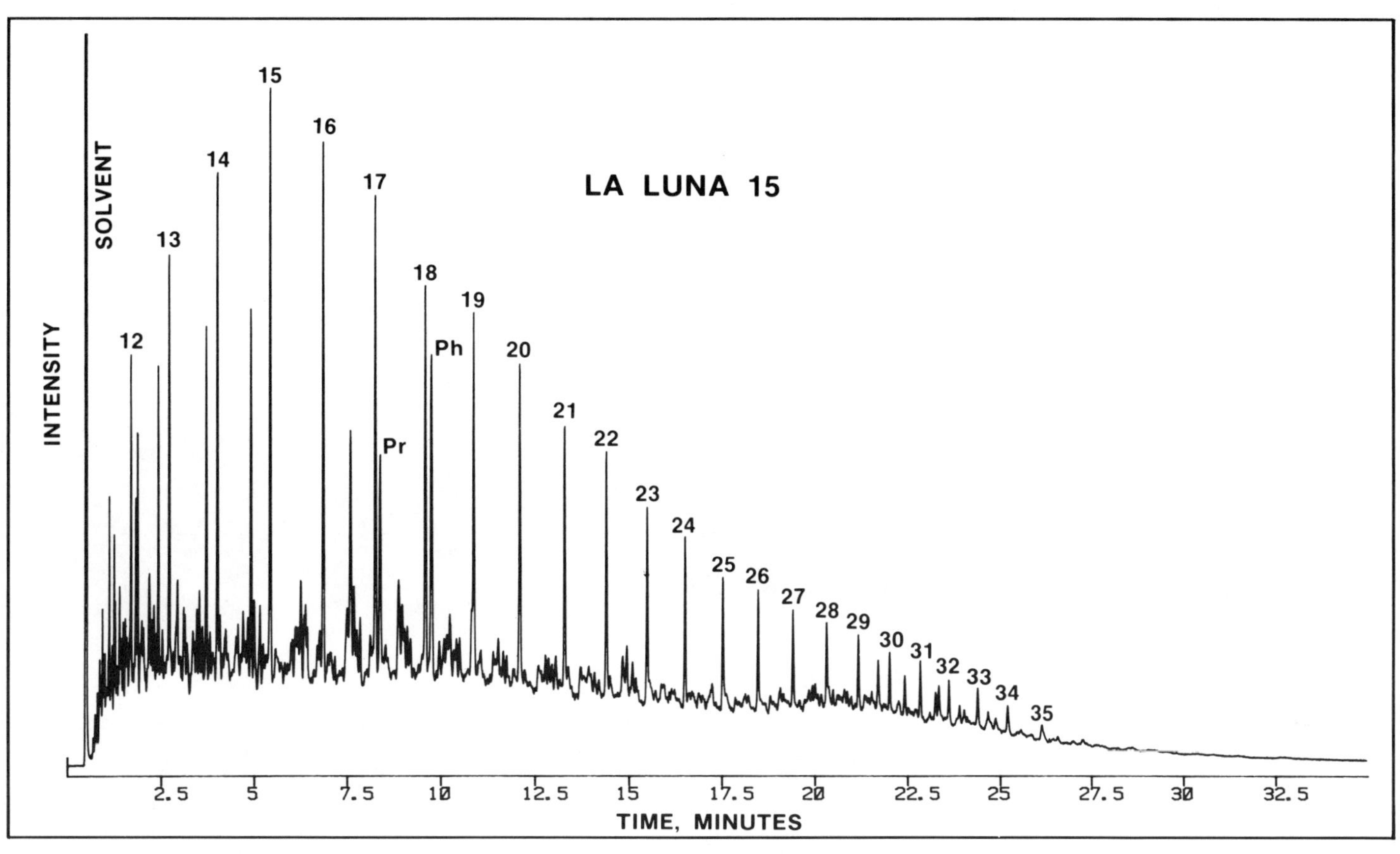

Figure 3 — Gas chromatogram of the aliphatic-hydrocarbon fraction of a Salada Member sample (LL15) from the La Luna Formation. Numbered peaks represent n-alkanes; Pr and Ph represent pristane and phytane, respectively.

average −27.5 ± 0.4‰ and −27.1 ± 0.4‰, respectively (relative to the PDB standard), for the Pujamana and Salada samples. These values are typical for oils of marine origin as reported by Sofer (1984). Oils produced from the Payoa, La Salina, and Corazon fields in the Middle Magdalena Valley (~30 km [18.6 mi] northwest of La Luna outcrop locations; Fig. 1) average −27.6 ± 0.2‰ and − 27.2 ± 0.1‰ for the aliphatic- and aromatic-hydrocarbon fractions, respectively (Zumberge, 1980). The ratios of the oils in the reservoirs are essentially identical to the La Luna extractable hydrocarbons—i.e., the proposed source rock.

The distributions of the geochemical fossils (biomarkers) known as steranes (tetracyclic alkanes) and terpanes (tricyclic and pentacyclic alkanes) are important in geochemical correlations. The tricyclic terpanes are especially useful in correlating nondegraded to biodegraded crude oils (such as are commonly found in the Middle Magdalena Valley) because they appear to be relatively unaffected by even severe microbial attack (e.g., Reed, 1977). Also, sterane and terpane biomarkers are often useful in determining source-rock depositional environments (e.g., Tissot and Welte, 1978). The sterane, diterpane, and triterpane distributions of a representative La Luna extract (LL15) are illustrated in Figure 4, and identified components are listed in Table 3. The C_{27} sterane stereoisomers (peaks 6–9, Fig. 4A) are more abundant than the C_{29} steranes (peaks 14–17), indicating a predominance of aquatic microorganisms (relative to higher land-derived plants; e.g., Tissot and Welte, 1978) that contributed to the La Luna sediments. Low concentrations of rearranged C_{27} steranes (peaks 1–5, Fig. 4A) are present relative to regular steranes, perhaps indicating a nonacidic, carbonate depositional environment (e.g., Sieskind, Joly, and Albrecht, 1979). The roughly equal abundance of the 20S and 20R component pairs of the 5α(H),14α-(H),17α(H) stereoisomers (peak pairs 6 and 9, 10 and 13, 14 and 17, Fig. 4A) is characteristic of maturities sufficient to generate significant quantities of hydrocarbons (Seifert and Moldowan, 1981; Mackenzie et al, 1980).

The most abundant tricyclic diterpane compound (i.e., C_{23}, MW 318, Fig. 4B) in the La Luna samples is also the most abundant in the oils from the Payoa, La Salina, and Corazon fields (Zumberge, 1980). The relative lack of C_{19} and C_{20} diterpanes (compounds generally believed to be derived from terrestrial plants; e.g., Simoneit, 1977, Palmer, 1981) in both La Luna rocks and Middle Magdalena Valley oils again suggests a marine source. The range of tricyclic terpanes extends out to the C_{30} molecular-weight range (MW 416, Fig. 4B and 4C).

The pentacyclic triterpane components in the La Luna source rocks (Fig. 4C) consist primarily of the ubiquitous hopane series, with the C_{29} norhopane (peak D) and C_{30} hopane (peak F) predominating. The hopanes are believed to be derived from bacteria (e.g., Ensminger et al, 1974; van Dorsselaer, Albrecht, and Ourisson, 1977). The thermodynamically

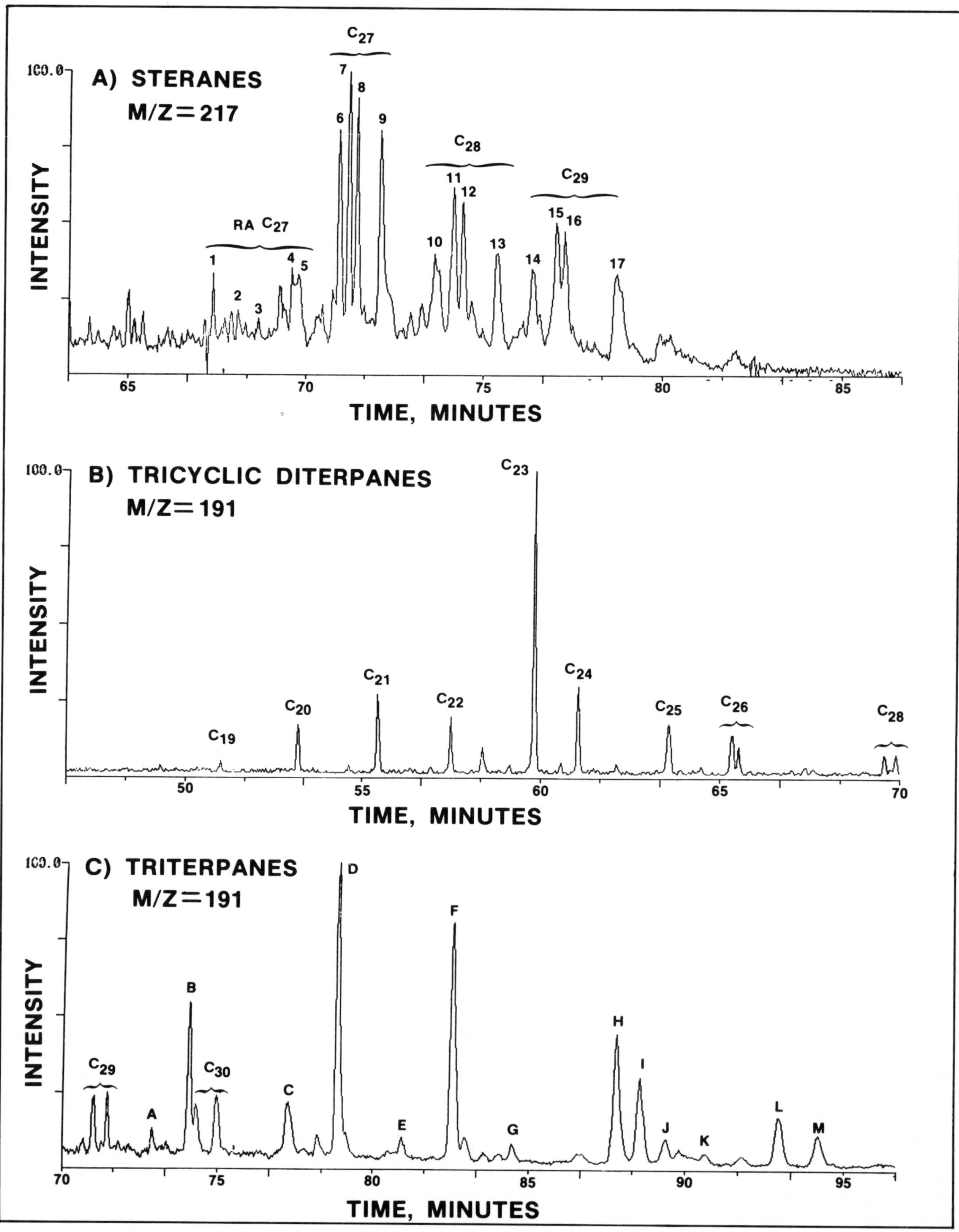
A) STERANES
M/Z=217
B) TRICYCLIC DITERPANES
M/Z=191
C) TRITERPANES
M/Z=191
INTENSITY
TIME, MINUTES
100.0
RA C27
C27
C28
C29
C19
C20
C21
C22
C23
C24
C25
C26
C28
C29
C30
A
B
C
D
E
F
G
H
I
J
K
L
M

less stable moretane series (i.e., the 17β(H),21α(H) hopane stereoisomers, peaks E, G, and K) are present in minor amounts, which is consistent with the other maturity indicators. Not shown in Figure 4C are traces of C_{32+} hopane 22S and 22R pairs.

CONCLUSIONS

The following listing is a summary of the significant results and inferences based on the organic-geochemistry analyses of the Upper Cretaceous La Luna Formation from the Middle Magdalena Valley, Colombia.

1. La Luna marlstones ($\sim$42% carbonate minerals) of the Pujamana and Salada Members average 4.3% total organic carbon. The carbonate-poor Galembo samples ($\sim$2% carbonate minerals) average less than 0.9% total organic carbon.

2. Microscopic and elemental analyses of La Luna kerogen suggest a marine origin, with a present maturity well within the main phase of oil generation.

3. C_{15+} extractable hydrocarbons of relatively carbonate-rich La Luna samples average almost 2,500 ppm, which indicate good possible oil source rocks.

4. Gas-chromatographic results of the aliphatic fractions, including pristane/phytane values less than 1, abundant C_{15} to C_{19} n-alkanes with respect to C_{20+} alkanes, and a predominance of cycloalkanes, indicate that La Luna organic matter is derived from marine plankton and microorganisms. These results agree with the kerogen analyses.

5. Stable-carbon-isotope ratios of La Luna hydrocarbon extracts are identical to those of crude oils produced from Tertiary reservoirs within the Middle Magdalena Valley, which suggests a genetic relationship.

6. Geochemical fossil distributions also indicate a marine depositional environment (evidenced by abundant C_{27} steranes relative to C_{29} steranes, relatively low quantities of C_{19} and C_{20} diterpanes, and the presence of bacterially derived hopanes) with molecular maturation parameters consistent with the generation of oil.

Figure 4 — Geochemical fossil composition of a La Luna hydrocarbon extract (LL15). A. Sterane distribution as determined by the m/z = 217 mass fragmentogram; numbered peaks are identified in Table 3; RA = rearranged steranes. B. Tricyclic diterpane distribution (m/z = 191); molecular weights were determined by monitoring m/z = 262 (C_{19}), 276 (C_{20}), 290 (C_{21}), 304 (C_{22}), 318 (C_{23}), etc. C. Pentacyclic triterpane distribution (m/z = 191). Lettered peaks are identified in Table 3.

ACKNOWLEDGMENTS

I thank T. Murray, J. Heard, S. Sellers, C. Schiefelbein, and P. Hoffmann for technical analyses, and M. Draughon for typing the manuscript. I also thank L. Baie, M. Leenheer, S. Palmer, and C. Sutton for reviewing the manuscript and providing helpful comments and suggestions. Special thanks go to R. S. Williams (Cities Service, Western Region) and to R. Mejia, A. Erazo, and G. Velasco (Colcito, Colombia) for sample collection and geological information and advice.

REFERENCES CITED

Ensminger, A., et al, 1974, Pentacyclic triterpanes of the hopane type as ubiquitous geochemical markers: Origin and significance, *in* B. Tissot and F. Bienner, eds., Advances in organic geochemistry 1973: Paris, Éditions Technip, p. 245–260.

Hedberg, H.D., 1931, Cretaceous limestone as petroleum source rock in northwestern Venezuela: AAPG Bulletin, v. 15, p. 229–246.

Hunt, J.M., 1979, Petroleum geochemistry and geology: San Francisco, W.H. Freeman, 617 p.

Mackenzie, A.S., et al, 1980, Molecular parameters of maturation in the Toarcian shales, Paris Basin, France: I. Changes in the configurations of acyclic isoprenoid alkanes, steranes and triterpanes: Geochimica et Cosmochimica Acta, v. 44, p. 1709–1721.

Palmer, S.E., 1981, Organic facies of the lacustrine Elko Formation (Eocene/Oligocene), northeastern Nevada (abs.): GSA Abstracts with Programs, v. 13, p. 525.

Reed, W.E., 1977, Molecular compositions of weathered petroleum and comparison with its possible source: Geochimica et Cosmochimica Acta, v. 41, p. 237–247.

Seifert, W.K., and J.M. Moldowan, 1981, Paleoreconstruction by biological markers: Geochimica et Cosmochimica Acta, v. 45, p. 783–794.

Sieskind, O., G. Joly, and P. Albrecht, 1979, Simulation of the geochemical transformations of sterols: superacid effect of clay minerals: Geochimica et Cosmochimica Acta, v. 43, p. 1675–1679.

Simoneit, B.R., 1977, Diterpenoid compounds and other lipids in deep-sea sediments and their geochemical significance: Geochimica et Cosmochimica Acta, v. 41, p. 463–476.

Sofer, Z., 1980, Preparation of carbon dioxide for stable carbon isotope analysis of petroleum fractions: Analytical Chemistry, v. 52, p. 1389–1391.

———, 1984, Stable carbon isotope compositions of crude oils: application to source depositional environments and petroleum alteration: AAPG Bulletin, v. 68, p. 31–49.

Tissot, B., and D.H. Welte, 1978, Petroleum formation and occurrence: Berlin, Springer–Verlag, 538 p.

van Dorsselaer, A., P. Albrecht, and G. Ourisson, 1977, Identification of novel (17αH)-hopanes in shales, coals, lignites, sediments, and petroleum: Bulletin de la Société Chimique de France, v. 58, p. 1243–1252.

Young, A., P.H. Monaghan, and R.T. Schweisberger, 1977, Calculation of ages of hydrocarbons in oils—Physical chemistry applied to petroleum geochemistry I: AAPG Bulletin, v. 61, p. 573–600.

Zumberge, J.E., 1980, Oil-oil and oil–source rock correlations of bacterially degraded oils and Cretaceous outcrops from Colombia, South America (abs.): Paris, International Geological Congress, 26th, p. 806.

Occurrence of Indigenous Biogenic Gas in Organic-Rich, Immature Chalks of Late Cretaceous Age, Eastern Denver Basin

Dudley D. Rice
U.S. Geological Survey, Denver Federal Center, Denver, Colorado

Natural gas is produced at shallow depths from chalk beds of the Upper Cretaceous Niobrara Formation in northeastern Colorado and northwestern Kansas. The depth of the gas-productive trend increases to the northwest from 250 to 850 m (820–2,790 ft). The chalks are fine-grained limestone, consisting mainly of calcareous nannofossils and other microfossils, that are characterized by high porosity (30–45%) and low permeability (about 1 md). Core samples from the productive area average 30% acid-insoluble residue. Most of the insoluble residue, which consists of clay minerals and organic matter, is concentrated in alternating laminations.

The gases are methane-rich ($C_1/C_{1-5} > 0.98$), are enriched in the light isotope ^{12}C ($\delta^{13}C_1$ values range from -65 to -55 ppt), and become isotopically heavier with increasing depth across the trend. The shallow gas in the Niobrara is interpreted to be of biogenic rather than thermogenic origin because of its chemical and isotopic composition and of source-rock studies indicating that Upper Cretaceous rocks in this region are immature with respect to thermogenic hydrocarbon generation. To the northwest, however, these rocks are mature and were capable of generating oil at the time of maximum burial and/or heat flow.

The biogenic gas was generated early in the burial history of the chalks by microbial degradation of organic matter in an anaerobic, presumably sulfate-free environment. *In-situ* gas generation is indicated, because low permeability inhibited long-range migration and because organic-rich laminae provided an adequate source for the gas. Organic-carbon values average 3.2%, and the organic matter consists primarily of hydrogen-rich sapropelic kerogen (type II), typical of an open-marine environment. The chalks are overlain by a thick section of shale containing many bentonite beds in the lower part that served as a seal for gas after the reservoirs were naturally fractured later in the burial history.

INTRODUCTION

Natural gas was found more than 60 years ago in shallow chalk reservoirs of the Upper Cretaceous Niobrara Formation in northwestern Kansas. Because of low flow rates and gas prices, the wells were plugged and abandoned, or the small amounts of gas were utilized for local industrial and/or domestic purposes. Increases in gas prices and advances in recovery technology in the 1970's have resulted in the development of more than 60 small gas fields mainly in northeastern Colorado and northwestern Kansas, as shown in Figure 1. Recently, new fields have been discovered in southwestern Nebraska, and the potential for development exists northward into North and South Dakota. Gas reserves of 1.2 tcf (0.34×10^{12} m^3) have been estimated for the current area of exploration from the shallow chalk reservoirs (National Petroleum Council, 1980).

The gas fields lie mainly along a northwest–southeast trend with present-day depths of burial increasing from about 250 to 850 m (820–2,790 ft) to the northwest (Fig. 1). The chalk reservoirs at these shallow depths of burial are characterized by high porosity and low permeability, and they require hydraulic stimulation to provide economic flow rates. Details of the production are provided by Lockridge (1977), Lockridge and Scholle (1978), Brown, Crafton, and Golson (1981), and Smagala (1981).

The shallow gas fields producing from Niobrara chalks are along the eastern side of the Denver basin. The basin includes approximately 120,000 km^2 (45,330 mi^2) in eastern Colorado, southeastern Colorado, southeastern Wyoming, and southwestern Nebraska. It is strongly asymmetric, with a broad, gently sloping eastern flank and an abrupt western flank along the mountain front (Fig. 2).

The Niobrara gas fields are situated on low-relief, structural closures. The folding of these structures has resulted in fracturing of the brittle chalk reservoirs, which enhances the reservoir quality. The structures are modified by normal faults that may be listric and do not extend below the Upper Cretaceous marine shales (Smagala, 1981).

The Denver basin is an extensively explored province that has produced more than 400 million barrels (1.5×10^6 m^3) of oil and 1 tcf (0.28×10^2 m^3) of mainly associated gas. Most of the oil and associated gas is produced from stratigraphic traps in sandstones of Cretaceous age. Fields developed in these traps also lie along the eastern flank of the basin but occur at greater depths than do the gas fields producing from the chalk reservoirs.

The existence of nonassociated gas at Wattenberg field and in the area of shallow gas fields producing from Niobrara chalks is a notable exception to the hydrocarbon production throughout most of the basin, which is predominantly oil and associated gas. The Wattenberg field lies along the axis of the basin (Fig. 2), and the gas is produced from depths ranging from 2,300 to 2,600 m (7,500–8,500 ft). Reservoirs are low-permeability sandstones of Cretaceous age that require massive hydraulic stimulation to provide economic flow rates.

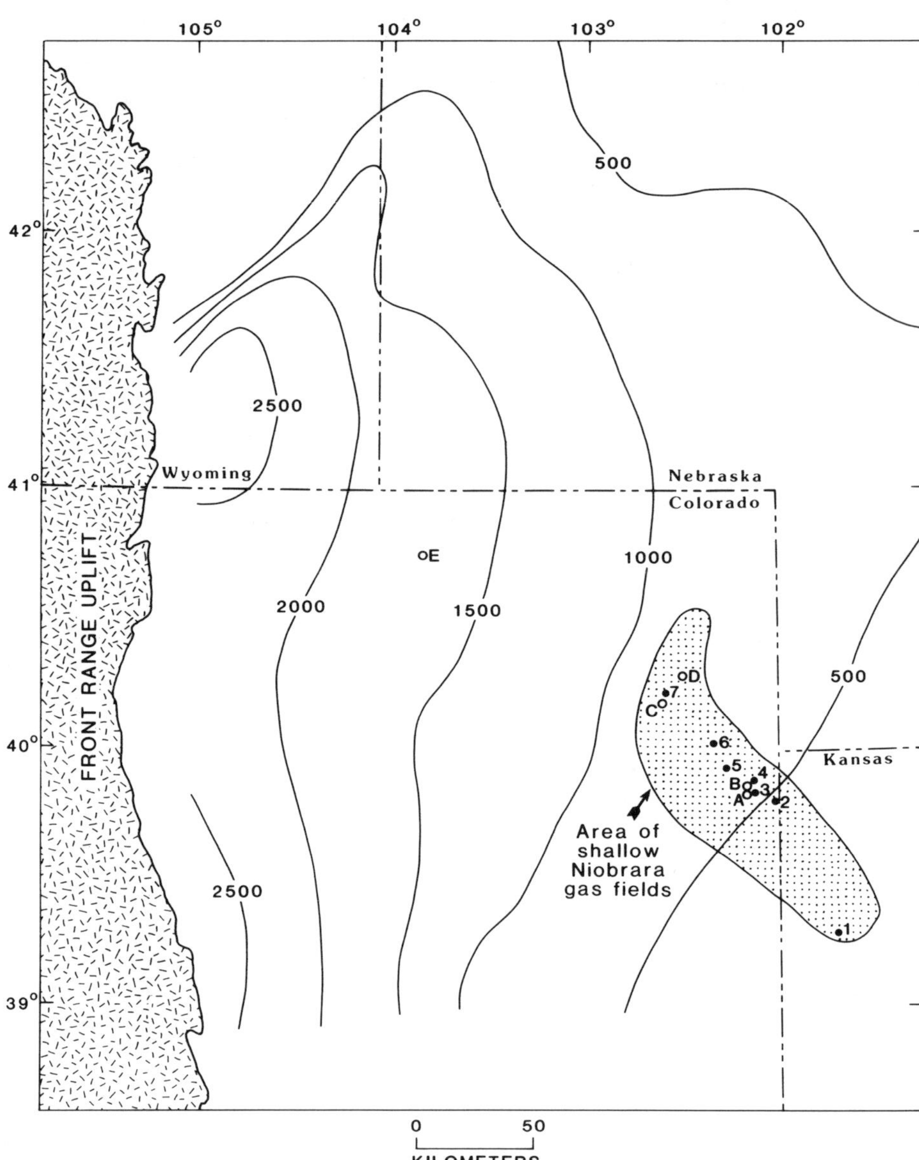

Figure 1 — Index map of Denver basin showing present-day depth of burial to top of Niobrara Formation (modified from Shurr, 1980). Contour interval is 500 m (1,640 ft). Stipple pattern indicates area of shallow biogenic-gas production from Niobrara chalks. Wells from which gas samples were analyzed are labeled by numbers (Table 1) and from which core samples were analyzed are labeled by letters (Table 2).

Reserves in the field are estimated at 1.5 tcf (0.42×10^{12} m^3) (Matuszczak, 1973).

This paper deals with the occurrence of nonassociated gas in shallow chalk reservoirs in the eastern part of the Denver basin. The objectives of the paper are to (1) describe the reservoir rock and adjacent strata, (2) discuss the characteristics and origin of the gas, and (3) characterize the organic matter in the chalks.

STRATIGRAPHY

The Cretaceous section in the Denver basin consists predominantly of siltstone and shale with minor amounts of sandstone and carbonate. Sediments that form this fine-grained sequence were deposited along and in an epicontinental seaway. The seaway invaded the area north from the Gulf of Mexico and south from the Arctic Ocean and at times formed a continuous north–south–trending body of water during Late Cretaceous time. The dominant source area, particularly for coarse–grained clastic sediments, was highlands to the west.

Sandstones were deposited along a fluctuating shoreline during regressive events and are the primary hydrocarbon-bearing reservoirs. The "J" and "D" sandstones (Fig. 3) account for more than 90% of the cumulative oil and gas production in the basin. The finer grained sediments, which form the Skull Creek, Huntsman (of subsurface usage), Graneros, and Carlile Shales (Fig. 3), were deposited offshore during more widespread invasions of the sea. These strata are probably the major source rocks for thermally generated oil and associated gas in the basin (Clayton and Swetland, 1980).

During maximum transgressions, carbonates, primarily chalk, formed in the central and eastern parts of the seaway, away from the influence of the dominantly westerly derived clastic sediments. The Greenhorn Limestone and Niobrara Formation (Fig. 3) are examples of carbonates formed during major transgressions.

The shallow gas in the eastern Denver basin occurs in carbonates assigned to the Niobrara Formation of Late Cretaceous (Coniacian to Campanian) age (Fig. 3). The carbonates form tongues that

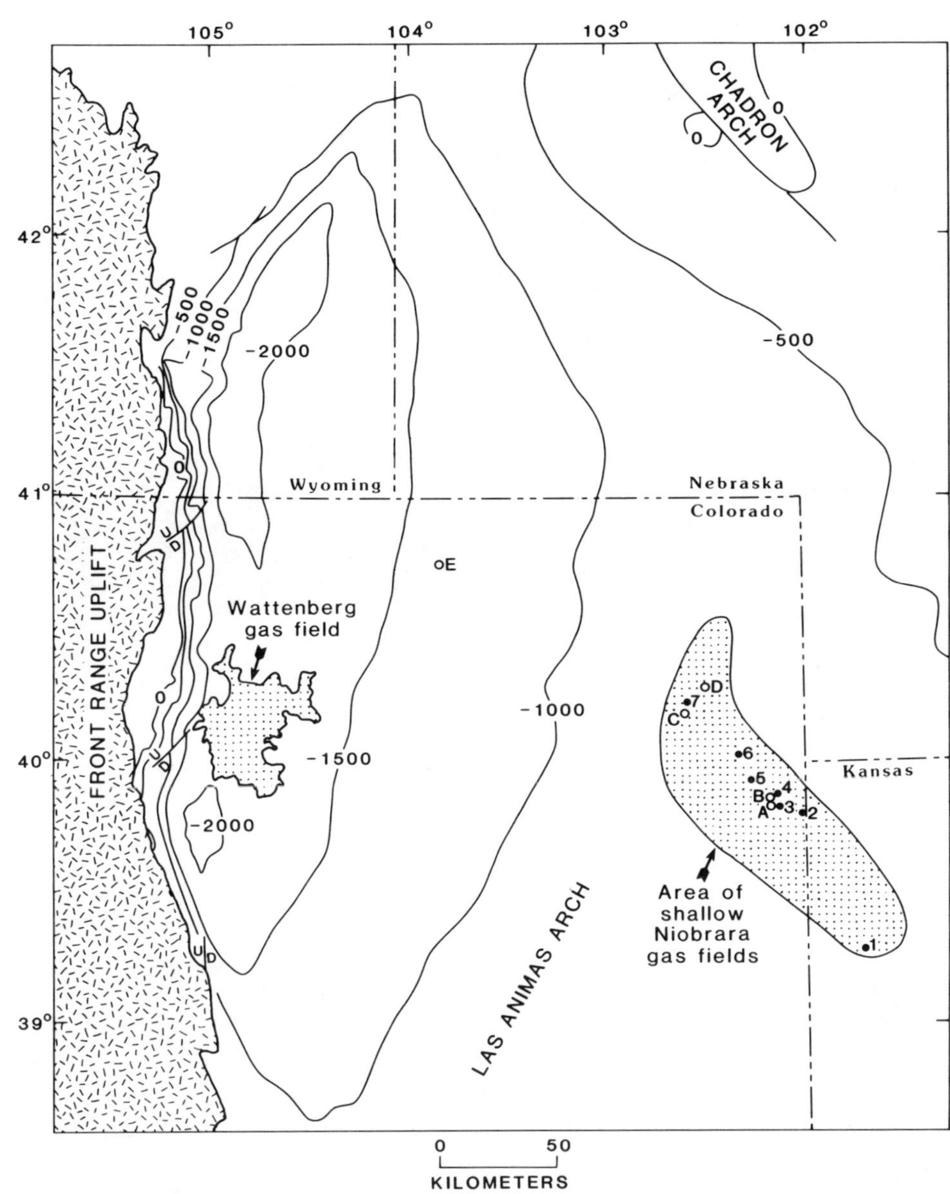

Figure 2 — Basement structure-contour map of Denver basin. Contour interval is 500 m (1,640 ft). Stipple pattern indicates area of shallow biogenic-gas production from Niobrara chalks. Wells from which gas samples were analyzed are labeled by numbers (Table 1) and from which core samples were analyzed are labeled by letters (Table 2).

grade north and west of the productive area into calcareous shales (Shurr and Sieverding, 1980). The calcareous shales in succession grade to noncalcareous shales and finally to sandstones along the western shoreline in central Utah (McGookey, 1972). The exact location of the eastern shoreline is unknown.

The Niobrara Formation is underlain and overlain by marine shales assigned to the Carlile and Pierre Shales, respectively (Fig. 3). In the eastern Denver basin, the Niobrara can be divided into two members. The lower member, the Fort Hays Limestone, consists of 10–25 m (33–82 ft) of relatively clean, well-indurated chalk. The upper Smoky Hill Member consists of 150–200 m (490–650 ft) of impure shaly chalk with locally massive chalk beds. Production is from a clean chalk bed at the top of the Niobrara informally named the Beecher Island zone (Lockridge, 1977).

NATURE OF RESERVOIR

Studies by scanning-electron microscopy (SEM) indicate that the Niobrara Formation in the area of shallow gas production is a relatively unusual type of carbonate called chalk (Scholle, 1977a, 1977b; Lockridge and Scholle, 1978). Chalks are very fine-grained limestones composed primarily of coccoliths, rhabdoliths, foraminifers, and inoceramid prisms. Figure 4 is an SEM photomicrograph showing the internal makeup of a typical shallow, gas-bearing chalk in eastern Colorado.

The Niobrara chalks are impure carbonates, and the amount of HCl-insoluble residue varies by stratigraphic member and geographic location (Scholle, 1977a). Runnels and Dubins (1949) reported 2–12% insoluble residue in the Fort Hays Limestone Member in Kansas. Farther to the west at Pueblo, Colorado, samples from the Fort Hays average 18% insoluble residue (Scholle, 1977a).

In contrast, the overlying Smoky Hill Member contains larger quantities of insoluble residue. Scholle (1977a) reported average values of 30% in south-

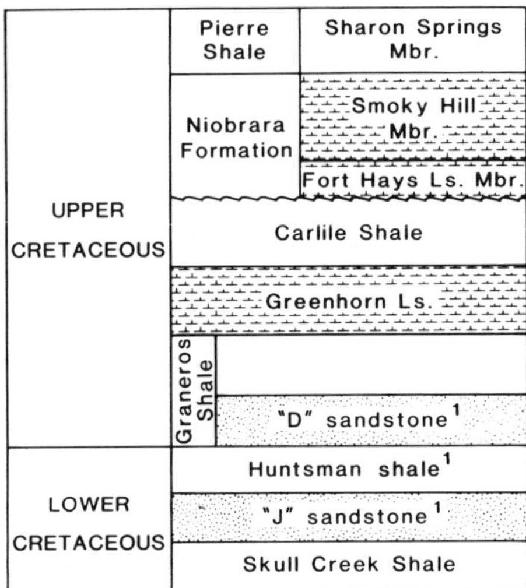

Figure 3 — Correlation chart of selected Cretaceous rocks in Denver basin.

Figure 4 — Scanning-electron photomicrograph of a gas-productive Niobrara chalk showing coccoliths and rhabdolith spines. ×6,000.

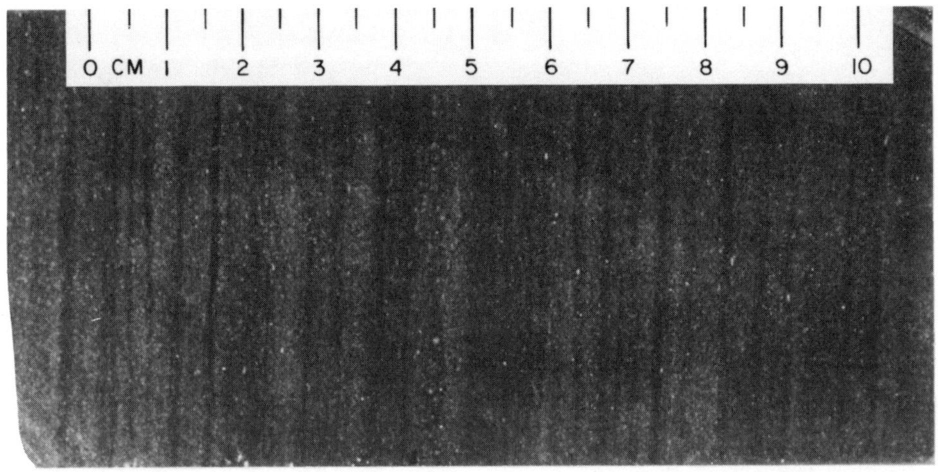

Figure 5 — Core photograph showing rhythmic laminations of Smoky Hill Member of Niobrara Formation.

central Colorado and 28% in central and western Kansas. Core samples from the upper part of the Smoky Hill Member in northeastern Colorado contain amounts of insoluble residue that range from 5 to 80% (R. M. Pollastro, written communication, 1981). The insoluble residue of core samples from five wells shown in Figure 1 averages 30%. In addition, the averages for individual wells are fairly constant, indicating that the chalk has not graded into a shale facies.

Analyses by x-ray diffraction show that the insoluble residue consists of 40–60% clay minerals, 20–30% quartz, 10–25% pyrite, and minor amounts of feldspar and gypsum (Pollastro, 1981a). This insoluble residue also includes significant amounts of organic carbon, which is discussed later.

On a larger scale, the Smoky Hill Member is distinctly laminated at a millimeter scale (Fig. 5). There are rhythmic alternations between lighter colored, carbonate-rich laminae and darker colored, organic-rich and clay-rich laminae. This alternation of laminae probably reflects shorter cycles of low-oxygen conditions during the overall "oceanic anoxic event" (M. A. Arthur, written communication, 1981).

Because of their very small grain size, chalks are deposited with high porosity (70–80%) and with extremely low permeability (Scholle, 1977a, 1977b). These properties provide the chalks with ample storage space but inhibit the movement of hydrocarbons. The initial porosity and permeability of chalk are rapidly and systematically reduced with increasing depth of burial, primarily owing to mechanical and chemical compaction produced by the addition of overburden (Scholle, 1977a, 1977b). Other factors associated with porosity and permeability loss are pore-water chemistry, primary mineralogy, tectonic stress, presence of hydrocarbons, and pore-fluid pressure (Scholle, 1977a, 1977b).

In the eastern Denver basin, the thickness of overburden on the gas-bearing chalk reservoirs gradually increases to the northwest from 250 to 850 m (820–2,790 ft) (Fig. 1). Figure 6 is a plot of porosity calculated from density logs versus present-day depth of burial for the Niobrara chalks. The plot shows that the porosities progressively decrease from about 45% at 300 m (980 ft) to about 10% at 1,800 m (5,900 ft). There is a corresponding decrease in permeability values.

However, the Niobrara chalks have lower porosity values at similar present-day depths of burial than do comparable European chalks (Lockridge and Scholle, 1978). Lockridge and Scholle (1978) suggested that the difference is because the Niobrara chalks (1) were subjected to greater burial depths than present day and (2) contain larger amounts of impurities. Depositional and erosional patterns of younger Cretaceous and Tertiary rocks shown by Shurr (1980) and Shurr and Sieverding (1980) indicate that there was probably more overburden during Late Cretaceous and Tertiary time. The occurrence of more overburden or higher geothermal gradients also is suggested by the degree of maturation of organic matter, discussed below.

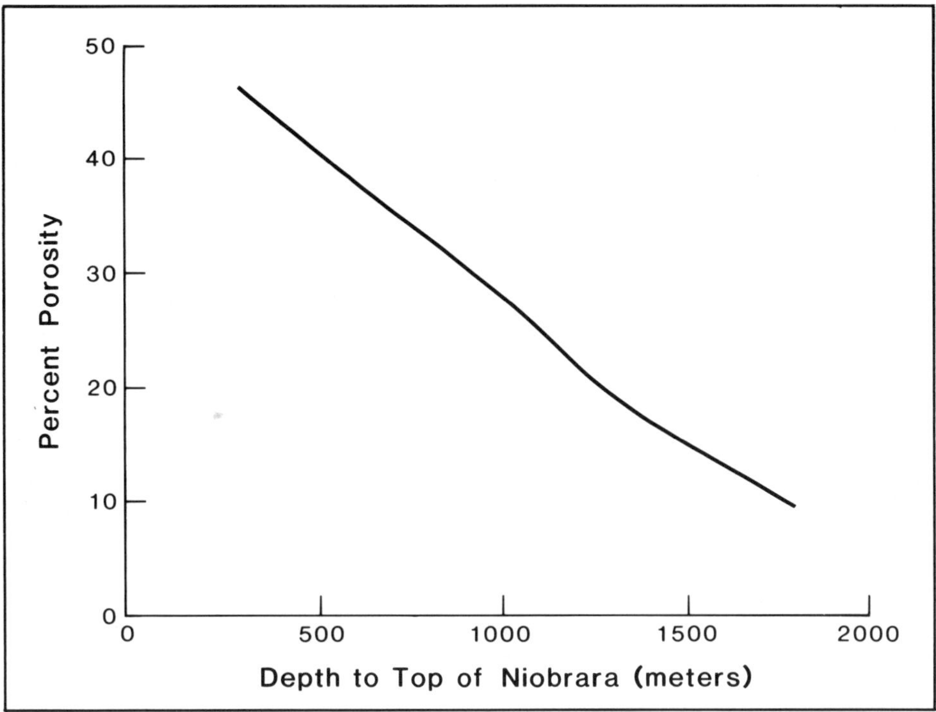

Figure 6 — Porosity versus depth plot for shallow Niobrara chalk reservoirs of eastern Denver basin (after Lockridge and Scholle, 1978). 1 m = 3.3 ft.

CHARACTER AND ORIGIN OF NATURAL GAS

Gas samples were collected from seven wells along a southeast–northwest belt that parallels the productive trend (Fig. 1). Present-day depths of burial to the Niobrara chalk reservoirs in these wells increase to the northwest from 328 to 842 m (1,076–2,763 ft). Analyses of the natural gases are summarized in Table 1 and are presented by well in order of increasing depth to the top of the pay zone. The volume percentage of selected components is reported, together with the proportion of methane in the hydrocarbon fraction (C_1/C_{1-5}) and the stable carbon-isotope ratio ($\delta^{13}C$) of the methane.

All the gases are uniformly dry, with methane generally accounting for more than 98% of the hydrocarbon fraction (Table 1). In addition, the methane component of the gases is enriched in the light isotope ^{12}C, and $\delta^{13}C_1$ values are lighter than −55 ppt. As shown in Figure 7, the gases from the Niobrara Formation become isotopically heavier with increasing depth; $\delta^{13}C_1$ values range from −65 ppt at 328 m (1,076 ft) to −55 ppt at 842 m (2,763 ft).

The chemical and isotopic compositions of the natural gases being produced from the Niobrara Formation suggest that they are dominantly of microbiological or biogenic origin (Fuex, 1977). Biogenic gas is generated during the immature stage of

Figure 7 — Plot showing changes with depth of $\delta^{13}C_1$ of natural gases from Niobrara Formation in Colorado and Kansas. Dots and numbers correspond to wells shown in Figure 1 and described in Table 1. 1 m = 3.3 ft.

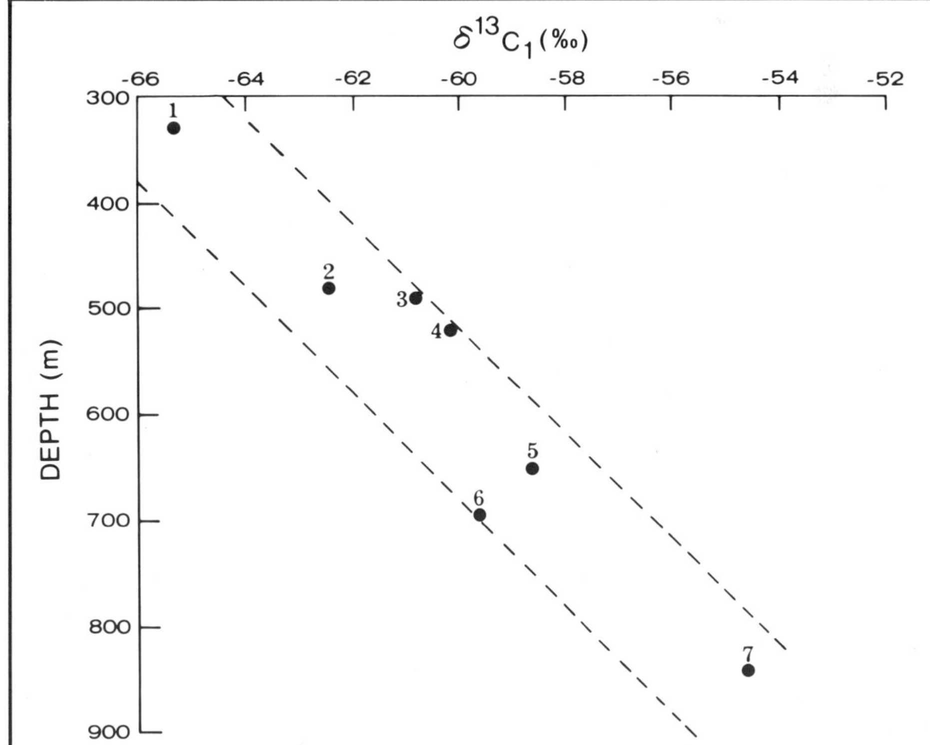

Table 1

Well Name	Location	Pay Zone (m)	N_2 and (or) Air	C_1	CO_2	C_2H_6	C_3H_8	$i\text{-}C_4H_{10}$	$n\text{-}C_4H_{10}$	$i\text{-}C_5H_{12}$	$n\text{-}C_5H_{12}$	$C_1/C_{1\text{-}5}$	$\delta^{13}C_1$ (ppt)
Kansas													
1. Mountain Petroleum 1-10 Topliff	NW 1/4 SE 1/4 sec. 10, T9S, R40W	322–328	5.18	93.91	0.25	0.67	tr	tr	tr	tr	tr	.993	−65.4
Colorado													
2. Kansas–Nebraska Natural Gas 1-10 Engel	SW 1/4 SW 1/4 sec. 10, T3S, R42W	478–482	6.83	91.18	0.13	1.24	0.39	0.06	0.06	0.02	0.01	.981	−62.5
3. Mountain Petroleum 2-34 Ekberg	NW 1/4 NW 1/4 sec. 34, T2S, R43W	481–491	5.08	92.9	0.29	1.16	0.41	0.07	0.06	tr	tr	.982	−60.8
4. Mountain Petroleum 1-16 State	NE 1/4 SW 1/4 sec. 16, T2S, R43W	513–518	5.46	92.53	0.23	1.06	0.48	0.09	0.08	0.03	0.02	.981	−60.1
5. Kansas–Nebraska Natural Gas 1-29 Allison	SE 1/4 SE 1/4 sec. 29, T1S, R44W	642–647	5.72	92.1	0.24	1.07	0.55	0.13	0.11	0.04	0.03	.979	−58.8
6. Kansas–Nebraska Natural Gas 1-27 Fonte	SW 1/4 NE 1/4 sec. 27, T1N, R45W	688–695	6.6	91.4	0.2	0.91	0.54	0.12	0.12	0.05	0.04	.981	−59.7
7. Kansas–Nebraska Natural Gas 1 State	NE 1/4 NE 1/4 sec. 16, T3N, R47W	838–842	5.12	92.85	0.43	1.05	0.34	0.09	0.07	0.03	0.02	.976	−54.7

1 m = 3.3 ft.

Table 1 — Chemical and isotopic composition of natural gas from selected wells shown in Figure 1.

thermal alteration of organic matter prior to the onset of major oil and gas generation by thermal-cracking processes (Fig. 8). The gas is generated in association with the breakdown of organic matter by microorganisms in anaerobic sediments at low temperatures (Mah et al, 1977).

Biogenic gas generally can be distinguished from thermally generated gases by both its chemical and isotopic composition. It consists predominantly of methane that results from the reduction of CO_2 by hydrogen (electron donor) produced by anaerobic oxidation of the organic matter. The occurrence of as much as 2% of higher alkanes (ethane, propane, butane, and pentane) in ancient accumulations, such as in the Niobrara, can be attributed to an early thermogenic contribution (Rice and Claypool, 1981).

Another important criterion for distinguishing biogenic gas from thermogenic gas is stable carbon-isotope ratios (Fuex, 1977). During the immature stage of hydrocarbon generation, the ^{12}C-enrichment in the methane is mainly the result of a large kinetic isotope effect associated with microbial decomposition of organic matter (Rosenfeld and Silverman, 1959). The general trend over the depth range of active biogenic-gas generation is for the methane to become isotopically heavier with increasing depth of burial because of more rapid removal than addition of isotopically light, organically derived CO_2, which is the main source for methane formation (Rice and Claypool, 1981). The range of $\delta^{13}C_1$ values of typical biogenic gases from DSDP sediments is −90 to −65 ppt (Claypool, 1974).

In addition, the methane in the Niobrara gases probably becomes heavier than −65 ppt because of mixing of minor amounts of thermogenic methane with biogenic methane previously formed at shallower depths. Rice and Claypool (1981) assigned an average but arbitrary lower $\delta^{13}C_1$ value of −55 ppt for ancient biogenic-gas accumulations. The gas sample from the deepest Niobrara well has a $\delta^{13}C_1$ value of about −ppt (Table 1). This $\delta^{13}C_1$ value at the northwestern limit of the gas-producing area more or less corresponds to the lower limit of the occurrence of dominantly biogenic gas. This interpretation is supported by the fact that well E, northwest of the gas-productive trend, is an oil well (Fig. 1).

In contrast, gases from the "J" sandstone in the deeper Wattenberg field in the central part of the basin are distinctly different from those produced from the Niobrara in both chemical and isotopic composition (Rice and Threlkeld, 1982). Gases from the Wattenberg field contain significant amounts of heavier hydrocarbons (C_1/C_{1-5} values are about 0.87) and are isotopically heavier ($\delta^{13}C_1$ values are about −43 ppt). The chemical and isotopic compositions of the gases indicate that they are thermogenic in origin (Fuex, 1977) and were generated by thermal-cracking processes during intermediate stages of thermal maturity in the deeper part of the basin. This interpretation is consistent with level of maturation determined by Clayton and Swetland (1980).

Another possible explanation for the origin of the isotopically light, chemically dry gases in the Niobrara is fractionation of thermogenic gases generated at greater depths during migration to shallower reservoirs (Neglia, 1979). Laboratory experiments have suggested that isotopic fractionation in methane may take place during migration, but that it may be significant only when migrating small quantities of gas (Galimov, 1967; Gunter and Gleason, 1971; Lebedev and Syngayevski, 1971). Recent work by Fuex (1980) suggests that fractionation is negligible for commercial quantities of gas and that ^{13}C is depleted slightly when migrating small quantities. This conclusion is confirmed by field studies cited by Bernard, Brooks, and Sackett (1977), Coleman et al (1977), and Rice (1980). They observed insignificant changes in isotopic composition with migration but noted a difference in chemical composition; i.e., gases become chemically drier with increasing distance of migration.

Fractionation of deeper thermogenic gases by migration is discounted as a possible origin for shallow Niobrara gases for the following reasons: (1) laboratory experiments and field studies cited earlier do not indicate significant fractionation, (2) gas is entrapped in low-permeability reservoirs that inhibit long-range migration, and (3) sufficient organic matter and proper environmental conditions were probably present during the time of deposition of the Niobrara such that significant quantities of biogenic methane could have been produced and retained.

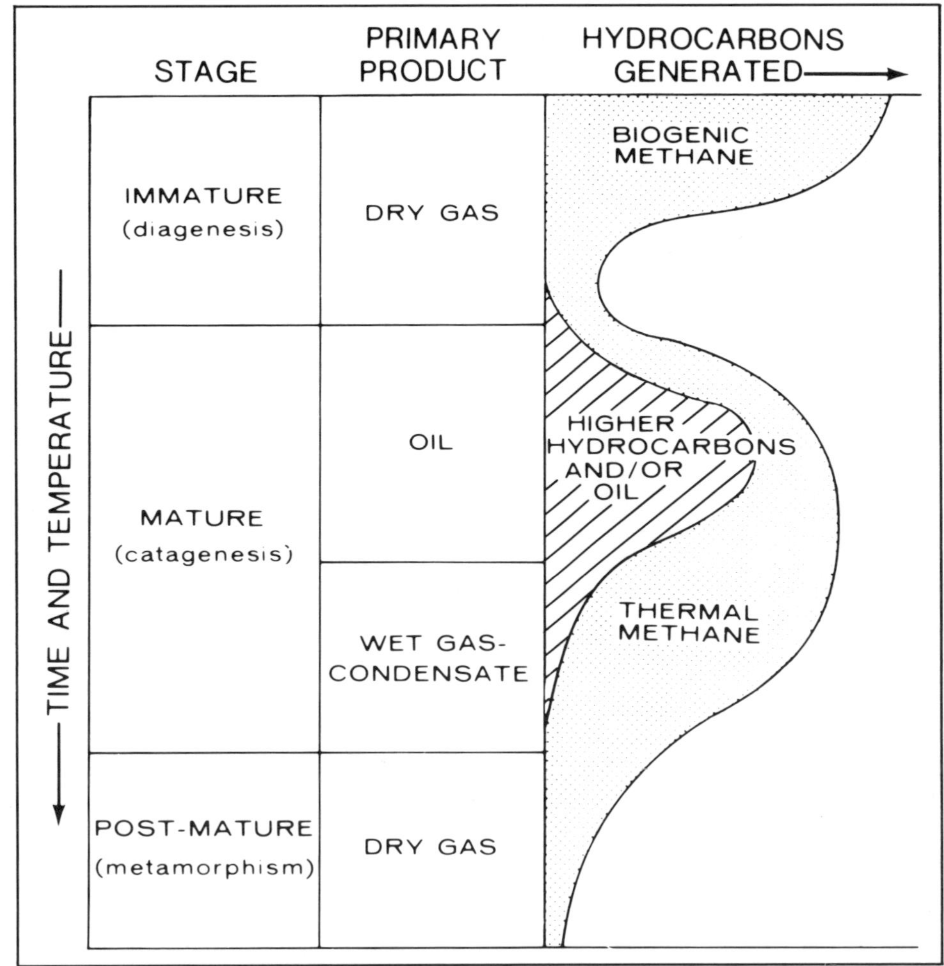

Figure 8 — Diagram showing generation of hydrocarbons with increasing temperature and time.

EVALUATION OF GAS SOURCE

The organic matter in the gas-bearing chalks of the Niobrara Formation is evaluated as a possible indigenous source of biogenic gas. An indigenous source is suggested by the fact that the chalks (1) have low permeability that inhibits long-range migration and (2) are overlain by a thick section of shale (Pierre Shale) containing many bentonite beds in the lower part. The shale served as a seal for the gas after the reservoirs were naturally fractured on low-relief structures late in the burial history. The fracturing resulted in enhancement of the reservoir quality and in localization of gas accumulation on the structures. Gas generation, however, is interpreted to have been widespread.

The amount, type, and level of maturation of the organic matter within the Niobrara chalks of the eastern Denver basin were evaluated by analysis of core samples from five wells in northeastern Colorado. The locations of the wells are shown in Figure 1, and details of the wells and cored intervals are given in Table 2. Four of the wells are situated within the gas-producing area; three of these are gas producers and the other is a dry hole because it is located off-structure. The fifth well, the Excelsior 1 Alice G. Nay, is northwest of the gas-productive area and is a shut-in oil well. Most of the core (67 m [220 ft] of a total of 74 m [243 ft]) is from the Smoky Hill Member, which contains the gas-productive Beecher Island zone. The remaining core (7 m [23 ft]) is from the Sharon Springs Member of the Pierre Shale in the Kansas–Nebraska 1-32 Whomble well.

Organic-Matter Content

Total-organic-carbon (TOC) analyses were conducted on 53 samples using a wet-oxidation method modified from Bush (1970) (Table 3). The TOC content for 46 samples from the Smoky Hill Member of the Niobrara Formation ranges from 0.43 to 5.8% and averages 3.2%.

Table 2

Well Name	Location	Status of Well	Top of Niobrara (m)	Pay Interval (m)	Cored Interval (m)	Maximum Recorded Bottom-hole Temperature (°C)
A. Kansas–Nebraska Natural Gas 1-32 Whomble	NE 1/4 NE 1/4 sec. 32, T2S, R43W	Gas well	454	457–466	447–463	36
B. Mountain Petroleum 1-29 State	SW 1/4 NE 1/4 sec. 29, T2S, R43W	Gas well	443	451–454	445–452	32
C. Kansas–Nebraska Natural Gas 1-33 Powell	NE 1/4 SW 1/4 sec. 33, T3N, R47W	Dry and abandoned	853		853–871	47
D. J–W Operating 1 Brophy	SE 1/4 SE 1/4 sec. 29, T4N, R46W	Gas well	836	839–842	842–863	46
E. Excelsior Oil 1 Alice G. Nay	NW 1/4 SE 1/4 sec. 14, T9N, R58W	Oil well	1,646	1,681–1,687	1,655–1,667	74

1 m = 3.3 ft.
°C = (°F − 32)/1.8.

Table 2 — Details of wells from which core samples were analyzed.

Seven samples from the overlying Sharon Springs Member had TOC contents ranging from 5.2 to 10.3%, with an average of 7.7%.

The organic matter in the Niobrara exceeds the 0.5% value used as minimum for significant methane generation by microbial activity in modern sediments (Claypool and Kaplan, 1974; Rashid and Vilks, 1977). The present TOC values for the Niobrara are minimal because some organic matter was consumed by biological activity prior to, during, and after methane genesis. Therefore, the Niobrara undoubtedly contained sufficient organic matter to have supported methane generation and other microbial activities.

The TOC content of the Niobrara is comparable to that of the Upper Cretaceous Austin Group in southeastern Texas. The Austin contains an average of 3.5% TOC (Grabowski, 1981). In contrast to the Niobrara, the Austin has been buried more deeply and is the source of indigenous oil in naturally fractured reservoirs.

Clayton and Swetland (1980) analyzed crude oils and shales from the main part of the Denver basin in order to determine oil–source-rock relations. They concluded that most of the oil in the basin was derived from the Carlile Shale, Greenhorn Limestone, Graneros Shale, and Huntsman interval. The average TOC content for 27 samples analyzed from these intervals is 2.1%. This value is considerably lower than the average value obtained for the Niobrara in the present study and indicates that the Niobrara contains sufficient TOC to generate significant quantities of thermogenic hydrocarbons when subjected to the depths of burial and temperatures necessary for thermal-cracking processes.

Organic-carbon contents reported by Clayton and Swetland (1980) are lower for samples of the Niobrara Formation from the central part of the basin (average 2.2%) than for samples from the eastern side of the basin as reported in the present study. The lower values suggest that (1) TOC values have decreased in the deeper, central part of the basin owing to thermal-cracking processes, and/or (2) original organic-matter content was lower in the central part of the basin because of a facies change from chalk to shale.

As stated earlier, the Smoky Hill Member of the Niobrara Formation is rhythmically laminated. The organic matter is concentrated in the darker colored laminae that have higher amounts of insoluble residue. These organic-rich laminae were probably formed during cycles of low-oxygen conditions during an overall "oceanic anoxic event" (Arthur and Schlanger, 1979).

Type of Organic Matter

The distinction among various types of kerogen generally is critical for source-rock evaluation because different types of organic matter yield thermogenic hydrocarbons of varying molecular weight—i.e., liquid versus gas (Hunt, 1979). However, Rice and Claypool (1981) showed that significant reserves of ancient biogenic gas have been generated from both marine and nonmarine sources. Therefore, the evaluation of the type of organic matter is not critical to the determination of the source of the biogenic gas. However, marine-derived organic matter generally is considered to contain a higher proportion of the tissues susceptible to degradation by microorganisms.

Visual Kerogen Analysis

Kerogen was classified by microscopic examination into four categories—algal-amorphous, herbaceous, woody, and coaly (inertinite) (Hunt, 1979). Results of the analyses are presented in Table 4 for 6 selected samples from 5 different wells. There is one Niobrara sample from each well, in addition to a single Sharon Springs sample from the Kansas–Nebraska 1-32 Whomble well.

Amorphous-type kerogen makes up the largest percentage (44–50) of the Niobrara samples. The rest of the kerogen is distributed among herbaceous, woody, and coaly, with decreasing amounts of each. Amorphous kerogen is the dominant type in most carbonates. The fact that the other types of kerogen collectively occur in amounts ranging from 50 to 56%, and the occurrence of

Table 3

Well Name	Depth of Sample (m)	TOC (wt%)	Volatile Hydrocarbons S_1 (ppm)	Pyrolitic Hydrocarbons S_2 (wt%)	T_{max} (°C)	Hydrogen Index (mg HC/g C_{org})	Oxygen Index (mg CO_2/g C_{org})	Production Index $S_1/S_1 + S_2$
Kansas-Nebraska Natural Gas 1-32 Whomble	447	6.93	64	31.73	391	458	30	0.02
	448	7.29	97	37.75	394	517	22	.03
	449	7.07	116	38.43	400	544	47	.03
	450	7.62	102	43.72	403	574	33	.02
	451	10.26	131	55.23	412	538	29	.02
	452	5.19	61	25.56	—	493	44	.02
	453	8.28	21	47.92	400	581	43	.04
	454	5.80	35	24.32	—	419	54	.01
	455	3.74	57	20.48	418	548	141	.03
	456	3.18	51	9.06	403	285	331	.05
	457	3.29	86	16.26	400	494	88	.05
	459	1.96	43	10.76	405	549	116	.04
	460	2.68	47	11.47	403	428	90	.06
	461	1.99	51	10.47	403	526	111	.04
Mountain Petroleum 1-29 State	445	4.09	47	20.42	—	499	63	.02
	446	4.26	34	24.66	399	579	55	.01
	447	2.79	53	10.76	—	386	103	.05
	448	2.69	22	15.09	397	561	100	.01
	449	2.49	31	14.52	—	583	113	.02
Kansas-Nebraska Natural Gas 1-33 Powell	854	2.25	66	10.15	406	451	69	.06
	856	1.55	52	5.04	418	325	131	.09
	857	3.19	88	14.26	—	447	69	.06
	858	2.54	47	12.08	397	476	76	.04
	862	5.31	84	27.51	400	518	44	.22
	863	4.98	138	28.53	406	573	54	.45
	864	3.65	40	20.4	403	559	46	.19
	865	4.78	70	24.46	403	512	53	.03
	866	3.11	76	19.74	406	635	68	.04
	867	2.36	97	15.58	—	660	85	.06
	868	3.48	77	26.82	406	770	54	.03
	870	3.49	70	16.99	—	487	81	.04
	871	3.42	116	23.56	394	689	66	.05
J-W Operating 1 Brophy	844	3.83	158	60.25	403	—	180	.03
	847	2.82	100	24.08	406	854	93	.04
	848	2.82	107	34.3	403	—	84	.03
	849	4.66	108	36.04	409	773	77	.03
	851	3.40	114	28.44	397	836	75	.04
	856	3.72	86	23.31	409	627	68	.04
	857	3.98	25	12.9	409	324	80	.02
	860	3.04	42	12.87	412	423	79	.03
Excelsior Oil 1 Alice G. Nay	1,655	.43	68	.2	421	45	29	.78
	1,656	2.66	785	8.78	415	330	25	.47
	1,657	1.44	913	4.15	—	288	52	.69
	1,658	2.02	1,019	5.98	412	296	20	.63
	1,659	2.00	614	6.14	411	307	32	.50
	1,660	1.74	636	4.8	415	276	26	.57
	1,661	4.72	617	16.81	418	356	15	.27
	1,662	2.56	759	9.96	421	389	20	.43
	1,663	2.8	787	8.72	421	311	29	.47
	1,664	4.53	354	13.57	427	300	21	.21
	1,665	4.01	289	12.30	430	307	29	.19
	1,666	13.76	4	12.45	425	331	—	.31
	1,667	0.57	50	.34	424	60	46	.59
	1,668	2.30	792	7.11	424	309	23	.53

1,000 ppm is equivalent to 1 mg HC/g rock relative to dry-rock weight or to 1 kg HC/metric ton of rock. First seven samples are from Sharon Springs Member of Pierre Shale; others are from Smoky Hill Member of Niobrara Formation. 1 m = 3.3 ft. °C = (°F − 32)/1.8.

Table 3 — Total-organic-carbon (TOC) and pyrolysis data for 53 samples.

Table 4

Well Name	Depth of Sample (m)	Type of Kerogen (%) Amorphous	Herbaceous	Woody	Coaly	Atomic H/C
1. Kansas–Nebraska Natural Gas 1-32 Whomble	450	40	30	20	10	1.19
2. Kansas–Nebraska Natural Gas 1-32 Whomble	461	44	34	11	11	1.19
3. Mountain Petroleum 1-29 State	447	50	26	12	12	1.19
4. Kansas–Nebraska Natural Gas 1-33 Powell	867	44	22	22	11	1.26
5. J–W Operating 1 Brophy	848	50	26	12	12	1.14
6. Excelsior Oil 1 Alice G. Nay	1,664	44	22	22	11	1.02

First sample is from Sharon Springs Member of Pierre Shale; others are from Smoky Hill Member of Niobrara Formation. Visual kerogen analyses were run by Geochem Laboratories, Inc., Houston, Texas. 1 m = 3.3 ft.

Table 4 — Visual kerogen and atomic hydrogen–carbon (H/C) ratios for six selected samples.

Table 5

Well Name	Depth of Sample (m)	TAI	R_o
1. Kansas–Nebraska Natural Gas 1-32 Whomble	450	1.9	0.32
2. Kansas–Nebraska Natural Gas 1-32 Whomble	461	1.7	0.35
3. Mountain Petroleum 1-29 State	447	1.6	0.32
4. Kansas–Nebraska Natural Gas 1-33 Powell	867	1.9	0.34
5. J–W Operating 1 Brophy	848	1.9	0.34
6. Excelsior Oil 1 Alice G. Nay	1,664	3.1	0.69

First sample is from Sharon Springs Member of Pierre Shale; others are from Smoky Hill Member of Niobrara Formation. 1 m = 3.3 ft.

Table 5 — Thermal-alteration-index (TAI) and vitrinite reflectance (R_o) data for six selected samples.

considerable amounts of insoluble residue composed of detrital clay minerals and quartz, indicates the strong influence of a land source.

In comparison, analysis of a single sample of the Sharon Springs Member of the Pierre Shale (sample 1, Table 4) indicates lesser amounts of sapropelic kerogen and greater amounts of herbaceous, woody, and coaly-type kerogens.

Elemental Analysis

Another classification system for kerogen was presented by Tissot et al (1974), using the elemental composition and evolution path on a van Krevelen diagram. The van Krevelen diagram involves the atomic hydrogen-to-carbon (H/C) and oxygen-to-carbon (O/C) ratios of demineralized kerogen samples. Three chemical types of organic matter (I, II, and III) can be determined using this diagram, but they do not necessarily correlate on a one-to-one basis with the four types recognized by visual kerogen analysis. In general, type I is comparable to amorphous; type II is a mixture of amorphous, herbaceous, and woody; and type III is composed of woody and coaly kerogens (Hunt, 1979). Usually, there is a decrease in the H/C and an increase in the O/C in the composition of the original organic matter in a comparison of types I, II, and III.

Atomic hydrogen-to-carbon ratios of kerogens from the same six samples described under visual kerogen analysis are shown in Table 4. The ratios of the samples from the Niobrara and overlying Sharon Springs show that the kerogens are hydrogen rich (H/C range from 1.02 to 1.26). Although the O/C are not available, the kerogens probably fall within the range of type II, which is comparable to the classification derived from visual kerogen analysis.

Pyrolysis (Rock-Eval)

Rock-Eval pyrolysis was conducted on 53 samples from the 5 wells listed in Table 2. The data for these samples are tabulated in Table 3. "Hydrogen-index" and "oxygen-index" values of samples from three wells are plotted on a modified van Krevelen diagram shown in Figure 9. These samples are considered to be typical of those from all five wells. All the samples shown in Figure 9 cluster along the type II path, which is consistent with interpretation made from visual kerogen analysis.

In summary, a mixture of kerogen types is present in the Niobrara chalks, although amorphous material is dominant. Amorphous kerogen is consistent with deposition in an open-marine environment, whereas a mixture of the other types suggests some contribution from terrestrial sources. The Niobrara chalks contain more than a sufficient amount of organic matter to have been the source of biogenic gas while the sediments were being deposited. In addition, this dominant type of kerogen was and still is capable of generating significant quantities of thermogenic oil and gas if subjected to suitable temperature conditions.

Thermal Maturity

Biogenic gas is generated at low temperatures (0°–75°C [32°–167°F]) prior to the generation of the full range of hydrocarbons by thermal degradation and cracking reactions with increased temperatures (Rice and Claypool, 1981). Several types of analyses were used to determine the level of thermal maturation of organic matter and associated clays in the Niobrara chalks. Thermal maturity is relevant to biogenic-gas occurrence in two main respects: (1) recognition of biogenic gas undiluted by thermogenic hydrocarbons and (2) preservation of reservoir properties in chalks.

The level of maturation in the Niobrara increases to the northwest. This interpretation is consistent with the following facts: (1) present-day depth of burial increases to the northwest, (2) porosity and permeability values in the chalks decrease to the northwest, (3) carbon-isotope compositions of the biogenic gases become heavier within the producing trend to the northwest, and (4) oil is produced from the Niobrara northwest of the gas-productive area.

Thermal-Alteration Index (TAI)

TAI values for six samples studied for this report range from 1.6 to 3.1 (Table 5). Samples from wells A–D (Fig. 1; Table 2), which occur in the gas-producing area, have TAI values between 1.6 and 1.9 and are interpreted to be thermally immature (Staplin, 1969). In contrast, the sample from the Excelsior 1 Alice G. Nay well has a TAI value of 3.1 and is considered to be thermally mature with respect to liquid-hydrocarbon generation.

Vitrinite Reflectance

Vitrinite-reflectance (R_o) values show the same trend as do TAI values (Table 5). In the gas-producing area, R_o values for all five samples are about 0.3%, which indicates immaturity (Dow, 1977). In western Canada, Hacquebard (1977) concluded that large quantities of biogenic gas occur at an R_o value of 0.4%, which is similar to values obtained from biogenic-gas-producing chalks of the Niobrara. In comparison, the R_o value for a sample from the Excelsior 1 Alice G. Nay well is considerably higher, at 0.69%, and represents the mature stage.

Clay Transformation

Investigations of shales from the Gulf Coast have demonstrated that highly expansible, mixed-layer clays undergo a temperature-controlled burial diagenesis (Burst, 1959; Hower et al, 1976). These changes have been related by Foscolos and Powell (1980) to petroleum generation.

According to Pollastro (1981b), clay-mineral assemblages in the insoluble-residue fraction and associated bentonite beds of the Niobrara in the five cored wells (Fig. 1; Table 2) show a similar transformation. In wells A and B, the clay fraction of the insoluble-residue fraction is characterized by randomly ordered, mixed-layer smectite–illite containing approximately 25–35% illite-type layers. A transition phase occurs in wells C and D in which illite-type layers account for approximately 55–65%. In the deepest well, E, ordered mixed-layer smectite–illite layers are present with approximately 70–80% illite-type layers. The associated bentonite beds progress through a similar transformation. This transformation occurs at approximately 100°C (212°F) (Hoffman and Hower, 1979), which is comparable to the temperature required for initiation of thermogenic hydrocarbon generation in Upper Cretaceous rocks (Hunt, 1979).

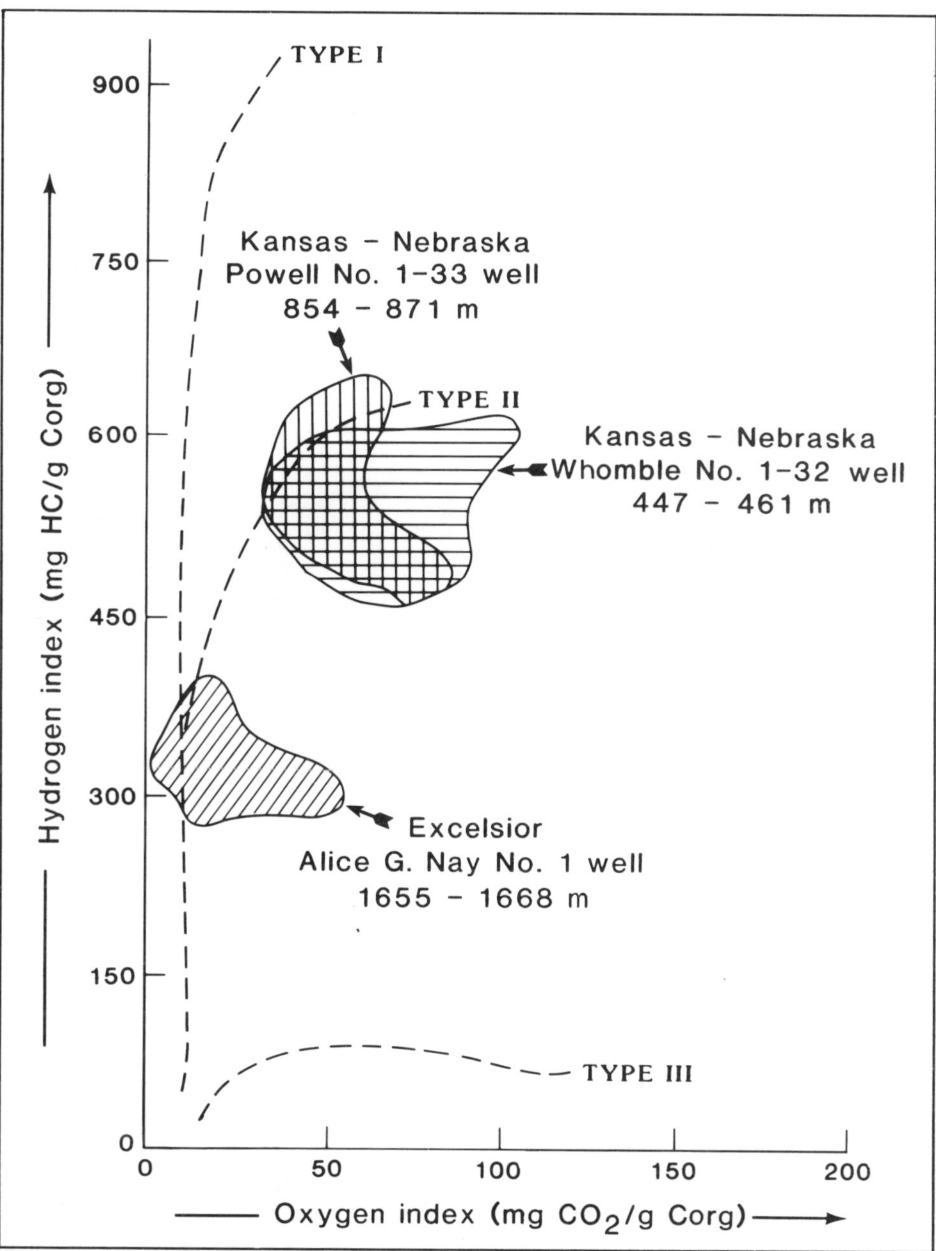

Figure 9 — Modified van Krevelen diagram showing type and maturity of kerogen in core samples from three selected wells. Pyrolysis data for all five wells are presented in Table 3.

Pyrolysis (Rock-Eval)

In addition to organic-matter type, pyrolysis data can be used to determine thermal maturity by three methods. The first method involves the use of a modified van Krevelen diagram. Maturation level is determined by the position of samples on the evolutionary path of a particular type of kerogen. At the beginning of the immature stage, a rapid decrease in the "oxygen index" occurs while the "hydrogen index" remains nearly constant (Espitalié et al, 1977). During the mature stage, only the "hydrogen index" decreases.

In Figure 9, the "hydrogen-index" and

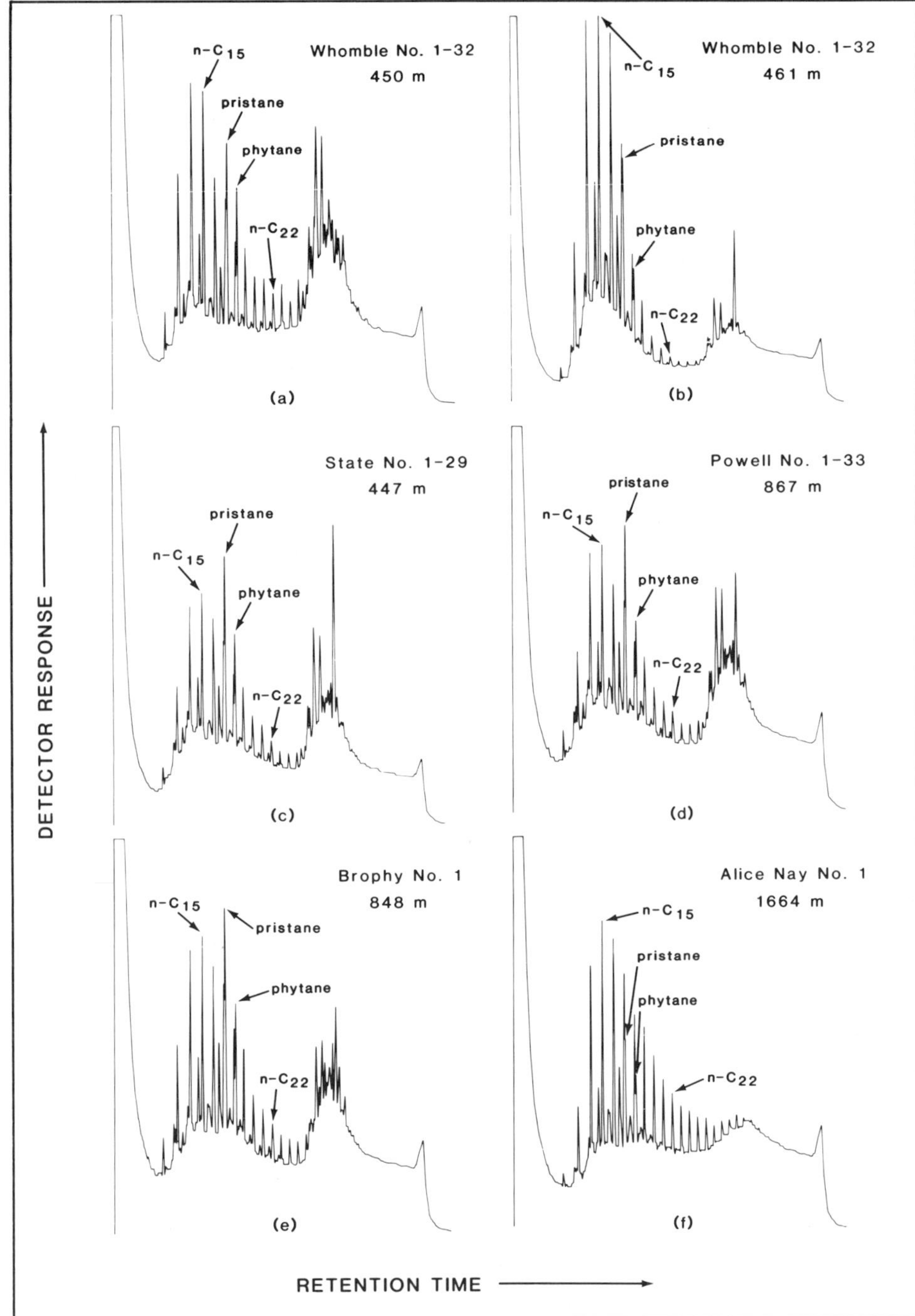

Figure 10 — Gas chromatograms of C_{12+} saturated hydrocarbons of extracted rock samples from Table 6. Column, 3% Se-30 on gas-chrom Q: 2 m, 6.25 mm; 80–300°C; temp. gradient 12°C/min for 10 minutes, 10°C/min for 10 minutes; flow rate 30 cc He/min.

"oxygen-index" data are plotted for samples from three wells. As discussed earlier, the kerogens for samples from all five wells are classified as type II. Samples from the Kansas–Nebraska 1-32 Whomble and 1-33 Powell wells plot on the immature portion of the evolutionary path. Samples from these two wells are considered to be in the same stage of maturation as samples from Mountain Petroleum 1-29 State and J–W Operating 1 Brophy wells. Kerogen from samples of the Excelsior 1 Alice G. Nay well has a substantially decreased "hydrogen index" and is interpreted to be in mature stage.

A second method for evaluating maturation level concerns the temperature at which maximum quantities of hydrocarbons are generated by pyrolysis (T_{max}). As a general rule, T_{max} increases with depth and temperature for a fixed kerogen type/mineral assemblage. Samples from wells A through D, which are in the gas-productive zone, have average T_{max} values of 403°, 398°, 404°, and 406°C (757°, 748°, 759°, 763°F), respectively (Table 3), and are considered typical for the immature stage. The average T_{max} for samples from well E is considerably higher at 420°C (788°F) (Table 3) and indicates a corresponding higher degree of thermal maturity.

A third parameter is the "production index," which is expressed by the ratio $S_1/S_1 + S_2$. The value of this index increases with maturation because the S_1 peak increases and the S_2 peak decreases with increasing burial depth (Espitalié et al, 1977). Samples from the four shallow wells (wells A–D) have production indices that range from 0.02 to 0.07 (Table 3) and are characteristic of immature rocks. The "production index" for the Excelsior 1 Alice G. Nay well averages 0.47. This higher value indicates advancement to a stage of thermal maturity at which petroleum generation is possible (Espitalié et al, 1977). This interpretation is confirmed by the fact that the well has produced oil.

Elemental Analysis

Elemental analysis is used in the same way for evaluating maturation as is pyrolysis data. The atomic H/C of kerogens from shallow wells A–D range from 1.14 to 1.26 and indicate a low level of maturation (immature) for this type of organic matter (Table 4). Because previous studies indicate that the kerogens from all wells are classified as type II, a decrease in the atomic H/C (1.02) for the sample from well E is due to the generation of thermogenic hydrocarbons and advancement to a higher level of thermal maturity.

Solvent Extraction and Elution Chromatography

Results of solvent extraction and elution chromatography are consistent with other maturation indices. Gas chromatograms of C_{12+} saturated hydrocarbons are presented in Figure 10. Samples 1 through 5, which represent samples from the shallow wells in the gas-producing area, display a strong bimodal size distri-

Table 6

Well Name	Depth of Sample (m)	TOC (wt%)	EOM (ppm)	HC (ppm)	Non-HC (ppm)	Saturated/Aromatic Ratio	HC/EOM (%)	Pristane/Phytane	Phytane/n-C_{18}	Sat HC/TOC (%)	EOM/TOC (%)	HC/TOC (%)
1. Kansas–Nebraska Natural Gas 1-32 Whomble	450	9.01	2,155	1,227	928	1.7	57	1.4	1.7	0.9	2.4	1.4
2. Kansas–Nebraska Natural Gas 1-32 Whomble	461	3.12	1,686	768	918	1.2	46	2.4	0.8	1.3	5.4	2.5
3. Mountain Petroleum 1-29 State	447	3.34	1,393	591	802	0.7	42	1.7	1.6	0.7	4.2	1.8
4. Kansas–Nebraska Natural Gas 1-33 Powell	867	4.78	1,865	1,101	764	1.1	59	1.9	1.3	1.2	3.9	2.3
5. J–W Operating 1 Brophy	848	3.52	1,218	692	526	1.0	57	1.8	1.3	1.0	3.5	2.0
6. Excelsior Oil 1 Alice G. Nay	1,664	4.65	10,113	7,862	2,251	1.7	78	1.6	0.5	10.7	21.7	16.9

First sample is from Sharon Springs Member of Pierre Shale; others are from Smoky Hill Member of Niobrara Formation. 1 m = 3.3 ft.

Table 6 — Organic-geochemical data for six selected samples.

bution on the chromatogram. Large amounts of both low-molecular-weight hydrocarbons (C_{13}–C_{17}) in addition to unresolved cycloalkanes (steranes and triterpanes) in the region of C_{28}–C_{30} are present. Sterane and triterpane precursors are synthesized by living organisms, and their occurrence in amounts predominating over normal alkanes is observed frequently in ancient immature sediments buried at shallow depths. In the low-molecular-weight range, normal alkanes are subordinate to the isoprenoids, pristane and phytane. Pristane and phytane are derived from the phytol chain of chlorophyll and generally occur in large amounts relative to n-alkanes in immature sediments (Tissot and Welte, 1978).

In contrast to the immature samples previously discussed, the sample from the Excelsior 1 Alice G. Nay well contains a relatively smooth distribution of alkanes with a significant decrease in the amount of higher molecular-weight molecules (C_{26}–C_{32}). There is also a marked decrease in the amount of the isoprenoids, pristane and phytane, relative to the n-alkanes. This distribution is typical of mature source rocks and is caused by the addition of hydrocarbons generated from kerogen (Tissot and Welte, 1978).

It is important to recognize that interpretation of maturity based on composition of extractable organic matter (EOM) may be affected by migration of oil in the Excelsior 1 Alice G. Nay well. However, several measurements discussed earlier, such as R_o, TAI, and clay transformation, are not affected by migration hydrocarbons and indicate that the rocks in this well are thermally mature..

Quantitative determinations of EOM for six selected samples are presented in Table 6. The data generally support the interpretation that rocks from the gas-producing area are immature and those from the Excelsior 1 Alice G. Nay well are more mature. Immature sediments usually have extractable hydrocarbon concentrations that are less than 1% of the TOC. The fact that relatively immature rocks in the study area have values slightly greater than 1–2% probably indicates that these rocks were deposited with a high content of labile nonkerogen hydrocarbon precursors, i.e., that they have greater than normal hydrocarbon-generating potential because of the hydrogen-rich nature of the organic matter. Mature source rocks are present in the Excelsior 1 Alice G. Nay well as indicated by (1) high contents of total EOM and hydrocarbons relative to TOC, (2) 78% of EOM composed of hydrocarbons, (3) a relatively high ratio of saturated to aromatic hydrocarbons (1.7), and (4) a phytane/n-C_{18} of less than 1 (0.5).

Burial Temperatures

Uncorrected bottom-hole temperatures for wells in the gas-productive area range from 32° to 46°C (90°–115°F) with a general increase to the northwest (Table 2). The uncorrected bottom-hole temperature for the Excelsior 1 Alice G. Nay well is significantly higher at 74°C (165°F). All data indicate that chalks from the latter well are thermally mature. Rocks that undergo a typical burial–subsidence history with average thermal gradients (30°C/km [2.6°F/100 ft]) and sedimentation rates (50–100 m/m.y. [164–328 ft/m.y.]) require temperatures greater than 100°C (212°F) to generate significant amounts of liquid hydrocarbons (Waples, 1980). The required existence of higher temperatures than are now present in this well suggests that either a higher geothermal gradient or more overburden was present during Tertiary time.

In summary, the level of thermal maturation in Niobrara chalks increases to the northwest in northeastern Colorado. In the biogenic-gas-productive area, the chalks are thermally immature but have generated and preserved significant amounts of biogenic gas because of their environment of deposition and subsequent burial history. Northwest of the gas-productive area, in the Excelsior 1 Alice G. Nay well, the chalks have produced oil and are thermally mature. Rocks at this locality probably also generated biogenic gas, but deeper burial and subsequent thermogenic hydrocarbon generation would make its recognition

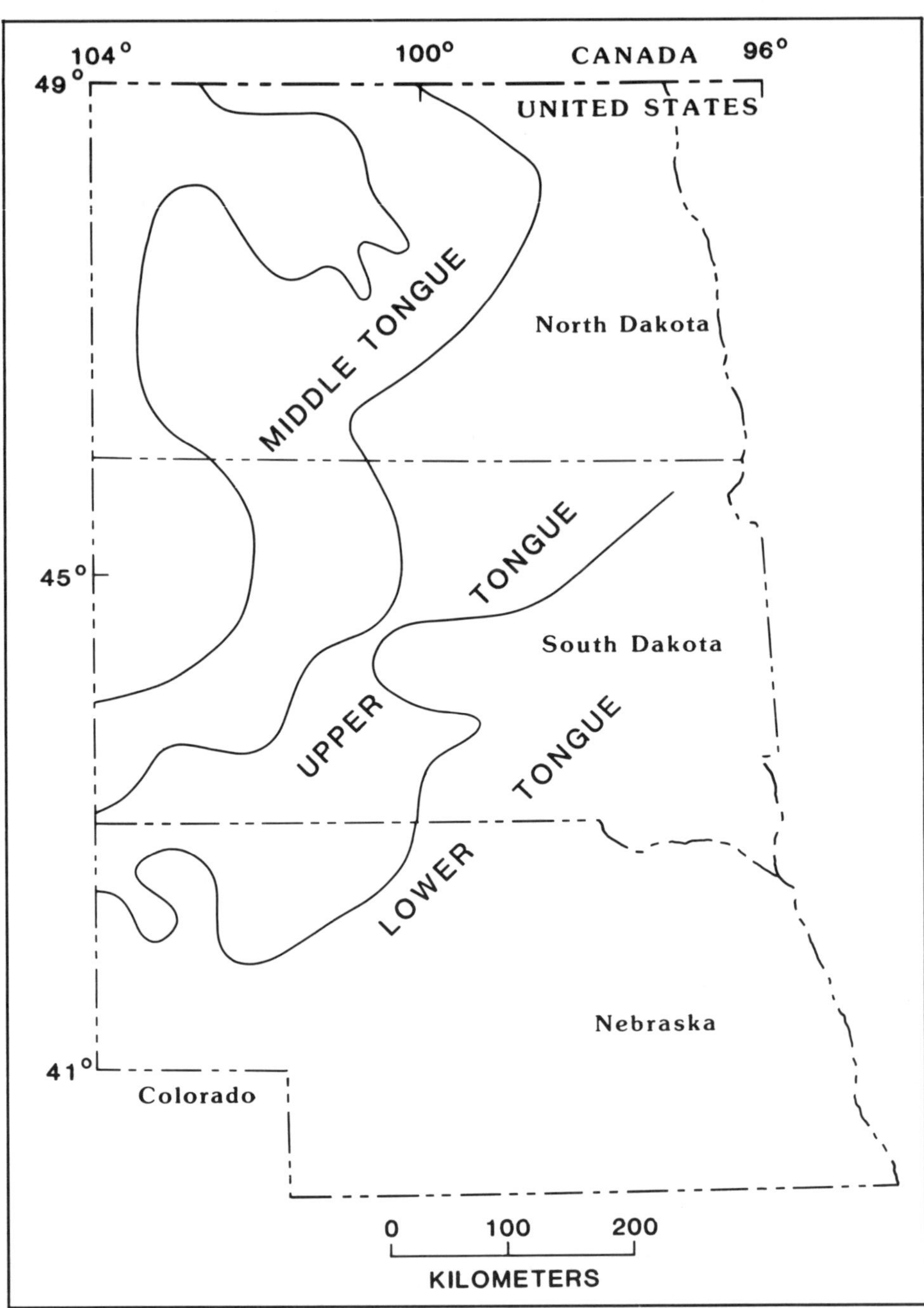

Figure 11 — Map showing northwest extent of chalk tongues of Niobrara Formation in central and northern Great Plains (after Shurr and Sieverding, 1980).

difficult. Also, loss of reservoir properties generally prevents successful completion of wells.

SUMMARY AND CONCLUSIONS

Isotopically light ($\delta^{13}C$ values ranging from -55 to -65 ppt), methane-rich ($C_1/C_{1-5} > 0.98$) gas is produced at shallow depths from chalks of the Upper Cretaceous Niobrara Formation in the eastern Denver basin. This biogenic gas was generated by decomposition of organic matter at low temperatures by anaerobic microorganisms in sediments deposited during widespread "oceanic anoxic events." The biogenic gas probably was generated *in situ*, as suggested by (1) low permeability of the chalks, which inhibited long-range migration; (2) organic-rich laminae with type II kerogen within the chalks, which is a product of the anoxic environment in which the gas was formed; and (3) a thick section of shale, with many bentonites, that overlies the chalks and which served as a seal for the gas accumulations.

In the eastern Denver basin, the degree of thermal maturity of the chalks increases to the northwest. In the gas-productive area, the chalks are immature with respect to thermogenic hydrocarbon generation. At a well about 100 km (62 mi) northwest of the gas-productive area, the Niobrara has produced small quantities of oil and is thermally mature. Present-day bottom-hole temperatures suggest that geothermal gradients were higher and/or more overburden was present to account for the generation of liquid hydrocarbons in this area.

Chalk beds of the Niobrara Formation are widespread on the eastern side of the Western Interior seaway in the central and northern Great Plains. The chalks form northwest-extending tongues (Fig. 11) that have minimal overburden, which is critical for the preservation of adequate reservoir properties. The potential for biogenic-gas accumulations exists throughout the area where the chalk tongues are developed (Rice and Shurr, 1980). Numerous gas shows have been documented in the Niobrara chalks of South Dakota (Bretz, 1981).

ACKNOWLEDGMENTS

R. M. Pollastro, U.S. Geological Survey, provided information on insoluble residues and clay mineralogy, which are a critical part of this study. I also thank C. N. Threlkeld, T. A. Daws, M. Pawlewicz, and C. M. Lubeck for analytical assistance.

REFERENCES CITED

Arthur, M.A., and S.O. Schlanger, 1979, Cretaceous "oceanic anoxic events" as causal factors in development of reef-reservoired giant oil fields: AAPG Bulletin, v. 63, p. 870–885.

Bernard, B., J.M. Brooks, and W.M. Sackett, 1977, A geochemical model for characterization of hydrocarbon gas sources in marine sediments: Ninth Annual Offshore Technology Conference Proceedings, v. 3, p. 435–438.

Bretz, Richard, 1981, List of natural gas occurrences in South Dakota by county: South Dakota Geological Survey Open-File Report Z–BAS, 148 p.

Brown, C.A., J.W. Crafton, and J.G. Golson, 1981, The Niobrara gas play: exploration and development of a low pressure, low permeability gas reservoir: AIME Society of Petroleum Engineers Paper 10304, 28 p.

Burst, T.F., Jr., 1959, Postdiagenetic clay mineral environmental relationships in the Gulf Coast Eocene, *in* Ada Swineford, ed., Clays and clay minerals: Berkeley, California, Sixth National Conference on Clays and Clay Minerals, 1957, Proceedings, p. 327–341.

Bush, P.R., 1970, A rapid method for determination of carbonate carbon and organic carbon: Chemical Geology, v. 6, p. 59–62.

Claypool, G.E., 1974, Anoxic diagenesis and bacterial methane production in deep sea sediments: University of California at Los Angeles unpublished Ph.D. thesis, 276 p.

——— and I.R. Kaplan, 1974, The origin and distribution of methane in marine sediments, *in* Natural gases in marine sediments: New York, Plenum Press, p. 99–139.

Clayton, J.L., and P.J. Swetland, 1980, Petroleum generation and migration in Denver Basin: AAPG Bulletin, v. 64, p. 1613–1633.

Coleman, D.D., et al, 1977, Isotopic identification of leakage gas from underground storage reservoirs—a progress report: Illinois State Geological Survey Petroleum 3, 10 p.

Dow, W.G., 1977, Kerogen studies and geological interpretations: Journal of Geochemical Exploration, v. 7, p. 79–99.

Espitalié, J. et al, 1977, Source rock characterization method for petroleum exploration: Ninth Annual Offshore Technology Conference Proceedings, v. 3, p. 439–444.

Foscolos, A.E., and T.G. Powell, 1980, Mineralogical and geochemical transformation of clays during catagenesis and their relation to oil generation, *in* Facts and principles of world petroleum occurrence: Canadian Society of Petroleum Geologists Memoir 6, p. 153–172.

Fuex, A.N., 1977, The use of stable carbon isotopes in hydrocarbon exploration: Journal of Geochemical Exploration, v. 7, p. 155–188.

———, 1980, Experimental evidence against an appreciable isotopic fractionation of methane during migration, *in* Advances in organic geochemistry 1979: New York, Pergamon Press, p. 725–732.

Galimov, E.M., 1967, ^{13}C enrichment of methane during passage through rocks: Geochemistry International, v. 4, p. 1180–1181.

Grabowski, G.J., Jr., 1981, Source-rock potential of the Austin Chalk, Upper Cretaceous, southeastern Texas: Gulf Coast Association of Geological Societies Transactions, v. 31, p. 105–113.

Gunter, B.D., and J.D. Gleason, 1971, Isotope fractionation during gas chromatographic separations: Journal of Chromatographic Science, v. 9, p. 191–192.

Hacquebard, P.A., 1977, Rank of coal as an index of organic metamorphism for oil and gas in Alberta, *in* The origin and migration of petroleum in the western Canadian sedimentary basin, Alberta: Geological Survey of Canada Bulletin 262, chap. 3, p. 11–22.

Hoffman, J., and J. Hower, 1979, Clay mineral assemblages as low grade metamorphic geothermometers: application to the thrust faulted disturbed belt of Montana, U.S.A.: SEPM Special Publication 26, p. 55–79.

Hower, J., et al, 1976, Mechanism of burial metamorphism of argillaceous sediment: 1. Mineralogical and chemical evidence: GSA Bulletin, v. 87, p. 725–737.

Hunt, J.M., 1979, Petroleum geochemistry and geology: San Francisco, W.H. Freeman, 617 p.

Lebedev, V.S., and E.D. Syngayevski, 1971, Carbon isotope fractionation in sorption processes: Geochemistry International, v. 8, p. 460.

Lockridge, J.P., 1977, Beecher Island field, Yuma County, Colorado, *in* H.K. Veal, ed., Exploration frontiers of the central and southern Rockies: Denver, Colorado, Rocky Mountain Association of Geologists, p. 271–279.

——— and P.A. Scholle, 1978, Niobrara gas in eastern Colorado and northwestern Kansas, *in* J.D. Pruit and P.B. Coffin, eds., Energy resources of the Denver Basin: Denver, Colorado, Rocky Mountain Association of Geologists, p. 35–49.

Mah, R.A., et al, 1977, Biogenesis of methane: American Revue of Microbiology, v. 31, p. 309–341.

Matuszczak, R.A., 1973, Wattenberg field, Denver Basin, Colorado: Mountain Geologist, v. 10, p. 99–105.

McGookey, D.P., 1972, Cretaceous System, *in* Geologic atlas of the Rocky Mountain region: Denver, Colorado, Rocky Mountain Association of Geologists, p. 190–228.

National Petroleum Council, 1980, Denver Basin, *in* Unconventional gas sources, tight gas reservoirs, pt. 2, p. 15-1–15-39.

Neglia, S., 1979, Migration of fluids in sedimentary basins: AAPG Bulletin, v. 63, p. 573–597.

Pollastro, R.M., 1981a, Authigenic kaolinite and associated pyrite in chalk of

Cretaceous Niobrara Formation, eastern Colorado: Journal of Sedimentary Petrology, v. 51, p. 553–562.

——, 1981b, Clay-mineral diagenesis within a fine-grained, marine, hydrocarbon-bearing, carbonate sequence: evidence from the Cretaceous Niobrara Formation (abs.): Clay Minerals Society, 1981 Program and Abstracts, p. 13.

Rashid, M.A., and G. Vilks, 1977, Environmental controls of methane production in Holocene basins in eastern Canada: Organic Geochemistry, v. 1, p. 53–59.

Rice, D.D., 1980, Chemical and isotopic evidence of the origins of natural gases in offshore Gulf of Mexico: Gulf Coast Association of Geological Societies Transactions, v. 30, p. 203–213.

—— and G.W. Shurr, 1980, Shallow, low-permeability reservoirs of northern Great Plains—assessment of their natural gas resources: AAPG Bulletin, v. 64, p. 969–987.

—— and G.E. Claypool, 1981, Generation, accumulation, and resource potential of biogenic gas: AAPG Bulletin, v. 65, p. 5-25.

—— and C.N. Threlkeld, 1982, Occurrence and origin of natural gas in ground water, southern Weld County, Colorado: U.S. Geological Survey Open-File Report 82–496, 6 p.

Rosenfeld, W.D., and S.R. Silverman, 1959, Carbon isotope fractionation in bacterial production of methane: Science, v. 130, p. 1658–1659.

Runnels, R.T., and I.M. Dubins, 1949, Chemical and petrographic studies of the Fort Hays chalk in Kansas: Kansas Geological Survey Bulletin 82, pt. 1, p. 1–36.

Scholle, P.A., 1977a, Chalk diagenesis and its relation to petroleum exploration—Oil from chalks, a modern miracle?: AAPG Bulletin, v. 61, p. 982–1009.

——, 1977b, Current oil and gas production from North American Upper Cretaceous chalks: U.S. Geological Survey Circular 767, 51 p.

Shurr, G.W., 1980, Geologic setting of the Pierre Shale (Upper Cretaceous) in the northern Great Plains: U.S. Geological Survey Open-File Report 80–675, 8 p., 4 plates.

—— and J.L. Sieverding, 1980, Preliminary synthesis of subsurface stratigraphy of the Niobrara Formation (Upper Cretaceous) in the northern Great Plains: U.S. Geological Survey Open-File Report 80–1266, 19 p., 6 plates.

Smagala, Tom, 1981, The Cretaceous Niobrara play: Oil and Gas Journal, v. 79, no. 10, p. 204–218.

Staplin, F.L., 1969, Sedimentary organic matter, organic metamorphism, and oil and gas occurrence: Bulletin of Canadian Petroleum Geology, v. 17, p. 47–66.

Tissot, B., and D.H. Welte, 1978, Petroleum formation and occurrence: Berlin, Springer–Verlag, 538 p.

Tissot, B., et al, 1974, Influence of nature and diagenesis of organic matter in formation of petroleum: AAPG Bulletin, v. 58, p. 499–506.

Waples, D.W., 1980, Time and temperature in petroleum formation: application of Lopatin's method to petroleum exploration: AAPG Bulletin, v. 64, p. 916–926.

Environment of Deposition of Middle Miocene (Alcanar) Carbonate Source Beds, Casablanca Field, Tarragona Basin, Offshore Spain

Gerard Demaison and F. T. Bourgeois
Chevron Overseas Petroleum, Inc., San Francisco, California

The oil in the Casablanca field, offshore Spain, is present in Jurassic limestones unconformably overlain by thick Tertiary sediments that provide both source and seal. The source beds for the Casablanca oil have been identified as parts, but not the whole, of a carbonate unit at the very base of the Tertiary transgressive sequence, the Alcanar Formation of middle Miocene age. The Alcanar oil-source beds are dark-brownish-gray, well-lithified, pelagic marlstones to limestones containing foraminifers. The mineral carbonate content ranges from 38 to 75%. The organic-carbon content in the source facies ranges from 3.2 to 4.7%. The hydrocarbon yield from whole-rock pyrolysis is high (5 kg/ton or more than 1 gal/ton) and is compatible with the identified kerogen type (type II).

Oriented thin sections from conventional cores in the Alcanar source facies show no evidence of large burrowing infauna and only sparse benthic foraminifers. Rare microscopic, horizontal worm burrows are present. Most, but not all, thin sections in the source facies show microscopic laminations. These combined characteristics are indicative of past low oxygen concentrations above sea bottom typical of dysaerobic environments. These conditions, when combined with high sedimentation rates, favor oil-source-bed deposition.

INTRODUCTION

The Casablanca field (Watson, 1981) is the largest of four producing oil fields in the Spanish Mediterranean Sea (Garcia–Sineriz et al, 1980), some 50 km south of Tarragona (Fig. 1). The crude has a density of 33.7° API, a sulfur content of 0.2%, and a gas–oil ratio of 155 cf/bbl. Reservoir temperature is high, close to 300°F (150°C), but effective-heating time is short, since the bulk of the section overlying the reservoir is late Miocene to Pliocene in age. The Casablanca crude closely resembles oil in the nearby Castellon field, which is entirely different from the oil in the Amposta field (Albaiges and Torradas, 1977; Seifert et al, 1983).

The reservoir, filled to the spill point (2,733 m or 8,966 ft), is a weathered and fractured Upper Jurassic carbonate ridge unconformably draped by a middle Miocene carbonate-rock unit, the Alcanar Formation, which is the source of the oil accumulation. Oil–source-rock correlations, both by isotope chemistry (see Fig. 2) and gas chromatography (Fig. 3), are sufficient to document this genetic association. Moreover, the Mesozoic rocks between the Tertiary unconformity and the metamorphic basement have been recognized geochemically as organic poor and without hydrocarbon-source potential. The bulk of the oil charge migrated from a short distance, as the mature Alcanar generative depression ("kitchen") is adjacent to the oil accumulation itself (see Fig. 1). The Alcanar and younger Mio–Pliocene shales and clays must provide efficient seals for the Casablanca oil accumulation, since it is full to spill point.

The objective of this study was to investigate the environment of deposition of the Alcanar Formation, by sedimentological and paleontological methods, in order to compare it with petroleum-source potential as evaluated by geochemical methods. Worldwide observations, for both modern and ancient sediments, show that hydrogen-rich, "oil-prone" types I and II kerogens (Tissot and Welte, 1978) are preferentially preserved in sediments deposited under anoxic or, at least, oxygen-poor conditions at the benthic boundary (Didyk et al, 1978; Demaison and Moore, 1980; Maynard, 1981; Summerhayes, 1981a; Pelet, 1983; Barrows and Cluff, 1984). Therefore, our sedimentological and paleontological studies were aimed at measuring characteristics relating to past oxygen concentration at the benthic boundary, namely, degree of bioturbation and the relative abundance and types of benthic foraminifers (see Table 3). The geochemical analyses (Table 1) consisted of kerogen typing by Rock-Eval pyrolysis and organic-carbon measurements. Pyrolysis data were confirmed by conventional studies such as elemental analysis of kerogen, vitrinite reflectance, and analyses of extracted bitumen (see Table 2) in 1977 for the Casablanca–5 and –6 wells.

SEDIMENTOLOGICAL AND PALEONTOLOGICAL METHODS OF STUDY

This investigation was made possible by the availability of conventional cores of the Alcanar Formation from the Casablanca–3, Casablanca–4, Casablanca–5, and Casablanca–6 wells. Since these cores are uniformly dense, hard, and well lithified, they were amenable to examination in thin sections that were cut perpendicular to bedding. A split corresponding to each thin section was submitted for Rock-Eval pyrolysis and for organic-carbon and carbonate analyses.

Thin sections, while providing information about lithology and sedimentary features such as laminations and burrows, created problems in the identification of foraminiferal taxa, and to some degree in the methods by which populations were evaluated. For paleoecological

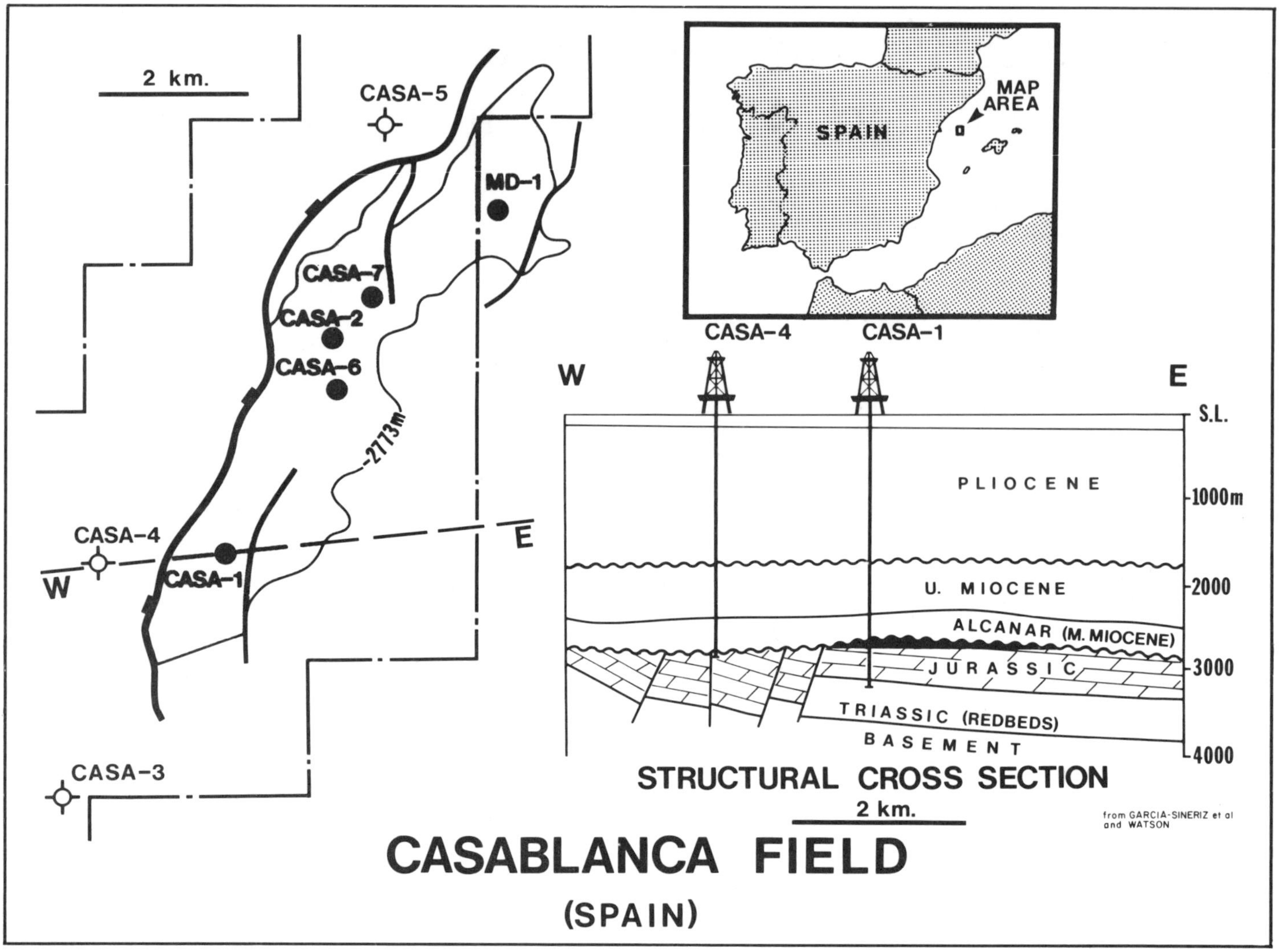

Figure 1 — Geologic setting of Casablanca oil field, Tarragona basin, offshore Spain.

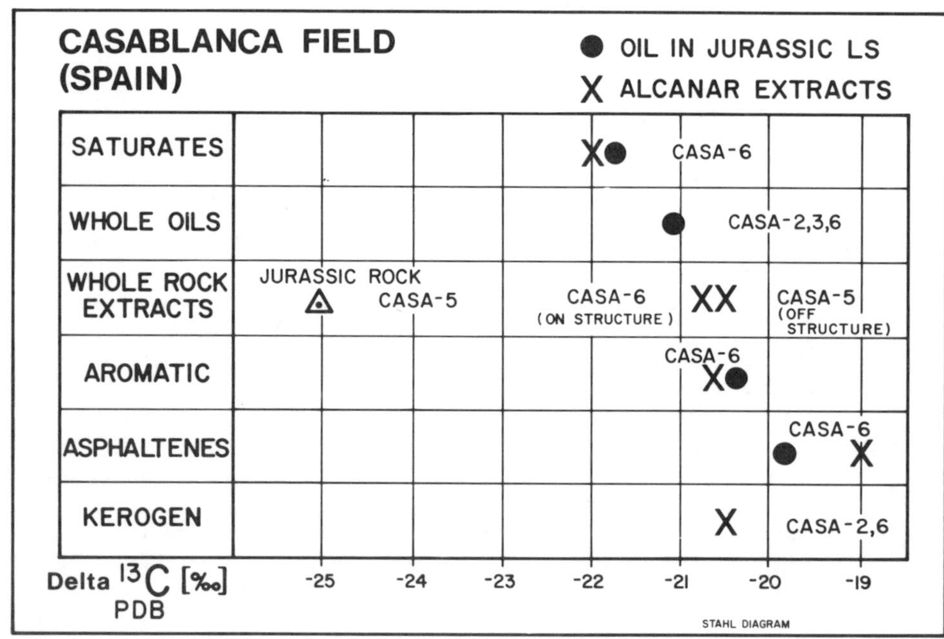

Figure 2 — Oil–rock correlations by stable-isotope chemistry. Oil entrapped in the Jurassic reservoir matches the kerogen and bitumen extracted from the Alcanar Formation.

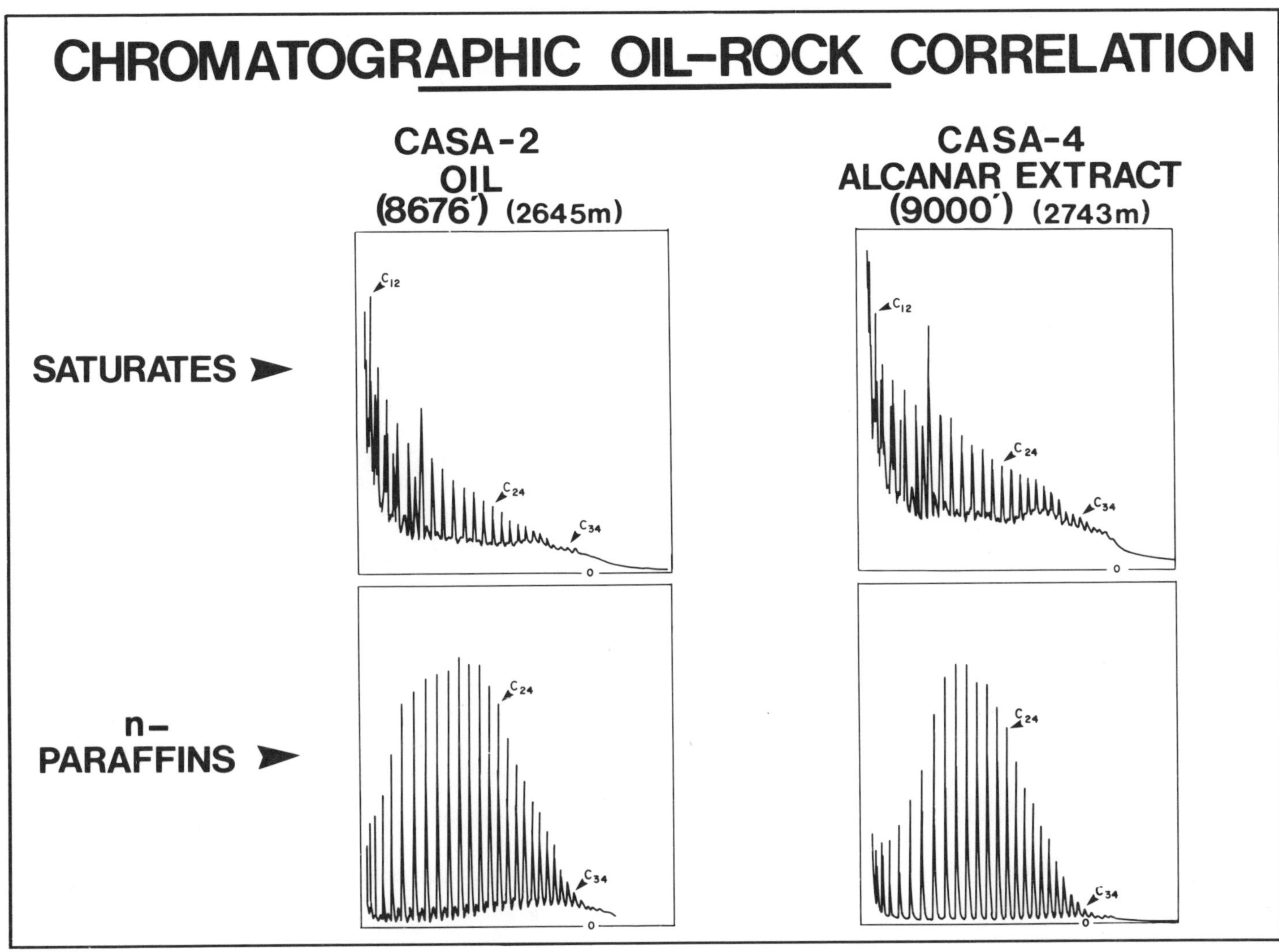

Figure 3 — Chromatograms of oil from Casablanca–2 well in the Jurassic reservoir and Alcanar Formation extract from the Casablanca–4 well. The latter well is off structure outside the area of oil accumulation (see Fig. 1).

analyses of sea-bottom samples in modern unconsolidated sediments, statistical studies of species diversity, abundance, and planktic-to-benthic ratios are used to describe living fauna. The reliability of these measurements depends on large volumes of washable sediments sampled by gravity cores. However, in thin sections of ancient sediments, both the number of foraminiferal tests on a slide and the volume of rock observed are obviously much smaller.

The basis of our method is to measure the density of benthic foraminifers per unit area in thin sections and to determine approximately how many species of benthic foraminifers are present on each slide. This is only a semiquantitative technique, since it is neither possible to observe the total number of specimens nor to identify all the species present in a slide. Recognition of species was difficult with some of the microfauna present in the Alcanar Formation, such as *Uvigerina*, *Nodosaria*, *Dentalina*, *Cibicides*, *Gyrodina*, *Cassidulina*, and *Anomalina*. Many of these microfossils lack distinctive characteristics in thin sections. Thus, minor errors in species counts can be made, depending upon the operator and the care taken. However, overall population counts are reasonably accurate, especially in slides with fewer microfossils. Furthermore, all shell fragments, echinoid fragments, and ostracods were counted. A size distinction also was made when counting echinoid debris, since one facies characteristically contained very small, delicate echinoid spines and extremely thin, porous plates.

It is recognized that these numerical results generated from thin sections leave much to be desired when compared to classical paleoecological techniques on modern unconsolidated sediments. Nevertheless, benthic-foraminiferal statistics for the Alcanar exhibit meaningful correlation trends in relation to other data such as rock textures, lithofacies, burrow size, amount of echinoid debris, kerogen types, and hydrocarbon-source potential. Hence, we suggest that this method is a useful technique for oxygen-level, paleoenvironmental studies in carbonate rocks.

OBSERVATIONS AND DISCUSSIONS

Energy Level of the Depositional Environment

Material studied in the cores of the Alcanar Formation from the Casablanca–

Table 1

Depth (ft)	S_1 (mg/g)	S_2 (mg/g)	S_3 (mg/g)	T_{max} (°C)	S_2/S_3	HI (S_2/TOC)	TOC (wt%)	Total Carbonate (wt%)
Casa–3								
8,989.5	0.3	0.9	0.3	441	3	36	2.5	79
8,991	.1	.1	.3	442	.33	9.09	1.1	81
8,992	.1	.1	.3	442	.33	9.09	1.1	82
8,997.5	.1	.1	.3	442	.33	9.09	1.1	83
9,000.5	.1	.1	.3	437	.33	10.75	.93	80
9,004	.1	.1	.3	442	.33	20	.5	68
9,008	.1	.2	.2	446	1	31.75	.63	71
9,011	.2	.3	.2	442	1.5	33.3	.9	74
9,012.5	.1	.2	.2	445	1	11.83	1.69	82
9,026	.4	.5	.3	440	1.7	20	2.5	88
9,037	.6	.4	.6	433	.67	23.3	1.72	70
9,045	.4	.3	.3	447	1	23.1	1.3	86
9,059.5	.6	2.3	.3	441	7.7	137.7	1.67	
9,063.5	.6	1.6	.2	443	8	65.3	2.45	82
9,074.5	1.1	3.3	.3	444	11	107.84	3.06	40
Casa–4								
8,978	1.2	2.4	.5	448	4.8	84.8	2.83	49
8,981	1.6	4.1	.5	449	8.2	111.7	3.67	50
8,983	1.5	3.2	.6	449	5.3	90.4	3.54	42
8,986	1.8	5.2	.6	448	8.7	125	4.16	38
8,990	1.5	3	.5	447	6	117.2	2.56	47
8,993	1.9	4.4	.6	447	7.3	111.1	3.96	50
8,997	1.5	2.4	.6	448	4	94.1	2.55	58
8,999	1.5	2.9	.5	446	5.8	109.8	2.64	32
9,000	1.7	4.5	.7	450	6.4	124.3	3.62	44
9,005	1	.6	.3	444	2	51.28	1.17	55
9,172	2.1	2.9	.8	449	3.6	69.5	4.17	64
Casa–5								
9,958	2.6	5.1	.4	438	12.8	116.2	4.39	66
9,961	2.8	4.4	.5	438	8.8	131.7	3.34	75
9,964	3.1	5.4	.5	439	10.8	120	4.5	61
9,967	3.5	6.8	.6	437	11.3	151.1	4.5	59
9,970.3	3.8	7.8	.5	437	15.6	179.7	4.34	54
9,973	4.0	6.1	.6	439	10.2	133.2	4.58	54
9,975.9	3.7	6.4	.6	439	10.7	153.8	4.16	54
Casa–6								
9,151.8	2.5	9.1	.5	435	18.2	192.4	4.73	54
9,162.0	2.1	8.4	.5	435	16.8	195.8	4.29	59
9,172.8	1.9	8	.4	432	20	198.5	4.03	50
9,185.6	2.7	9.9	.6	436	16.5	231.9	4.27	58
9,196.4	2.2	10.3	.5	434	20.6	262	3.93	54
9,203.3	2.1	8.5	.6	437	14.2	224.3	3.79	44
9,218.1	1.6	5.9	.4	434	14.8	178.2	3.31	52
9,227.6	1.6	5.8	.4	434	14.5	177.9	3.26	58
9,242.0	1.9	7.4	.4	434	18.5	194.2	3.81	54

Rock-Eval and TOC data by Exploration Logging (1980). TOC data independently cross checked by Global Geochemistry (1980). Conventional core material only.

Table 1 — Geochemical data obtained from sample splits that are microscopically described in Table 3.

3, –4, –5, and –6 wells is fairly representative of the formation as a whole. The Alcanar can be subdivided into two distinct lithofacies each of which represents part of a textural and compositional continuum reflecting different energy levels close to the past sediment surface.

"Alcanar Marl" Lithofacies

The rocks in the marl facies are fine grained, consisting of microbioclastic carbonate silt with less than 15% planktonic foraminifers in a clay matrix. They contain very finely crystalline pyrite and dolomite and fall close to the shale–carbonate boundary, the range of carbonate content varying between 38 and 75%. An accurate description of this rock is planktonic foraminiferal dolomitic calcisiltite. The energy level of the depositional environment is interpreted as low, with only feeble currents having acted intermittently.

"Alcanar Chalk" Lithofacies

The chalk lithofacies is a carbonate-rich unit (68 to 88% total carbonate content) that can be subdivided into two subunits (1 and 2).

The rocks in subunit 1 are slightly glauconitic and phosphatic planktonic foraminiferal wackestones to packstones. Their outward macroscopic appearance is that of a chalky limestone, but the dolomite content was found to be fairly high by analysis (up to 70% dolomite). Shell fragments are absent, and echinoid fragments rare. The lithofacies of subunit 1, as indicated by sorting characteristics of planktonic foraminifers, were deposited in an environment in which weak to moderate bottom currents prevailed.

"Alcanar chalk" subunit 2 rocks are echinoid–planktonic-foraminiferal packstones ("chalks") with a micritic to slightly argillaceous matrix. Shell fragments are present but rare. Echinoid fragments are abundant. Chemical analyses indicate that these rocks also are highly dolomitic. The dolomite is mostly very finely crystalline, but all the echinoidal material has been dolomitized by coarser crystals. Some fine to medium quartz sand is present (about 5%), and sporadic rock fragments (including limestone and volcanic rocks) are seen. About 3–5% phosphate occurs, within fossils and as a replacement of fossil debris. the environment of deposition is interpreted as having a moderate energy level with relatively stronger bottom currents than in subunit 1.

Oxygen Level of the Depositional Environment

Criteria used by sedimentologists to evaluate degree and type of bioturbation, and therefore past oxygen levels at the benthic boundary, are extensively discussed by Rhoads and Morse (1971), Byers (1977), and Savrda, Bottjer, and Gorsline (1984). They are also discussed in reference to stratigraphic interpretation by Morris (1980), Hallam (1981),

Demaison et al (1983), and Pratt (1984).

The sedimentological criteria we chose to use in our study are useful and significant, according to these authors, yet these criteria were observable on the core material at our disposal, namely (1) absence, presence, and/or type of laminations; and (2) type, size, and orientation of worm burrows.

We also added to our investigation a statistical micropaleontological study, in thin section, to evaluate (1) relative numbers of individual benthic foraminifers and (2) relative numbers of species of benthic foraminifers.

Using these sedimentological and paleontological criteria in the Alcanar Formation permitted us to reach the following conclusions about the "Alcanar marl" and the "Alcanar chalk" lithofacies.

Table 2

Depth (ft)	Kerogen (Atomic H/C)	R_o	EOM (ppm)	HC (ppm)	Saturates (%)	Aromatics (%)	Asphaltics + Non eluted (%)
Casa–5							
9,966.0	ND	0.59	4,498	2,739	22.6	38.3	39.1
9,969.6	ND	0.56	5,250	3,449	21.4	44.3	34.3
9,974.0	1.03	0.62	6,582	4,041	22.8	38.6	38.6
Casa–4							
8,978.6	0.96	0.46	ND	ND	ND	ND	ND
8,989.8	0.95	0.46	2,657	1,674	21.6	41.4	37.0
8,999.9	0.90	0.52	2,313	1,564	31.0	36.6	32.4

Data by CORELAB (1977) under the supervision of G. W. Ruth. Solvent composition: 91% benzene, 9% methanol. Conventional core material only. ND = no data. R_o = vitrinite reflectance in oil. EOM = extractable organic matter. HC = hydrocarbons.

Table 2 — Additional geochemical data on sample splits described in Table 3.

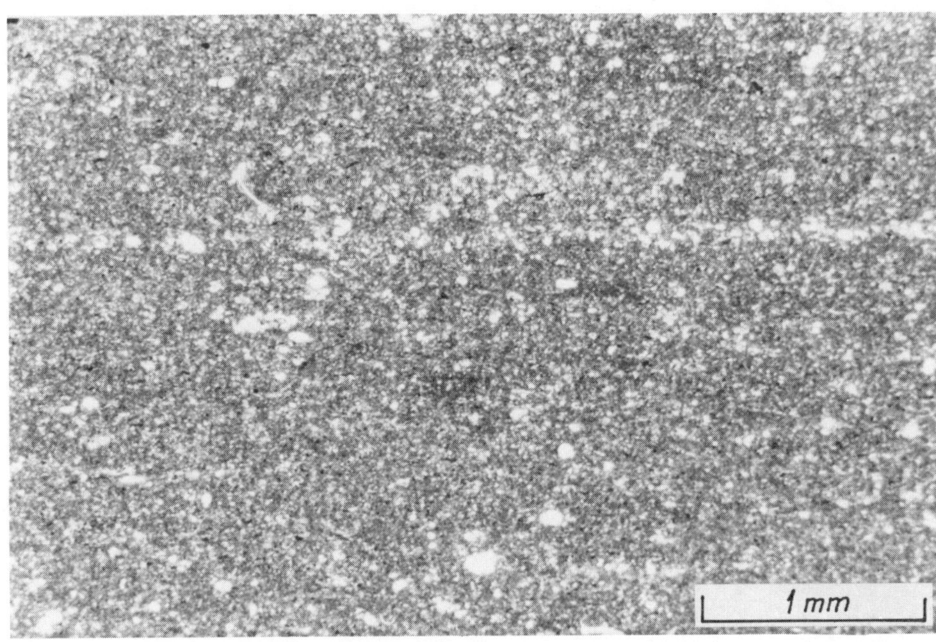

Figure 4 — Photomicrograph of rock from Casablanca–6 well, 9,218 ft (2,810 m). Strongly dysaerobic "Alcanar marl" lithofacies with calcisilt lamina (approximately 0.2 mm thick). Disruption by microbioturbation is seen in left half of lamina. Note also preferred orientation of particles throughout rock. Some of this may be due to compaction, but most particle orientations indicate nondisrupted bedding. Total carbonate, 52%; total organic carbon, 3.31%; hydrogen index, 178; S_1, 1.6 mg/g.

"Alcanar Marl" Lithofacies

This facies is characterized by the presence of compositional laminations ranging in width from 0.2 to about 1.5 mm. These laminations can be observed microscopically owing mainly to differences in the amount of carbonate silt (Fig. 4). In places the layering is vague and gradational; elsewhere laminations are sharp, especially where the carbonate-silt content is high. Concentrations of planktonic foraminifers also form laminations, but these are not common and are always much thinner than the silty laminae. Some planktonic foraminifers and very fine elongate planktonic fossils are usually found lying on bedding planes. This preferred orientation is also seen in the orientation of coarse silt-sized grains and flakes of organic matter (kerogen).

Laminations are commonly disrupted by microbioturbation, as shown by the presence of very small horizontal burrows, by the dispersal of planktonic foraminifers from certain laminae, and by swirl fabrics or other features which indicate that sediment has been pushed aside by minute burrowing organisms. These microscopic burrows are frequently seen in the "Alcanar marl" facies of the Casablanca–4 well (Fig. 5) but are rare in the "Alcanar marl" facies of the Casablanca–5 and –6 wells (Fig. 4). Most burrows are of the clay-fill type, but some are filled by silt. The burrows are always horizontal or subhorizontal and are microscopic in size (0.2–0.3 mm in diameter; the size range is approximately 0.1–0.4 mm). These microscopic horizontal burrows are interpreted to have resulted from the past activity of very small, smooth wire worms, probably nematodes. There is no evidence of burrowing by large segmented worms, bivalves, or holothurians.

Benthic foraminifers are always rare in the "Alcanar marl" lithofacies. Typically, there are 1 to 8 benthic-foraminiferal specimens per cm^2 in the slides. Our counts also indicate that there is a range of 2 to 16 species of benthic foraminifers per slide in the "Alcanar marl." This contrasts with a range of 11 to 35 species in the "Alcanar chalk." Thus the number of benthic organisms tends to be lower both in the number of species and in the number of distinct individuals in the "Alcanar marl" facies, as compared to the "Alcanar chalk" facies (Table 3).

Table 3

Depth (ft)	Individual Benthic Foraminifers Per Slide	No. of Benthic Foraminifer Species Per Slide	Individual Benthic Foraminifers/ cm^2	No. of Species of Benthic Foraminifers/ cm^2	Laminations	Burrow Size (mm) Vertical (ϕ) Horizontal (θ)
Casa–3						
8,989.5	108	25	12.77	2.97	No	0.8 ϕ
8,991	72	19	9.3	2.47	No	1.5 ϕ
8,992	51	14	7.5	2.07	No	0.8 ϕ
8,997.5	78	23	9.63	2.83	Trace	0.8 ϕ
9,000.5	109	29	13.4	3.57	No	1.6 ϕ
9,004	40	16	4.83	1.93	No	1.3 ϕ
9,008	55	15	7.8	2.13	No	1.6 ϕ
9,011	133	35	17.6	4.63	No	No
9,012.5	75	11	9.23	1.37	No	No
9,026	78	17	10.6	2.3	No	No
9,037	90	16	20.03	3.57	No	No
9,045	153	11	17.7	1.27	Trace	No
9,059.5	27	19	3.73	2.63	No	0.6 ϕ
9,063.5	49	16	6.83	2.23	No	0.6 ϕ
9,074.5	54	21	7.17	2.8	No	0.4 ϕ
Casa–4						
8,978	29	11	4	1.5	No	0.48ϕ
8,981	21	6	2.7	.77	No	No
8,983	21	4	2.8	.53	Yes	0.8 θ
8,986	21	6	3.2	.9	Yes	ND
8,990	8	5	1	.63	Yes	0.4 θ 0.8 ϕ
8,993	43	16	5.4	2	No	0.8 ϕ
8,997	12	8	1.53	1.03	Yes	2.4
8,999	25	11	4.57	2	Yes	0.4 θ
9,000	14	8	3.17	1.83	Yes	0.25θ
9,005	14	7	1.87	.93	No	1.6 ϕ
9,172	46	9	6.77	1.33	No	0.3 θ
Casa–5						
9,958	28	10	4.93	1.77	Yes	Trace
9,961	19	8	3	2.1	No	0.3 θ
9,964	17	7	5.27	2.17	Yes	Trace
9,967	44	8	7.87	1.43	Yes	No
9,970.3	24	7	4.33	1.27	Yes	No
9,973	13	5	2.63	1	No	No
9,975.9	25	8	6.57	2.1	No	Trace
Casa–6						
12,790.2	16	8	3.3	1.63	Yes	Trace
12,793.3	4	3	.83	.63	No	No
12,796.6	22	10	4.53	2.07	Yes	No
12,800.5	27	12	5.67	2.5	Yes	No
12,803.8	20	7	4.1	1.43	Yes	0.4 θ
12,805.9	10	6	2.13	1.27	Yes	Trace
12,810.4	16	5	3.3	1.03	Yes	0.4 θ
12,813.3	3	2	.73	.5	Yes	Trace
12,817.7	10	4	2.43	.97	Yes	0.4 θ

Paleontological and sedimentological investigations by F. T. Bourgeois.

Table 3 — Micropaleontological and sedimentological data from oriented thin sections from conventional cores.

In summary, the "Alcanar marl" lithofacies is characterized by (1) microscopic laminations within the sediment, (2) a low density of benthic-foraminiferal tests, (3) low numbers of benthic species, (4) the presence of microscopic horizontal burrows, and (5) the absence of large and/or vertical burrows. Apart from the microscopic burrows described above, there are no signs of other infaunal elements such as, for instance, bivalves: what shell debris is seen in the "Alcanar marl" facies consists of transported microfragments brought in with carbonate silt from other environments.

Using sedimentological criteria referenced earlier in this chapter, we conclude that most of the "Alcanar marl" lithofacies was deposited under dysaerobic conditions (Byers, 1977), with an oxygen level fluctuating between perhaps 1.0 and 0.1 mL/L of oxygen in water. There was no permanent anoxia in the bottom water; the anoxic boundary was probably oscillating from just above the sediment surface at times to just below it. Overall low oxygen content was persistent enough to inhibit active metazoan scavenging at the benthic boundary as well as heavy bioturbation of the sediment. Weak bioturbation by microscopic wire worms appears to have been the only metazoan benthic activity in the dysaerobic "Alcanar marl" facies.

"Alcanar Chalk" Lithofacies, Subunits 1 and 2

Both subunits are heavily and coarsely bioturbated: when viewed under a low-powered microscope, these rocks have a mottled fabric, and rarely show any kind of bedding. Clay-filled burrows are commonly present, which are filled with micropelletal material; some of these show backfill structures. They range from 0.4 to 1.6 mm in diameter but commonly are about 0.8 mm. Burrow orientation is strictly random (Fig. 6). The worm populations are interpreted as being different from those observed in the "Alcanar marl" facies. Worms that inhabited the "Alcanar chalk" facies were probably large segmented worms (polychaete?) that respired oxygen above the benthic boundary rather than the small wire worms that dwelled within the anaerobic muds of the "Alcanar marl" facies.

The "Alcanar chalk" facies is interpreted, according to sedimentological criteria referenced earlier in this chapter, as having been deposited under water with high oxygen concentrations, greater than 1.0 mL/L. This allowed for vigorous benthic activity and heavy burrowing by large metazoans. Benthic-foraminiferal density is high in the chalk facies, ranging from 5 to 14 specimens per cm^2 in subunit 1 and from 10 to 20 specimens per cm^2 in subunit 2, in contrast to a

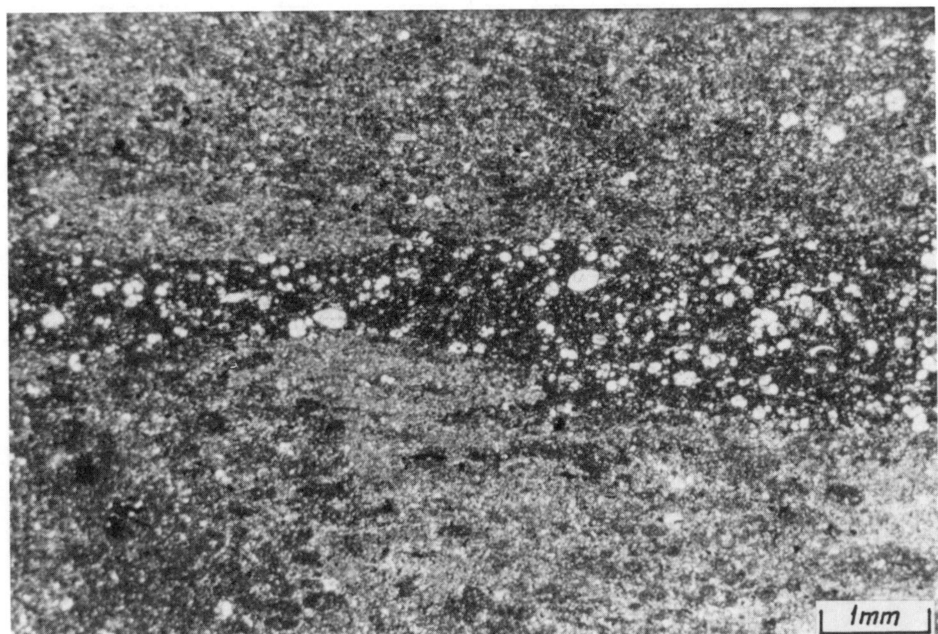

Figure 5 — Photomicrograph of rock from Casablanca-4 well, 8,999 ft (2,743 m). Weakly dysaerobic "Alcanar marl" lithofacies with rich calcisilt content and lamina with planktic foraminifers. The calcisilt layer in lower part of micrograph has many small *horizontal* burrows about 0.2 mm in diameter. These are common at left, where they border a darker bioturbated area. Small *horizontal* burrows also are seen in the upper silty layer, but they contrast less and therefore are not so easily recognized. Total carbonate, 32%; total organic carbon, 2.64%; hydrogen index, 109; S_1, 1.5 mg/g.

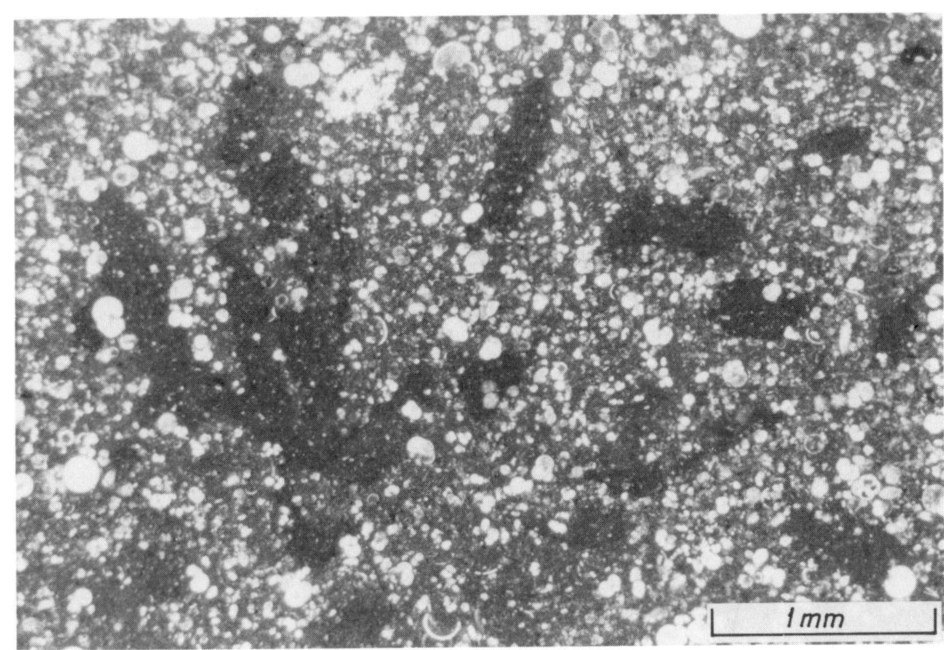

Figure 6 — Photomicrograph of rock from Casablanca-3 well, 8,992 ft (2,741 m). Strongly aerobic, coarsely bioturbated "Alcanar chalk" lithofacies (subunit 1) with branching burrows, which range from 0.4 to 0.8 mm in diameter and display random orientations; some show backfilling structures and pellety fabrics. Total carbonate, 82%; total organic carbon, 1.1%; hydrogen index, 9; S_1, 0.1 mg/g.

range of only 1 to 8 specimens in the dysaerobic "Alcanar marl" facies. Twenty-nine species of benthic foraminifers per slide were observed in subunit 1 and 35 in subunit 2 of the chalk facies, in contrast to a maximum of only 16 species in the dysaerobic "Alcanar marl" facies.

In summary, a positive correlation between decreasing degree of bioturbation (which relates to past oxygen levels at the benthic boundary) and decreasing density of the benthic population is apparent in the investigated Alcanar rocks. This correlation is expressed in Figure 7. The least oxygenated environment ("Alcanar marl") is also that with the lowest energy level, the finest grained sediments, and the highest clay content.

For the purpose of this study, this correlation gives us confidence in using benthic-foraminiferal statistics as an indicator of past oxygenation during Alcanar deposition. Furthermore, our observations on the Alcanar formation concur with those made by Ingle, Keller, and Kolpack (1980) for recent sediments of the southern Peru–Chile trench area. Ingle and his coworkers observed that high benthic-foraminiferal species and abundance correlate with waters high in oxygen, whereas low species diversity and abundance correlate with waters low in oxygen.

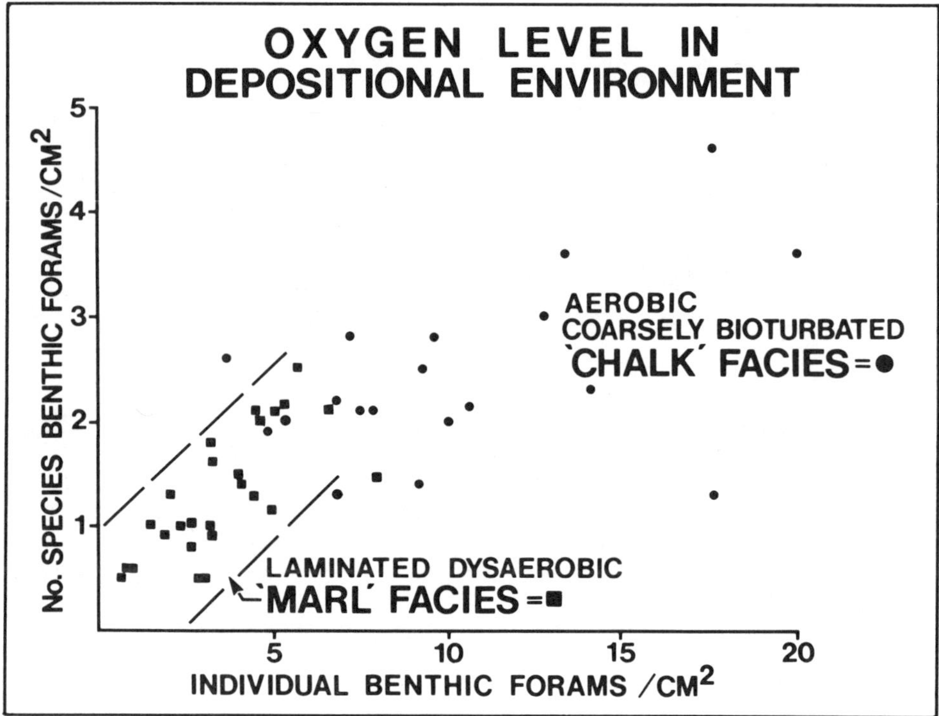

Figure 7 — Oxygen level in the depositional environment. The data identified as black squares are those in the dysaerobic "marl" facies in which laminations and only rare microscopic horizontal worm burrows are present. Both number of benthic-foraminifer species and number of benthic-foraminifer specimens are significantly lower in the dysaerobic "marl" facies than in the aerobic, coarsely bioturbated "chalk" facies.

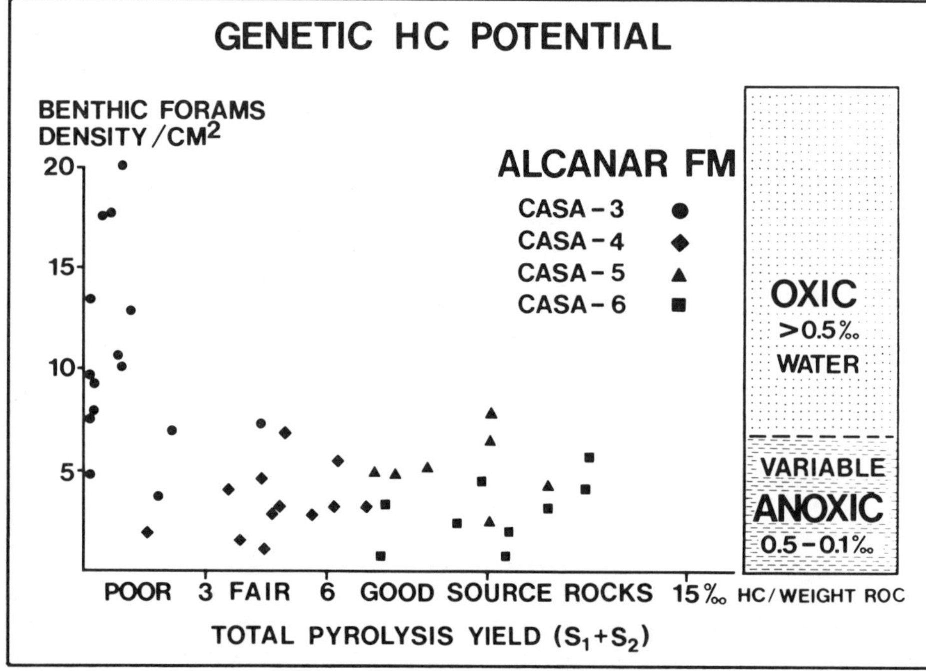

Figure 8 — Genetic hydrocarbon potential in relation to past degrees of oxygenation at benthic boundary (as expressed by benthic-foraminifer densities). Note poor genetic potential for rocks with a benthic-foraminifer density higher than 7–8 specimens per cm^2.

KEROGEN TYPE, OIL-SOURCE POTENTIAL, AND DEPOSITIONAL ENVIRONMENT

Sample splits from all the thin sections were analyzed by conventional Leco and Rock-Eval techniques (Tissot and Welte, 1984). The data are listed in Table 1, in reference to the two recognized lithofacies and specific wells. They lead to the following observations in terms of petroleum source potential as related to past oxygen content above the benthic boundary, which themselves control the degree of bioturbation:

1. There is a distinct correlation between the environment of deposition and hydrocarbon source potential in the Alcanar Formation (Fig. 8). The hydrocarbon genetic potential ($S_1 + S_2$) is highest in the weakly bioturbated, dysaerobic "Alcanar marl" facies (6–12 kg of hydrocarbons per metric ton of rock); conversely, it is generally low to very low in the highly bioturbated, aerobic "Alcanar chalk" facies. There is a complete gradation in terms of paleoenvironment and petroleum potential between these two extremes.

2. Samples from the Casablanca–3 well (strongly bioturbated "Alcanar chalk" facies) are characterized by low-quality, hydrogen-poor, type III kerogens (see Figs. 9, 10). The organic-carbon content varies from less than 0.5 to slightly more than 3%, merely reflecting high sedimentation rates in an oxic environment (Demaison et al, 1983). S_1 values (free hydrocarbons) are consistently low (0.1 to 1.1 kg of hydrocarbons per metric ton of rock).

3. Samples from the Casablanca–4 well (mainly weakly bioturbated "Alcanar marl" facies) feature low-quality, hydrogen-depleted type II kerogens, derived from plankton, with only marginal potential for oil generation. These are so-called "degraded type II kerogens" (J. Espitalié, 1981, personal communication). Organic-carbon contents range from about 2.5 to 4%. Characteristically, Alcanar thin sections from this Casablanca–4 well (Fig. 5) display a significantly higher density of microscopic worm burrows than thin sections from the less bioturbated and more "oil-prone" facies in the Casablanca–5 and –6 wells (Fig. 4), in which even microscopic burrows are rare.

4. Samples from the Casablanca–5 and –6 wells (least bioturbated "Alcanar

marl" facies) are characterized by fair- to good-quality type II "oil-prone" kerogens (Figs. 9, 10) derived from plankton. The S_1 values are the highest recorded in the Alcanar: 1.0 to 4.0 kg of hydrocarbons per metric ton of rock. The organic-carbon content is also the highest and varies from about 3 to 5%.

Other observations have been made with respect to interpretation of pyrolysis data:

1. All Alcanar rocks were sampled at nearby depths and indicate (R_o and T_{max}) similar degrees of maturation. Furthermore, these sediments, now at their maximum burial depth, with identical burial histories, are exposed to near-identical present-day temperatures. Hence the respective positions of the studied kerogens on path II (Fig. 11) relate to different degrees of oxygenation in the early depositional environment rather than to different degrees of thermal maturation. This point is made by the observation that the Casablanca–5 samples (R_o = 0.59 to 0.62) are higher on the hydrogen-index path than the less mature (but more oxygenated) Casablanca–4 samples (R_o = 0.46 to 0.52). Similar observations are commonly made by the authors and other investigators (J. Espitalié, 1981, personal communication) with rocks of different age and in other sedimentary basins. Thus it is suggested that paths I and II on the HI versus OI diagram (Fig. 11) are partly influenced by thermal maturation (hydrogen loss during deep catagenesis) and partly by early diagenetic oxidation of the organic matter (hydrogen loss during shallow sedimentation). Thus the HI–OI diagram is capable of yielding useful paleoenvironmental insights that can help in geologically mapping source-bed distribution patterns.

2. Mineral carbon from the carbonate matrix (32–88% of the Alcanar Formation) statistically has no significant influence on Rock-Eval S_2 and S_3 data, as can be seen in Table 1. Moreover, the S_2/S_3 ratios and hydrogen indices (S_2/TOC) are compatible with each other (Table 1) and lead to identical interpretations concerning kerogen type, although derived from different measurements (S_3 and TOC). These conclusions are based on results obtained from a calibrated instrument in proper working order.

3. There is no evidence of an artificial dependence of S_2 in relation to organic carbon. Alcanar rocks containing poor-

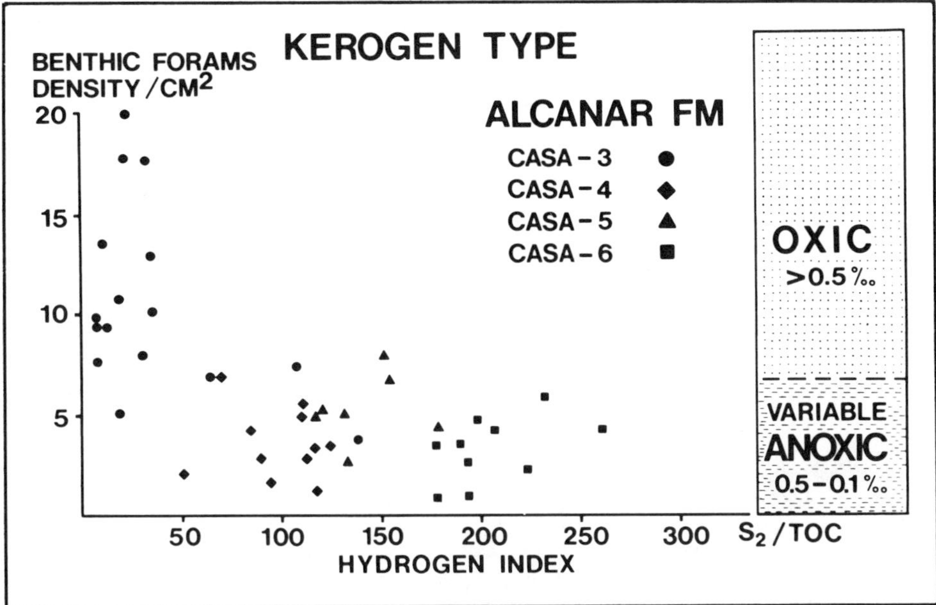

Figure 9 — Hydrogen index (S_2/TOC) in relation to past degree of oxygenation at the benthic boundary (as expressed by benthic-foraminifer densities). Note that virtually all hydrogen indices below 50 coincide with benthic-foraminifer densities higher than about 8 specimens per cm^2.

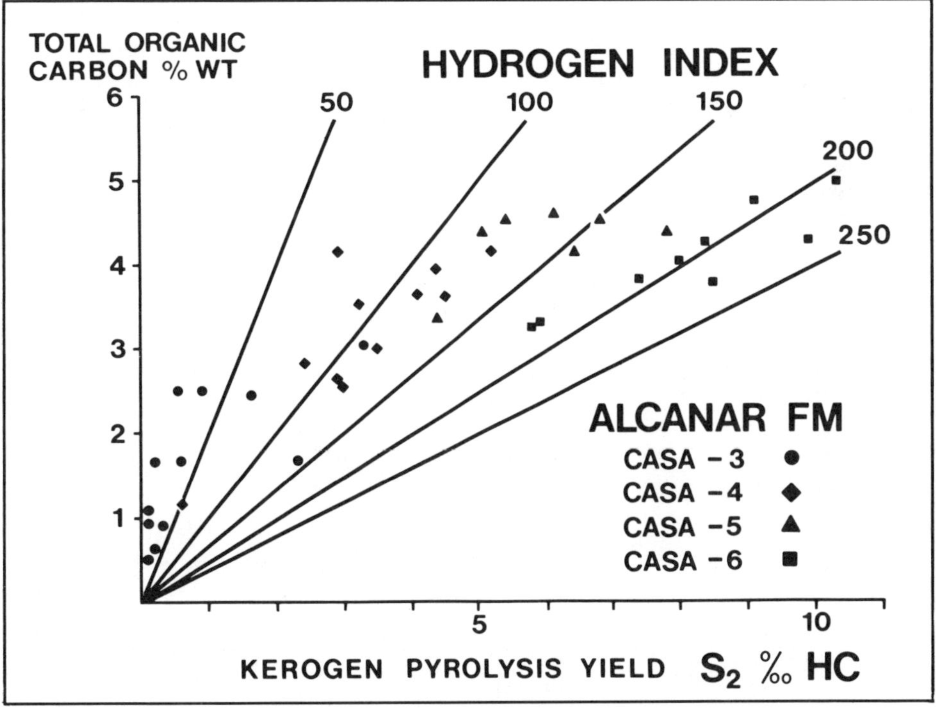

Figure 10 — Hydrogen-index display. Alcanar rocks containing type III kerogen (HI = 9 to 36) range in organic-carbon content from 0.36 to 2.5%. These rocks were deposited in the high-energy "chalk" facies under well-oxygenated conditions. Beyond 2.5% OC, the sharp rise of S_2 represents the onset of oxygen-depleted conditions in the low-energy "marl" facies (the geochemically demonstrated oil source for the Casablanca field).

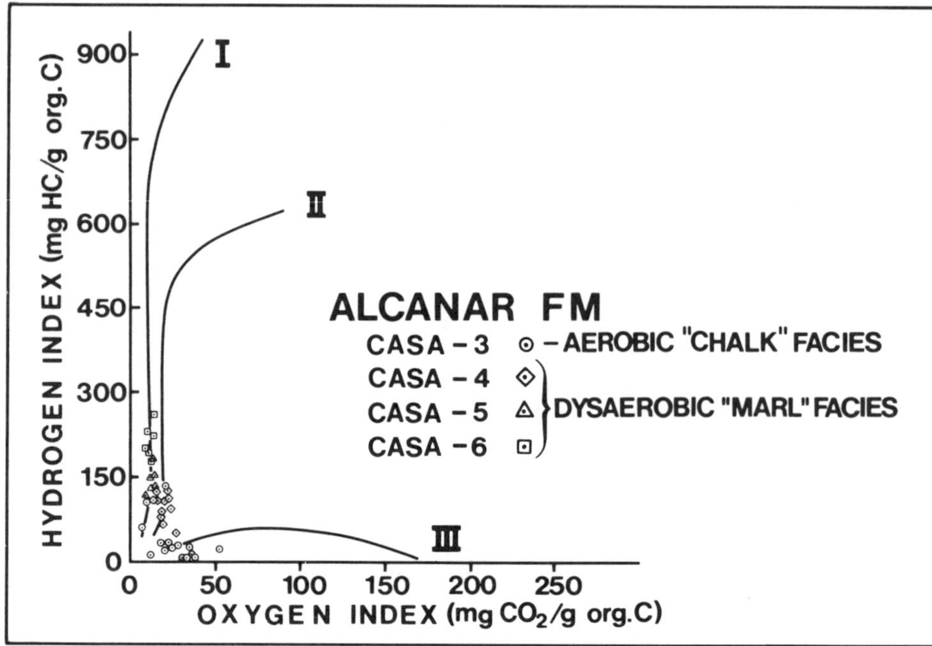

Figure 11 — Diagram of hydrogen index versus oxygen index. Note a significant anomaly: the more mature (R_o = 0.62) Casablanca–5 well samples are higher on path II than the less mature (R_o = 0.52), but more oxygenated, Casablanca–4 well samples. This and other similar observations led us to propose that the relative positions of the studied kerogens on path II mainly reflect different degrees of oxygenation in the early environment rather than different degrees of maturation. The overall clustering of the data points, however, reflects a fairly high level of thermal maturation for all the rocks under investigation.

quality kerogens with low hydrogen indices (HI = 9 to 36) in the Casablanca–3 well range in organic-carbon contents from 0.63 to 2.5%. These geochemical characteristics are compatible with fairly high sedimentation rates in an oxic environment. Above 2.5% organic carbon, the sharp rise of S_2 with respect to total organic carbon (Fig. 10) represents the onset of oxygen-depleted conditions, leading to better preservation of hydrogen in the organic matter and thus higher hydrogen indices. Please note that the S_2-to-TOC relationship for the Alcanar Formation is in excellent concurrence with our paleoenvironmental data relating to oxygen levels at the benthic boundary, as shown in Table 3.

The following remarks also can be made about using sedimentological and micropaleontological methods as a guide to source potential.

There are no rocks with good hydrocarbon-source potential in which benthic-foraminiferal densities in the thin sections exceed 6 to 8 specimens per cm^2. However, there are a few samples with very low benthic-foraminiferal densities (as low as 2 foraminifers per cm^2) that are *not* source rocks by geochemical criteria. *Thus the absence of benthic foraminifers is not necessarily an indication of source potential.* Sedimentological features relating to bioturbation, specifically absence or presence of laminations, present the same ambiguities as benthic-foraminiferal statistics: most, *but not all*, demonstrated carbonate source rocks show microscopic laminations, while all coarsely bioturbated rocks are nonsource. So the absence of laminations is not necessarily an indication of *lack* of source potential. Conversely, the useful point is that coarse bioturbation (heavy burrowing by large worms) and high benthic-foraminiferal densities are, together, associated with lack of oil-source potential.

CONCLUSIONS

1. Oil-source potential in the Alcanar Formation relates primarily to degree of oxygenation at the benthic boundary during early diagenesis of the carbonate rocks. The fine-grained, low-energy, weakly bioturbated, commonly laminated marlstones and limestones deposited in a low-oxygen "dysaerobic" environment are the demonstrated Alcanar source beds. The fine-grained marlstones deposited under somewhat more oxygenated conditions have only marginal source potential. The heavily bioturbated "chalks" deposited under well-oxygenated bottom water, in a higher energy environment, are essentially without oil-source potential, although their organic-carbon contents are commonly greater than 1%.

2. Marked vertical variations in kerogen type and thus in oil-source potential within the cored Alcanar section, as well as lateral organic-facies changes, can be explained by the interplay of a fluctuating oxgyen-minimum layer in the water column with respect to past sea-bottom configuration (Demaison et al, 1983). The Casablanca–3 well in the heavily bioturbated chalk facies is on the southern plunge of the Casablanca structure, where, apparently, fairly high-energy (and probably shallow-water) carbonate sediments accumulated (see Fig. 1). The Casablanca–4, a southwest-flank well drilled through a marl facies, recovered a lesser quality type II kerogen than that of the "Alcanar marl" facies of the Casablanca–6, a northerly crestal well. The Casablanca–5 well, which penetrated the most oil-prone facies, is off-structure, on the northwest flank. This organic-facies distribution leads us to believe that oxygen-depleted conditions at the onset of the middle Miocene transgression near the Casablanca structure were associated with an oxygen-minimum layer that intersected *only the northern half of the structure, perhaps in deeper water.* The vertical and lateral organic-facies variability observed in the Alcanar Formation, as well as the borderline dysaerobic rather than strongly anoxic character of the source beds, does point out a risk in assuming widespread regional stratigraphic occurrence of this source unit. This practical conclusion, reached by integration of regional structure, geochemistry, sedimentology, and paleo-oceanography, helps in the prediction of stratigraphic source-bed distribution. This, in turn, can help in the evaluation of exploration risk for yet undrilled prospects in the studied area.

ACKNOWLEDGMENTS

We thank the management of Chevron Overseas Petroleum, Inc., for permission to publish. We also thank our colleagues at Chevron Oil Co. of Spain and Chevron Oil Field Research, La Habra, California, for their help. F. Tamo drafted the figures.

This paper was read at the annual convention of the Geological Society of America in Atlanta, Georgia, in November 1980.

REFERENCES CITED

Albaiges, J., and J. Torradas, 1977, Geochemical characterization of the Spanish crude oils, *in* R. Campos and J. Goni, eds., Advances in organic geochemistry 1975: Madrid, Enadimsa, p. 99–115.

Barrows, M.H., and R.M. Cluff, 1984, New Albany Shale Group (Devonian–Mississippian) source rocks and hydrocarbon generation in the Illinois basin, *in* Gerard Demaison and R.J. Murris, eds., Petroleum geochemistry and basin evaluation: AAPG Memoir 35, p. 111–138.

Byers, C.W., 1977, Biofacies pattern in euxinic basins: A general model: SEPM Special Publication 25, p. 5–17.

Demaison, G.J., and G.T. Moore, 1980, Anoxic environments and oil source bed genesis: AAPG Bulletin, v. 64, p. 1179–1209.

Demaison, G.J., et al, 1983, Predictive source bed stratigraphy: A guide to regional petroleum occurrence: Chichester, John Wiley, 11th World Petroleum Congress Proceedings, p. 17–29.

Didyk, B.M., et al, 1978, Organic geochemical indicators of paleoenvironmental conditions of sedimentation: Nature, v. 272, p. 216–222.

Garcia–Sineriz, B., et al, 1980, A new hydrocarbon province in the Western Mediterranean: London, 10th World Petroleum Congress Preprint, Panel Discussion, p. 1–7.

Hallam, A., 1981, Facies interpretation and the stratigraphic record: Oxford and San Francisco, W.H. Freeman, 291 p.

Ingle, J.C., O. Keller, and R.L. Kolpack, 1980, Benthic foraminiferal biofacies, sediments and water masses of the southern Peru–Chile Trench area, southeastern Pacific Ocean: Micropaleontology, v. 26, p. 113–150.

Jones, R.W., and G.J. Demaison, 1981, Organic matter characteristics near shelf–slope boundary (abs.): AAPG Bulletin, v. 65, p. 942.

Maynard, J.B., 1981, Some geochemical properties of the Devonian–Mississippian shale sequence, *in* R.C. Kepferle and J.B. Roen, Chattanooga and Ohio Shales of the southern Appalachian basin, *in* T.G. Roberts, ed., GSA Cincinnati '81 field trip guidebooks: Falls Church, Virginia, American Geological Institute, v. 2, p. 336–343.

Morris, K.A., 1980, Comparison of major sequences of organic-rich mud deposition in the British Jurassic: Geological Society of London Journal, v. 137, p. 157–170.

Pelet, R., 1983, Preservation and alteration of present-day sedimentary organic matter, *in* M. Bjorøy et al, eds., Advances in organic geochemistry 1981: Chichester, John Wiley, p. 241–250.

Pratt, L.M., 1984, Influence of paleoenvironmental factors on preservation of organic matter in middle Cretaceous Greenhorn Formation, Pueblo, Colorado: AAPG Bulletin, v. 68, p. 1146–1159.

Rhoads, D.C., and J.W. Morse, 1971, Evolutionary and ecologic significance of oxygen-deficient marine basins: Lethaia, v. 4, p. 413–428.

Savrda, C.E., D.J. Bottjer, and D.S. Gorsline, 1984, Development of a comprehensive oxygen-deficient marine biofacies model: Evidence from Santa Monica, San Pedro, and Santa Barbara basins, California continental borderline: AAPG Bulletin, v. 68, p. 1179–1192.

Seifert, W.K., R.M.K. Carlson, and J.M. Moldowan, 1983, Geomimetic synthesis, structure assignment, and geochemical correlation application of monoaromatized petroleum steroids, *in* M. Bjorøy et al, eds., Advances in organic geochemistry 1981: Chichester, John Wiley, p. 710–724.

Summerhayes, C.P., 1981a, Organic facies of Middle Cretaceous black shales in deep North Atlantic: AAPG Bulletin, v. 65, p. 2364–2380.

———, 1981b, Oceanographic controls on organic matter in the Miocene Monterey Formation, offshore California, *in* R.E. Garrison and R.G. Douglas, eds., The Monterey Formation and related siliceous rocks of California: SEPM, Pacific Section, Special Publication, p. 213–219.

Tissot, B., and D.H. Welte, 1978, Petroleum formation and occurrence, 2nd edition: Berlin, Springer–Verlag, 538 p.

Watson, H.J., 1981, Casablanca field, offshore Spain, a paleogeomorphic trap (abs.): AAPG Bulletin, v. 65, p. 1005.

Comparison of Carbonate and Shale Source Rocks

R. W. Jones
Chevron Oil Field Research Company, La Habra, California

As with shales, the source potential of carbonate rocks depends primarily upon the organic facies rather than the mineral matrix. Where the depositional and early diagenetic environment is highly oxygenated, the total-organic-carbon (TOC) content is low. The remaining kerogen is highly oxygenated, with a negligible generative capacity for hydrocarbons, despite a relatively high hydrocarbon/TOC ratio in the immature state. An anoxic depositional early diagenetic environment can result in the deposition of organic-rich, fine-grained carbonate sediments that are excellent potential source rocks.

Excellent oil-prone source rocks, whether with carbonate- or clay-mineral matrices, have many characteristics in common. Both form in anoxic environments, are generally laminated and heterogeneous, have moderate to high TOC, and contain high-quality organic matter (OM). The latter is exemplified by atomic H/C ratios ≥ 1.2 near a vitrinite reflectance of 0.50% R_o. Although they constitute a small percentage of all carbonate rocks, organic-rich, fine-grained carbonate rocks are widespread in both time and space and are the probable source of 30–40% or more of the petroleum reserves of the world.

Gas-prone organic facies are rare in carbonate rocks because they are usually dominated by terrestrial organic matter deposited in a dominantly clay matrix. However, gas-prone organic facies may occur in carbonate rocks as a result of turbidite deposition or by a mixture of kerogen types II and IV. Most carbonate rocks contain nongenerative organic facies, as do most siliceous rocks. Oxygen-rich depositional environments for carbonates are found from sea level (reefs) to the ocean depths (*Globigerina* ooze).

Despite the basic relationship between organic-rich oil-prone carbonate and shale source rocks, some significant differences exist. Oils derived from carbonate rocks are often richer in cyclic hydrocarbons and sulfur compounds than oils derived from shales, owing to the dearth of terrestrial-plant waxes in the OM and less iron in the pore water. In addition, the generally earlier decrease of porosity and permeability and the greater contrast between the physical properties of the OM and the rock matrix in carbonate source rocks often result in different primary migration characteristics.

INTRODUCTION

The literature unfortunately has placed carbonates and shales in antagonistic positions regarding source-rock evaluation. Different criteria have been used to evaluate their source potential (Tissot and Welte, 1978; Hunt, 1967, 1979). This approach requires discussion. Why not also compare carbonates and siltstones, or carbonates and sandstones? Clearly, one missing ingredient here is the original grain size of the rock. Veber and Gorskaya (1965) noted that the variations in organic content of carbonate rocks with grain size are similar to those in terrigenous facies. Because the density of oil-prone organic matter (OM) is very close to that of water, oil-prone source rocks are almost invariably fine grained. Thus, to arrive at a meaningful comparison of carbonate and shale source rocks, we must compare the shales with fine-grained carbonates (e.g., micrites) deposited in quiet water. Most carbonates are not pertinent except to understand why they are not source rocks, just as siliceous siltstones and sandstones are not.

The source potential of any sediment depends primarily on its organic facies (Jones and Demaison, 1982). The latter is determined by the amount and type of the original OM input, the time the OM spends in an oxygenated water column, and the oxygen content of the water at and near the sediment–water interface during and shortly after deposition. The hydrologic factors that control the grain size of the rock matrix also influence the organic facies by their effect on the size and density of the organic particles deposited, by the rate of oxygen diffusion, and by the rate of replacement of the bottom water and, thereby, the input of oxygen to the bottom waters. Other important secondary factors that influence the organic facies preserved in the sediment include the sedimentation rate, the thickness of the overlying oxygenated water column, the near-surface organic productivity, and the varying amounts of terrestrial OM input (Jones, 1983).

This paper stresses the similarities between oil-prone source rocks that range in composition from nearly all clay and organic matter to nearly all carbonate and organic matter. Emphasis is on the independence of organic facies and overall petroleum-generating capacity from the composition of the enclosing rock matrix. Thus, carbonates, marls, and shales are placed in a continuum, with each term occupying one-third of the distance from 100 to 0 percent carbonate. Where additional refinement is available or necessary to make a point, numbers or modifiers are used. Also included is a discussion of TOC requirements and catalytic effects, two areas in which the literature suggests substantial differences between carbonate and shale source rocks; but I disagree. The paper concludes with a discussion of one area of major differences: the relative efficiency of different migration mechanisms in carbonate and shale source rocks.

ORGANIC FACIES

General

In this paper an organic facies is defined as a mappable rock unit, distinguishable from adjacent rock units by the character of its OM *without regard to the inorganic aspects of the sediment*. Thus, organic facies are defined similarly to other facies such as biofacies. All organic facies occur in both carbonates and shales, although in different proportions.

Previous Work

The origin of the concept of organic facies is obscure, but it clearly lies with the coal petrographers. They first recognized that the different coal macerals (types of OM) commonly occur in specific groupings with unique chemical and physical properties. They also recognized that the different microlithotypes, as organic facies are called in coal terminology, were determined by both the types of OM available in the coal swamp and the early diagenetic environments (Stach et al, 1982).

In the early 1960s, Krejci–Graf (1963, 1964a, 1964b) distinguished four main groups of organic substances in the earth's crust that were determined by the oxygen content of the depositional environment. The less oxygen in the depositional environment, the more oil-prone the OM. At the same time, Bitterli (1963a, 1963b) made an extensive study of the bituminous rock sequences in Europe and noted with moderate surprise that a major controlling factor in their deposition was not water depth but the anoxia of the depositional environment. The two authors mentioned carbonate content only in passing and clearly did not consider it to be a controlling parameter. Other studies relating depositional environment and/or organic-matter type to oil-generating capacity were made by Bass (1963), Breger and Brown (1963), Tissot et al (1974), Parparova and Neruchev (1977), Maksimov et al (1976), Cornelius (1978), Larskaia (1978), and many others. With the exception of Cornelius, who used carbonate content as a parameter to complete his Eh–pH diagrams, the carbonate content of the rock did not enter into their evaluations.

In the American literature the emphasis has been on differentiating a dominantly gas-generative, "terrestrial" organic facies from a dominantly oil-generative "marine organic facies" (Breger and Brown, 1963; Rogers and Koons, 1971; Dow, 1977; Barker, 1978, 1979; and Rogers, 1980). This dichotomy has not been conspicuously successful, despite its wide use. The distribution of gas and oil in reservoirs depends not only on the original organic facies of the source rock but also on the maturity level of the OM and on the relative ease of migration of gas and oil in different geologic situations (Spillers, 1965; Weber and Daukow, 1975; Tissot and Welte, 1978; Durand and Oudin, 1980). In addition, the degree of oxidation of both the terrestrial- and marine-derived OM and initial variations within each group have a profound effect on the amounts and types of generated products. For example, waxy terrestrial-plant debris is probably a major source contributor to some oils (Hedberg, 1968); plant resins have been touted as a source for some condensates (Snowdon and Powell, 1982); vitrinite yields dominantly gas to reservoirs despite the generation of abundant liquids in vitrinitic coals (Tissot and Welte, 1978); and highly oxidized terrestrial OM is essentially inert (Tissot and Welte, 1978). Although marine OM initially has a higher, more homogeneous oil-generating capacity than terrestrial OM, it has the same problems of varying degrees of oxidation in the water column and during early diagenesis as the terrestrial OM. Thus, its generating capacity also runs the full spectrum from highly oil prone to inert. Clearly some tie between generating capacity and comparatively easily measured chemical parameters has been needed.

Without explicitly using the concept of organic facies, Tissot et al (1974) used the atomic H/C ratio of kerogen to define three kerogen types with different hydrocarbon-generating capacities. The kerogen types were subsequently related to Rock-Eval pyrolysis data (Espitalié et al, 1977). Building on the concepts developed by the French investigators, Jones and Demaison (1982) explicitly defined different organic facies[1] in terms of atomic H/C ratios and Rock-Eval data at a maturation defined by a vitrinite reflectance of $\cong 0.50\%$ R_o—i.e., near the initiation of significant generation of liquid hydrocarbons. It is important to recognize that, as defined by nature, the different organic facies are completely gradational, and the assigned boundaries arbitrary. The basic effort of Jones and Demaison (1982) tied generating capacity, organic chemistry, early diagenetic environment, sedimentology, and original OM input into a concept understandable to and usable by a petroleum geologist (Tables 1, 2). The independence of organic facies and rock matrix is stressed in the following discussion of four different organic facies that occur in both carbonate-rich and carbonate-poor rocks.

Organic Facies A

In both carbonates and shales, organic facies A is characterized by deposition in a highly anoxic environment. The anoxic environment can be either preexisting or created by the high input of OM (Demaison and Moore, 1980; Calvert, 1983). The associated rock is usually laminated and heterogeneous, the TOC contents are moderate to very high, and atomic H/C ratios are ≥ 1.4 at $\cong 0.5\%$ R_o (Tables 1, 2). The Rock-Eval S_2 response curve is usually rather narrow. T_{max}, the temperature of maximum hydrocarbon generation from the kerogen, is unusually high for a given maturation. The narrow S_2 and high T_{max} probably reflect the high degree of polymerization and homogenization of the OM.

Chevron data indicate that organic facies A occurs more frequently in carbonate rocks than in shales. This association is probably due to the dearth of terrestrial and previously degraded or thermally altered OM in most carbonate sequences of organic facies A. Microscopically, the OM is usually amorphous, clumpy, and quite homogeneous. It is presumed to be of algal/bacterial origin (Tissot and Welte, 1978). However, a strong case for hydrogenation and chemical homogenization by high alkalinity has been made for the Eocene Green River Formation (Smith and Lee, 1982; Smith, 1983). A preferred depositional environment for documented cases is shallow-water saline lakes, as exemplified by the Green River Formation and some other Tertiary lakes in the Great Basin. However, in some marine carbonate sequences, most notably some carbonate source rocks in the Middle East, the input of clay and terrestrially derived OM is negligible, the atomic H/C ratios of immature OM slightly exceed 1.4, and the existence of organic facies A can be inferred (Tissot and Welte, 1978). In

[1]The concepts of kerogen type and organic facies overlap and seem similar, but they are quite different. For example, organic facies D, commercially nongenerative, can be composed of a variety of kerogen types, including (1) highly oxidized terrestrial OM oxidized above the water table (e.g., terrestrial red beds), (2) OM from a variety of sources oxidized by oxygen-rich water at any water depth, and (3) redeposited kerogen of any type that was overmatured in a previous thermal cycle.

Table 1

Organic Facies	Products	Depositional Environment	Internal Structures	Organic Matter
A	Oil	Anoxic (saline), lacustrine; rare marine	Finely laminated	Algal; amorphous; rare terrestrial
B	Oil	Anoxic; marine	Laminated, well bedded	Algal; amorphous; common terrestrial
B–C	Oil–gas	Variable; deltaic	Poorly bedded	Mixed marine, terrestrial
C	Gas	Mildly oxic; shelf/slope; coals	Poorly bedded; bioturbated	Terrestrial, mostly "vitrinites"; partially degraded algal
D	Dry gas	Highly oxic; anywhere	Massive; bioturbated	Highly oxidized; reworked

Table 1 — Some generalized sedimentary characteristics of organic facies A–D.

Table 2

Organic Facies	Dominant Products	Atomic H/C at $\%R_o \cong 0.5$	Pyrolysis Yield[a]	
			HI	OI
A	Oil	≥1.4	700–1000+	10–40
B	Oil	1.2–1.4	350–700	20–60
B–C	Oil–gas	1.0–1.2	200–350	40–80
C	Gas	0.7–1.0	50–200	50–150
D	Dry gas	0.4–0.7	<50	20–200

[a]Derived from Rock-Eval pyrolysis data, where HI = hydrogen index = mg hydrocarbons generated/g TOC; OI = oxygen index = mg CO_2 generated/g TOC.

Table 2 — Some generalized geochemical characteristics of organic facies A–D.

carbonate-poor sequences containing organic facies A, discrete algae (e.g., *Botryococcus*, *Tasmanites*) usually dominate (Hutton, Kantsler, and Cook, 1980).

Organic facies A is likely to be more laterally and vertically extensive in carbonate rocks than in noncarbonates because of the greater stability of the depositional environment and the dearth of terrestrial input. Unfortunately, these factors, conducive to development of excellent potential source rocks, often carry with them a limited accessibility to reservoir rocks unless substantial tectonic activity has severely fractured the carbonates (Fouch, 1983b). Because everything has to be just right to form organic facies A, it is rare in the geologic record and has not been a major source of oil, except perhaps locally in the Middle East, whether it was deposited in a carbonate or a clay matrix.

Organic Facies B

Organic facies B is the source of most of the world's oil. The carbonate content of the rock matrix ranges from <1 to >99%. Irrespective of carbonate content, the TOC contents are usually >1.5 wt%, and the OM dominantly amorphous with an atomic H/C ratio ≥1.2 at $\%R_o \cong 0.5$. The sediments are usually laminated or banded and either were deposited under anoxic water or else the pore water became anoxic soon after deposition. The Rock-Eval pyrograms of carbonate and shale organic-rich source rocks of organic facies B are similar. However, the S_2 peak is usually broader in shales and the hydrogen-index averages somewhat lower. The differences reflect a greater diversity of OM in the shales.

Conditions favorable for the deposition of organic facies B can be found in several scenarios, including the following:

1. Rapid deposition of OM can create local anoxia in a generally oxic environment, as occurs on the shelf off southwestern Africa (Calvert, 1983; Calvert and Price, 1971).

2. Deposition of OM within an oxygen minimum layer (OML) where it impinges on the upper slope can create local anoxia, as suggested by Summerhayes (1981) for the Monterey Formation in southern California.

3. Thermal or salinity layering can cause a persistent density stratification that precludes replenishment of the oxygen supply, even if oxygen is only slowly consumed by the oxidation of the incoming OM. In this situation, anoxia first develops in the basin's topographic lows and gradually expands upward until oxygen insertion by water mixing produces an upper limit.

Despite increasing recognition of the occurrences of scenarios 1 and 2 and their combination, most carbonate source rocks were deposited within the constraints of scenario 3.

Density stratification owing to salinity differences in a developing evaporitic regime is often a prelude to the deposition of carbonate-rich source rocks in a situation in which the anoxia builds upward from the sediment–water interface. Examples include (1) the basinal source rocks of Silurian pinnacle reefs in the Michigan basin (Gardner and Bray, 1984); (2) the lagoonal source rocks for the Lower Cretaceous rudist mounds in south Florida (Palacas, 1983); (3) the deep-water organic-rich Permian Lamar and Bone Spring limestones underlying the Castile evaporites in the Delaware basin of west Texas–New Mexico (Fig. 1); (4) the inferred interreef carbonate source beds in the Devonian of the Rainbow–Zama area in western Canada (McCames and Griffith, 1967); (5) the deep-water Rudeis Formation marls in the Miocene of the Gulf of Suez (Heybroek, 1965; Hassan and El-Dashlouty, 1970); (6) parts of the middle Paleozoic of the Russian platform (Zhuze, 1972); (7) the Cambrian of the southeastern Siberian platform (Bakhturov, 1981); (8) parts of the Tethys Ocean in the Mesozoic, particularly the Upper Jurassic of the Middle East (Murris, 1980); (9) the Jurassic of the Gulf Coast (Oehler, 1984); and (10) the precursors of the main deposition of the Permian Zechstein evaporites in western Europe (Fig. 2). There may be a source contribution from the "Stinkdolomit" and "Stinkkalk," but the primary source is probably the Stinkschiefer, which is a laminated carbonate-rich (70–90%) rock that contains the highest TOC contents (Botz and Müller, 1981; Botz et al, 1981).

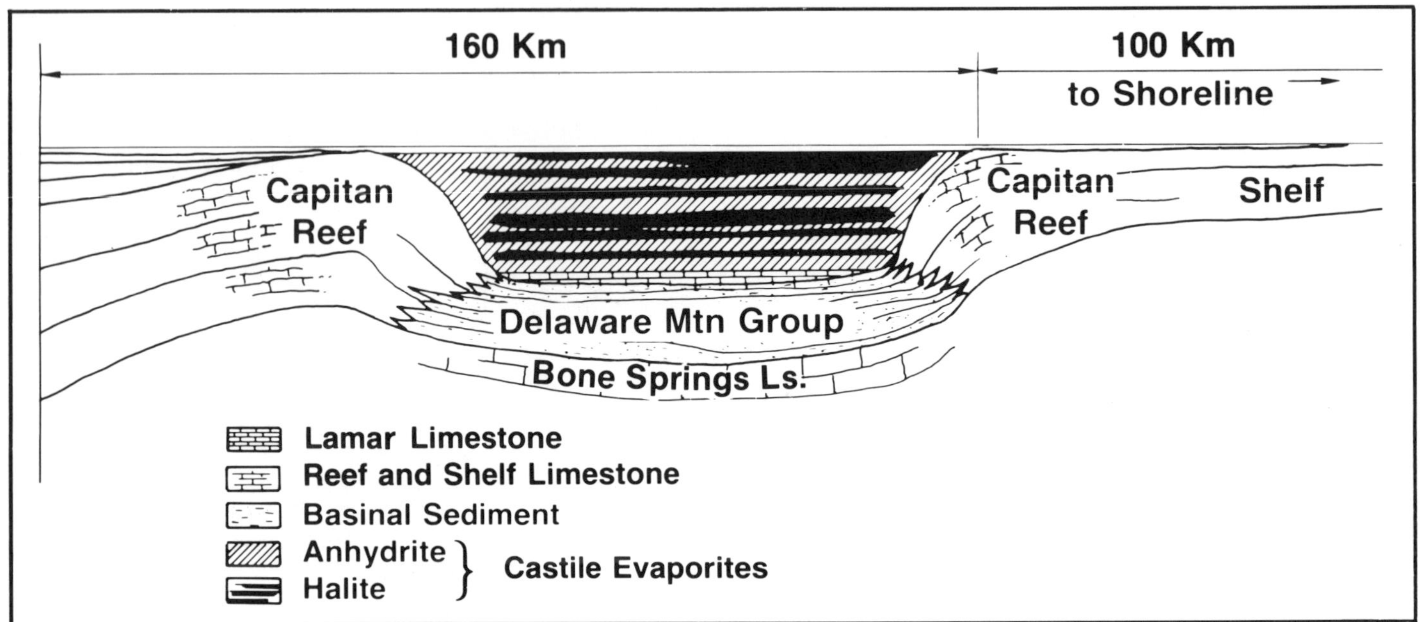

Figure 1 — Diagrammatic stratigraphic cross section, Late Permian time, Delaware basin, west Texas and southeastern New Mexico, showing the propitious geometrical relationships between the source rocks of the Lamar and Bone Springs Limestones, deposited in the basin center, and the adjacent reservoirs in the reef and on the shelf.

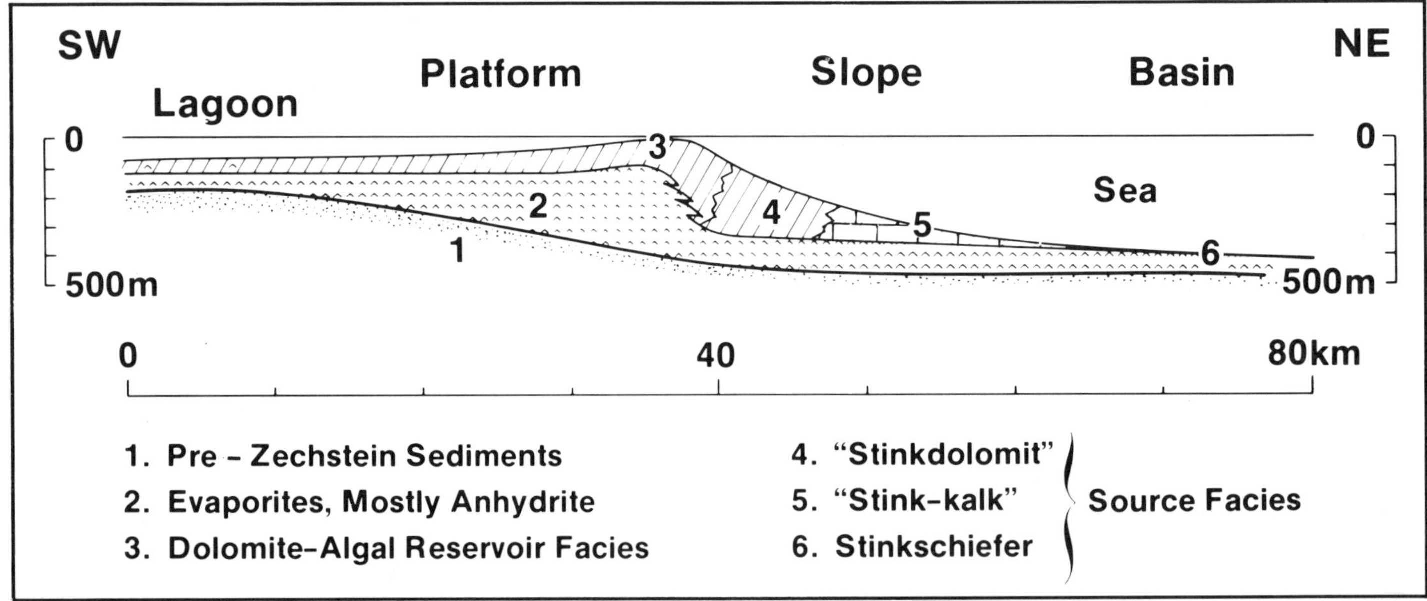

Figure 2 — Diagrammatic stratigraphic cross section of a part of East Germany showing the relationships between source and reservoir rocks in the lower part of the Permian Zechstein Formation prior to deposition of the succeeding evaporitic sequence. Cross section modified from Botz and Müller (1981).

Where carbonate rocks with good source potential are interbedded in evaporite sequences, migration problems often have precluded effective drainage. Any thermally generated hydrocarbons may remain where they originated or very close by. Thus, accumulations are likely to be small, as are those having a source in the Cretaceous Binga Formation carbonates in the Cuanza basin of Angola (Brognon and Verrier, 1966). The organic-rich calcareous shales interbedded with evaporite deposits in the Pennsylvanian Paradox Formation of the Paradox basin, Utah, still retain almost all of their hydrocarbons. Where they were deposited around the porous algal mounds at the southern end of the basin, they have been the source of locally prolific production (Peterson and Hite, 1969).

Many organic-rich carbonate source rocks have been deposited in lake systems. On the average they have been

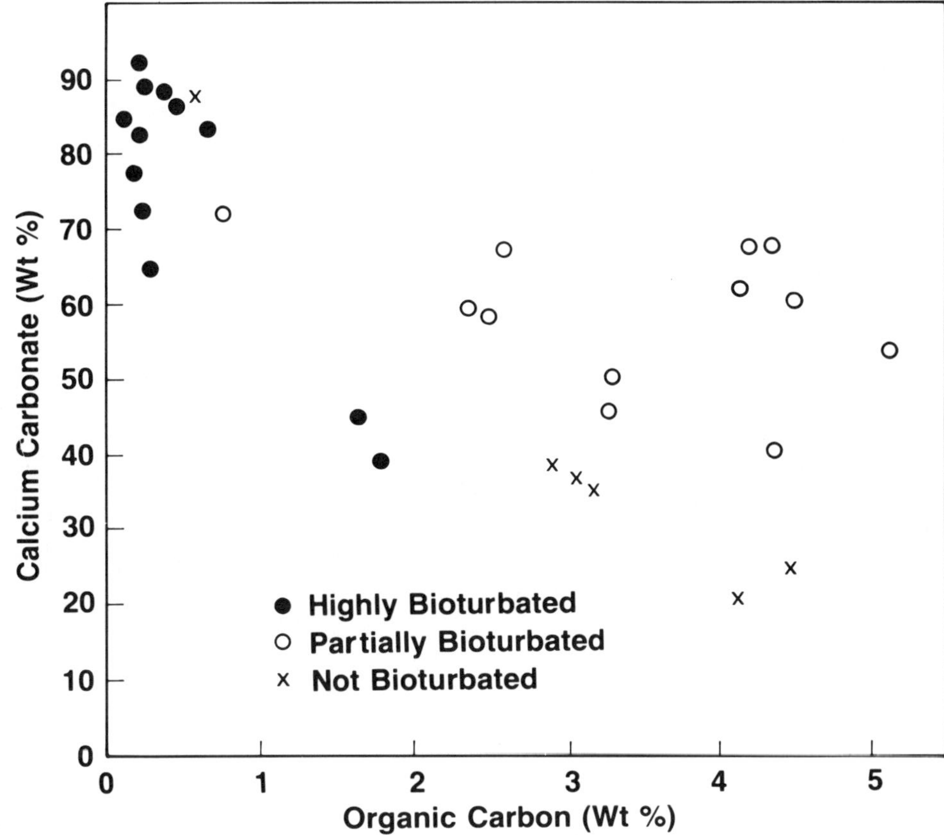

Figure 3 — Correlation of calcium carbonate and organic-carbon contents in a 30-m (98-ft) core from the middle Cretaceous Greenhorn Formation, Colorado. Data from Pratt (1982).

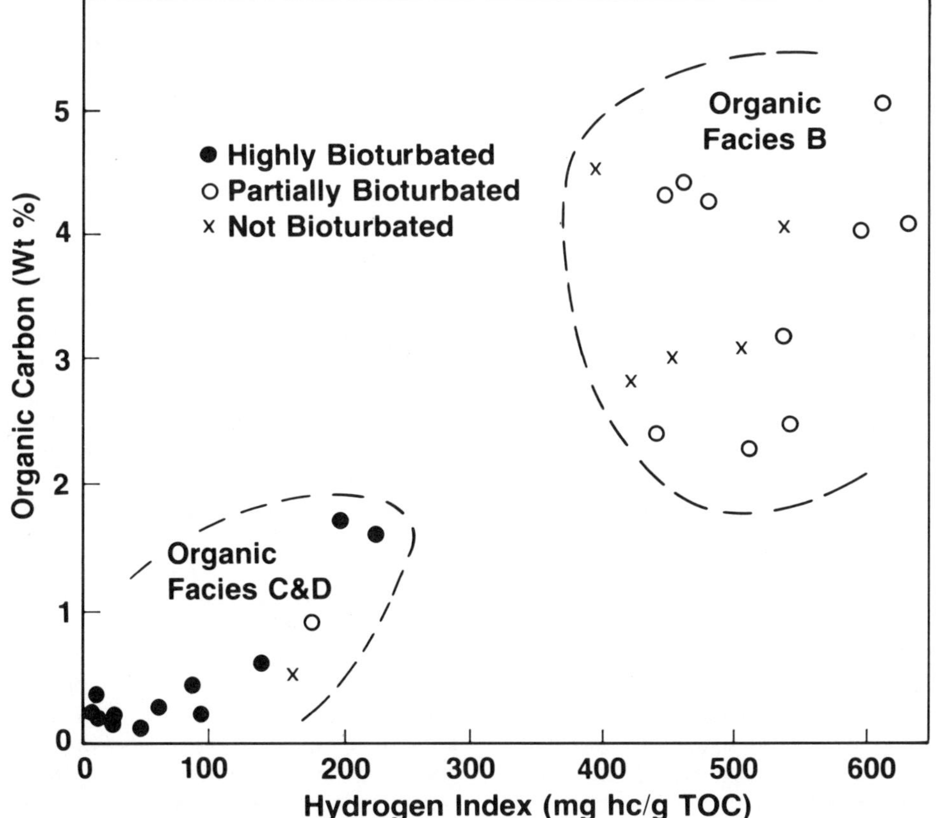

Figure 4 — Correlation of organic carbon and hydrogen index in a 30-m (98-ft) core from the middle Cretaceous Greenhorn Formation, Colorado. Data from Pratt (1982). The very low hydrogen indices of some of the high carbonate samples may be due in part to mineral-matrix effects during pyrolysis (Espitalié, Madec, and Tissot, 1980; Dembicki, Horsfield, and Ho, 1983).

responsible for less commercial oil than their often less organic-rich shale counterparts because of poorer access to reservoirs (Fouch, 1983a, 1983b).

Density stratification caused by salinity differences also can be created by an ingress of fresh water at the top of the water column. Documented cases in the geologic record in which carbonate source rocks resulted from this type of density stratification are rare and difficult to identify. However, the flow of fresh water through deltaic systems into Lake Uinta must have contributed to the density stratification there that led to the formation of organic-rich dolomitic marls in the Green River Formation. Ozimic (1982) attributed formation of the carbonate-rich oil shales of the Lower Cretaceous in the Eromanga basin, Australia, to anoxia created by the input of fresh water into a silled marine basin.

Perhaps the best documented illustration of how density stratification created by fresh-water input into a marine environment can affect the carbonate and organic content of the underlying sediments is provided by an excellent study by Pratt (1982) of a 30-m (98-ft) core from the middle Cretaceous Greenhorn Formation near Pueblo, Colorado. Figure 3, generated from Pratt's tabular data, shows two distinct groupings of sediments: one that is carbonate-rich and organic-carbon-poor, and the other with a moderate carbonate content and a moderate to high organic-carbon content. Pratt explained the two sediment types as products of variable terrestrial and fresh-water input from highlands to the west. During times of high fresh-water input, enhanced density stratification and bottom-water anoxia resulted in inhibited growth of calcareous algae and preservation of organic-rich sediment.

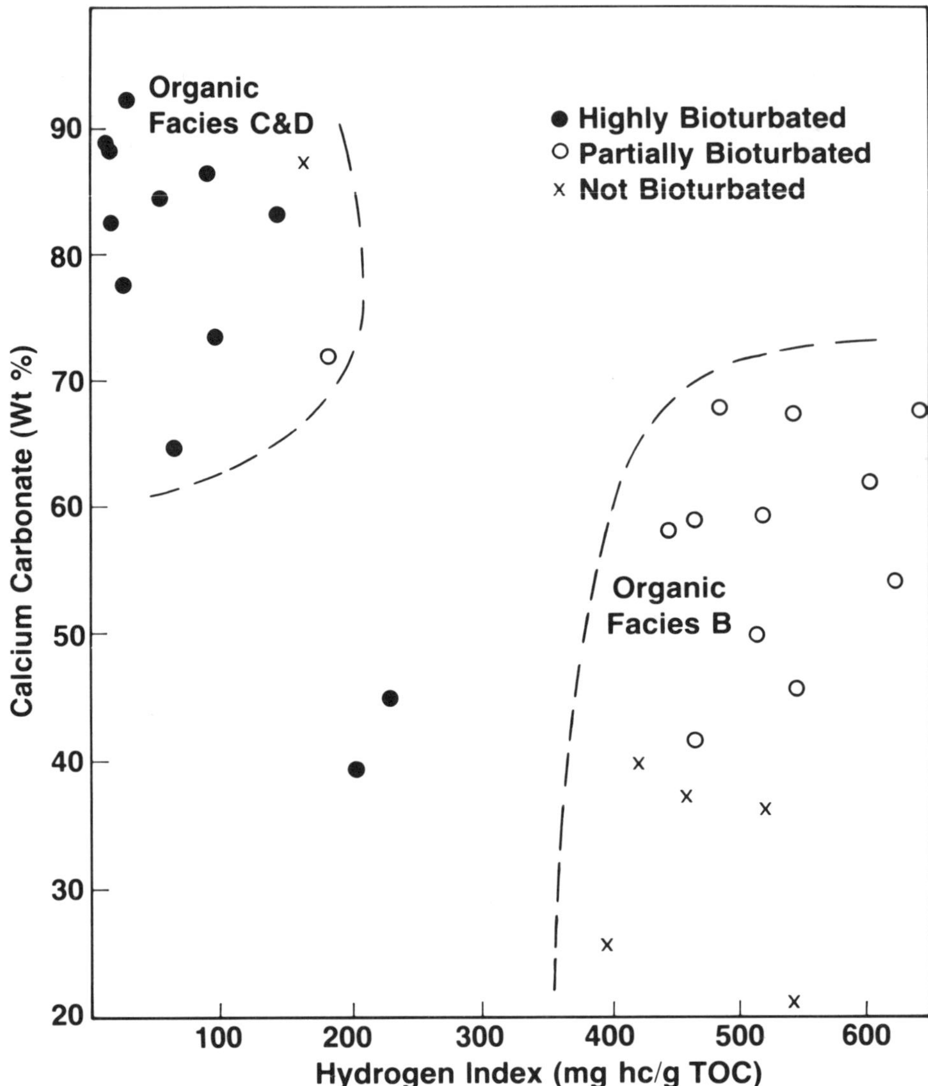

Figure 5 — Correlation of calcium carbonate and hydrogen index in a 30-m (98-ft) core from the middle Cretaceous Greenhorn Formation, Colorado. Data from Pratt (1982).

Pyrolysis and chromatographic data indicate that the OM in the Greenhorn Formation is at the beginning of the oil window. Therefore, the different organic facies can be recognized by using the hydrogen index (Jones and Demaison, 1982). Figure 4 shows a well-defined organic facies B, and a gradation from organic facies C to organic facies D, as defined by the hydrogen-index data of Pratt (1982). As the TOC decreases, the OM becomes less oil prone, reflecting the increased proportion of bacteria-resistant, residual, nongenerative OM. The plot of hydrogen index versus percentage of carbonate (Fig. 5) indicates the same separation of organic facies as shown in Figure 4. These results show the effects of increased bioturbation recorded by Pratt (1982) in the high-carbonate rocks.

Three other less clearly defined points can be made:

1. Five of the six nonbioturbated samples have a lower carbonate percentage and a slightly lower hydrogen index than the higher carbonate samples of organic facies B. This may reflect the increased input of terrestrial OM that Pratt (1982) observed in the more clay-rich samples.
2. Two highly bioturbated samples with 40–50% carbonate content occupy a distinct niche in Figure 5. Stratigraphically, they occur as thin marls within a limestone sequence and probably reflect an influx of fresh water and associated clay insufficient to create complete anoxia. Their hydrogen indices are at the extreme upper end of organic facies C and demonstrate that marls, despite a dominant marine-algal input, can, if partly degraded, contain organic facies C.
3. The anomalous, nonbioturbated sample with a high carbonate content is from a thin limestone bed within a marl. Thus, it may be surmised that the length of time of normal marine salinity at the surface necessary to create the limestone was insufficient to disturb the anoxia at the water–sediment interface. In addition, the modest TOC of the sample, albeit the highest of the high carbonate samples, suggests that most of the original OM was not derived from calcareous algae. The carbonate was a diluent.

Dilution is common to many potential carbonate source rocks. The Holocene sediments of the Black Sea are a case in point. Anoxia was created by density stratification, this time caused by influx of saline marine water into a fresh-water lake. However, the most organic-rich sediments usually contain the least carbonate, a fact in part caused by the com-

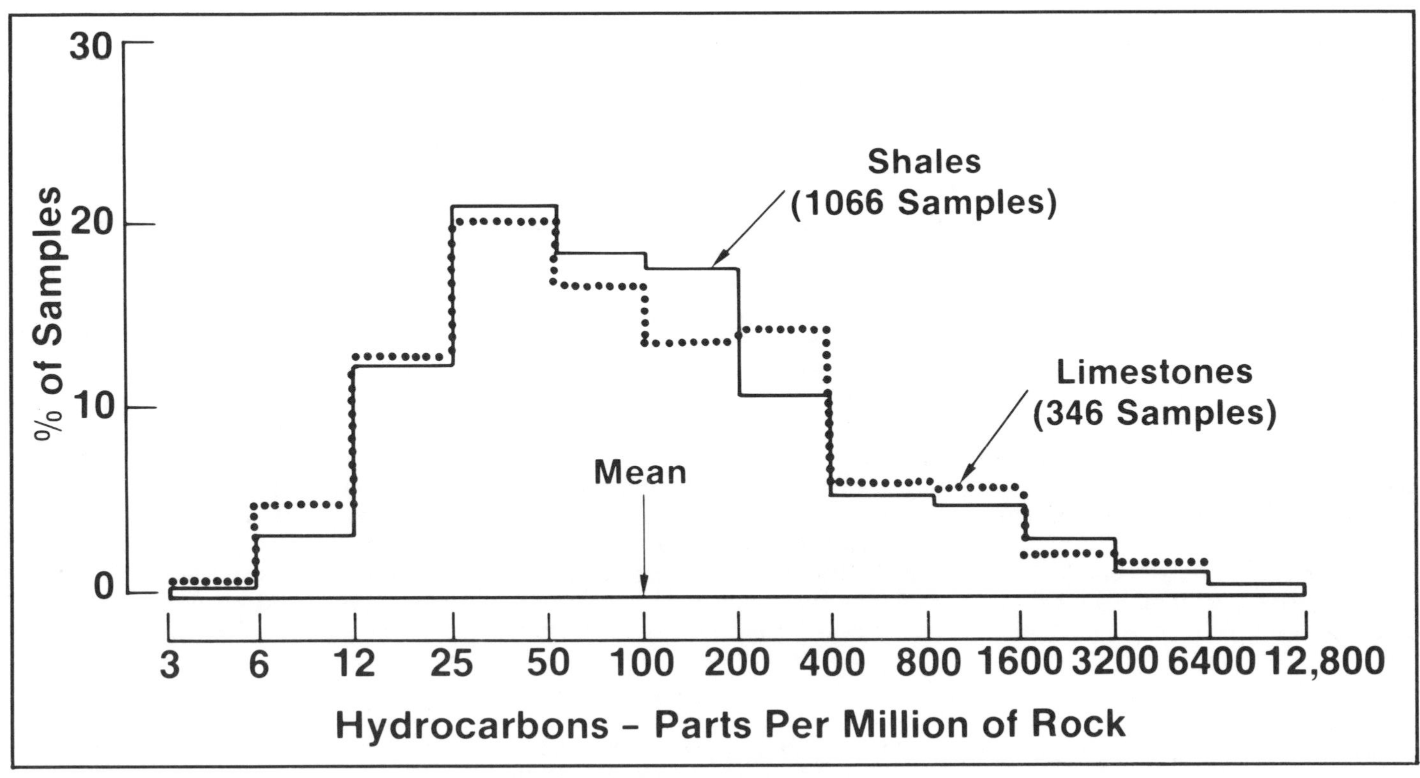

Figure 6 — Distribution of hydrocarbons in Cambrian to Late Tertiary shales and limestones. Figure from Gehman (1962).

bination of a roughly constant input of OM superimposed on a varying rate of coccolith deposition (Degens, 1974; Degens et al, 1978; Jones, 1983).

Carbonate dissolution caused by a low pH at or near the sediment–water interface can also increase the OM/carbonate ratio. This process also contributed to the strong inverse correlation between OM and carbonate content in the Black Sea (Degens and Stoffers, 1976). Dean, Gardner, and Cepek (1981) stress variations in the carbonate compensation depth and the resulting periodic dissolution of carbonate in Cretaceous–Eocene sediments of the northwest Atlantic as a major control on varying organic and carbonate content. Similar effects were postulated as one of the controls in the deposition of organic-rich, middle Cretaceous limestones in the equatorial Pacific (Dean, Claypool, and Thiede, 1981).

Sometimes the transition from carbonate-poor to carbonate-rich organic facies B can be very gradual. An excellent example is the Eagle Ford Shale to Austin Chalk transition in the Cretaceous of the East Texas basin. Grabowski (1981, 1984) showed that the organic facies in this interval remained essentially constant as the Cretaceous transgression proceeded. As the locus of deposition became farther removed from terrigenous input, the carbonate content of the sediments increased. Only when the water became shallower and the anoxia at the water–sediment interface was destroyed did the organic facies become less oil prone, although the carbonate content remained high.

Many of the world's prolific oil source rocks were deposited under conditions in which the density stratification necessary for anoxia to develop and persist at the ocean bottom was created by temperature gradients in widespread warm, temperate climates. Water turnover owing to temperature-induced density inversions did not occur, and oxygen injection near the water–sediment interface was minimal because of a lack of strong mixing. Such periods typically developed oxygen-deficient bottom water worldwide, occurring during quiet but persistent worldwide transgressions (Tissot, 1979), and extended the oxygen-deficient conditions well onto the shelf, particularly into broad shelf basins. These conditions led to deposition of carbonate-, siliceous-, and clay-rich organic facies B source rocks.

The most widespread transgressive source rocks are usually clay rich. Examples include the Upper Devonian–Lower Mississippian Bakken Shale–Woodford Shale–Chattanooga Shale unit in the Mid-Continent province of the United States and the Kimmeridgian source rocks of the North Sea and Grand Banks. Deeper water Woodford equivalents (Arkansas Novaculite) are also organic rich but are less so than the Woodford, whose organic content was not diluted by major siliceous input. The siliceous Lower Cretaceous Mowry Shale is widespread in the Rocky Mountain area and is a major oil source. The Upper Cretaceous Niobrara carbonate and equivalents of organic-rich organic facies B are also widespread in the Rocky Mountains; however, their contribution to oil production has been limited by maturity and reservoir-access problems. The very prolific Upper Jurassic source rocks in the Middle East also were deposited in relatively shallow water during a widespread transgression in a mild climate (Murris, 1980; Ayres et al, 1982). These rocks are virtually all carbonate and OM. The climate-induced density stratification was probably aided by salinity stratification, as indicated by associated evaporites in the Gotnia basin to the north and pseudomorph structures after halite in the broad lagoonal areas on the Arabian Jurassic shelf (Ayres et al, 1982).

Many shallow-water carbonate depositional environments can include a variety of local conditions that permit the accumulation and preservation of a significant amount of OM (Malek-Aslani, 1980). Local vertical salinity gradients can exist on a carbonate shelf, particularly in areas of irregular topography. Because of the low iron content of many carbonate environments, carbonate sediments tend both to generate and to preserve H_2S. The H_2S can create an anoxic environment at lesser sediment thicknesses than for clay muds (Lindbloom and Lupton, 1961). The combination of high organic productivity, high carbonate input, and low clay input can create an H_2S–O_2 interface at the water–sediment interface in shallow water, as postulated by Abed and Amireh (1983) for the Upper Cretaceous oil shales of Jordan. Anoxia can develop in shallow-water sediments at/near the sediment–water interface in fine-grained sediments of any type of sediment matrix if the input of organic matter is sufficiently rapid to deplete the available oxygen. A large amount of organic matter is destroyed to permit the preservation of the remainder (Jones, 1983). Examples include certain areas of noncarbonate muds in the Baltic Sea (P. J. Müller, written communication), the siliceous sediments in areas of upwelling off Peru and southwest Africa (Müller and Suess, 1979; Calvert and Price, 1971), and thin organic-rich layers interbedded with Ordovician shelf carbonates and evaporites in the Williston basin (Kohm and Louden, 1979). Blue-green algal (stromatolitic) mats containing thin, organic-rich laminae have existed throughout much of geologic time. However, their restricted lateral and vertical extent at any particular time has limited their significance as source rocks of organic facies B. The best documented exception to this generalization is the probable algal-mat source for the substantial hydrocarbon accumulations in the Upper Jurassic Smackover Formation of the Gulf Coast (Oehler, 1984).

Despite the overlapping and intertwined nature of carbonate-, siliceous-, and clay-dominated source rocks of organic facies B demonstrated above, some differences exist. Although most OM in most organic facies B rocks is marine derived in both clay- and carbonate-rich rocks, the clay-rich rocks usually contain more terrestrially derived OM. Fortunately the more waxy and hydrogen-rich macerals of terrestrial plants (e.g., spores, pollen, cuticle, and the more hydrogen-rich varieties of vitrinite) are the most resistant to bacterial attack. Thus, if the terrestrial input is minor, it is likely to be dominated by hydrogen-rich OM similar in oil-generating capacity to the marine OM. This is true for the Kimmeridgian of the North Sea and the better Cretaceous source rocks of the Rocky Mountain region. Of course, as the percentage of terrestrial OM increases, the overall quality decreases, resulting in organic facies B–C (e.g., Tertiary of the Gulf Coast) and ultimately the primarily gas-generative organic facies C, characterized by a dominant terrestrial input.

Statistical differences exist between the oils generated in carbonate- and clay-dominated rocks of organic facies B (Gransch and Posthuma, 1974; Tissot and Welte, 1978; Breger, 1980; Tissot and Roucaché, 1980). Although they both yield a naphthenic–paraffinic oil, the paraffin content increases with maturity and terrestrial OM content. Oils derived from source rocks with iron-rich clays are usually low in sulfur; conversely, carbonate and siliceous rocks of organic facies B typically yield moderate- to high-sulfur oils. In addition, the high sulfur content of the kerogens in some carbonate/siliceous oil-prone source rocks can result in the formation of low-gravity, sulfur-rich oils formed at less than normal thermal stress (unpublished Chevron data). Such oils primarily reflect the lower activation energy required to break sulfur–carbon than carbon–carbon bonds. The higher sulfur content can persist into the condensate stage of maturation. For example, in western Wyoming, condensates derived from the Phosphoria Formation (Permian) are distinctly higher in sulfur than those of the same API gravity derived from essentially carbonate-free Cretaceous sources.

Typically, giant oil fields derived from organic facies B in either carbonate-rich (e.g., Arabia) or clay-rich (e.g., North Sea) rocks have low gas–oil ratios unless the maturity of the OM is high. This reflects both the high TOC contents necessary for prolific oil production from either rock type and oil migration prior to reaching the main stage of gas generation.

Other differences exist, particularly in dominant migration mechanisms, as discussed later.

Organic Facies C

The gas-prone organic facies C is typically composed of partially oxidized terrestrial OM deposited in a carbonate-poor clay matrix. Tremendous thicknesses of this rock–OM association were deposited on continental margins during extensive episodes of the Mesozoic and Cenozoic Eras, as the oil industry discovered to its consternation during exploration on continental shelves during the past two decades.

Organic facies C can occur in carbonate rocks. However, it is volumetrically unimportant and usually distributed in thin layers of limited areal extent. Several processes can be involved. Limited amounts of terrestrially derived OM can be introduced into a carbonate environment, but because the terrestrial OM is almost always associated with clay the carbonate environment will usually be overwhelmed if the influx of terrestrial OM becomes dominant. The OM that initially formed in carbonate environments is dominated by the potential for organic facies A–B; however, the variation in oxygen content near the water–sediment interface can sometimes stabilize the degrading OM in the geochemical characteristics of organic facies C (Fig. 4; Dean, Gardner, and Cepek, 1981; Pratt, 1982). Another possible scenario for the development of organic facies C in carbonate rocks is the side-by-side existence of highly degraded OM and higher quality OM protected from degradation by inclusion in carbonate skeletal remains. In such cases, the TOC is usually quite low, and a small amount of late-stage gas migration is the best that can be hoped for. Basically, organic facies C is economically unimportant in carbonate rocks.

Organic Facies D

Unfortunately, most carbonate rocks contain nongenerative organic facies D, a fact that has undoubtedly postponed the recognition of some carbonates as excellent oil-source rocks. The primary reason for the dominance of organic facies D in carbonate rocks is straightforward: most carbonates were deposited in highly oxygenated environments (Larskaia, 1977). This environment existed in carbonate banks and reefs, in the shelf and ramp carbonates that are important volu-

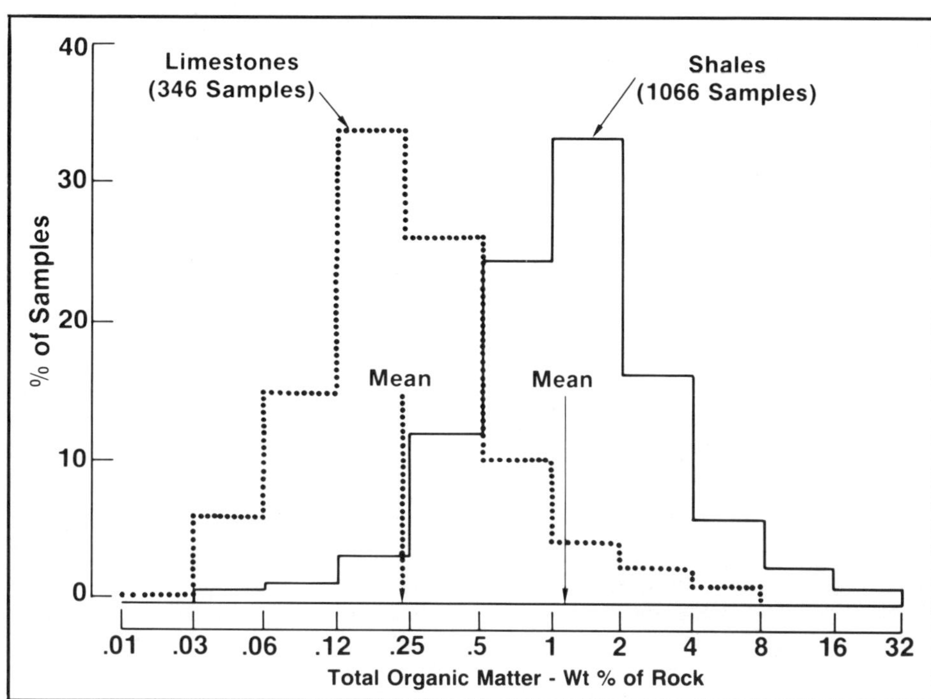

Figure 7 — Distribution of total organic matter in Cambrian to Late Tertiary shales and limestones. Figure from Gehman (1962).

metrically, in most of the chalks deposited in intermediate water depths, and in deep-sea oozes deposited above the carbonate compensation depth. These are the rocks with low TOC contents but high hydrocarbon/TOC ratios that have occasionally been touted as possible source rocks (Hunt, 1967). I consider their generating capacity negligible. The epitome of organic facies D in carbonate rocks was reached in most of the Jurassic back-reef carbonates of the East Coast of the United States. There, a highly oxidizing environment was also subjected to an input of residual nongenerative OM derived from erosion of the strongly thermally altered Paleozoic rocks of the Appalachian Mountains.

Of course, organic facies D is also dominant in carbonate-poor rocks. Nonmarine rocks, except for most coals and some lake sediments, are composed of organic facies D, as are most shelf sediments, particularly the siltstones and sandstones. Red beds contain organic facies D, whether their highly oxidized kerogen was originally derived from structured terrestrial OM deposited in flood-plain deposits or nonstructured marine OM deposited in the ocean depths.

Table 3

Total Organic Matter (wt%)	Carbonate (wt%)	Bitumen/TOM	Number of Samples
0–0.1	95.7	23.8	63
0.1–0.2	89.8	14.4	77
0.2–0.4	84.5	12.1	45
0.4–0.6	75.4	6.8	23
0.6–0.8	42.1	6.6	16
0.8–1.0	33.9	5.5	18
>1.0	27.2	2.2	37

[a]Data from Uspenskii and Chernysheva (1951).

Table 3 — Relationships between total-organic-matter (TOM), bitumen, and carbonate contents for 279 samples from the USSR.[a]

THE TOC PROBLEM

The repeated suggestions in the literature that carbonates may be effective oil source rocks at lower TOC contents than shales are largely based on interpretations by others of Gehman's (1962) excellent pioneer work. Thus, Gehman's study is worth reexamining despite the lack of both thermal-maturation data and the pairing of TOC and hydrocarbon data on the same samples. Gehman's study is summarized in Figures 6 and 7. For the 346 limestone and 1,066 shale samples analyzed, the data show that (1) the average total-organic-matter (TOM) content of the shales ≅ 4 times that of the limestones; (2) the absolute amounts and distribution of hydrocarbons in limestones and shales are nearly identical; and, consequently, (3) the ratio of hydrocarbons to TOM ≅ 4 times greater in limestones than in shales. Gehman's data, particularly item 3, have been used to argue that carbonates do not require as much organic carbon as shales to be source rocks. A minimum figure of 0.3 wt% TOC for carbonates has been suggested (Hunt, 1967, 1979; Tissot and Welte, 1978).

Other interpretations of Gehman's data are possible. The lower TOM and higher hydrocarbon/TOM ratio in carbonates could, in part, simply reflect the greater preferential destruction of nonhydrocarbon OM in highly oxygenated and bioturbated sediments. Although primarily concerned with the effects of oxygenated meteoric waters, Gehman (1962) suggested this interpretation. Unfortunately, Gehman did not break out the hydrocarbon/TOM ratios as a function of both TOM and carbonate percentage. However, others have.

Although using bitumen rather than

Table 4

Total Organic Matter (wt%)	Carbonate Rich ($CaCO_3 \geq 92$ wt%) Carbonate (wt%)	Bitumen/TOM	Number of Samples	Relatively Carbonate Poor ($CaCO_3 \leq 71.5$ wt%) Carbonate (wt%)	Bitumen/TOM	Number of Samples
0.1	98.7	24.2	56	71.5	21.4	7
0.1–0.2	98.1	14.3	61	51.7	14.7	16
0.2–0.4	95.8	13.2	34	49.4	9.6	11
0.4–0.6	92.7	5.8	14	48.6	8.5	9
0.6–0.8	95.9	5.6	6	9.7	7.2	10
0.8–1.0	93.1	6.1	5	11.2	5.2	13
>1.0	92.1	3.9	2	23.5	2.3	35

[a]Data from Uspenskii and Chernysheva (1951).

Table 4 — Correlation of total organic matter (TOM) with bitumen/TOM for 178 carbonate-rich (>90% $CaCO_3$) and 101 carbonate-poor (<72% $CaCO_3$) samples from the USSR.[a]

hydrocarbon data on 279 samples from Paleozoic–Tertiary carbonates of the USSR, Uspenskii and Chernysheva (1951) showed conclusively that the basic correlation is between low TOM and high bitumen/TOM ratios, not between carbonates and high bitumen/TOM ratios (Tables 3, 4). Table 3 shows a strong positive correlation between carbonate percentage and the bitumen/TOM ratio. The data also show an equally strong inverse correlation between the carbonate percentage and the TOM and between the TOM and the bitumen/TOM ratio. What underlying relationship controls these correlations? Uspenskii and Chernysheva (1951) resolved this problem by dividing each TOM group into high and low (relatively) carbonate subgroups (Table 4). These data clearly show that the primary controlling factor of the bitumen/TOM ratio is the amount of TOM, not the carbonate percentage. Uspenskii and Chernysheva (1951, p. 104) explicitly attributed the increased bitumen/TOM ratio in carbonate rocks to a faster rate of decomposition of the OM. Their explanation is consistent with the dominant control of both the TOM percentage and the bitumen/TOM ratio being the oxygen content of the depositional and early diagenetic environment rather than the chemical composition of the rock matrix.

Bordovskiy et al (1974) provided an unusual example of the preferential destruction of nonhydrocarbon OM in the bottom sediments of the Kuril–Kamchatka trench. They showed that as OM moved through the alimentary canal of a bottom feeder, the total amount of oils (≅ hydrocarbons) remained essentially constant, whereas the TOC dropped by ≅50% and the bitumen content by nearly an order of magnitude. Thus, the hydrocarbons were preferentially preserved, and the hydrocarbon/TOC ratio went up as the TOC dropped.

Despite the overriding inverse correlation between TOM and the bitumen/TOM ratio for all rock matrices, there is a definite tendency, particularly at very low TOMs, for the bitumen/TOM and hydrocarbon/TOM ratios to be higher in carbonate rocks. Bordovskiy and Takh (1978) compared the bitumen/TOM ratios for Holocene sediments defined as carbonate and carbonate-free sediments in the Caspian Sea (Fig. 8). The shapes of the two curves are essentially identical, confirming the basic TOM to bitumen/TOM correlation. However, the two curves show a persistent separation between the carbonate and carbonate-free sediments, a separation that clearly increases with decreased TOC. Do the substantially higher hydrocarbon/TOC ratios at low TOCs for carbonate rocks really matter when we are considering rocks that are potentially the source of significant amounts of petroleum? I think not. The total amount of bitumen–hydrocarbons in these low TOC rocks is negligible compared to what is necessary to produce significant accumulations. The total generating capacity is what is of interest, and the ratio of bitumen–hydrocarbons/TOC in immature low-TOC rocks, are not pertinent. For example, if one believes that in immature sediments, such as those described by Bordovskiy and Takh (1978), the higher the bitumen/TOC ratio the better the source potential, one is forced to prefer an infinitesimal TOC, a *reductio ad absurdum*.

Figure 9 shows TOC and hydrocarbon/TOC data compiled from the literature by Tissot and Welte (1978, p. 95) for 20 areas of Holocene clastic sedimentation in five different depositional environments. They also included analyses from two areas of carbonate muds. Although many unknown factors are operating, three conclusions pertinent to this discussion are apparent from Figure 9: (1) overall, an inverse correlation exists between the TOC and the hydrocarbon/TOC ratio; (2) the values for the carbonate samples are surrounded by samples from a variety of clastic environments; and (3) the highest hydrocarbon/TOC ratios are from clastic environments with low TOCs.

At least two effects might contribute to the increasing spread between the bitumen/TOC ratio for carbonates and shales that occur with decreasing TOC (Fig. 8). In rocks that are almost all carbonate and have a low TOC, the ratio may be inflated by loss of a substantial portion of the OM in the necessarily severe oxidation process. In shales with low TOC, a substantial portion of the OM may be land-derived inertinite with negligible bitumen content. Thus the bitumen/TOC ratio would be lower than for shales with a higher TOC but the same background inertinite content (Hood and Castaño, 1974; Leythaeuser et al, 1980).

Various reasons have been advanced to explain the differences (and similarities) that often exist in hydrocarbon/TOC and bitumen/TOC ratios and average TOC between carbonate-rich and carbonate-poor rocks at all values of TOC reported by Gehman (1962), Bordovskiy and Takh (1978), Tissot and Welte (1978), and others. None of the possible explanations has been accompanied by a suffic-

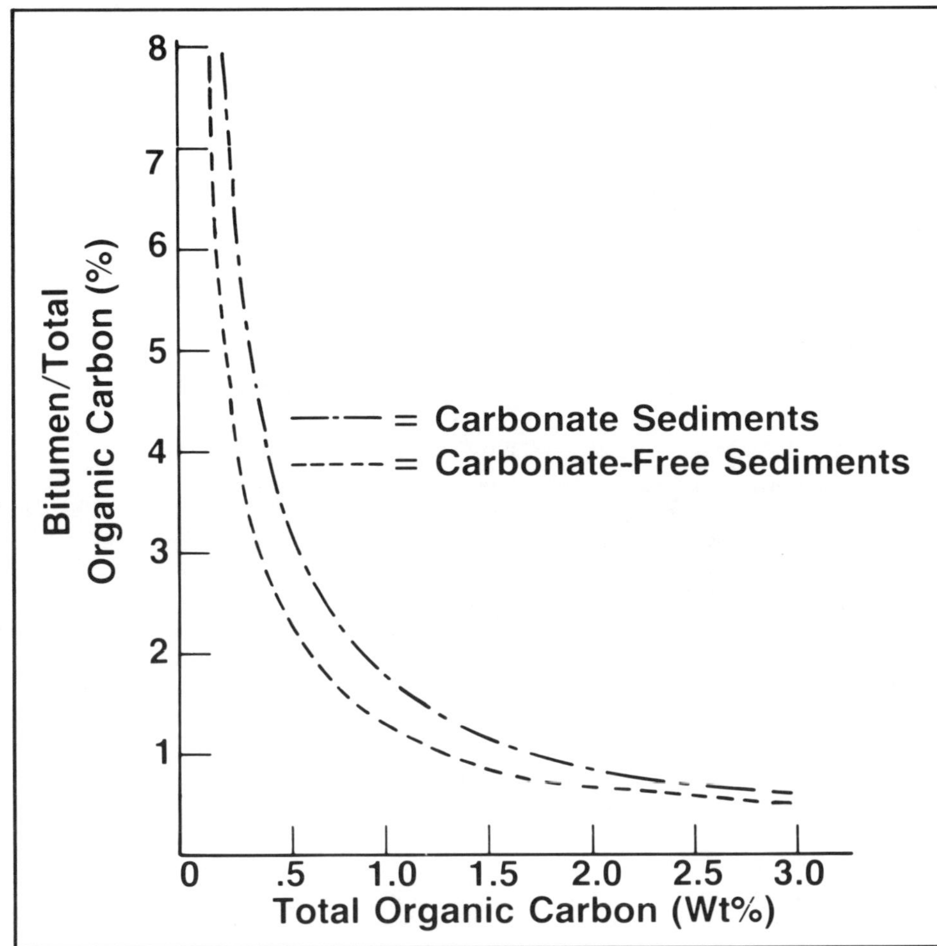

Figure 8 — Correlation of bitumen/total organic carbon with total organic carbon for carbonate and carbonate-free bottom sediments of the Caspian Sea, USSR. Figure from Bordovskiy and Takh (1978).

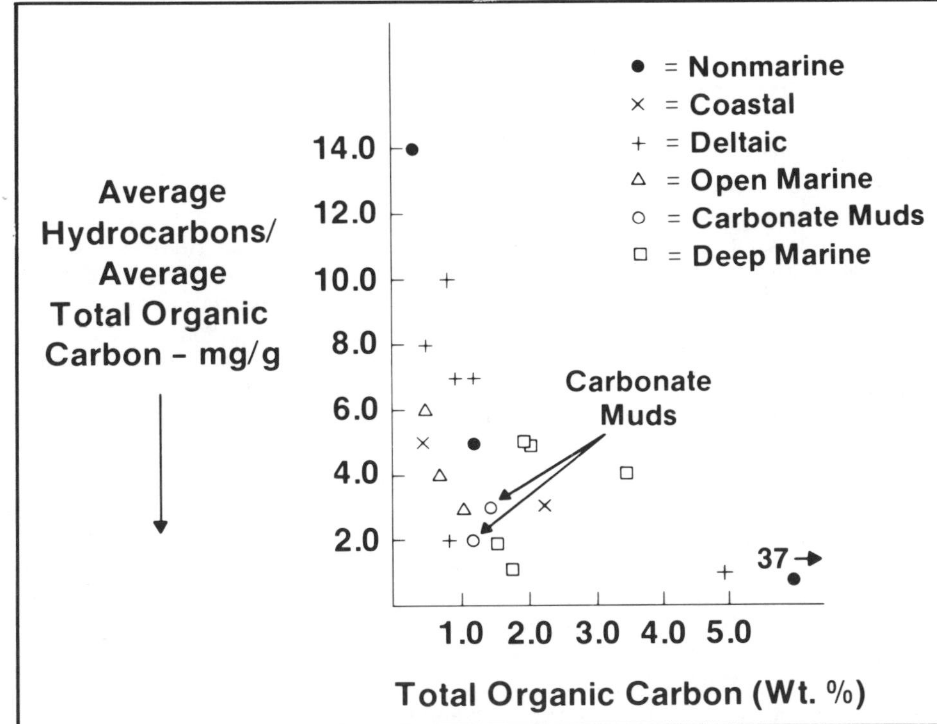

Figure 9 — Correlation of hydrocarbon/total organic carbon with total organic carbon for Holocene sediments from five siliciclastic environments and two carbonate-mud environments. Data from Tissot and Welte (1978, p. 95).

ient amount of data on the many factors that could influence the numbers to be completely convincing.

The quality (type) of the OM could be significant (Hunt, 1979), as suggested by an average greater algal input to carbonate rocks and a corresponding slightly higher H/C ratio. However, the H/C ratio difference in oil-prone source rocks is small, and the sulfur often incorporated in kerogens in carbonate rocks probably decreases the hydrocarbon-generating capacity of a given H/C ratio. Also, Figure 9, although containing no explicit data regarding kerogen quality, gives no support to the thesis that the types of OM in Holocene carbonate muds contain higher percentages of hydrocarbons than do their siliciclastic counterparts.

Differences in adsorption capacities between clays and carbonates for hydrocarbons and bitumens are a possible explanation. However, despite the excellent observations and experimental work on carbonates by Shearman and Skipwith (1965), Chave (1965), and Suess (1970, 1973), and many studies on clays (summarized in Hunt, 1979), the quantitative significance of adsorption effects of carbonates and shales on the hydrocarbon/TOC ratio at different maturation levels is not known. For example, most clays presumably have greater catalytic effects regarding conversion of various bitumen components to hydrocarbons; yet carbonates, statistically, have a higher hydrocarbon/TOC ratio.

I suggest (also with no proof) that the hydrocarbon–bitumen–TOC relationships being discussed are primarily caused by the overall longer exposure of OM in carbonate rocks to an oxygenated environment. At least four separate effects might contribute within such a framework: (1) higher average organic productivity in carbonates accompanied by a preferential destruction of nonhydro-

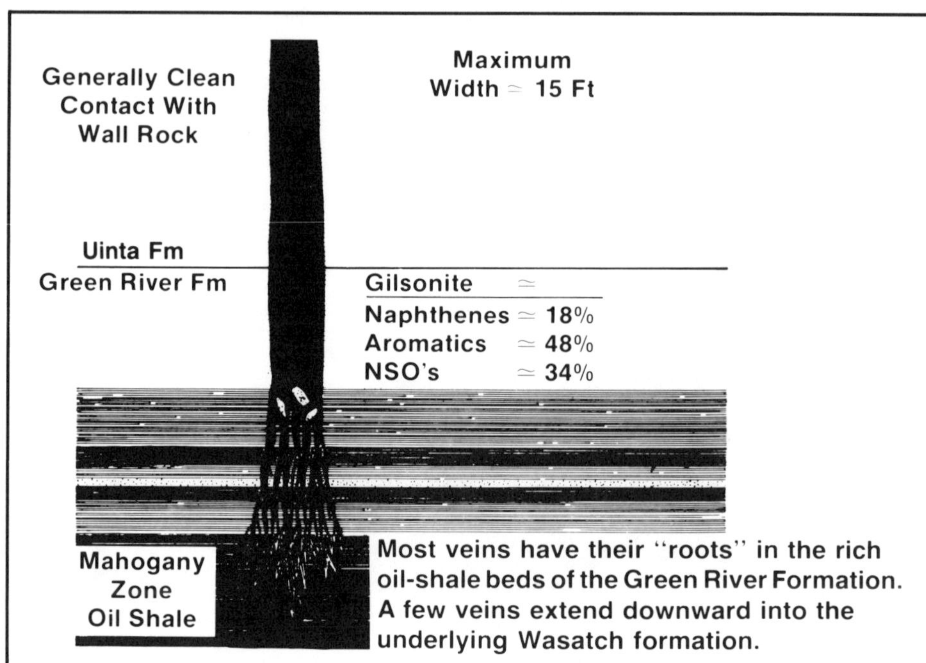

Figure 10 — Generalized cross section of a gilsonite mine, Uinta basin, Utah. After Eldridge (1901).

carbon OM could result in a higher hydrocarbon/TOC ratio than for shales with a similar TOC; (2) as a group, carbonates are deposited in a shallower, more agitated, and more oxygenated environment than shales and, everything else being equal, lower TOC contents should result; (3) most carbonates have a larger average grain size than shales, and diffusion of oxygen into such carbonates should be greater than for shales; (4) lower sedimentation rates and a larger average grain size for carbonates than for shales could result in the preservation of an oxic early diagenetic environment for a longer period of time for carbonates than for shales. Again, everything else being equal, lower TOC contents should result from the lower sedimentation rates (Toth and Lerman, 1977; Müller and Suess, 1979; Müller and Mangini, 1980).

In summary, the data presented and discussed above do not indicate that carbonate-rich rocks have a lower minimum TOC requirement to be source rocks than carbonate-poor rocks. This relationship simply reflects the lower TOC contents and more oxygenated depositional and early diagenetic environment of most carbonates analyzed. Only one of the carbonate source rocks reported on in this volume contains TOC values close to the 0.3 wt% minimum suggested by Tissot and Welte (1978) and Hunt (1979). Gardner and Bray (1984) report an average TOC of 0.35 wt% for the Silurian source rocks of the Michigan basin. However, many individual values are higher, and it is possible to speculate that the deeper and sparsely sampled parts of the depositional basin contain higher TOC contents. In all the other examples discussed in this volume the TOC contents are 1.5+ wt%, indistinguishable from the TOC contents of shales that were the source of similar amounts of oil.

EFFECTS OF VARIATIONS IN ROCK MATRIX ON GENERATION

Catalytic Effects

Arguments about the relative catalytic effects of carbonates and clays on hydrocarbon generation have been around for a long time. In my opinion, their longevity is due to the following reasons: (1) most experiments have been performed on hydrocarbons or organic acids (Shimoyama and Johns, 1972) with extremely high surface-to-volume ratios, not on kerogens with a much smaller surface-to-volume ratio; (2) any catalytic effects in the subsurface during the generative process in organic-rich rocks are minor and so are difficult to demonstrate conclusively as a significant subsurface effect; and (3) effects that have been demonstrated (Dembicki, Horsfield, and Ho, 1983) are strongest in low-TOC rocks with comparatively low surface-to-volume ratios of the kerogen. Such demonstrated alterations are also more directly related to catalytic effects as hydrocarbons move through the source rock than to the initial generation process. The lack of a rock-matrix effect in the more popular of the thermal-maturation models (Hood, Gutjahr, and Heacock, 1975; Waples, 1980) is a clear indication that geochemists in the oil industry have not perceived a significant difference between the catalytic effects of clay- and carbonate-rock matrices.

In excellent oil-prone source rocks, only a small percentage of the potentially generative OM is in contact with either clays or carbonates during generation, and generation is primarily a temperature effect. This problem is being increasingly recognized (Johns, 1982; Halpern and Kaplan, 1982). Undoubtedly, some differential catalytic effects occur as the bitumen moves through and out of an organic-rich source rock. However, if the movement is primarily through fractures (Momper, 1978), this effect is probably also small. Saturation occurs quickly. A possible exception to this generalization is clay-catalytic effects where hydrocarbons move substantial distances in overpressured high-porosity shales (Davis, 1982; Davis and Stanley, 1983). Even this possibility as a quantitatively significant effect can be questioned. Savin (1982) used isotopic studies to show that most water moving out of and through shales moves through cracks.

Thus the arguments go on. However, as stated previously, the surface-to-volume ratio of the kerogen particles in organic-rich oil-prone rocks strongly suggests that the catalytic effects of different rock matrices regarding generation are insignificant.

Delay in Generation

Nevertheless, there is some evidence that generation is delayed in fine-grained, tight, oil-prone carbonate source rocks (Palacas, 1984). Differences in generation timing within both carbonate and clay rocks of organic facies B having identical thermal histories probably reflect whether the system is open or closed. For example, in pyrolysis bomb experiments, generation reactions reach equilibrium at a constant pressure and temperature if the generation products are not removed from the bomb (S. R. Silverman, personal communication). In most stratigraphic sections, local pressure buildups tend to dissipate into the regional pressure gradient. A normal generation versus depth (temperature) profile results, as in the Douala basin, Cameroons (Albrecht, Vandenbroucke, and Mandengue, 1976).

However, in a relatively fine-grained carbonate sequence, particularly if it is argillaceous or organic rich, even a moderate amount of hydrocarbon generation is likely to build up and maintain the local pressure sufficiently to retard the chemical reactions necessary for additional generation. The retardation can also occur in tight or overpressured shales but is probably more common in fine-grained carbonates owing to their lower porosity and permeability for a given thermal history. This effect has not been discussed adequately in the source-rock literature, although it has been observed occasionally. It is probably the explanation for the liquid hydrocarbons observed by Price, Clayton, and Rumen (1981) in the subsurface at high temperatures; the higher percentage of saturate hydrocarbons in open pores than in closed pores in a Triassic limestone sequence in Hungary (Brukner and Vetö, 1983); the dramatic retardation in maturation of the OM in a Triassic calcareous shale overlying a reservoir at 200°C (392°F) in the AGIP Canoica well in Italy (Neglia, 1980); and an anomalous immaturity in overpressured shale in the Miesbach 1 well in West Germany (Jacob and Kuckelhorn, 1977). As expected, fine-grained carbonate rocks dominate the above examples.

In some carbonate source rocks the retardation of generation by local pressure buildup creates a bomb that is ready to explode if the system is driven to high enough temperatures or is tectonically fractured. This observation leads into the next section: migration similarities and contrasts in carbonate-rich and carbonate-poor, oil-prone source rocks.

MIGRATION

The same suite of possible migration mechanisms exists in all rocks. However, its relative quantitative significance depends not only on the organic characteristics but also on the rocks' chemical and, particularly, physical properties. Although the chemical differences between carbonate and shale source rocks are substantial, their effects are more evident on oil characteristics than on the efficiency or type of migration. On the other hand, effective shale and carbonate source rocks usually are quite different in their physical properties. In particular, carbonate source rocks are usually more brittle, more heterogeneous, and tighter than shale source rocks. Consequently they are more prone to fracture and emit hydrocarbons more readily than shales in a variety of geologic settings.

Fracturing on various scales has recently achieved considerable recognition as a principal cause of primary migration from organic-rich, oil-prone source rocks (Momper, 1978; Tissot and Welte, 1978; Hunt, 1979). However, Momper (1978), in an otherwise perspicacious paper, denigrated the significance of carbonate source rocks. He believed that their general dearth of clay would not permit the episodic buildup of pressure from hydrocarbon generation and its subsequent release by microfracturing that he considered the *sine qua non* of commercially significant expulsion and migration of hydrocarbons (Momper, 1978, p. B22–B24). In support of this view, Momper cited the lack of documented high pressures in carbonate sequences and a generally poor documentation of carbonate source rocks. The question of documentation has been answered by several papers in this volume and by Murris (1980) and Ayres et al (1982) on the Jurassic of the Arabian Peninsula.

The pressure-buildup question is more involved. Certainly, carbonate sections are leaky where they contain little clay or organic matter. For example, there is strong empirical support for not drilling below the top of massive carbonate sequences, such as the Ellenburger Formation in the southern Mid-Continent, if tests at the top are water wet (Cardwell, 1977). However, fine-grained carbonate rocks that contain considerable clay and/or organic matter are not leaky to fluids beneath them. In the process of normal compaction, such source rocks become overpressured with respect to both the immediately overlying and underlying sections. Thus, they can emit petroleum upward and downward. They can be a source for overlying accumulations with a separate cap, as with the Hanifa–Hadriya carbonate source rock in the Middle East (Murris, 1980; Ayres et al, 1982), or be both a source and cap for underlying accumulations, as with the Oxfordian carbonate source rocks in the Middle East and the Miocene marls that cap and act as the source for reservoirs in offshore southern Spain (Demaison and Bourgeois, 1984).

Organic-rich carbonate source rocks are typically deposited in an anoxic, salinity-stratified environment that is the prelude to evaporite deposition. If such a cycle is carried to completion, it is immaterial whether the emission of hydrocarbons is episodic or continuous; the evaporite cap is available at all times to channel the hydrocarbons into any available porosity. An example of this situation is the Jurassic Smackover section along the northeastern coast of the Gulf of Mexico, where the source and reservoir directly underlie an anhydritic caprock. There, it is easy to visualize a heterogeneous, finely laminated, shallow-water, stromatolitic source that underwent continuous compression, microfracturing, and hydrocarbon migration as it moved to greater depths and higher temperatures (Oehler, 1984). Additional overpressuring owing to hydrocarbon generation only improved an already felicitous situation.

The possibility for "explosive" migration from carbonate source rocks exists in certain geologic situations. If carbonate source rocks are dominantly composed of fine-grained carbonate and organic matter, they will, upon adequate burial, become impermeable, and almost all the noncarbonate space in the rock will be occupied by organic matter. Increasing burial will place the organic matter under near-lithostatic stress, and only a slight change in the regional stress system can result in the injection of large amounts of the more mobile fraction of the organic matter into the adjacent rocks. A dramatic illustration of this situation is given in Figure 10. The ejection of organic matter from the Mahogany Zone of the dolomitic Green River Formation (Eocene) and the injection of the thermally immature gilsonite into both the overlying and underlying section took place when the immature organic matter under lithostatic stress was buried ≅2,000 m (6,560 ft) and the Uinta basin area in Utah was subjected to relatively mild regional warping that created tensile fracturing (Jones, 1981). Similar injections of other types of bitumens from different parts of the Green River Formation also occurred (Hunt, Stewart, and Dickey, 1954; Hunt, 1963). The most paraffinic, and presumably most mature, bitumens originated in the lower parts of the Green River Formation, which presumably were under the greatest thermal stress at the time of migration.

If an organic-rich carbonate source rock is subjected to high temperatures in a passive tectonic regime, petroleum migration may be limited. However, when tectonic fracturing finally occurs, the ejected and injected OM may be producible oil. Perhaps the Ghawar field in Arabia, the world's largest producible-petroleum accumulation, was partially charged with hydrocarbons by this mechanism.

The recent intensive study of the Miocene Monterey Formation in southern California has revealed many examples of "explosive" migration. The Monterey Formation is a heterogeneous, brittle rock whose rock matrix is composed of various silica phases, limestone, dolomite, a variable clay content, and OM (Isaacs, 1980, 1983). Although the siliceous aspects of the Monterey have received the most attention in the literature (e.g., Garrison and Douglas, 1981), the OM content shows an inverse correlation with silica wt%, whereas a positive correlation exists with calcite wt% (Isaacs, 1983). Organic contents reach 24 wt% and average 13 wt% in the phosphatic carbonaceous-marl member of the Monterey Formation along the Santa Barbara coast (Isaacs, 1983). After being subjected to tectonism, the Monterey was highly fractured locally, and the fractures were filled with its own oil, apparently in an explosive manner (Redwine, 1981; Surdam and Stanley, 1981; Roehl, 1981; Belfield et al, 1983). This type of fracturing commonly creates regions of lower pressure. The pressure gradients thus formed can draw the oil from the rock matrix to the fractures. Similar effects have been described in the Russian literature (Veber and Gorskaya, 1965).

Many, but by no means all, of the oils having the Monterey as their source are low gravity and are immature by geochemical criteria. They contain unusually high contents of sulfur and asphaltenes. Such oils originated under thermal conditions less severe than those usually associated with "normal" generation and migration, a possibility not recognized early on by geochemists and managers of all oil companies.

The migration of immature oils within and from carbonate-rich source rocks is not confined to the Monterey Formation. De Groot (1980) suggested that oils in the Eastern Mediterranean are immature, early-expulsion products. Palmer and Zumberge (1981) showed that petroliferous marls in the upper Miocene evaporites of Sicily emitted asphaltic-seep oils at low levels of maturation. Powell (1984) described heavy immature oils and bitumens in the Devonian of Canada. These occurrences have been tied to the existence of a low-molecular-weight kerogen rich in nitrogen, sulfur, and oxygen (NSO) compounds that was deposited under highly euxinic conditions (Powell and Snowdon, 1983). The source rocks are organic rich and interlayered with other rocks of quite different physical properties. When tectonically deformed, these rocks tend to mobilize large quantities of heavy oil.

One possible reason for the delay in recognition of carbonates as important source rocks is that, until recently, migration hypotheses had been dominated by those involving water. Continuous-phase oil migration from carbonate source rocks is clearly a preferred mechanism. Various concepts of gas flushing that have, in part, replaced the water-dominated mechanisms originally proposed for thick Tertiary delta systems cannot explain migration from carbonate source rocks. It is likely that the low porosity and permeability of carbonate source rocks would preclude gas from having access to significant amounts of liquid hydrocarbons. There is no need to invoke the solution of oil in either water or gas as the primary migration mechanism responsible for any of the carbonate source–oil correlations reported in this volume. Continuous-phase primary oil migration is the most reasonable explanation for the oil accumulations discussed herein that originated in the Austin Chalk (Grabowski, 1984), Sunniland carbonates (Palacas, Anders, and King, 1984), Alcanar Formation (Demaison and Bourgeois, 1984), Salina A-1 carbonates (Gardner and Bray, 1984), Smackover alginites (Oehler, 1984), and the La Luna Formation (Zumberge, 1984).

SUMMARY AND CONCLUSIONS

Whether dominated by a carbonate or shale rock matrix, excellent oil-prone source rocks share many characteristics. Both (1) were formed in anoxic environments, (2) are generally finely laminated, (3) contain moderate to high TOC contents (>1.5–$20+$ wt%), and (4) contain high-quality OM (atomic H/C ≥ 1.2 at $\%R_o \cong 0.5$).

All of the organic facies exist in both carbonate and shale rocks, although they occur in different proportions and with slightly different organic characteristics in some cases.

Organic facies A is preferentially found in carbonate rocks primarily owing to the dearth of terrestrial organic input to marine carbonates, but perhaps also owing to inorganic hydrogenization in lakes under highly alkaline conditions (e.g., Green River Formation).

Organic facies B has been the source of hundreds of billions of barrels of oil from both carbonate and shale source rocks. Differences in oils derived from organic facies B in carbonate and shale source rocks primarily reflect the dearth of iron in carbonate systems and the variations in algal and terrestrial organic input to the OM.

Organic facies C is more important in shales, owing to the dominance of terrestrial input descending through an oxic water column, but this facies does occur in carbonates where partially degraded OM is preserved.

Organic facies D is the dominant facies in carbonate rocks, simply because most carbonates were deposited slowly in shallow, oxic water.

There is no convincing evidence that TOC requirements are less for carbonates than for shales, or that catalytic effects are more pronounced for shales with respect to subsurface generation from kerogen.

Substantial amounts of generally non-producible bitumen (dominantly NSO compounds with a small percentage of hydrocarbons) can be mobilized early in organic-rich, laminated, heterogeneous carbonate rocks after they have been tectonically fractured.

There is no convincing evidence that shale and carbonate source rocks have different thermal requirements to produce moderate- to high-gravity oils.

ACKNOWLEDGMENTS

Thanks are expressed to D. Leythaeuser, G. T. Moore, P. Müller, J. G. Palacas, K. E. Peters, and an anonymous reviewer for helpful reviews of the manuscript at various stages. The responsibility for the occasional controversial opinions expressed in the manuscript is solely that of the author.

REFERENCES CITED

Abed, A.M., and B.S. Amireh, 1983, Petrography and geochemistry of some Jordanian oil shales from north Jordan: Journal of Petroleum Geology, v. 5, p. 261–274.

Albrecht, P., M. Vandenbroucke, and M. Mandengue, 1976, Geochemical studies on the organic matter from the Douala Basin (Cameroon) I. Evolution of the extractable organic matter and the formation of petroleum: Geochimica et Cosmochimica Acta, v. 40, p. 791–799.

Ayres, M.G., et al, 1982, Hydrocarbon habitat in main producing area, Saudi Arabia: AAPG Bulletin, v. 66, p. 1–9.

Bakhturov, S.F., 1981, Bituminous carbonate rocks of the Tinnovskii Formation of the periphery of the Patom Highlands: Geologiya i Geofiziks, v. 22, p. 132–135.

Barker, C., 1978, Plate tectonics, organic matter and basin evaluation (abs.): AAPG Bulletin, v. 62, p. 493.

——, 1979, Organic geochemistry in petroleum exploration: AAPG Continuing Education Course Note Series 10, 159 p.

Bass, N.W., 1963, Composition of crude oils in northwestern Colorado and northeastern Utah suggests local sources: AAPG Bulletin, v. 47, p. 2039–2064.

Belfield, J.M., et al, 1983, Monterey fractured reservoir, Santa Barbara Channel, California (abs.): AAPG Bulletin, v. 67, p. 421–422.

Bitterli, P., 1963a, Aspects of the genesis of bituminous rock sequences: Geologie en Mijnbouw, v. 42, no. 6, p. 183–201.

——, 1963b, On the classification of bituminous rocks from Western Europe: Sixth World Petroleum Congress Proceedings, v. 2, p. 155–165.

Bordovskiy, O.K., and N.I. Takh, 1978, Organic matter in the Recent carbonate sediments of the Caspian Sea: Oceanology, v. 18, p. 673–678.

Bordovskiy, O.K., et al, 1974, Evaluation of the role of bottom fauna in the transformation of organic matter in sediments (with specific reference to the deep-sea detritus feeders in the Kuril–Kamchatka Trench): Oceanology, v. 14, p. 128–132.

Botz, R., and G. Müller, 1981, Mineralogie, Petrographie, anorganische Geochemie und Isotopen-Geochemie der Karbionatgesteine des Zechstein 2: Geologisches Jahrbuch, Reihe D, v. 47, p. 3–112.

Botz, R., et al, 1981, Kriterien und Bewertung des Zechstein-Stinkschie-fers im Hinblick auf sein Erdöl- und Erdgaspotential: Geologisches Jahrbuch, Reihe D, v. 47, p. 113–132.

Breger, I.A., 1980, Carbonate sedimentary rocks and the origin of heavy crude oils (abs.): Geological Society of America Abstracts with Programs, v. 12, p. 391.

—— and A. Brown, 1963, Distribution and types of organic matter in a barred marine basin: New York Academy of Science Transactions, v. 25, p. 741–755.

Brognon, G.P., and C.R. Verrier, 1966, Oil and geology in the Cuanza basin of Angola: AAPG Bulletin, v. 50, p. 108–158.

Brukner, A., and I. Vetö, 1983, Extracts from open and closed pores of an Upper Triassic sequence from W. Hungary: a contribution to studies of primary migration, *in* M. Bjorøy et al, eds., Advances in organic geochemistry 1981: Chichester, John Wiley, p. 175–182.

Calvert, S.E., 1983, Organic matter accumulation in the ocean: implication for the origin of sapropels, black shales, and petroleum source beds (abs.): Canadian Society of Petroleum Geologists News Letter, v. 10, no. 2, p. 1–2.

—— and N.B. Price, 1971, Recent sediments of the South-West African Shelf, *in* F.M. Delany, ed., Atlantic continental margins: London Institute of Geological Science, p. 175–185.

Cardwell, A.L., 1977, Petroleum source rock potential of Arbuckle and Ellenburger Groups, southern Mid-Continent, United States: Golden, Colorado, Colorado School of Mines Quarterly, v. 72, no. 3, 134 p.

Chave, K.E., 1965, Carbonates: association with organic matter in surface seawater: Science, v. 148, p. 1723–1724.

Cornelius, C., 1978, The role of source rock facies in the origin of petroleum: Erdoel-Erdgas–Zeitschrift, v. 94, no. 3, p. 90–94. (In German.)

Davis, J.B., 1982, Catalytic effects of smectitic clays in hydrocarbon generation (abs.): AAPG Bulletin, v. 66, p. 1442.

—— and J.P. Stanley, 1983, Catalytic effect of smectite clays in hydrocarbon generation revealed by pyrolysis-gas chromatography: Journal of Analytical and Applied Pyrolysis, p. 227–240.

Dean, W.E., J.V. Gardner, and P. Cepek, 1981, Tertiary carbonate-dissolution cycles on the Sierra Leone Rise, eastern equatorial Atlantic Ocean: Marine Geology, v. 39, p. 81–101.

—— G.E. Claypool, and J. Thiede, 1981, Origin of organic carbon-rich Mid-Cretaceous limestones, mid-Pacific Mountains and southern Hess Rise: Deep Sea Drilling Project, Initial Report, v. 62, p. 877–890.

Degens, E.T., 1974, Cellular processes in Black Sea sediments, *in* E.T. Degens and A.A. Ross, eds., The Black Sea—geology, chemistry and biology: AAPG Memoir 20, p. 296–307.

—— and P. Stoffers, 1976, Stratified waters as a key to the past: Nature, v. 263, p. 22–27.

—— et al, 1978, Varve chronology: estimated rates of sedimentation in the Black Sea deep basin: Deep Sea Drilling Project, Initial Report, v. 42, pt. 2, p. 499–508.

De Groot, K., 1980, Origin, migration and accumulation of hydrocarbons, *in* Panel discussion: Heyden, London,

10th World Petroleum Congress Proceedings, v. 2, p. 50.
Demaison, G.J., and G.T. Moore, 1980, Anoxic environments and oil source bed genesis: AAPG Bulletin, v. 64, p. 1179–1209.
——— and F.T. Bourgeois, 1984, Environment of deposition of middle Miocene (Alcanar) carbonate source beds, Casablanca field, Tarragona basin, offshore Spain, *in* J.G. Palacas, ed., Petroleum geochemistry and source-rock potential of carbonate rocks: AAPG Studies in Geology 18, this volume.
Dembicki, H., B. Horsfield, and T.T.Y. Ho, 1983, Source rock evaluation by pyrolysis gas chromatography: AAPG Bulletin, v. 67, p. 1094–1103.
Dow, W.G., 1977, Kerogen studies and geological interpretations: Journal of Geochemical Exploration, v. 7, p. 79–99.
Durand, B., and J.L. Oudin, 1980, Exemple de migration des hydrocarbunes dans une séries deltaique: Le delta de la Mahakam, Indonésie: 10th World Petroleum Congress Proceedings, v. 2, p. 1–12.
Eldridge, G.H., 1901, The asphalt and bituminous rock deposits of the United States: U.S. Geological Survey 22d Annual Report, 339 p.
Espitalié, J., M. Madec, and B. Tissot, 1980, Role of mineral matrix in kerogen pyrolysis: influence on petroleum generation and migration: AAPG Bulletin, v. 64, p. 59–66.
——— et al, 1977, Méthode rapide de charactérisation des roches mères, de leur potential pétrolier et de leur degré d'évolution: Revue de l'Institut Français du Pétrole, v. 32, p. 23–42.
Fouch, T.D., 1983a, Character of ancient petroliferous lake basins of the world (abs.): Canadian Society of Petroleum Geologists News Letter, v. 10, no. 3, p. 1–3.
———, 1983b, Lacustrine siliciclastic rocks and hydrocarbons (abs.): AAPG Bulletin, v. 67, p. 462.
Gardner, W.C., and E.E. Bray, 1984, Oils and source rocks of Niagaran reefs in the Michigan basin, *in* J.G. Palacas, ed., Petroleum geochemistry and source-rock potential of carbonate rocks: AAPG Studies in Geology 18, this volume.
Garrison, R.E., and R.G. Douglas, eds., 1981, The Monterey Formation and related siliceous rocks of California: SEPM, Pacific Section, Special Publication, 327 p.
Gehman, H.M., 1962, Organic matter in limestones: Geochimica et Cosmochimica Acta, v. 26, p. 885–894.
Grabowski, G.J., 1981, Source-rock potential of the Austin Chalk, Upper Cretaceous, southeastern Texas: Gulf Coast Association of Geological Societies Transactions, v. 31, p. 105–113.
———, 1984, Generation and migration of hydrocarbons in Upper Cretaceous Austin Chalk, south-central Texas, *in* J.G. Palacas, ed., Petroleum geochemistry and source-rock potential of carbonate rocks: AAPG Studies in Geology 18, this volume.
Gransch, J.A., and J. Posthuma, 1974, On the origin of sulfur in crudes, *in* B. Tissot and F. Bienner, eds., Advances in organic geochemistry 1973: Paris, Éditions Technip, p. 727–739.
Halpern, H.I., and I.R. Kaplan, 1982, Mineral–kerogen interactions in laboratory experiments—significance for petroleum genesis (abs.): AAPG Bulletin, v. 66, p. 1690.
Hassan, F., and S. El–Dashlouty, 1970, Miocene evaporites of Gulf of Suez region: AAPG Bulletin, v. 54, p. 1686–1696.
Hedberg, H.D., 1968, Significance of high-wax oils with respect to the genesis of petroleum: AAPG Bulletin, v. 52, p. 736–750.
Heybroek, F., 1965, The Red Sea Miocene evaporite basin, *in* Salt basins around Africa: London Institute of Petroleum, p. 17–40.
Hood, A., and J.R. Castaño, 1974, Organic metamorphism: Its relationship to petroleum generation and application to studies of authigenic minerals: United Nations ESCAP, CCOP Technical Bulletin 8, p. 85–118.
——— C.C.M. Gutjahr, and R.L. Heacock, 1975, Organic metamorphism and the generation of petroleum: AAPG Bulletin, v. 59, p. 986–996.
Hunt, J.M., 1963, Composition and origin of the Uinta Basin bitumens, *in* A.L. Crawford, ed., Oil and gas possibilities of Utah, re-evaluated: Utah Geological and Mineralogical Survey Bulletin 54, p. 249–273.
———, 1967, The origin of petroleum in carbonate rocks, *in* G.V. Chilingar, H.J. Bissell, and R.W. Fairbridge, eds., Carbonate rocks: New York, Elsevier, p. 225–251.
———, 1979, Petroleum geochemistry and geology: San Francisco, W.H. Freeman, 617 p.
———, F. Stewart, and P.A. Dickey, 1954, Origin of hydrocarbons of Uinta basin, Utah: AAPG Bulletin, v. 38, p. 1671–1698.
Hutton, A.C., A.J. Kantsler, and A.C. Cook, 1980, Organic matter in oil shales: Australian Petroleum Association Journal, v. 20, pt. 1, p. 44–68.
Isaacs, C.M., 1980, Lithostratigraphy of the Monterey Formation, Goleta to Point Conception, Santa Barbara coast, California: AAPG Field Guide 4, p. 9–24.
———, 1983, Compositional variation and sequence in the Miocene Monterey Formation, Santa Barbara coastal area, California, *in* D.K. Larue and R.J. Steel, eds., Cenozoic marine sedimentation, Pacific margin, U.S.A.: SEPM, Pacific Section, p. 117–132.
Jacob, H., and K. Kuckelhorn, 1977, The carbonization profile of the Miesback 1 well and its geological interpretation: Erdoel–Erdgas–Zeitschrift, v. 93, p. 115–124. (In German.)
Johns, W.D., 1982, Clay mineral catalysis and petroleum generation (abs.): AAPG Bulletin, v. 69, p. 1445.
Jones, R.W., 1981, Some mass balance and geological constraints on migration mechanisms: AAPG Bulletin, v. 65, p. 103–122.
———, 1983, Organic matter characteristics near the shelf–slope boundary, *in* D.J. Stanley and G.T. Moore, eds., The shelfbreak: critical interface on continental margins: SEPM Special Publication 33, p. 391–408.
——— and G.J. Demaison, 1982, Organic facies—stratigraphic concept and exploration tool, *in* A. Saldivar–Sali, ed., Proceedings of the Second ASCOPE Conference and Exhibition: Manila, p. 51–68.
Kohm, J.A., and R.O. Louden, 1979, Ordovician Red River (2)—Exploration success depends upon finding Ordovician Red River structures: Oil and Gas Journal, v. 77, no. 29, p. 89–94.
Krejci–Graf, K., 1963, Origin of oil: Geophysical Prospecting, v. 11, p. 244–275.
———, 1964a, Organic geochemistry: Naturwissenschaftliche Rundschau, v. 16, p. 175–186.
———, 1964b, Geochemical diagenesis of facies: Yorkshire Geological Society Proceedings, v. 34, pt. 4, no. 23, p. 469–521.

Larskaia, E.S., 1977, Disseminated organic matter of carbonate rocks and the oil forming process: Akademiya Nauk SSSR, Seriya Geologicheskaya, no. 12, p. 90–98. (In Russian.)

——, 1978, Distribution, balance, and type of disseminated organic matter in the Paleozoic deposits of the Russian Platform, depending on conditions of sedimentation: Lithology and Mineral Resources, v. 12, p. 323–333.

Leythaeuser, D., et al, 1980, Hydrocarbon generation in source beds as a function of type and maturation of their organic matter: a mass balance approach: 10th World Petroleum Congress Proceedings, v. 2, p. 31–42.

Lindbloom, G.P., and M.D. Lupton, 1961, Microbial aspects of organic geochemistry: Developments in Industrial Microbiology, v. 2, p. 9–22.

Maksimov, S.P., et al, 1976, On aspects of facies-genetic types of dispersed organic matter and its role in process of oil and gas generation: Moscow, Nauka, Issledovaniya organicheskogo veshchestva sovremennyki i iskopaemykh osadkov, p. 168–175. (In Russian.)

Malek-Aslani, M., 1980, Environmental and diagenetic controls of carbonate and evaporite source rocks: Gulf Coast Association of Geological Societies Transactions, v. 30, p. 445–458.

McCames, J.G., and L.S. Griffith, 1967, Middle Devonian facies relationships, Zama area, Alberta: Bulletin of Canadian Petroleum Geology, v. 15, p. 434–467.

Momper, J.A., 1978, Oil migration limitations suggested by geological and geochemical consideration, *in* Physical and chemical constraints on petroleum migration: AAPG Short Course Notes, v. 1, p. B1–B60.

Müller, P.J., and E. Suess, 1979, Productivity, sedimentation rate, and sedimentary organic matter in the oceans—I. Organic carbon preservation: Deep-Sea Research, v. 22A, p. 1347–1362.

—— and A. Mangini, 1980, Organic carbon deposition rates in sediments of the Pacific manganese nodule belt dated by 230Th and 231Pa: Earth and Planetary Science Letters, v. 51, p. 94–114.

Murris, R.J., 1980, Middle East: stratigraphic evolution and oil habitat: AAPG Bulletin, v. 64, p. 597–618.

Neglia, S., 1980, Migration of fluids in sedimentary basins: Reply to R.E. Chapman: AAPG Bulletin, v. 64, p. 1543–1547.

Oehler, J.H., 1984, Carbonate source rocks in the Jurassic Smackover trend of Mississippi, Alabama, and Florida, *in* J.G. Palacas, ed., Petroleum geochemistry and source-rock potential of carbonate rocks: AAPG Studies in Geology 18, this volume.

Ozimic, S., 1982, Depositional environment of the oil shale–bearing Cretaceous Toolebuc Formation and its equivalents, Eromanga basin, Australia, *in* J.H. Gary, ed., 15th Oil Shale Symposium Proceedings, p. 137–148.

Palacas, J.G., 1983, Carbonate rocks as sources of petroleum: Geological and chemical characteristics and oil–source correlations: World Petroleum Congress Proceedings, PD1, Origin, migration and accumulation of hydrocarbons, preprint, p. 1–13.

——, 1984, South Florida basin, a prime example of carbonate source rocks of petroleum, *in* J.G. Palacas, ed., Petroleum geochemistry and source-rock potential of carbonate rocks: AAPG Studies in Geology 18, this volume.

Palmer, S.E., and J.H. Zumberge, 1981, Organic geochemistry of upper Miocene evaporite deposits in Sicilian basin, Sicily, *in* J. Brooks, ed., Organic maturation studies and fossil fuel exploration: New York, Academic Press, p. 393–426.

Parparova, G.M., and S.G. Neruchev, 1977, Foundations of the genetic classification of dispersed organic matter in rock: Geologiya: Geofizika, v. 18, no. 5, p. 45–51.

Peterson, J.A., and R.J. Hite, 1969, Pennsylvanian evaporite–carbonate cycles and their relationship to petroleum occurrence, southern Rocky Mountains: AAPG Bulletin, v. 53, p. 884–908.

Powell, T.G., 1984, Some aspects of the hydrocarbon geochemistry of a Middle Devonian barrier-reef complex, western Canada, *in* J.G. Palacas, ed., Petroleum geochemistry and source-rock potential of carbonate rocks: AAPG Studies in Geology 18, this volume.

—— and L.R. Snowden, 1983, A composite hydrocarbon generation model: Erdöl und Kohle, Erdgas, Petrochemie, v. 36, no. 4, p. 167–175.

Pratt, L.M., 1982, A paleo-oceanographic interpretation of the sedimentary structures, clay minerals, and organic matter in a core of the Middle Cretaceous Greenhorn Formation drilled near Pueblo, Colorado: Princeton, New Jersey, Princeton University unpublished Ph.D. dissertation, 176 p.

Price, L.C., J.L. Clayton, and L.L. Rumen, 1981, Organic geochemistry of the 9.6 km Bertha Rogers No. 1 well, Oklahoma: Organic Geochemistry, v. 3, p. 59–77.

Redwine, L.E., 1981, Hypothesis combining dilation, natural hydraulic fracturing, and dolomitization to explain petroleum reservoirs in Monterey Shale, Santa Maria area, California, *in* R.E. Garrison and R.G. Douglas, eds., The Monterey Formation and related siliceous rocks of California: SEPM, Pacific Section, Special Publication, p. 221–249.

Roehl, P.O., 1981, Dilation breccia—a proposed mechanism of fracturing, petroleum expulsion and dolomitization in the Monterey Formation, California, *in* R.E. Garrison and R.G. Douglas, eds., The Monterey Formation and related siliceous rocks of California: SEPM, Pacific Section, Special Publication, p. 285–316.

Rogers, M.A., 1980, Application of organic facies concepts to hydrocarbon source rock evaluation: 10th World Petroleum Congress Proceedings, v. 2, p. 425–440.

—— and C.B. Koons, 1971, Generation of light hydrocarbons and establishment of normal paraffin preferences in crude oils: Origin and refining of petroleum: American Chemical Society, Advances in Geochemistry Series, p. 67–80.

Savin, S.M., 1982, Oxygen isotopic studies of diagenetic clay minerals: Implications for geothermometry, diagenetic reaction mechanisms, and fluid migration (abs.): AAPG Bulletin, v. 66, p. 1447.

Shearman, D.J., and P.A.d'E. Skipwith, 1965, Organic matter in recent and ancient limestones, and its role in their diagenesis: Nature, v. 208, p. 1310–1311.

Shimoyama, A., and W.D. Johns, 1972, Formation of alkanes from fatty acids in the presence of $CaCO_3$: Geochimica et Cosmochimica Acta, v. 36, p. 87–91.

Smith, J.W., 1983, The chemistry which created Green River Formation oil shale: American Chemical Society,

Symposium on Geochemistry and Chemistry of Oil Shales, preprint, p. 76–84.

——— and K.K. Lee, 1982, Chemistry and physical paleolimnology of Piceance Creek oil shales, *in* J.H. Gary, ed., 15th Oil Shale Symposium Proceedings: Golden, Colorado, p. 101–114.

Snowdon, L.R., and T.G. Powell, 1982, Immature oil and condensate—modification of hydrocarbon generation model for terrestrial organic matter: AAPG Bulletin, v. 66, p. 775–788.

Spillers, J.P., 1965, Distribution and hydrocarbons in south Louisiana by types of traps and trends (abs.): AAPG Bulletin, v. 49, p. 1749–1751.

Stach, E., et al, 1982, Stach's textbook of coal petrology, 3rd edition: Berlin, Gebrüder Borntraeger, 535 p.

Suess, E., 1970, Interaction of organic compounds with calcium carbonate—I. Association phenomena and geochemical implications: Geochimica et Cosmochimica Acta, v. 34, p. 157–168.

———, 1973, Interaction of organic compounds with calcium carbonate—II. Organo-carbonate association in Recent sediments: Geochimica et Cosmochimica Acta, v. 37, p. 2435–2447.

Summerhayes, C.P., 1981, Oceanographic controls on organic matter in the Miocene Monterey Formation, offshore California, *in* R.E. Garrison and R.G. Douglas, eds., The Monterey Formation and related siliceous rocks of California: SEPM, Pacific Section, Special Publication, p. 213–220.

Surdam, R.C., and K.O. Stanley, 1981, Diagenesis and migration of hydrocarbons in the Monterey Formation, *in* R.E. Garrison and R.G. Douglas, eds., The Monterey Formation and related siliceous rocks of California: SEPM, Pacific Section, Special Publication, p. 317–327.

Tissot, B., 1979, Effects on prolific source rocks and major coal deposits caused by sea-level changes: Nature, v. 277, p. 463–465.

——— and D.H. Welte, 1978, Petroleum formation and occurrence: Berlin, Springer–Verlag, 538 p.

——— and J. Roucaché, 1980, Principles of a genetic classification of crude oils, *in* A.D. Miall, ed., Facts and principles of world petroleum occurrence: Canadian Society of Petroleum Geologists Memoir 6, p. 209–217.

——— et al, 1974, Influence of nature and diagenesis of organic matter in formation of petroleum: AAPG Bulletin, v. 58, p. 499–506.

Toth, D.J., and A. Lerman, 1977, Organic matter reactivity and sedimentation rates in the ocean: American Journal of Science, v. 277, p. 465–485.

Uspenskii, V.A., and A.S. Chernysheva, 1951, The composition of the organic material from Lower Silurian limestones in the region of Chudovo City, *in* Contributions to geochemistry: Israel Program for Scientific Translations, 1965, p. 103–114.

Veber, V.V., and A.I. Gorskaya, 1965, Bitumen formation in sediments of carbonate facies: International Geology Review, v. 7, p. 816–825.

Waples, D.W., 1980, Time and temperature in petroleum formation: application of Lopatin's method to petroleum exploration: AAPG Bulletin, v. 64, p. 916–926.

Weber, K.J., and E. Daukow, 1975, Petroleum geology of the Niger Delta: World Petroleum Congress Proceedings, v. 2, p. 209–221.

Zhuze, N.G., 1972, Bitumen content of carbonate deposits of Upper Devonian and Lower Carboniferous of the Dnepr–Donets basin: International Geology Review, v. 15, p. 1212–1219.

Zumberge, J.E., 1984, Source rocks of the La Luna Formation (Upper Cretaceous) in the middle Magdalena Valley, Colombia, *in* J.G. Palacas, ed., Petroleum geochemistry and source-rock potential of carbonate rocks: AAPG Studies in Geology 18, this volume.

Use of Thiophenic Organosulfur Compounds in Characterizing Crude Oils Derived from Carbonate Versus Siliciclastic Sources

William B. Hughes
Research and Development Department, Phillips Petroleum Company, Bartlesville, Oklahoma

Several families of thiophenic organosulfur compounds (primarily benzo- and dibenzothiophenes) show promise for distinguishing oils derived from carbonate sources versus siliciclastic sources. This observation is based on analyses, using high-resolution gas chromatography coupled with a flame photometric detector, of several representative oils derived from carbonate sources (Smackover and Sunniland Formations, southeastern United States; Burgan and Minagish, Kuwait; and the Khatiyah Formation, Dubai) and, for comparison, of oils from siliciclastic sources (Ekofisk, North Sea; McKee, Texas; Muddy/Dakota, Wyoming; Beaverhill Lake, Alberta; and Klasaman, Indonesia). Some of the distinguishing features of carbonate-derived oils are an abundance of benzothiophenes, a fairly equal distribution among the dibenzothiophenes, and a distinctive distribution of methyldibenzothiophene isomers. Conversely, oils derived from siliciclastic sources commonly show low concentrations of benzothiophenes, decreasing amounts of dibenzothiophenes, and a different pattern in methyldibenzothiophene isomers.

The distinguishing features of carbonate-derived oils are applicable through the main phase of oil generation. Additional thermal maturity results in a significant reduction in the amounts of benzothiophenes and a redistribution of methyldibenzothiophene isomers, making the differences between carbonate- and siliciclastic-derived oils less distinctive. However, in these cases the distribution of mono-, di-, and trimethyl substituted dibenzothiophenes is generally relatively uniform in carbonate oils, while siliciclastic oils show decreasing amounts of dimethyl- and trimethyldibenzothiophenes relative to methyldibenzothiophenes.

Used in conjunction with such other geochemical parameters as total sulfur content, n-alkane compositions, and sterane/triterpane compositions, thiophenic organosulfur analysis provides a useful tool for oil–oil and oil–source-rock correlations as exemplified by a study of 20 oils from the Salawati basin, Indonesia.

INTRODUCTION

The sulfur-selective flame photometric detector (FPD), coupled with high-resolution gas chromatography (GC), provides a rapid, simple, and powerful tool for studying the organosulfur compositions of samples. As a result, the GC (FPD) technique can be applied routinely in geochemical studies. We have used this technique to analyze the organosulfur composition of the aromatic fraction of crude oils and rock extracts to determine if the relative distribution of various families of organosulfur compounds contains useful geochemical information. High-resolution gas chromatography/mass spectrometry (GC/MS) was also used to characterize the sulfur compounds and to analyze sterane compositions, thus providing additional insight into the origin and maturity level of a sample. Some results of this work are reported here, with particular emphasis on crude oils derived from carbonate source rocks.

Previous work on sulfur in petroleum includes the identification of compound types (Martin and Grant, 1965; Dean and Whitehead, 1967; Drushel, 1970; Rall et al, 1972; Orr, 1978), their origin (Obelentsev, 1967; Gransch and Posthuma, 1974; Ho et al, 1974; Orr, 1977), and their use in characterizing crude oils and rock extracts (Castex, Roucaché, and Boulet, 1974; Deroo, 1976). The use of the flame photometric detector with packed columns was described by Wenzel and Aiken (1979).

METHODS OF ANALYSIS

Crude oils or rock extracts were deasphalted and separated into a saturate, aromatic, and polar fraction by liquid chromatography on silica gel. The aromatic fraction was used directly for GC (FPD) analysis, and the saturate fraction, after removal of n-alkanes by urea adduction, was used for GC/MS sterane analysis.

Gas-chromatographic analyses were obtained on 30- and 60-m (98.5- and 197-ft) SE-54 WCOT glass capillary columns and a 60-m (197-ft) SE-54 fused-silica capillary column. The best resolution was given by the 60-m (197-ft) WCOT column. Typical analysis conditions were: injection temperature of 300°C (571°F), initial column temperature of 80°C (176°F), and heating at 15°C (28°F) per minute to 200°C (391°F) with a 5-minute hold followed by heating at 4°C (7°F) per minute to 280°C (536°F) with a 30-minute hold. All GC analyses were done on a Perkin–Elmer Sigma 1 gas chromatograph. The GC/MS data were obtained using a Perkin–Elmer gas chromatograph interfaced with a Finnegan 4000 or Kratos MS-25 mass spectrometer.

RESULTS

Typically, three classes of organosulfur compounds were found in the aromatics fraction of both carbonate- and siliciclastic-derived crude oils and rock

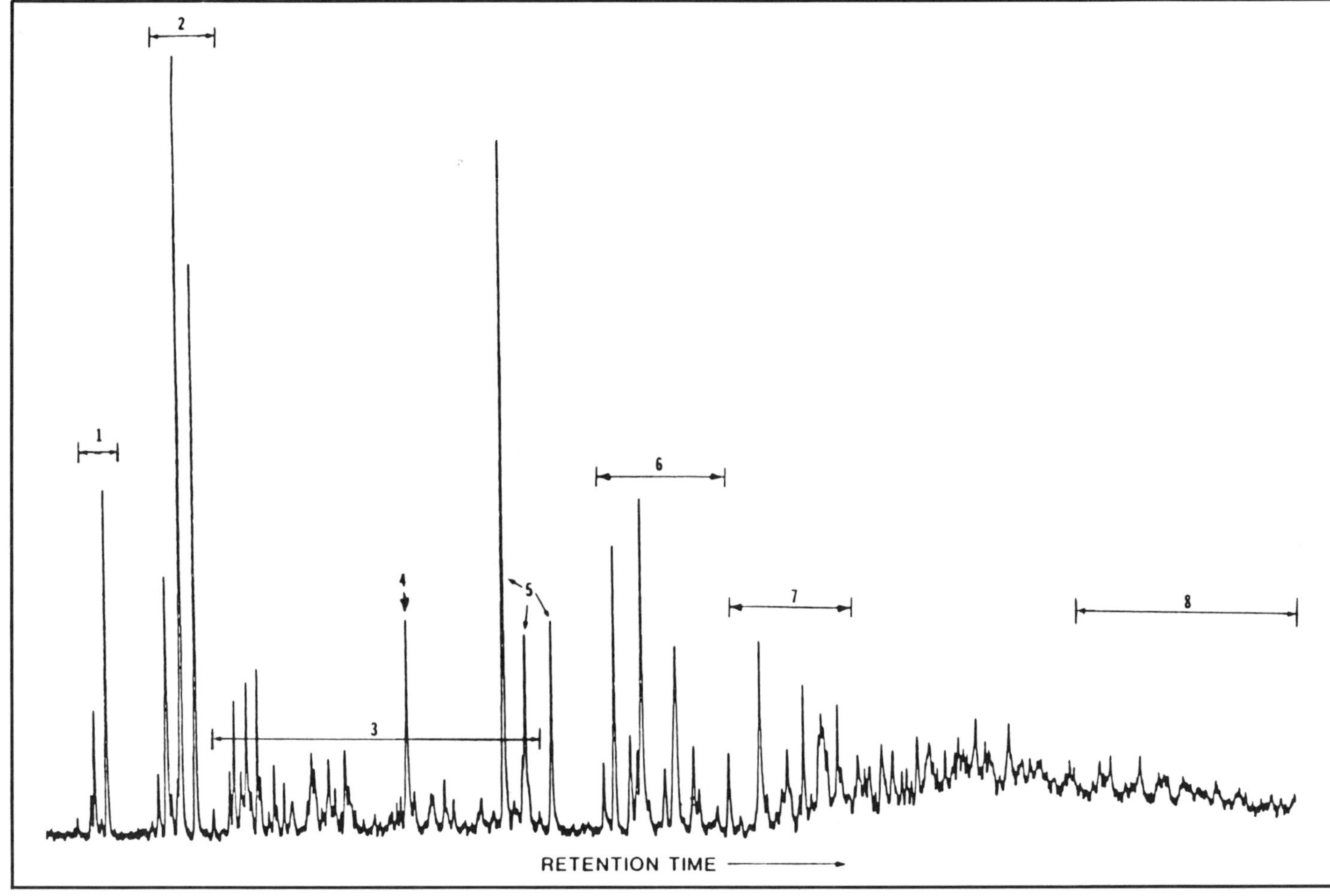

Figure 1 — Gas chromatogram of crude-oil aromatic fraction, using flame photometric detector (60-m, 197-ft, SCOT SE-54 glass capillary column).

extracts: the benzo-, dibenzo-, and benzonaphthothiophenes (Fig. 1; see Table 1 for peak-identification key). The benzothiophenes consist of three families: the two-carbon, three-carbon, and four-carbon and higher substituted derivatives. The substituents are believed to be mainly methyl groups. Benzothiophene itself and its one-carbon derivatives observed by Ho et al (1974) and by Rall et al (1972) probably were lost during solvent evaporation. The presence of substituted benzothiophenes of the proper mass was confirmed by 70 ev high-resolution field ionization mass spectrometry. Authentic samples were used to bracket the GC elution time ranges for the two- and three-carbon derivatives. The benzothiophenes labeled C_{4+} extend into the dibenzothiophene region.

The dibenzothiophenes consist primarily of dibenzothiophene itself and its methyl, dimethyl, and trimethyl derivatives. The span of retention times for the mono-, di-, and trimethyl derivatives was established by a comparison of the molecular-ion mass fragmentograms of the dibenzothiophenes with the FPD chromatogram of the same crude oil (Fig. 2). Evidence for the presence of C_4–C_7–substituted dibenzothiophenes was also obtained by GC/MS, but these are 3 to 10 times less abundant than the lower molecular-weight homologs.

A third family of compounds, benzonaphthothiophene and its C_1–C_5–substituted derivatives, was distinguished in a similar manner by comparing the GC/MS fragmentograms with the GC (FPD) trace. These components are generally present in much smaller amounts than the benzo- and dibenzothiophenes.

The various peaks distinguishable in the FPD trace within each C_n-substituted thiophenic family correspond primarily to structural isomers. For example, there are 15 possible isomers for dimethyldibenzothiophene. Identification of the individual isomers has not been made.

The retention times on a 60-m (98.5-ft) WCOT SE-54 column of the thiophenic sulfur compounds are compared with representative crude-oil alkanes and aromatics in Figure 3.

Carbonate-Derived Oils

To determine if carbonate source rocks impart distinctive organosulfur compositions, three groups of oils for which there is substantial evidence for a carbonate source were selected for study. One group was from the Jurassic Smackover Formation, western Alabama; a second group from the Lower Cretaceous Sunniland Limestone, South Florida basin; and a third group from Cretaceous reservoirs, Middle East (Burgan and Minagish sandstones, Kuwait, and Ilam Formation, Dubai).

Crude oil from reservoirs in the upper Smackover is believed to have its source in micritic carbonate mudstones of the lower Smackover. Geological support for this belief is based on the fact that no other source is reasonable. Neither the underlying Norphlet sand and Louann salt nor the overlying evaporites and clastics of the Haynesville and Cotton Valley Groups are likely to have been sufficiently organic rich to have served as source rocks. In addition to geological considerations, Oehler (this volume) has presented extensive geochemical data correlating Smackover oils to lower Smackover carbonate mudstones.

Palacas, Anders, and King (this volume) have presented geological and geochemical evidence that carbonate rocks are the indigenous source beds of the oils in Sunniland Limestone reservoirs. Murris and de Groot (1979), Murris (1980), and Ayres et al (1982) presented compelling geological evidence that most Middle Eastern oils have a carbonate source.

The Smackover, Sunniland, and Middle Eastern oils included in this study are listed in Table 2, along with some of their conventional properties and relevant geologic information.

The GC (FPD) chromatograms of the Gerry A-1 and Sunoco-Felda Sunniland oils, the Scott–Bolinger Smackover oil, and the Kuwait oils are shown in Figure 4. The other Sunniland oils in Table 2 have GC (FPD) traces similar to those shown in figure 4A, B. The shallowest Smackover oil (Scott–Bolinger), which is of comparable total sulfur content and API gravity to the Sunniland oils, is also similar in its GC (FPD) trace to these oils (Fig. 4C) but differs from the other Smackover crudes, which are discussed below.

The carbonate-derived oils are characterized by significant amounts of benzothiophenes. In the case of the Sunniland oils, the benzothiophenes predominate over the dibenzothiophenes; in the Smackover, Burgan, and Minagish oils the benzo- and dibenzothiophenes are comparable. Within the benzothiophenes, the C_3-derivatives are the major components in the Sunniland oils, and the C_{4+}-derivatives are most abundant in the Smackover, Burgan, and Minagish oils.

Another characteristic of carbonate-derived oils is the relative amount of the

Table 1

	Thiophenic-Sulfur Compounds
1	C_2–Benzothiophenes
2	C_3–Benzothiophenes
3	C_{4+}–Benzothiophenes
4	Dibenzothiophene
5	Methyldibenzothiophenes[1]
6	Dimethyldibenzothiophenes
7	Trimethyldibenzothiophenes
8	Benzonaphthothiophenes
	Steranes
9	13β(H),17α(H),20(R + S)-diacholestanes
10	5α(H),14α(H),17α(H),20R-cholestane + 24-ethyl-13β(H),17α(H),20R-diacholestane
11	24-methyl-5α(H),14α(H),17α(H),20R-cholestane
12	24-ethyl-5α(H),14α(H),17α(H),20S-cholestane
13	24-ethyl-5α(H),14β(H),17β(H),20(R + S)-cholestane
14	24-ethyl-5α(H),14α(H),17α(H),20R-cholestane

[1]By analogy with published data (Radke, Welte, and Willsch, 1982) obtained under similar GC conditions, the three peaks in this group, in order of elution time, are: 4-methyl-, 2- plus 3-methyl-, and 1-methyldibenzothiophene.

Table 1 — Peak-identification key for Figures 1, 2, 5, 7–10, 12–16.

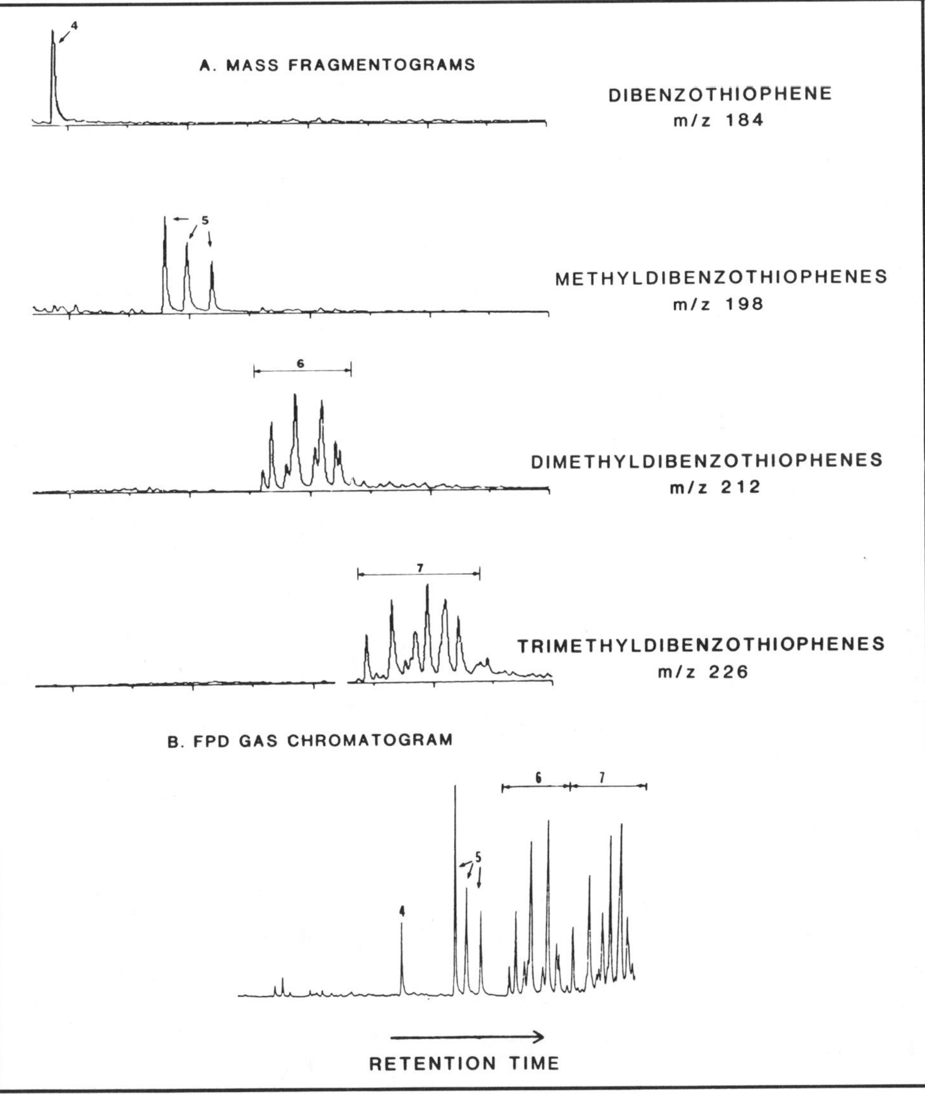

Figure 2 — Comparison of mass fragmentograms and FPD gas chromatogram of crude-oil aromatic fraction (see Table 1 for peak-identification key).

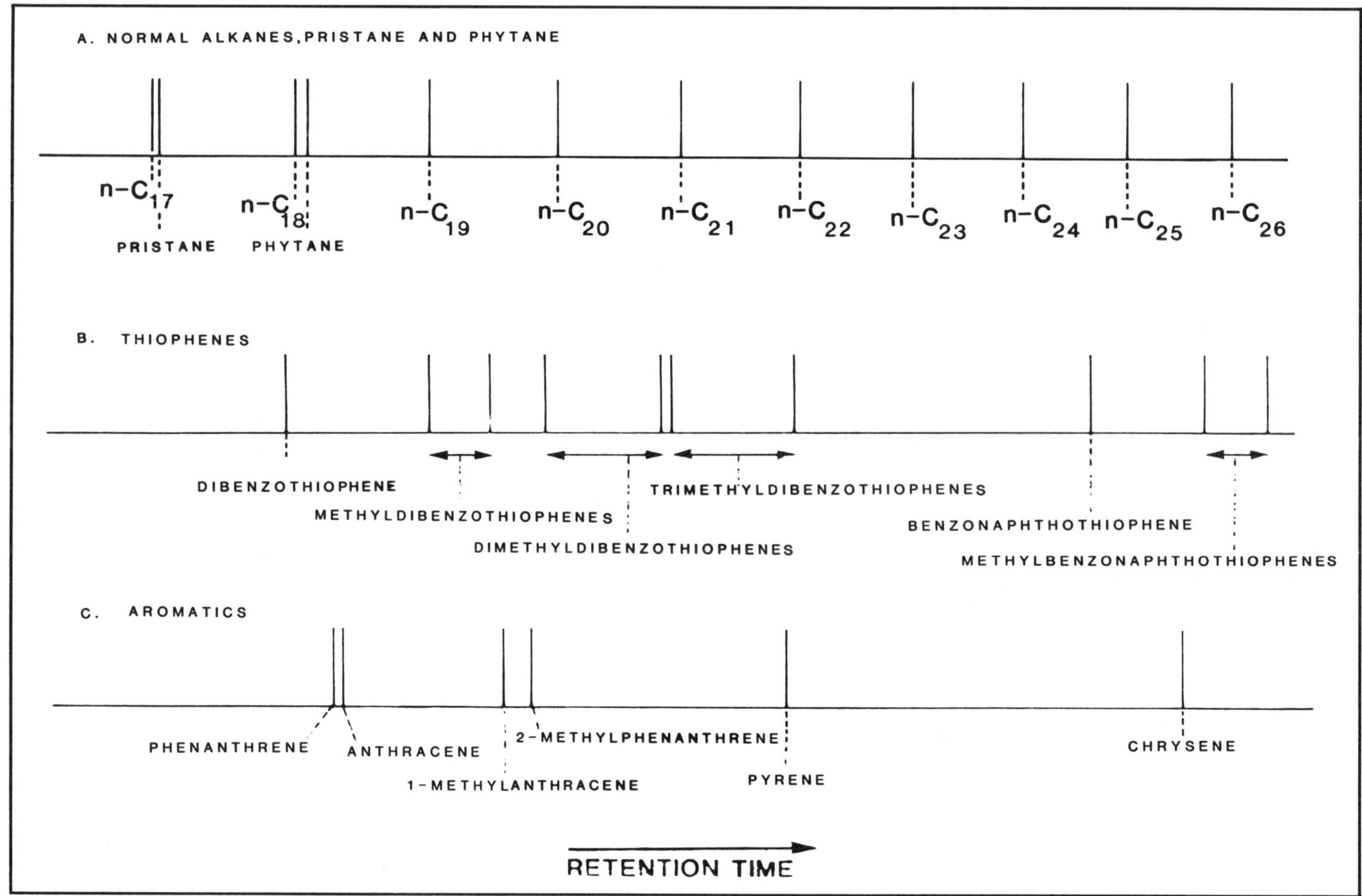

Figure 3 — Relative retention times of thiophenes, alkanes, and aromatics (60-m or 197-ft SCOT SE-54 glass capillary column).

Table 2

Field, Location	Well	Reservoir Formation	Age	Depth [1](m)	Gravity (°API)	Total Sulfur (wt%)	[3]OEP	Pristane / Phytane
Lake Trafford, Florida	1	Sunniland	Early Cretaceous	3621	25.4	4.20	0.88	0.49
Sunoco-Felda, Florida	32-1	Sunniland	Early Cretaceous	n.a.	22.2	3.61	0.96	0.58
West Felda, Florida	32-2	Sunniland	Early Cretaceous	n.a.	25.4	3.27	1.00	0.65
(Wildcat), Florida	Gerry A-1	Sunniland	Early Cretaceous	3517	19.2	4.19	0.99	0.53
Toxey, Alabama	Scott-Bolinger 4-3	Smackover	Late Jurassic	3203	19.0	3.60	0.84	0.46
Womack Hill, Alabama	WFU 10-14	Smackover	Late Jurassic	3447	39.0	1.04	0.88	0.62
Womack Hill, Alabama	WFU 15-1	Smackover	Late Jurassic	3451	40.6	1.02	0.89	0.52
N. Choctaw Ridge, Alabama	Ezell 29-5	Smackover	Late Jurassic	3565	37.7	1.27	0.89	0.65
Barrytown, Alabama	Phillips 12-12	Smackover	Late Jurassic	3611	42.4	0.80	0.93	0.65
W. Barrytown, Alabama	Abston 10-6	Smackover	Late Jurassic	3675	44.6	0.74	0.88	0.62
Sabiriyah, Kuwait	n.a.	Burgan	Cretaceous	[2]2470	32.1	1.82	0.98	0.79
Minagish, Kuwait	n.a.	Minagish	Cretaceous	[2]2942	33.2	2.12	0.97	0.55
SW Fateh, Dubai	n.a.	Ilam	Late Cretaceous	2351	26.9	2.30	0.97	0.66

[1]Midpoint of producing interval. [2]Midpoint of producing formation. [3]Average odd-even predominance. n.a. = not available.

Table 2 — Properties of carbonate-derived crude oils.

four isomeric methyldibenzothiophenes (peak 5, Fig. 4). These generally show a V pattern, with the first-eluting isomer (4-methyl) being largest and the last-eluting isomer (1-methyl) next largest. The middle peak of the V pattern contains two coeluting isomers (2- plus 3-methyl). Another characteristic feature is the comparable abundance of the methyl-, dimethyl-, and trimethyldibenzothiophenes (peaks 5, 6, 7).

The Smackover oils show changes in their sulfur fingerprint with reservoir depth (Fig. 5). The major change is the virtual disappearance of the ben-

Figure 4 — FPD gas chromatograms of aromatic fractions of carbonate oils (see Table for peak-identification key).

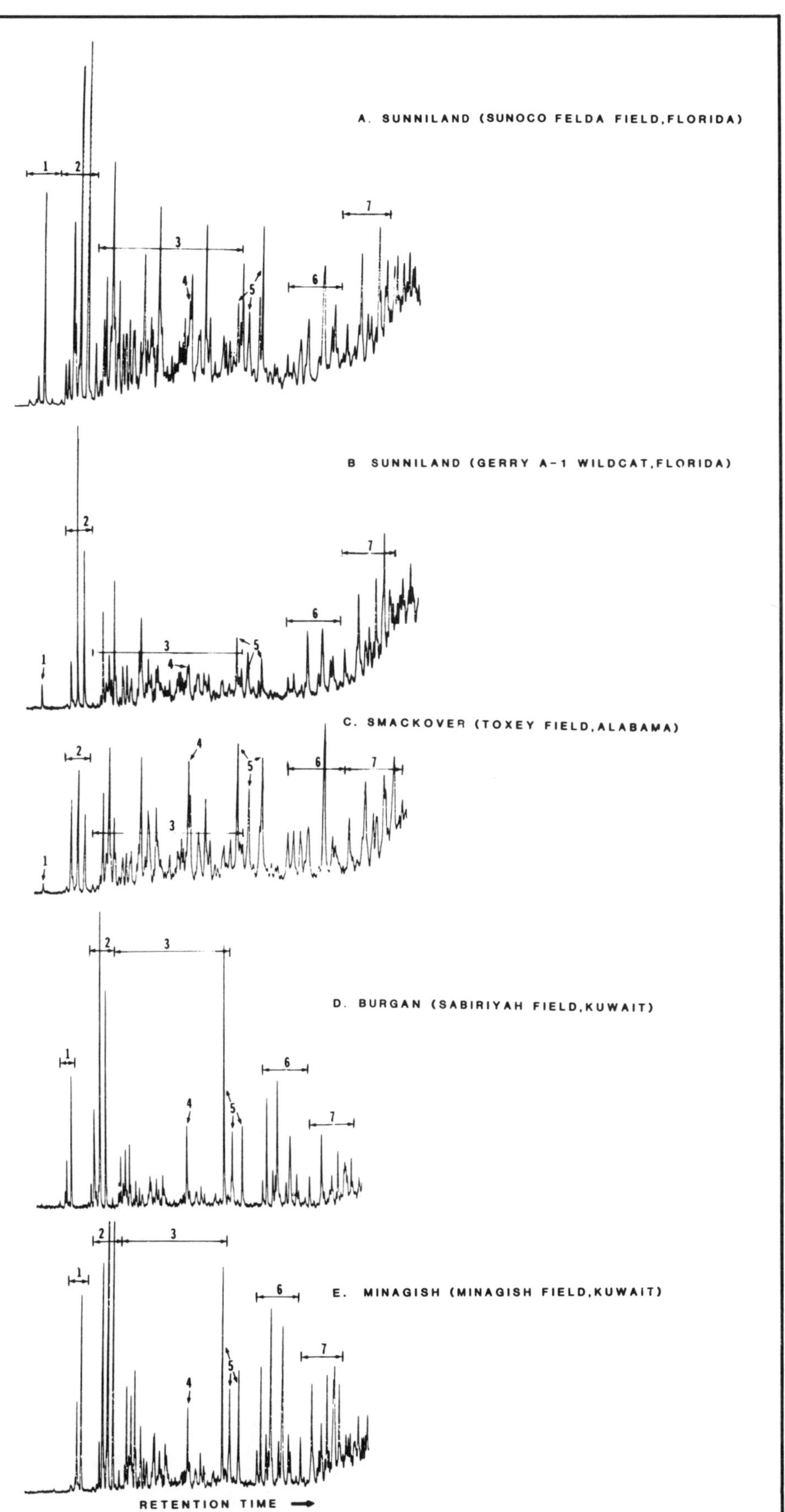

zothiophenes. In addition, there is a decrease in dibenzothiophene relative to its methyl derivatives, the last-eluting isomer of the methyldibenzothiophenes greatly decreases relative to the other three isomers, and there is an increase in the proportion of earlier-eluting isomers in both the di- and trimethyldibenzothiophenes.

These changes in the thiophenic sulfur derivatives parallel changes in the bulk properties of the oils. Total sulfur decreases from 3.6 wt% in the Scott–Bolinger well at 3,203 m (10,510 ft) to 0.74 wt% in the Abston 10-6 well at 3,675 m (12,058 ft). At the same time, API gravity changes from 19.0 to 44.6 and the percentage of the total crude oil in the C_5–C_{10} fraction increases from 14 to 47%.

The virtual disappearance of the benzothiophenes and the redistribution of methyldibenzothiophene isomers observed in the Smackover oils are most likely thermal effects. Based on sterane-isomer ratios (see below), the deeper Smackover oils are of advanced maturity (late main-phase oil generation). These results are consistent with the observations of Ho et al (1974), who found a decrease in benzothiophenes relative to dibenzothiophenes with increasing maturity of crude oils and with the changes in the distribution of methyldibenzothiophenes relative to dibenzothiophene with maturity observed by Radke, Welte, and Willsch (1982).

In addition to the dominant effect of maturity, there may be some source control with the Scott–Bolinger oil from a more carbonate-rich facies and the other Smackover oils from a more argillaceous facies. This is consistent with the higher proportion of rearrangement steranes in the Abston 10-6 oil relative to the Scott–Bolinger oil (Fig. 6).

Our results indicate that the distribution of methyldibenzothiophene isomers (peak 5, Fig. 5) tends toward what one would expect to be the most thermodynamically favorable, i.e., 4-methyl > 2- + 3-methyl > 1-methyl. This could be interpreted to mean that selective destruction of the less stable isomers or interconversion of isomers can occur.

To gain additional insight into the nature of the source and some informa-

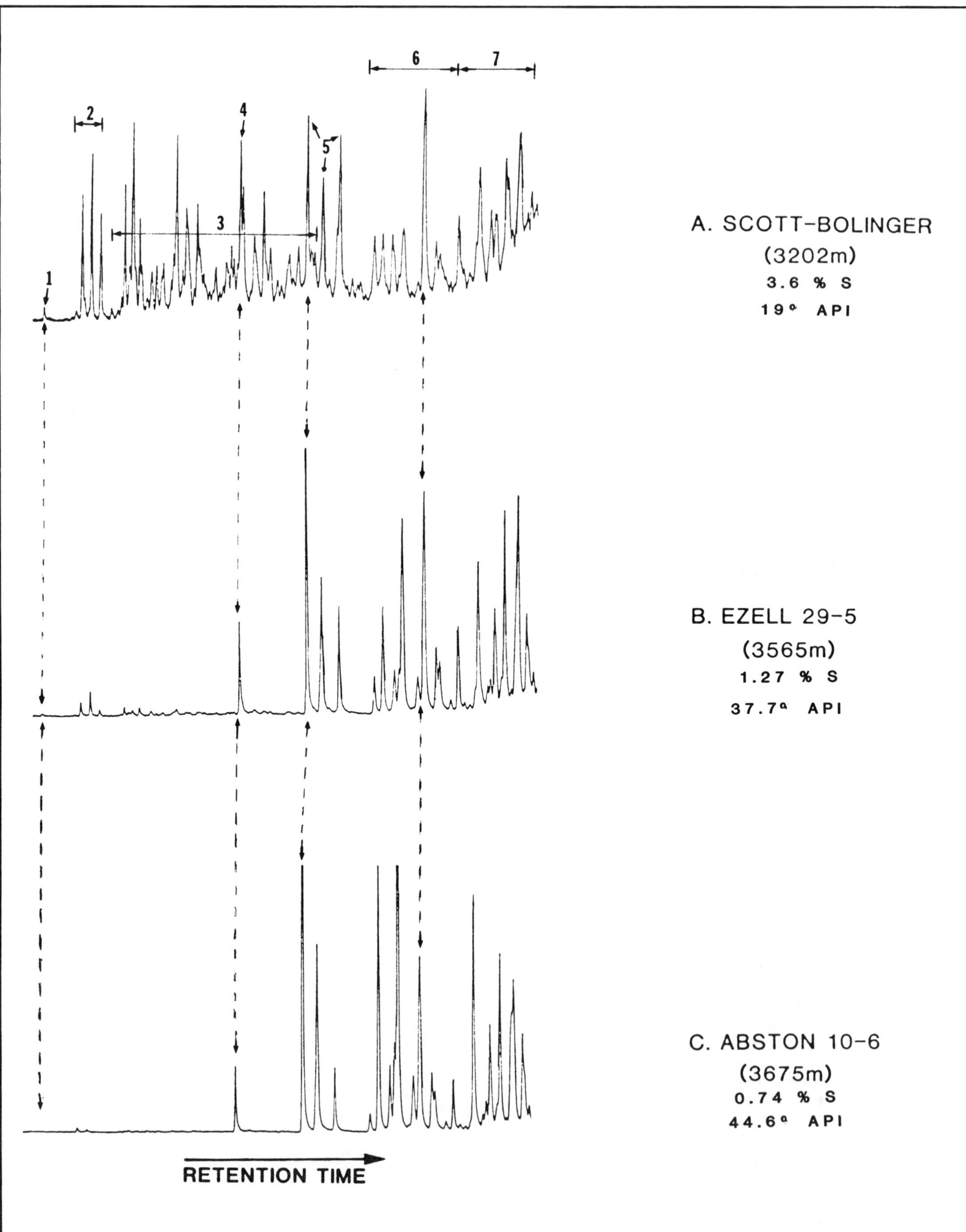
1
2
3
4
5
6
7
A. SCOTT-BOLINGER
(3202m)
3.6 % S
19° API
B. EZELL 29-5
(3565m)
1.27 % S
37.7° API
C. ABSTON 10-6
(3675m)
0.74 % S
44.6° API
RETENTION TIME

Figure 5 — Changes in FPD gas chromatograms of aromatic fractions with depth for Smackover crude oils (see Table 1 for peak-identification key).

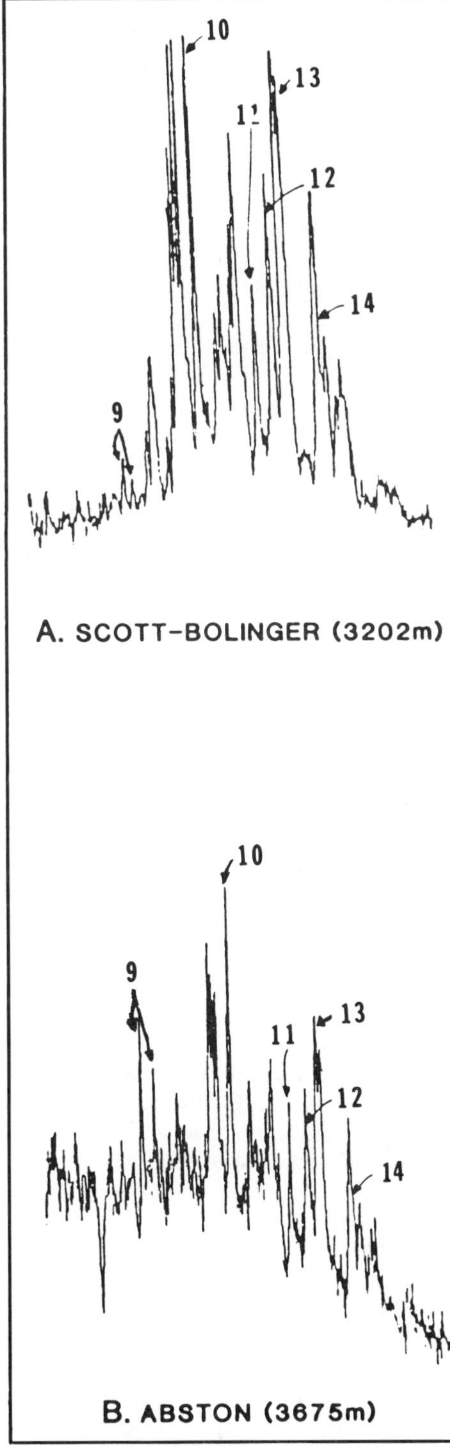

Figure 6 — Comparison of m/z 217 mass fragmentograms of saturate fractions of Scott–Bolinger and Abston 10-6 (Smackover) crude oils (see Table 1 for peak-identification key).

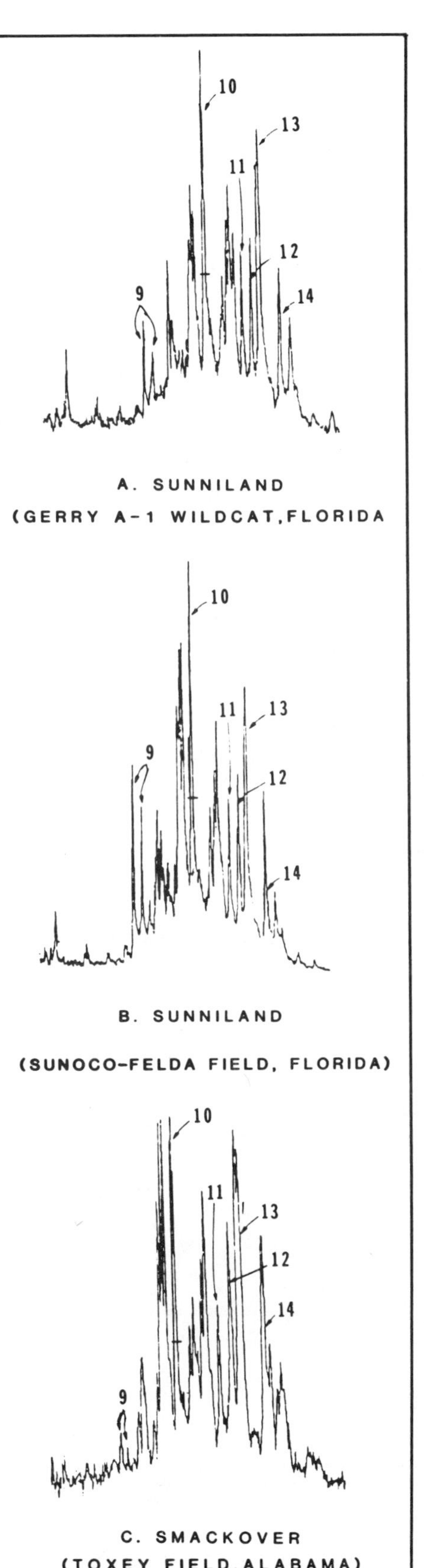

Figure 7 — Sterane m/z 217 mass fragmentograms of saturate fractions of carbonate oils (see Table 1 for peak-identification key).

tion about the maturity of the carbonate-derived oils, the sterane patterns of the Smackover and Sunniland samples were examined (Fig. 7). The steranes are characterized by nearly equal amounts of cholestane and 24-ethylcholestane, which predominate over 24-methylcholestane (the relative contributions of 5α(H),14α(H),17α(H),20R-cholestane and 24-ethyl-13β(H),17α(H),20R-diacholestane to peak 10 of Figure 5 were determined from the m/z 259 ion fragmentogram). The Smackover oils show a somewhat higher proportion of cholestane. The nearly equal amounts of C_{27} plus C_{28} steranes and C_{29} steranes indicate a mixed aquatic/terrestrial source for the organic matter that generated the oils. The relative peak areas of 24-ethylcholestane epimers—5α(H),14α(H),17α(H),20R; 5α(H),14α(H),17α(H),20S; and 5α(H),14β(H),17β(H),20(R+S)—indicate mature (main-phase oil generation) oils (Mackenzie et al, 1980).

The amounts of methyl-rearranged C_{27} steranes (peak 9, Fig. 7) are less than those of the unrearranged isomers in all the carbonate-derived oils in Table 2. The amount of rearrangement is particularly low in the Lake Trafford and Gerry Sunniland oils and the Scott–Bolinger Smackover oil. Palacas, Anders, and King (this volume), however, have noted larger amounts of diasteranes in some Sunniland oils. In contrast, oils from shale sources generally contain more of the methyl-rearranged than unrearranged C_{27} steranes. This likely reflects the higher concentration in shales of Lewis acid sites in clays to catalyze the methyl rearrangement (Rubenstein, Sieskind, and Albrecht, 1975; Sieskind, Joly, and Albrecht, 1979). Apparent exceptions can occur in calcareous shales, in which buffering by carbonate can suppress the rearrangement, or in carbonates with shale stringers, in which the generally higher organic content of the shales may contribute a disproportionate share to the total steranes.

Siliciclastic-Derived Oils

The aromatic-sulfur compositions of oils from noncarbonate siliciclastic sources were compared with those of the carbonate oils. The oils selected (Table 3) were from the Ekofisk and Tor chalks of the Ekofisk field, North Sea; the McKee sandstone of the Andector field, Texas; the Klasaman Formation of the Salawati

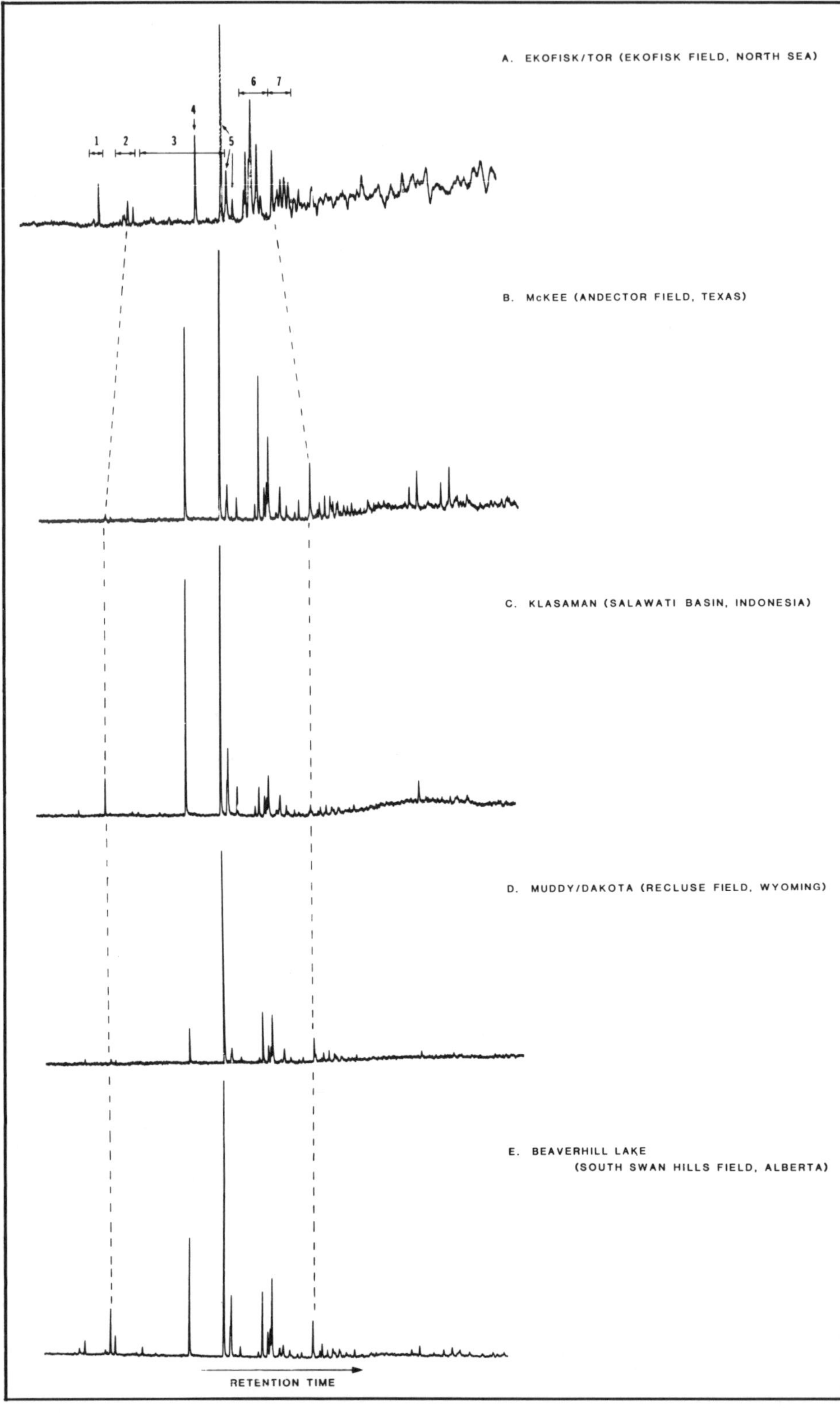

Figure 8 — FPD gas chromatograms of aromatic fractions of siliciclastic oils (see Table 1 for peak-identification key).

basin, Indonesia; the Muddy/Dakota sandstone of the Recluse field, Wyoming; and the Swan Hills carbonate member of the Beaverhill Lake Formation of the South Swan Hills field, Alberta.

Geochemical evidence has been presented for the Upper Jurassic Kimmeridge Clay as the source of the Ekofisk oil (Van den Bark and Thomas, 1980), Pennsylvanian shales as the source of the Andector Mckee oil (Cardwell, 1977), and the Lower Cretaceous Mowry Shale as the source of the Recluse Muddy/Dakota oil (Momper and Williams, 1979). The Swan Hills crude oil, although produced from carbonate reservoirs, is believed to have its source in shales of the Upper Devonian Beaverhill Lake Formation (Hitchon, 1961). Evidence for a shale source for the Salawati oil is circumstantial but nevertheless compelling, as the oil is produced from an isolated pinnacle reef completely encased in Plio–Pleistocene Klasaman shale.

The FPD gas chromatograms of the five siliciclastic oils are presented in Figure 8. In contrast to the carbonate oils (Fig. 4), these oils contain only minor amounts of benzothiophenes. In the dibenzothiophene series, dibenzothiophene and its methyl derivatives are enhanced relative to its di- and trimethyl derivatives, and the first-eluting isomer of the methyldibenzothiophenes greatly predominates over the other three isomers.

The range of maturities for the oils in Figure 8, based on the relative amounts of C_{29} sterane isomers, is comparable to that of the carbonate oils in Figure 4. Indeed, the Ekofisk oil, by this maturity measure, is less thermally mature than the Sunniland and Smackover oils in this study. Thus, the greatly reduced amounts of benzothiophenes and the different methyldibenzothiophene isomer pattern in the siliciclastic-derived oils are not primarily a thermal-maturity effect as observed in the deeper Smackover oils (see Fig. 5).

Source Versus Reservoir Effects

Evidence suggesting that the thiophenic-sulfur composition is controlled primarily by the source and not post-accumulation reservoir effects was obtained by comparing two oils generated from a common source but produced from reservoirs of different lithologies. Both oils are from the Andector field, Ector County, Texas; one is produced from the Ordovician McKee Formation, a quartzose sandstone, the other from the Ordovician Ellenburger Formation, a dolomite.

Geochemical evidence for a common source is provided by the essentially identical compositions in the C_5–C_{10}

Table 3

Field, Location	Well	Reservoir Formation	Age	Depth [1](m)	Gravity (°API)	Total Sulfur (wt%)	[3]OEP	Pristane / Phytane
Ekofisk, North Sea	2	Ekofisk/Tor	Danian/Maastrichtian	3171	36.0	0.20	0.98	1.47
Andector, Texas	Frank A-9	McKee	Ordovician	1444	42.0	0.08	1.02	1.04
Salawati, Indonesia	K-1X	Klasaman	Plio–Pleistocene	2735	30.0	0.33	1.24	5.4
Recluse, Wyoming	n.a.	Muddy/Dakota	Early Cretaceous	1479	38.6	0.06	1.04	1.38
South Swan Hills, Alberta	169	Beaverhill Lake	Devonian	2627	36.0	0.12	1.06	1.31

[1]Midpoint of producing formation. [2]Composite production from four wells. [3]Average odd–even predominance. n.a. = not available.

Table 3 — Properties of siliciclastic-derived oils.

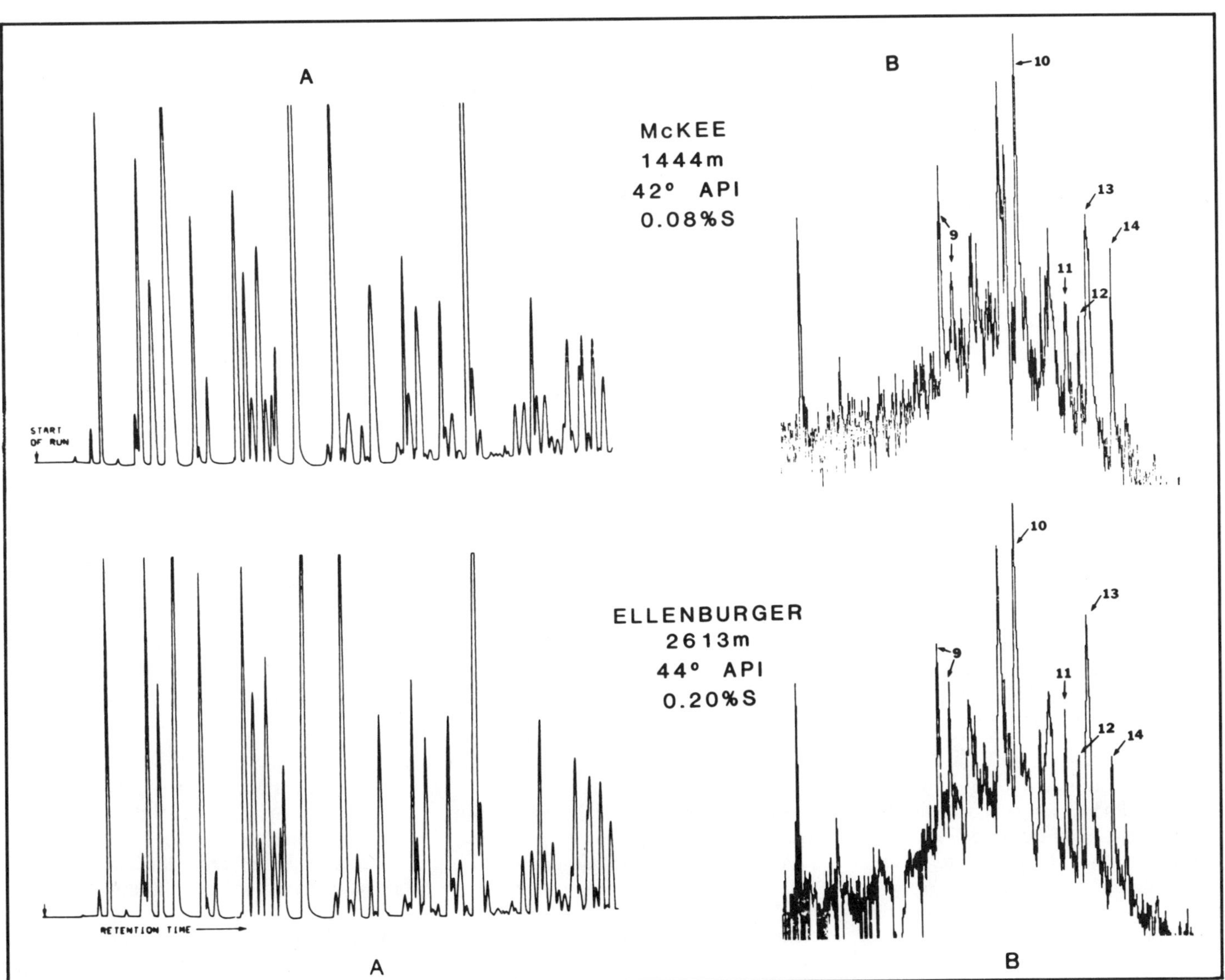

Figure 9 — Comparison of (A) high-resolution FID gas chromatograms of C_5–C_{10} fraction of whole crude oil and (B) m/z 217 mass fragmentograms of saturate fractions of crude oils, McKee and Ellenburger Formations, Andector field, west Texas (see Table 1 for peak-identification key).

molecular-weight range of the whole crude and in the steranes as shown by high-resolution GC (FID) (Fig. 9A) and GC/MS analyses (Fig. 9B), respectively. A study by Cardwell (1977) suggests a siliciclastic source in overlying Pennsylvanian shales.

Gas chromatograms (FPD) of the aromatic fraction of the two oils are compared in Figure 10. Although the McKee oil contains a somewhat higher proportion of dibenzothiophene, the overall

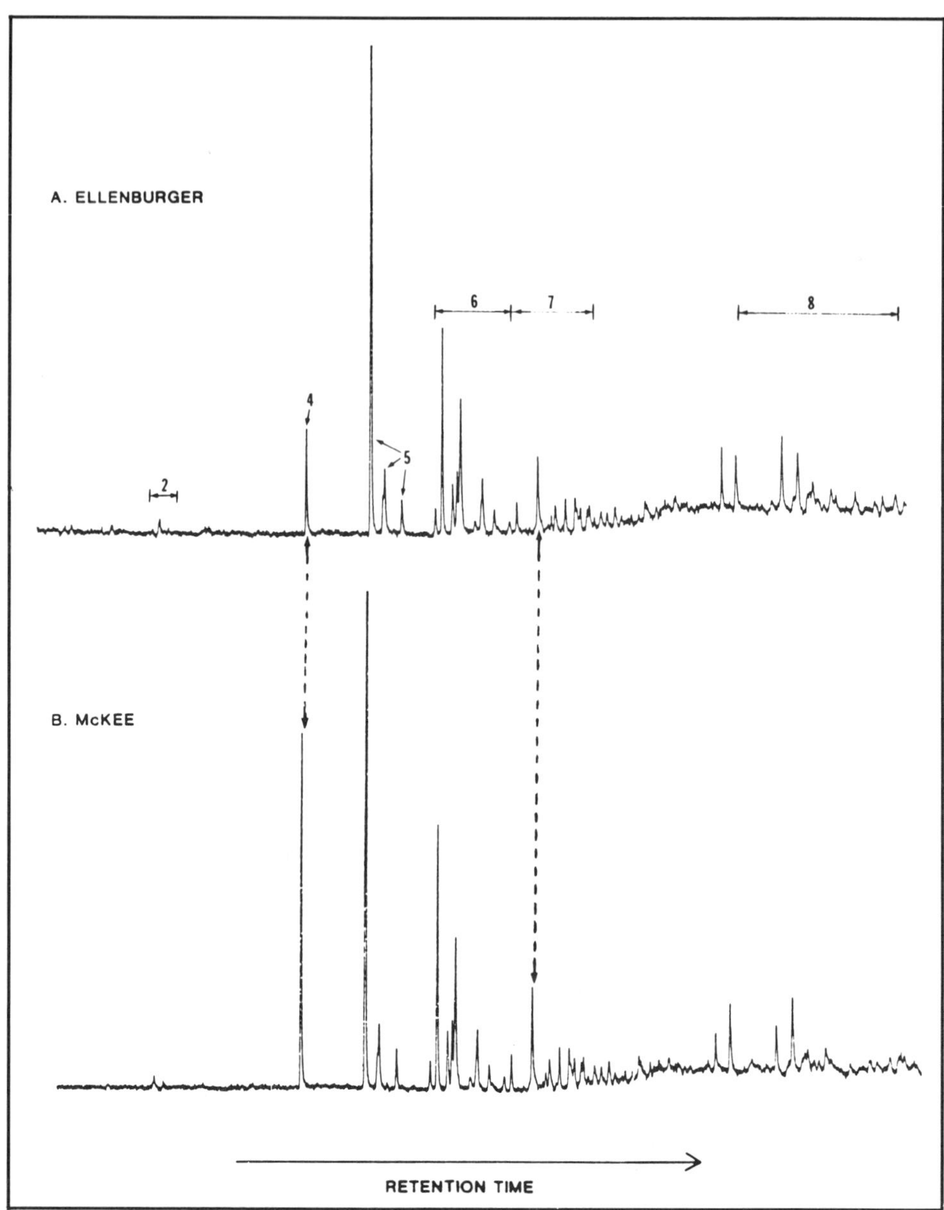

Figure 10 — Comparison of FPD gas chromatograms of crude-oil aromatic fractions, Ellenburger and McKee Formations, Andector field, west Texas (see Table 1 for peak-identification key).

thiophenic-sulfur compositions are essentially the same and typical of oils derived from siliciclastic sources (see Fig. 8).

Source control of the thiophenic-sulfur composition also is demonstrated by comparing the GC (FPD) trace of an oil with that of an extract from its presumed source rock. The oil is from the Upper Cretaceous Ilam Formation in the SW Fateh field of offshore Dubai, U.A.E. The extract is from a micritic limestone of the Cenomanian Khatiyah Formation, Fateh field of offshore Dubai.

The Khatiyah sample has a total organic-carbon content of 2.88 wt% and a carbonate-carbon content of 11.51 wt% (equivalent to 96% calcium carbonate). The Khatiyah extract has an odd–even predominance, OEP (Scalan and Smith, 1970) of 1.07 and a pristane/phytane ratio of 0.69, which are similar to an OEP of 0.97 and a pristane/phytane ratio of 0.66 for the oil. The similarity in these parameters and in sterane compositions (see Fig. 11B) is good evidence that the Khatiyah Formation is the source of the Ilam oil.

The FPD gas chromatograms of the aromatic fractions and the m/z 217 mass fragmentograms of the saturate fractions for the extract and oil are compared in Figure 11A. The distributions of thiophenic-sulfur compounds are similar and have many of the characteristics associated with oils from carbonate sources, in particular the V pattern in the methyldibenzothiophene isomers and the relatively high abundance of C_2-, C_3-, and C_4-substituted benzothiophenes, although these are somewhat reduced in the rock extract.

Additional support for source control is provided by extracts from the Upper Jurassic Kimmeridge Clay, the source of the Ekofisk oils (Van den Bark and Thomas, 1980), and from the Sunniland Formation, the source of the Sunniland oils (Palacas, Anders, and King, this volume). The thiophenic-sulfur compositions of the rock extracts are similar to the derived oils (compare Fig. 8A with

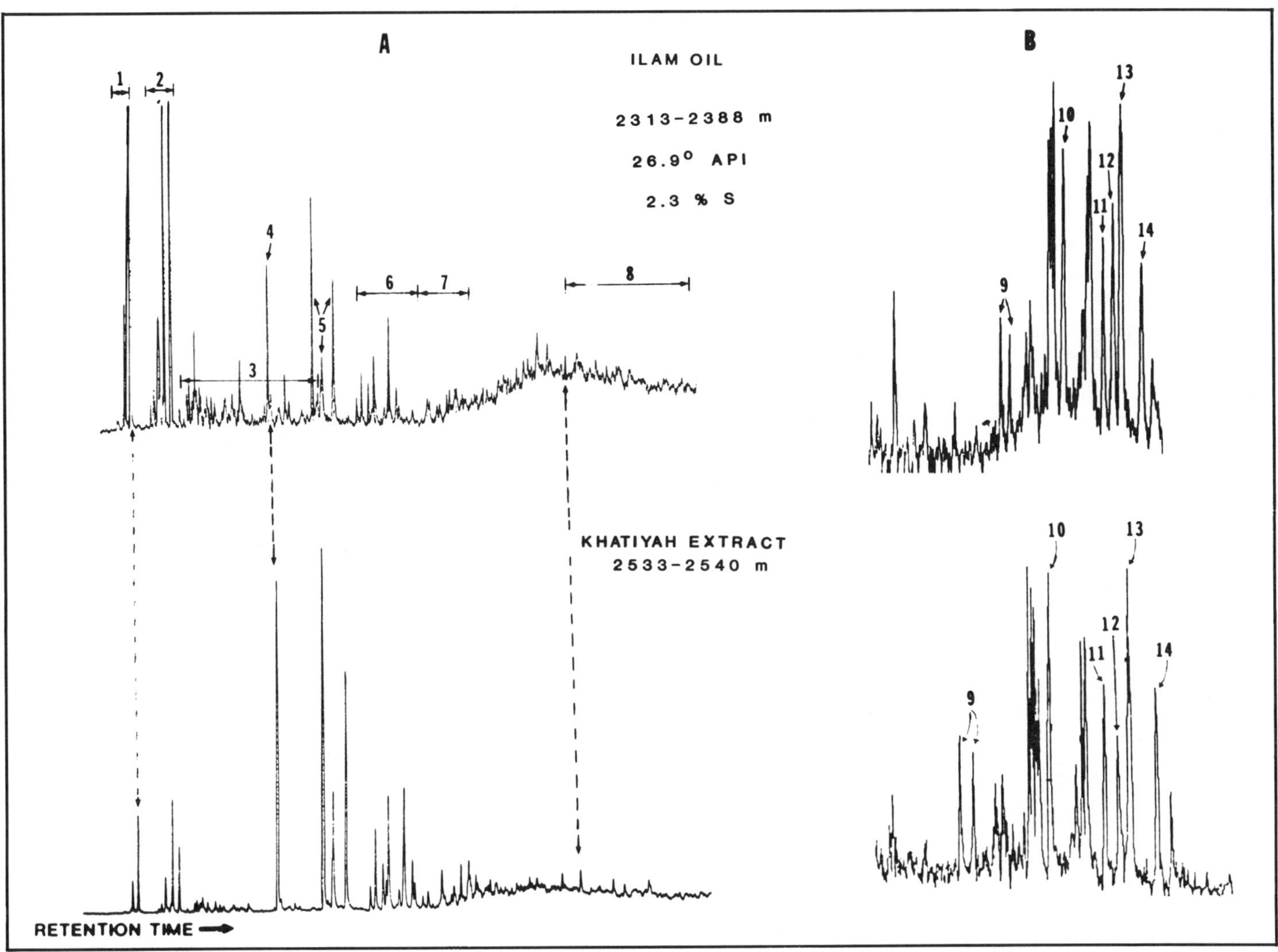

Figure 11 — Comparison of (A) FPD gas chromatograms and (B) m/z 217 mass fragmentograms of a crude oil, Ilam Formation, SW Fateh field, Dubai, and of a rock extract, Khatiyah Formation, Fateh field, Dubai (see Table 1 for peak-identification key).

Fig. 12A and Figs. 4A and 4B with Fig. 12B) and to the patterns expected for a siliciclastic source (Ekofisk) and a carbonate source (Sunniland).

Comparison of Other Chemical and Physical Properties of Carbonate and Siliciclastic Oils

In addition to typical thiophenic-sulfur compositions, oils from carbonate sources exhibit other characteristic features. Although the carbonate-derived oils in this study are mature (main-phase oil generation) as measured by the extent of epimerization of the C_{29} sterane, they have relatively low API gravities and high sulfur contents. The API gravity may increase and the total sulfur content may decrease with maturation, as shown by the Smackover oils discussed above. The implication here is that for oils of equal maturity, those from carbonate sources are generally higher in sulfur and lower in API gravity.

Oils derived from carbonate sources tend to show characteristic distributions in families of chemical compounds other than the thiophenes, which distinguish them from oils derived from siliciclastic sources. For example, the generally lower abundance of diasteranes relative to steranes has been mentioned. Other characteristics are an even predominance ($\overline{OEP} < 1$) in the C_{18}–C_{30} n-alkane range and relatively low (< 1) pristane/phytane ratios. Similar observations were made by Tissot and Welte (1978), Connan, Hussler, and Albrecht (1980), Powell (1984), and Palacas (1983). Connan, Hussler, and Albrecht (1980) pointed out that a predominance of even n-alkanes and unrearranged steranes and pristane/phytane ratios < 1 partly reflect an important microbial contribution, which is usually more pronounced in highly anoxic environments.

The characteristics of oils from carbonate sources are listed and compared with those of oils from siliciclastic sources in Table 4. These compositional signatures reflect the original organic input and environmental conditions during diagenesis of the source rock and not migration or in-reservoir alterations.

The characteristics listed in Table 4 for oils from carbonate sources may also be found in oils not usually considered to have a predominantly carbonate source lithology. For example, oil from the Wilmington field in California, which has

Table 4

Parameter	Carbonate Source	Siliciclastic Source
Total sulfur	High	Low
Gravity	Low to medium	Medium to high
n-alkanes (CPI or OEP)	Even to no predominance	Odd to no predominance
Pristane/phytane	Very low	Low to high
Steranes	High aquatic to mixed aquatic/terrestrial	High terrestrial to mixed aquatic/terrestrial
Diasteranes	Low (to medium)	Medium to high
Thiophenic sulfur	High in benzothiophenes	Low in benzothiophenes

[1]For non-biodegraded oils of comparable maturity.

Table 4 — Comparison of compositional parameters of oils from carbonate and siliciclastic sources.[1]

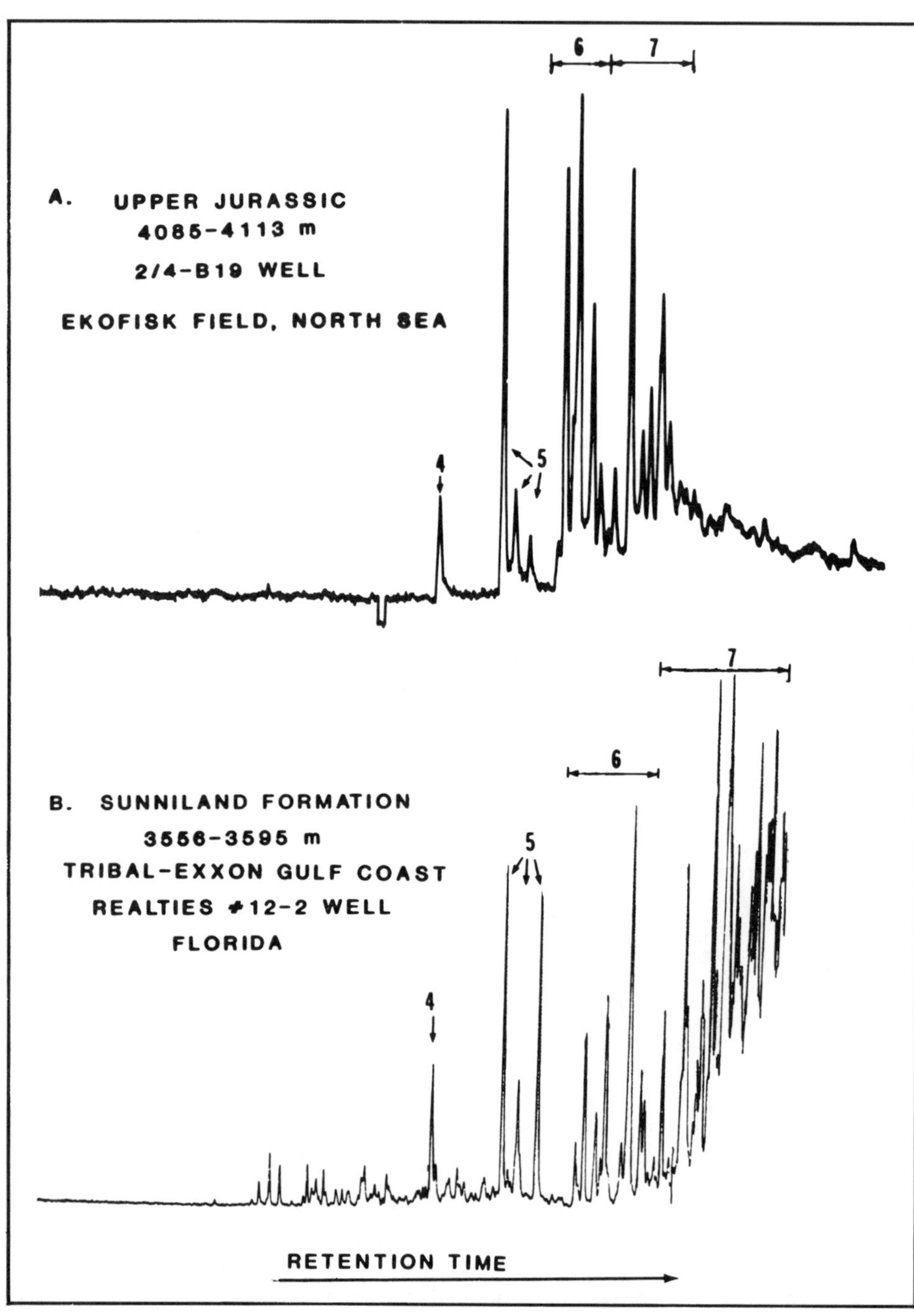

Figure 12 — FPD gas chromatograms of aromatic fractions of rock extracts of (A) Upper Jurassic Kimmeridge Clay, Ekofisk field, North Sea, and (B) Sunniland Formation, Florida (see Table 1 for peak-identification key).

its source in the upper Miocene Puente Shale (Philippi, 1965), shows a thiophenic-sulfur composition (Fig. 13) typical of oils from carbonate sources, especially in the abundance of benzothiophenes and the V pattern in the methyldibenzothiophenes. The API gravity (19.4°), sulfur content (1.59 wt%), OEP (0.97), and pristane/phytane ratio (1.13) also are typical of carbonate-derived oils. However, it should be noted that the upper Miocene D and E shales, which were established to be the source of Los Angeles basin oils, contain significant amounts of carbonate (Philippi, 1965).

Other examples may be the indigenous Monterey siliceous-zone oils from the Santa Maria basin Cat Canyon field, described by Milner, Rogers, and Evans (1977). Their data show these oils to have low API gravities (16°–21.8°), high sulfur contents (3.47–4.43 wt%), an even predominance in the C_{18}–C_{30} range of the n-alkanes, phytane more abundant than pristane, and benzothiophenes more abundant than dibenzothiophenes. Again, it should be noted that parts of the Monterey contain significant amounts of dolomite (Garrison and Douglas, 1981). It may well be that these oils, indeed, were derived from a predominantly carbonate–siliceous facies.

APPLICATION

An example to illustrate the use of thiophenic-organosulfur compositional patterns in exploration is given below. The objective was to use organosulfur compositions to establish whether a suite of crude oils from the Salawati basin, Indonesia, has a single, common source or multiple sources, and to characterize the source.

Twenty crude oils from reservoirs in the Miocene Kais Formation were analyzed by GC (FPD). The thiophenic-sulfur compositions of all the Salawati oils as a whole are similar; however, there are differences, primarily in the relative abundance of benzo- and dibenzothiophenes, which divide the oils into two groups, I and II (Fig. 14). Compared to group I oils, group II oils show a greater abundance of benzothiophenes, a higher proportion of C_2- and C_3-dibenzothiophenes, and less dibenzothiophene. These features of the group II Salawati oils are typical of carbonate-derived oils such as the shallow Smackover oils; in contrast, the group I oils in these features resemble siliciclastic-derived oils such as the Ekofisk oil (see Fig. 14).

The two groups of oils also differ in other geochemical parameters (Table 5). Consistent with the thiophenic-sulfur compositions, the group II oils have somewhat lower gravities, higher total sulfur contents, and lower proportions of rearranged steranes.

Thiophenic-sulfur compounds and sterane analyses indicate that the Salawati Kais oils were derived from a common source-rock facies characterized by slight differences in thermal maturity and by variable proportions of terrestrial and aquatic organic matter. These differences permit recognition of two subgroups of Kais oils. A plausible source is a variable carbonate/shale facies, perhaps a marlstone. Group II oils were generated in the less mature, more calcareous source facies, whereas group I oils were generated in the more mature, more shaly source facies.

Table 5

Parameter	Group I	Group II
Total sulfur, wt%	.07–0.7	0.7–1.2
API gravity, °	33–46	20–33
δC^{13} total crude, ‰	−18.8 to −21	−21 to −22.8
δC^{13} saturates, ‰	−18.8 to −21	−21 to −23.4
Ratio of rearranged steranes to unrearranged steranes	High	Low
n-alkane maximum, carbon number	C_{17}	C_{21}

Table 5 — Comparison of group I and group II Salawati Kais oils, Indonesia.

SUMMARY

The thiophenic-organosulfur compositions of crude oils, which can be conven-

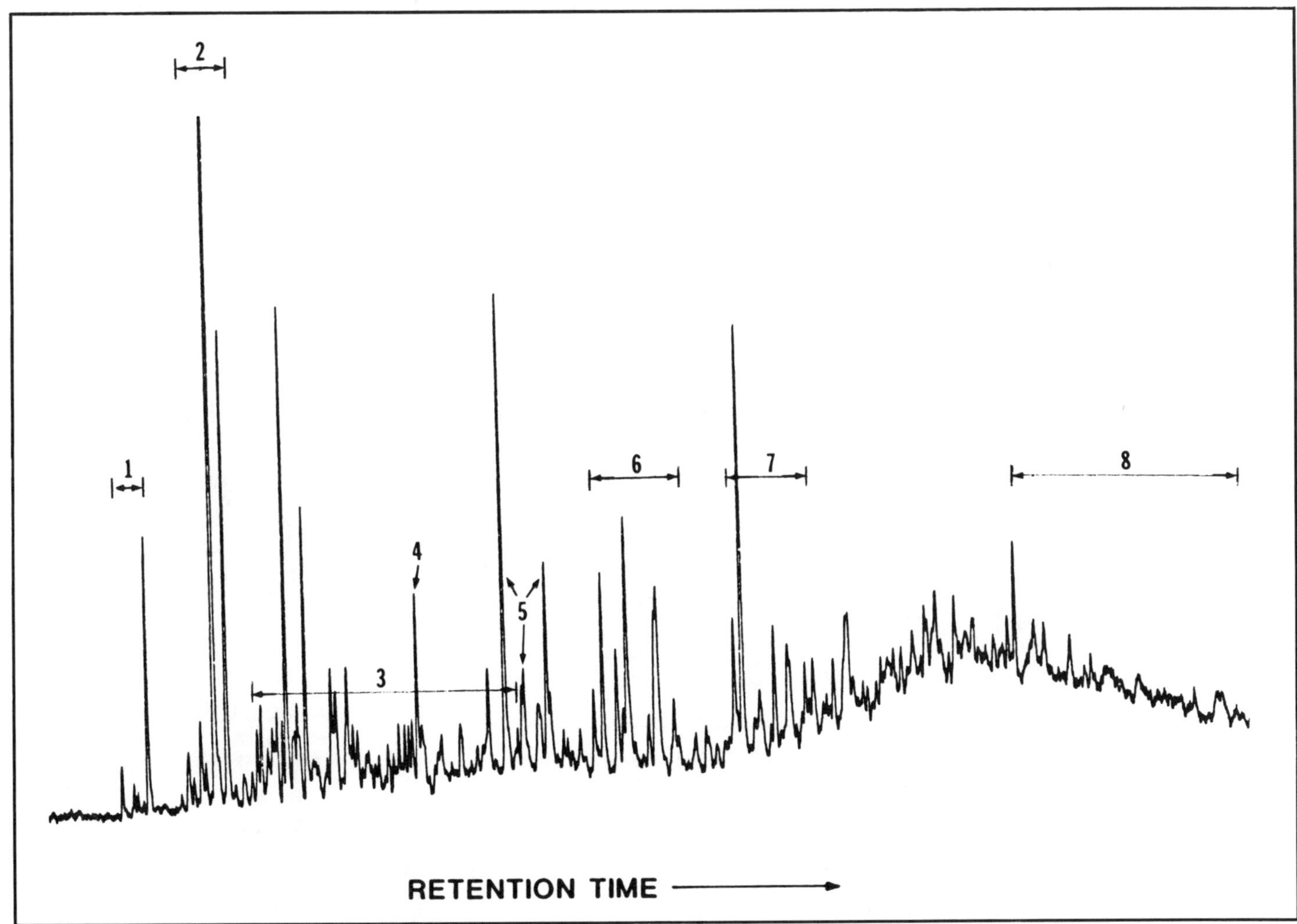

Figure 13 — FPD gas chromatograms of a crude-oil aromatic fraction, Wilmington field, California (see Table 1 for peak-identification key).

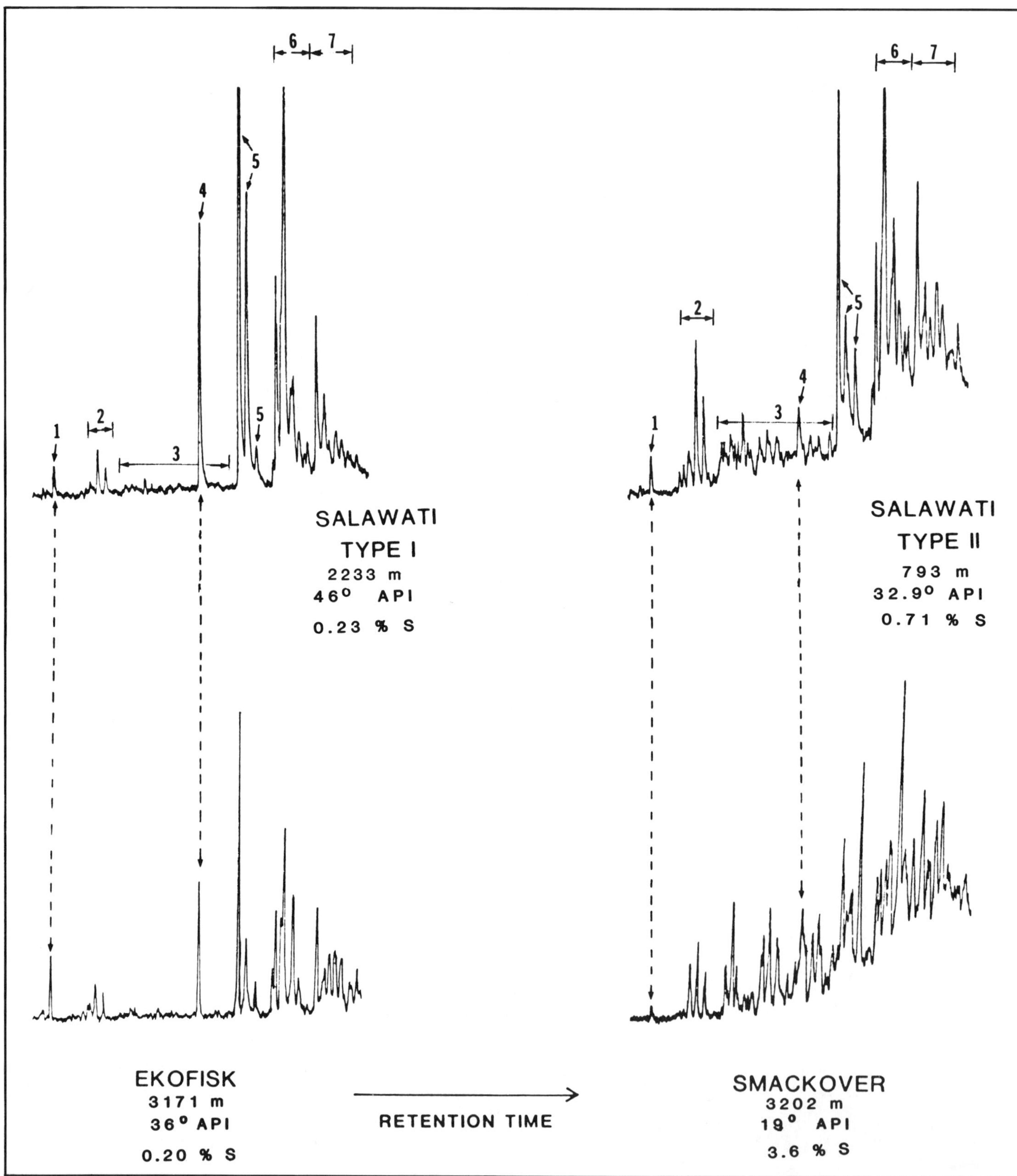

Figure 14 — Comparison of FPD gas chromatograms of aromatic fractions representative of group I and group II oils, Kais Formation, Salawati basin, Indonesia, with FPD gas chromatograms of oils from a typical carbonate source (Smackover) and a typical siliciclastic source (Ekofisk) (see Table 1 for peak-identification key).

iently determined by high-resolution gas chromatography using a flame photometric detector, show promise for characterizing oils from carbonate or, perhaps more generally, highly calcareous or nonsiliciclastic anoxic sources. Our observations are that the presence of a definite V pattern (4-methyl > 2- + 3-methyl < 1-methyl) in the methyldibenzothiophenes generally is associated with oils from predominantly carbonate source rocks. A stair-step pattern (4-methyl > 2- + 3-methyl > 1-methyl) in the methyldibenzothiophenes is associated with predominantly siliciclastic source rocks or advanced maturity (late– to post–oil window) oils from carbonate sources. The thiophenic-sulfur fingerprints of oils from carbonate sources also show an abundance of benzothiophenes and a fairly equal distribution of substituted dibenzothiophenes. Thermal maturation may modify this pattern so that the distribution of mono-, di-, and trimethyl-substituted dibenzothiophenes is generally relatively uniform in carbonate oils, while siliciclastic oils show decreasing amounts of dimethyl- and trimethyldibenzothiophenes relative to methyldibenzothiophenes.

The relationship between thiophenic-organosulfur composition and source-rock lithology is empirical but is useful in selecting between candidate formations of differing lithology. Used in conjunction with such other geochemical parameters as total sulfur, n-alkane, and sterane/triterpane compositions, thiophenic-organosulfur analysis provides a useful exploration tool. Indeed, their high degree of thermodynamic and kinetic stability and above-average resistance to microbiological degradation (Milner, Rogers, and Evans, 1977) suggest that thiophenes will be useful in oil–oil and oil–source-rock correlations in which other chemical markers fail.

REFERENCES CITED

Ayres, M.G., et al, 1982, Hydrocarbon habitat in main producing areas, Saudi Arabia: AAPG Bulletin, v. 66, p. 1–9.

Cardwell, A.L., 1977, Petroleum source rock potential of Arbuckle and Ellenburger Groups, southern Mid-Continent, United States: Golden, Colorado, Colorado School of Mines Quarterly, v. 72, no. 3, 134 p.

Castex, H., J. Roucaché, and R. Boulet, 1974, The thiophenic sulfur in petroleum and rock extracts analyzed by mass spectrometry and gas chromatography: Revue de l'Institut Français du Pétrole, v. 29, p. 3–40.

Connan, J., G. Hussler, and P. Albrecht, 1980, Geochemistry of crude oils and crude oil–source rock correlations in four carbonate basins (abs.): Geological Society of America Abstracts with Programs, v. 12, p. 405.

Dean, R.A., and E.V. Whitehead, 1967, Status of work in separation and identification of sulfur compounds in petroleum: Seventh World Petroleum Congress, Paper 7, Panel Discussion 23.

Deroo, G., 1976, Correlations of crude oils and source rocks in some sedimentary basins: Bulletin du Centre de Recherches de Pau, v. 10, p. 317–335.

Drushel, H.V., 1970, Sulfur compounds in petroleum—known and unknown: American Chemical Society, Division of Petroleum Chemistry, Preprints, v. 15, no. 2, C–13.

Garrison, R.E., and R.G. Douglas, eds., 1981, The Monterey Formation and related siliceous rocks of California: SEPM, Pacific Section, Special Publication, 327 p.

Gransch, J.A., and J. Posthuma, 1974, On the origin of sulfur in crudes, *in* B. Tissot and F. Bienner, eds., Advances in organic geochemistry 1973: Paris, Editions Technip, p. 727–739.

Hitchon, B., 1961, Formation fluids, *in* R.G. McCrosian and R.P. Glaister, eds., Geological history of western Canada: Alberta Society of Petroleum Geologists, p. 200–217.

Ho, T.Y., et al, 1974, Evolution of sulfur compounds in crude oils: AAPG Bulletin, v. 58, p. 2338–2348.

Mackenzie, A.S., et al, 1980, Molecular parameters of maturation in the Toarcian shales, Paris Basin, France: I. Changes in the configurations of acyclic isoprenoid alkanes, steranes and triterpanes: Geochimica et Cosmochimica Acta, v. 44, p. 1709–1721.

Martin, R.L., and J.A. Grant, 1965, Determination of thiophenic compounds by types in petroleum samples: Analytical Chemistry, v. 37, p. 649–657.

Milner, C.W.D., M.A. Rogers, and C.R. Evans, 1977, Petroleum transformations in reservoirs: Journal of Geochemical Exploration, v. 7, p. 101–153.

Momper, J.A., and J.A. Williams, 1979, Geochemical exploration in the Powder River basin: Oil and Gas Journal, v. 77, no. 50, p. 129–134.

Murris, R.J., 1980, Middle East: stratigraphic evolution and oil habitat: AAPG Bulletin, v. 64, p. 597–618.

——— and K. de Groot, 1979, Oil habitat of carbonate provinces: Mexico, Proceedings Congress Panama Ing. Petr., Sec. 1, Paper 5.

Obelentsev, R.D., 1967, Sulfo–organic compounds in oils of USSR: Seventh World Petroleum Congress Proceedings, v. 9, p. 109–116.

Oehler, J.H., 1984, Carbonate source rocks in the Jurassic Smackover trend of Mississippi, Alabama, and Florida, *in* J.G. Palacas, ed., Petroleum geochemistry and source-rock potential of carbonate rocks: AAPG Studies in Geology 18, this volume.

Orr, W.L., 1977, Geological and geochemical controls on the distribution of hydrogen sulfide in natural gas, *in* R. Campos and J. Goni, eds., Advances in Organic Geochemistry 1975: Madrid, Enadisma, p. 571–597.

———, 1978, Sulfur in heavy oils, oil sands and oil shales, *in* O.P. Strausz and E.M. Lown, eds., Oil sand and oil shale chemistry: New York, Verlag Chemie, p. 223–243.

Palacas, J.G., 1978, Preliminary assessment of organic carbon content and petroleum source rock potential of Cretaceous and Lower Tertiary carbonates, South Florida Basin: Gulf Coast Association of Geological Societies Transactions, v. 28, p. 357–381.

———, 1983, Carbonate rocks as sources of petroleum: Geological and chemical characteristics and oil–source correlations: London, 11th World Petroleum Congress Preprint, 13 p.

———, D.E. Anders, and J.D. King, 1984, South Florida basin, a prime example of carbonate source rocks of petroleum, *in* J.G. Palacas, ed., Petroleum geochemistry and source-rock potential of carbonate rocks: AAPG Studies in Geology 18, this volume.

Philippi, G.T., 1965, On the depth, time and mechanism of petroleum generation: Geochimica et Cosmochimica Acta, v. 29, p. 1021–1049.

Powell, T.G., 1984, Some aspects of the hydrocarbon geochemistry of a Middle Devonian barrier-reef complex, western Canada, *in* J.G. Palacas, ed., Petroleum geochemistry and source-rock potential of carbonate rocks: AAPG

Studies in Geology 18, this volume.

Radke, M., D.H. Welte, and H. Willsch, 1982, Geochemical study on a well in the Western Canada basin: relation of the aromatic distribution pattern to maturity of organic matter: Geochimica et Cosmochimica Acta, v. 46, p. 1–10.

Rall, H.T., et al, 1972, Sulfur compounds in crude oil: U.S. Bureau of Mines Bulletin 659, 187 p.

Rubenstein, I., O. Sieskind, and P. Albrecht, 1975, Rearranged steranes in a shale: occurrence and simulated formation: Jour. Chem. Soc. Perkin Trans. I, p. 1833–1836.

Scalan, R.S., and J.E. Smith, 1970, An improved measure of the odd–even predominance in the normal alkanes of sediment extracts and petroleum: Geochimica et Cosmochimica Acta, v. 34, p. 611–620.

Sieskind, O., G. Joly, and P. Albrecht, 1979, Simulation of the geochemical transformations of sterols: superacid effect of clay minerals: Geochimica et Cosmochimica Acta, v. 43, p. 1675–1679.

Tissot, B., and D.H. Welte, 1978, Petroleum formation and occurrence: Berlin, Springer–Verlag, 538 p.

Van den Bark, E., and O.D. Thomas, 1980, Ekofisk, first of the giant oil fields in western Europe, *in* M.T. Halbouty, ed., Giant oil and gas fields of the decade 1968–1978: AAPG Memoir 30, p. 195–224.

Wenzel, B., and R.L. Aiken, 1979, Thiophenic sulfur distribution in petroleum fractions by gas chromatography with a flame photometric detector: Journal of Chromatographic Science, v. 17, p. 503–509.

Compaction of Modern Carbonate Sediments: Implications for Generation and Expulsion of Hydrocarbons

Eugene A. Shinn and Daniel M. Robbin
U.S. Geological Survey, Fisher Island Station, Miami Beach, Florida

George E. Claypool
U.S. Geological Survey, Denver Federal Center, Denver, Colorado

Combined compaction and heating experiments conducted with *in-situ* modern carbonate sediments produced dissolved and undissolved tarry hydrocarbons. During heating at temperatures ranging from 100° to 200°C (212°–392°F) and under pressures of up to the equivalent of 13,000 ft (4,000 m) of burial, hydrocarbons with an immature, biologically related, bimodal carbon distribution were converted to hydrocarbons with a relatively more mature unimodal distribution. The new hydrocarbon distribution resembles those typical of Upper Cretaceous carbonate-rock extracts in south Florida. The data from these short-term experiments lend support to the idea that limestones can serve as hydrocarbon source beds and that compaction, both mechanical and chemical (i.e., pressure dissolution), could provide the "driving force" for expulsion into a carrier bed or directly into a reservoir. The suggestion is made that, at depths and temperatures at and within the oil-generation window, liquid hydrocarbons together with compaction waters escape during chemical compaction (akin to stylolitization). As temperature and pressure increase, liquid hydrocarbons escape through hydraulically formed microfractures. Finally, with continued burial and increased temperature, below the oil-generation window, the residual hydrocarbons and other hydrogen-rich materials (kerogen) are converted to and escape as gas condensates and gases.

INTRODUCTION

The primary purpose of this paper is to describe alteration of organic matter that resulted from combined compaction and heating experiments conducted on modern lime sediments. The secondary purpose is to discuss a possible mechanism for expulsion and migration of hydrocarbons from limestone source rocks.

Recent experimental work has shown that *in-situ* fine-grained lime sediment is compactable (Shinn et al, 1977; Shinn and Robbin, 1983). In those experiments, cores of *in-situ* lime sediment were compacted to between one-half and one-fourth of their original length under loads ranging from the equivalent of about 930–13,570 ft (300–4,100 m) of overburden. It was noted during those experiments that the resulting sedimentary and petrographic features strongly resembled limestones generally considered as source rocks. The suite of features produced by compaction includes (1) mashed burrow fillings (circular burrows were flattened into ovoid shapes); (2) obliteration of sea grasses and resultant production of thin, wavy organic seams superficially resembling stylolites; (3) obliteration of soft pellets but retention of pelletal texture when there was a slight degree of submarine cementation; (4) reorientation of randomly arranged fossils toward the horizontal (significantly, fossils such as shells and algal fragments were rarely broken, especially when surrounded by lime mud); and (5) chemical compaction (i.e., pressure dissolution), in which there was mutual penetration of grains without fracturing.

A typical example of features resulting from experimental compaction is shown in Figures 1 and 2. It should be pointed out that the "rock" shown in these figures is not cemented to form a true rock, even though it can be sawed and treated as though it were. When placed in water, the material quickly becomes friable and mushy. Initial heating experiments were conducted in an attempt to convert unstable aragonite sediment to calcite and to try to produce a true "rock" through recrystallization. In these experiments, cementation and alteration of aragonite to calcite was not achieved, even after heating to temperatures ranging from 100° to 200°C (212°–392°F). What was noted, however, was a tarry substance that accumulated on the exterior of the compacted and heated sediment cores (Fig. 1). The tarry substance readily fluoresces under ultraviolet light, and analyses confirmed its hydrocarbon composition and led to the further experimentation described herein.

EXPERIMENTAL METHODS

The devices and basic procedures for compaction of unconsolidated *in-situ* cores of modern sediment were described in detail by Shinn and Robbin (1983), and therefore much of the detail will not be repeated here. Basically, cores 30–50 cm long with diameters between 7.3 and 9.8 cm from various sedimentary environments were placed under either a mechanical or hydraulic press, and loads ranging from 933 to 14,416 psi (65.5–1,011.5 kg/cm^2) were applied. Fluid was allowed to escape through porous pistons placed at each end of the core tube. A porting system within the piston (see Shinn and Robbin, 1983) allowed fluid to be channeled through a flexible tube into a collector consisting of a glass test tube. A high-pressure valve was attached to the piston upstream from the flexible tubing.

Figure 1 — Compacted core of modern carbonate sediment (core 8), which was heated to 100°C (212°F) while being compacted for 30 days. See Tables 1 and 2 for additional data. The dark, tarry residue is a hydrocarbon substance that formed from disseminated organic matter during heating.

Table 1

Core No.	Location	Approximate Pressure in Chambers psi (kg/cm^2)	Original Length of Core (mm)	Percentage of Original Length	Time Under Load Days	Porosity Before Compaction	Porosity After Compaction	Heat Applied (°C)	Total Weight Dissolved Hydrocarbons Expelled with Water (g)
8	Totten Key	933 (66)	348	26	30.00	85	45	100	—
10	Totten Key	933 (66)	352	29	9.92	84	47	100	—
14	Totten Key	8,480 (595)	254	27	7.25	85	47	200	—
33	Crane Key	11,872 (833)	325	34	14.50	79	37	100	0.865
35	Rodriguez Key	13,568 (952)	355	39	34.00	75	39	—	0.005
36	Rodriguez Key	13,568 (952)	309	42	35.00	62	35	140	0.163
48	Cluett Key	5,000 (353)	182	45	28.00	—	—	200	—

Table 1 — Compaction data and expelled hydrocarbon weights for individual cores.

By closing the valve while force was being applied, both lithostatic and hydrostatic pressure was allowed to build in the pressure chamber. Under these conditions, sediment did not compact (i.e., dewater), allowing the equivalent of "geopressuring" to occur.

In these experiments, a heating jacket was placed around the pressure chamber, which consisted of a General Electric[1] "calrod" heating element wrapped tightly around the steel-walled pressure chamber and insulated by a glass wool jacket. The heating elements were moderated by a Fenwal Model 524[1] temperature-control device attached by a thermocouple embedded within the steel wall of the pressure chamber, which maintained temperatures to an accuracy of approximately 5%. Heating with the valves closed prevented rapid expulsion of water and allowed organic matter within the sediment to "cook" in a manner thought to be more similar to that which occurs in nature. When valves were open, water was quickly expelled (within

[1]Use of brand names does not constitute endorsement by the U.S. Geological Survey.

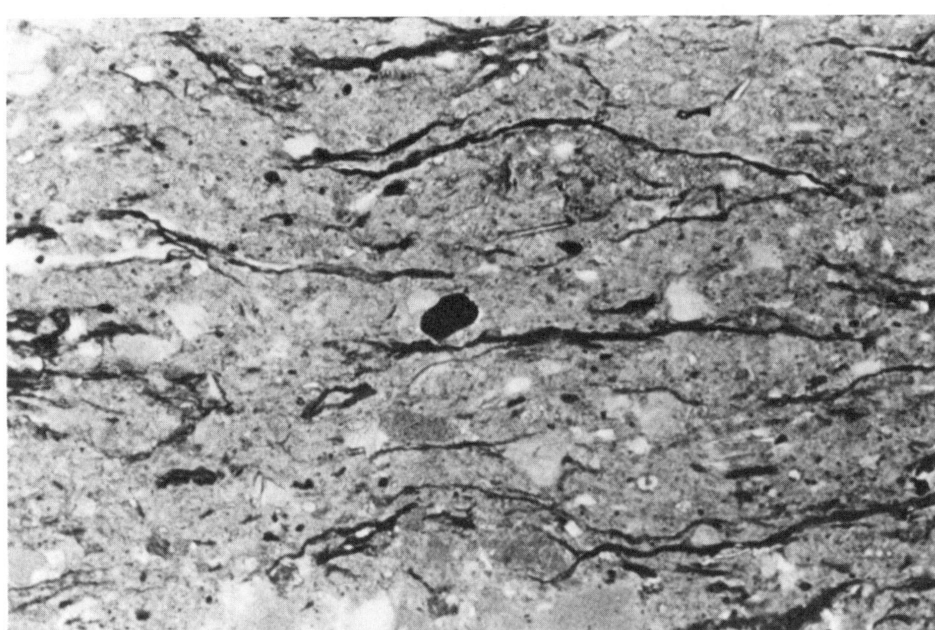

Figure 2 — Typical thin section of compressed modern sediment (core 8), showing the abundance of stylolite-like wisps of organic matter. The rock looks similar to typical ancient limestone source rocks. The organic wisps could be forerunners of true stylolites.

two days; see Shinn and Robbin, 1983), except for approximately 40% fluid, which was the average porosity of the cores after compaction. When heated, however, much of this water boiled off as steam, creating a condition not likely in a true oil-source bed. Tarry residue, such as shown in Figure 1, resulted when the water was allowed to boil off. By keeping the valves closed, however, boiling was prevented, and samples could be taken periodically by briefly opening the valves. Basic data, including temperature, pressure, and time, are given in Table 1.

Chemical analyses were made by the organic geochemistry laboratory of the U.S. Geological Survey, Denver, Colorado. The material analyzed consisted of (1) uncompacted and unheated lime sediment; (2) the heated sediment, including the tarry residue that accumulated on the outer surface of compacted and heated cores; and (3) fluids that were valved off periodically from compaction and heating experiments where fluid was retained within the sediment. Organic carbon was determined by a wet oxidation technique modified from Bush (1970). Organic matter (bitumen) was removed from the sediment samples by extraction with chloroform in a Soxhlet apparatus. The bitumen was separated by silica-gel column chromatography into three fractions: saturated hydrocarbons (heptane eluate), aromatic hydrocarbons (benzene eluate), and nonhydrocarbons or asphaltics (50:50 benzene–methanol eluate). Gas chromatograms of saturated-hydrocarbon distributions were run on a packed column coated with SE-30. Organic matter dissolved in water samples expelled during the compaction experiments was extracted with chloroform in a separatory funnel, and the extracts worked up as the bitumen above.

RESULTS

Tarry Residue

Figure 3 shows gas chromatograms of saturated-hydrocarbon distributions of the same sample (core 8) before and after being subjected to heating and compaction. The material analyzed after heating and compaction was a sample of the bulk compacted sediments with the tarry residue, as shown in Figure 1. The material in core 8 had been heated for 30 days at 100°C (212°F; see Table 1) under a load of 933 psi (65.5 kg/cm^2). Starting material contained approximately 2.9 wt% total organic carbon and 27 ppm extractable saturated hydrocarbons. Figure 3 shows that the saturated hydrocarbons in the unheated sample are characterized by a bimodal distribution typical of many modern and other immature sediments. Maximum concentration of saturated hydrocarbon molecules occurs in the region above n-C_{25}. In sharp contrast, heating and compaction of the sample yielded a unimodal saturated-hydrocarbon distribution with the maximum concentration of molecules in the n-C_{17} to n-C_{19} region. The amount of saturated hydrocarbons also showed a fivefold increase to 130 ppm.

Although the tarry material of core 8 was produced at a pressure of 933 psi (65.5 kg/cm^2) over a period of 30 days, experiments showed that similar effects (a tarry residue on the outside of the cores) could be produced in a shorter period of time. For example, tarry residues were observed in core 10 when heated at 100°C (212°F) for 9 days, 22 hours, and in core 14 when heated at 200°C (392°F) for just 7 days, 6 hours. It is not known whether pressure actually played a significant role in the alteration of organic matter.

Dissolved Hydrocarbons

Figures 4 through 6 summarize experiments in which valves were closed except for sampling of fluids. The sampling periods shown in these figures are represented by sudden changes in compaction rate that occurred during sampling. Dissolved total hydrocarbon concentrations (THC) in the fluids sampled are shown in these figures next to the abrupt deflections on the graphs. The total weights of dissolved hydrocarbons expelled with water during the compaction experiments were estimated by multiplying the volume of water times the hydrocarbon concentration for each of the sampling periods. The estimated total hydrocarbons expelled are listed in Table 1 for the three experiments summarized in Figures 4 through 6, along with the other compaction data. The expelled hydrocarbons can be expressed in terms of ppm concentration relative to the dry weight of sediment used in the experiments, yielding 9 ppm for core 35, 126 ppm for core 36, and 1,140 ppm for core 33. Figure 7 shows increases of hydrocarbons and other extractable organic matter by means of color changes that occurred during a typical experiment.

Core 33 was from an algal-mat-rich (supratidal stromatolite) area on Crane Key. The total-organic-matter content was undoubtedly much higher for this core, but it was not quantitatively characterized.

DISCUSSION

Tarry Residues

In the simple compaction and heating experiments, where fluid was allowed to escape continually, tarry residue accumulated on the core exterior as shown in Figure 1. The analysis not only showed a

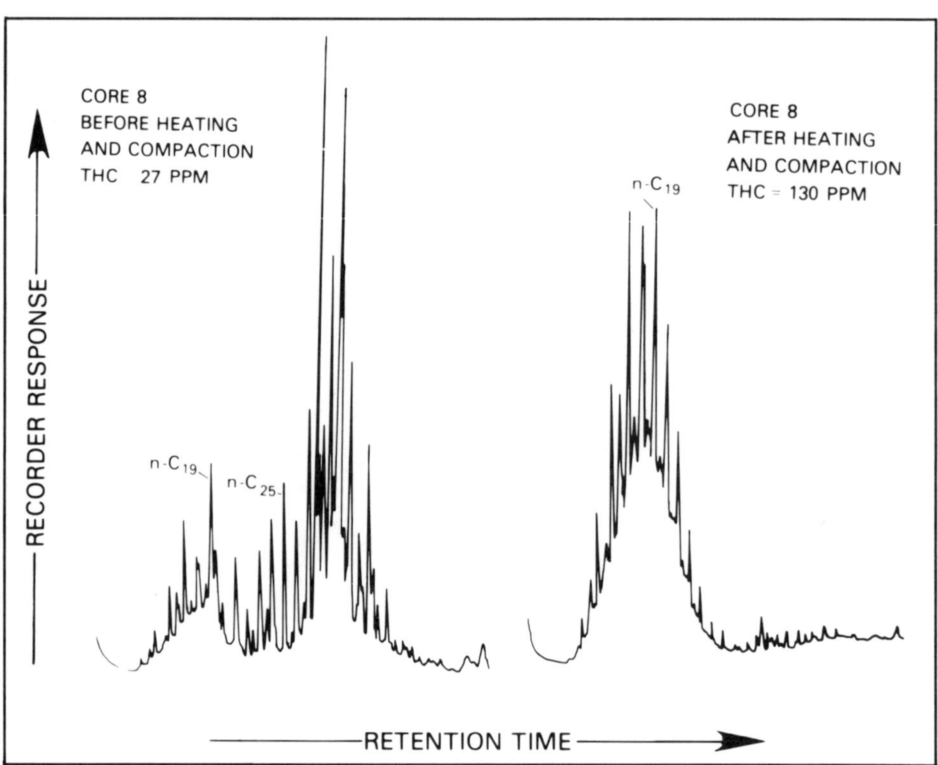

Figure 3 — Gas chromatograms of saturated hydrocarbons extracted from Florida Bay sediment (core 8) before and after heating and compaction. Note bimodal distribution before heating and unimodal distribution after heating. See Table 2 for additional data. THC = total hydrocarbon extracts.

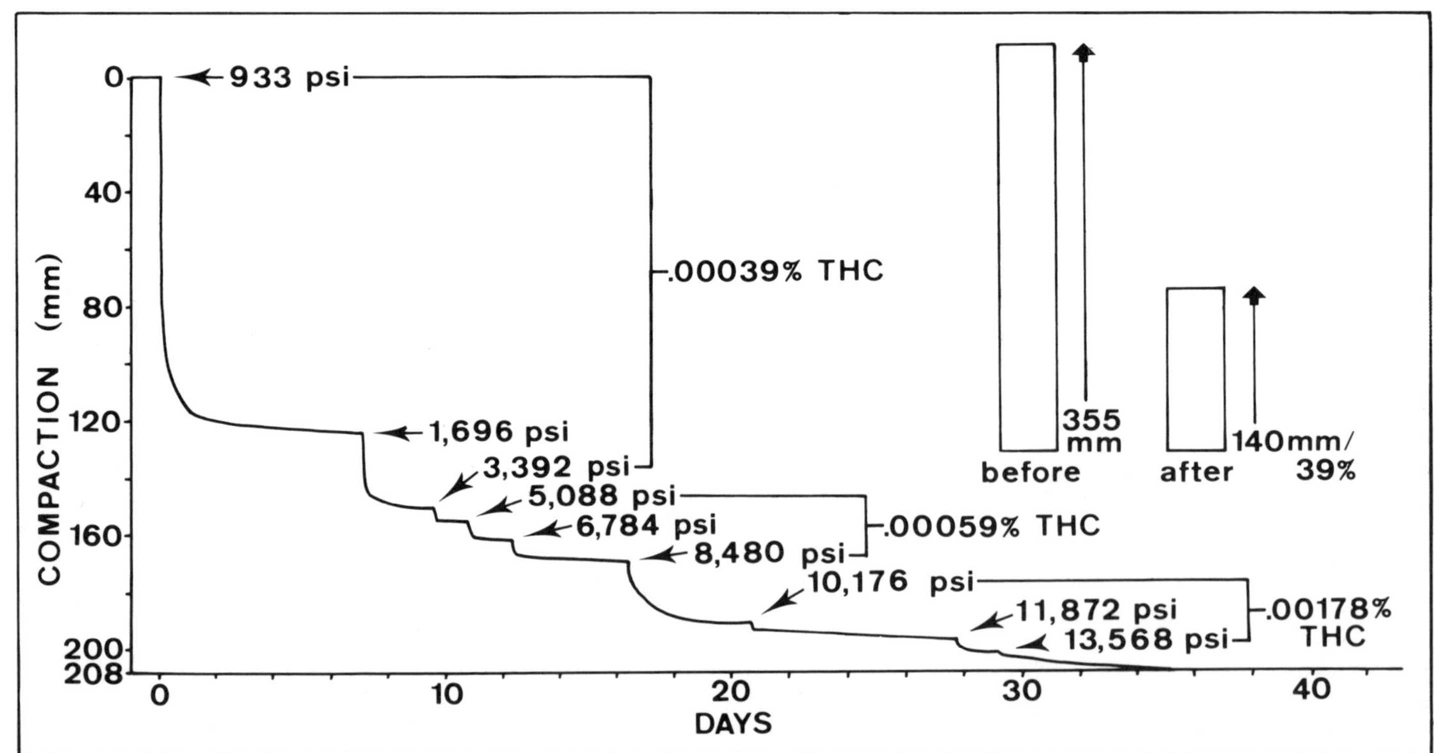

Figure 4 — Compaction graph of core 35, which was compacted without heating. Abrupt deflections in graph correspond to release in chamber pressure necessary to sample fluids. Pressure was increased after each sample, and the values are given in psi. Concentrations of total hydrocarbon (THC) extracted from the expelled fluid are given by each pressure value where samples were taken. Note gradual increase in THC values with time. Drawing in upper right corner shows the length of core before and after compaction and the porosity at the end of the experiment.

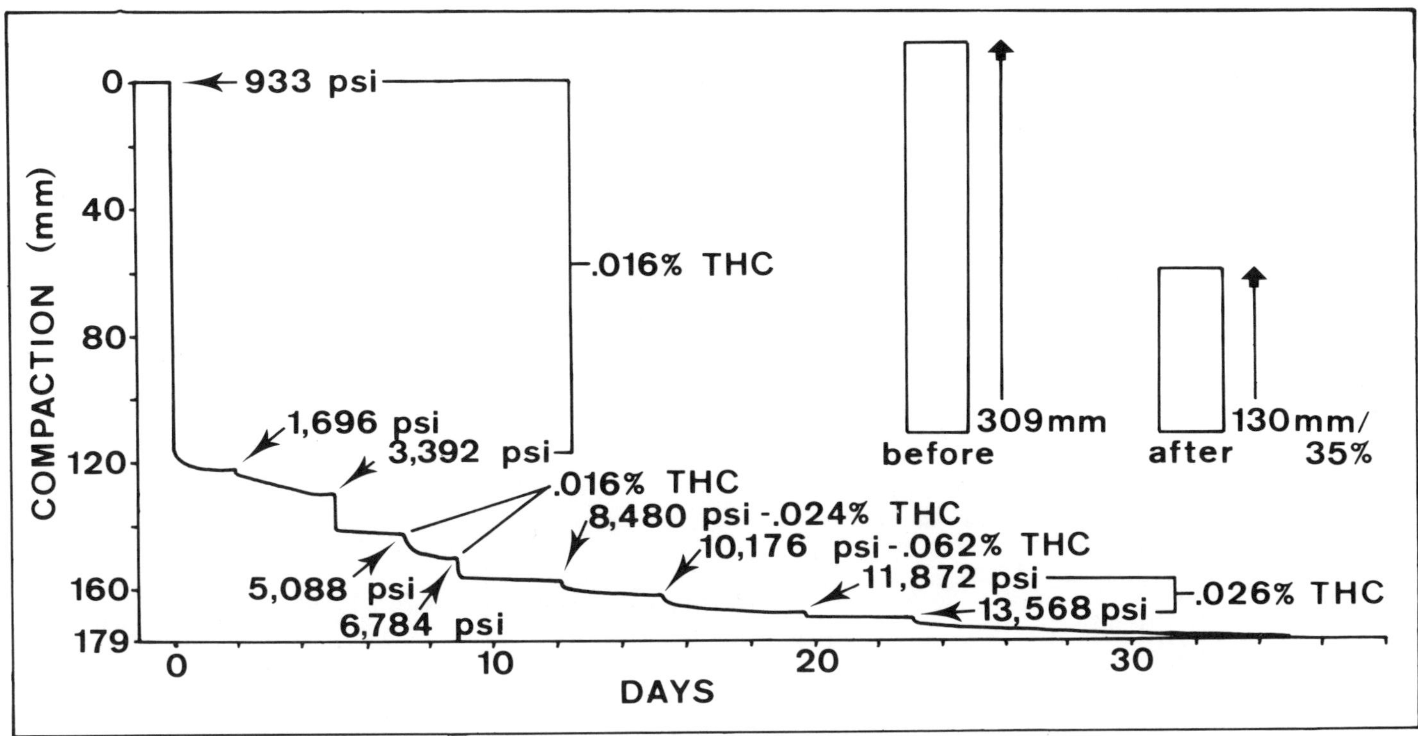

Figure 5 — Same type of graph as shown in Figure 4, except that the sediment was heated to 140°C (284°F) throughout the length of the experiment. Note that THC values are higher but still quite low.

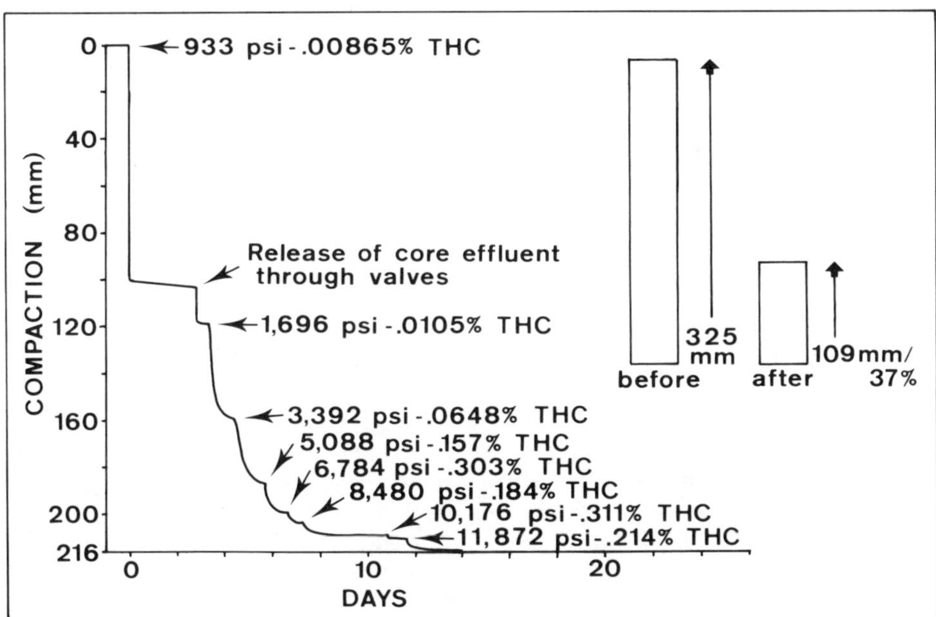

Figure 6 — Same type of graph as shown in Figures 4 and 5. Sediment was a supratidal algal-tich sediment from Crane Key in Florida Bay, and maximum temperature was 100°C (212°F). Greater amount of THC was probably a reflection of the high degree of leathery algal mats in the starting sediment. Pre-compaction values of organic matter for this sediment were not obtained.

significant increase in saturated hydrocarbons for core 8 (see Table 2) but also a clear change in carbon-number distribution (Fig. 3). In the untreated sample, saturated hydrocarbons in the n-C_{27} to n-C_{29} range predominate, whereas in the sample that was heated and compacted, maximum distribution occurs in the n-C_{17} to n-C_{19} range (Fig. 3). From the trace in Figure 3, one might conclude that the compounds in the n-C_{27} to n-C_{29} range fractionated during heating; however, it is likely that hydrocarbons in that range are simply masked by the generation of a relatively large concentration of hydrocarbons in the n-C_{17} to n-C_{19} range during heating (Fig. 3).

In Figure 8 a gas chromatogram of the experimentally produced tarry residue (core 48) is compared with natural immature hydrocarbons extracted from an Upper Cretaceous rock from a depth of 5,890 ft (1,795 m) in south Florida. The similarity is provocative and suggests that the heating experiments indeed have mimicked in a few weeks what took millions of years in nature.

Dissolved Hydrocarbons

Comparison of experimental data (Table 1) for core 35 (Fig. 4) and core 36 (Fig. 5) clearly shows that heat caused a significant increase in dissolved hydrocarbons in the expelled fluid. The sediment type of both cores appeared to be comparable, and hence the assumption is made that the organic-matter content was also comparable (total-organic-carbon content estimated to be less than 2–3%).

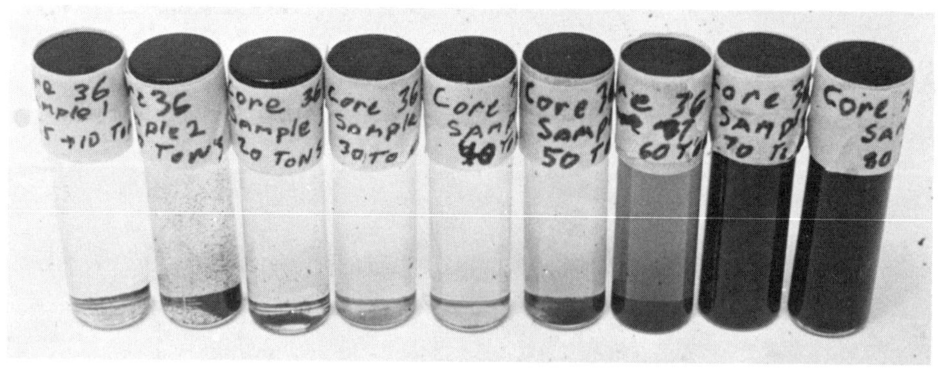

Figure 7 — Vials of fluid taken from experiment shown in Figure 4 and arranged in order of increasing time and pressure. Note progressive darkening of fluid.

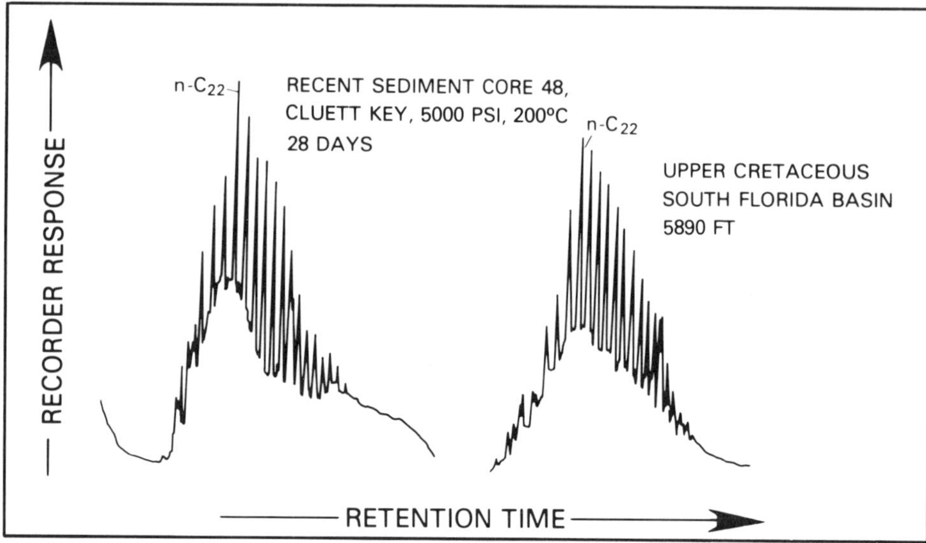

Figure 8 — Two gas chromatograms of saturated hydrocarbons. The trace at left is an extract from a recent sediment core (no. 48, Cluett Key) that was heated at 200°C (392°F) for 28 days under a pressure of 5,000 psi (353 kg/cm^2). The trace at right is from an extract of an Upper Cretaceous carbonate rock at a depth of about 5,890 ft (1,795 m) in a well in south Florida. Similarity between the two GC's suggests a similar degree of maturation.

The amount of dissolved hydrocarbons is mainly a function of the type and amount of organic matter in the starting material. A similar experiment was conducted (Fig. 6) on a core of supratidal sediment from Crane Key in Florida Bay. The sediment contains abundant organic-rich leathery algal mats, which appear similar to the well-known stromatolites of ancient limestones. The significantly higher concentration of generated hydrocarbons is attributed to the hydrogen-rich composition of the algae, which have an estimated organic-carbon content of 10% or more. Photographs of compacted algal mats are shown in Shinn and Robbin (1983).

Geologic Implications

These experiments, like most, cannot be translated directly to the "real" world owing to their short-term nature. One is struck, however, with the similarity of the produced rock (see Shinn and Robbin, 1983, for other examples) to ancient limestones, especially supposed lime source rocks. This similarity, plus the apparent conversion of organic matter to hydrocarbons in these experiments, supports the idea of some ancient limestones serving as source rocks. There are, however, several problems with applying the results of these experiments to ancient rocks and to the origin of migration of hydrocarbons.

In most of the compaction experiments, we have simulated burial depths equivalent to those associated with maturation of hydrocarbons in nature. In the experiments, however, the minimum porosity achieved after mechanical compaction was only a little less than 40%, whereas in nature fine-grained rocks at similar burial depths and temperatures probably would contain less than 5%. In addition, natural rocks originally containing aragonite would have been already converted to calcite. Further, in our experiments we applied high temperatures at simulated burial depths as shallow as 933 ft (285 m). Such temperatures ordinarily would not be encountered in nature until burial depths of 8,000–10,000 ft (2,440–3,050 m) are reached. The data nevertheless do have some useful implications.

In a natural situation, aragonitic or high-Mg limestones would be converted to calcite and porosities reduced to the range of a few percent long before reaching the temperatures attained in the experiments. Reduction of porosity in nature can be attributed to near-surface early diagenetic effects, such as submarine and subaerial lithification or shallow phreatic cementation in addition to later deeper diagenetic events, such as mechanical and chemical compaction (see Shinn and Robbin, 1983, for further discussion). The universal occurrence of stylolites in limestones buried to a few thousand feet or more is evidence of chemical compaction, a process that can occur only in the presence of water. Both mechanical and chemical compaction requires the expulsion of water or other fluids. Since heat is probably the single most important factor in hydrocarbon maturation, it is likely that fluids expelled early during burial compaction (both mechanical and chemical) would contain low concentrations of hydrocarbons.

The heat required to initiate and continue hydrocarbon maturation probably would not be obtained until burial depth has reached a point where porosity is low and fluid content is in the 5–10% range. At that time and depth, the ratio of organic matter to pore water would be high. Given an increase in temperature as burial depth increases, the amount of oil relative to water would be much greater than that simulated in our experiments. Dunnington (1954, 1967) proposed a mechanism for oil expulsion from limestones that involved pressure dissolution (chemical compaction). We believe that our experiments and the conclusions of previous work (Shinn and Robbin, 1983) support and refine the proposal of Dunnington (1954, 1967).

Expulsion of Oil from Limestone Source Rocks

As limestone porosity is reduced through chemical compaction, a point must be reached where hydrocarbon content is equal to or greater than the amount of water. For example, if the starting sediment contained 3% organic carbon by weight, this would correspond to about 5% organic matter by weight, or about 12% organic matter by volume. About half of the organic matter in typical carbonate rocks can be converted to hydrocarbons, so about 6% by volume of the rock is ultimately convertible to oil. If the porosity is reduced eventually to 3% through mechanical and chemical compaction prior to significant hydrocarbon generation, then the pore space would be inadequate to accommodate the residual fluids (oil and water). With continued chemical compaction, possibly enhanced by H_2O and CO_2 derived by thermal cracking of the organic matter, a point would be reached where both water with dissolved hydrocarbons, or liquid hydrocarbons, would be supporting the overburden. Under such conditions, if fluids and dissolved hydrocarbons are unable to escape through tiny pore throats to a carrier bed, then microfracturing is likely to occur. The process envisioned is roughly similar to the process generally proposed for expulsion from shale source rocks (Hunt, 1979, p. 216). Since chemical compaction of limestones requires water, the proposed model is self limiting. When all water has been expelled, chemical compaction stops. As temperature rises, any remaining hydrocarbons would probably be expelled as a continuous oil phase and later as gas.

In summary, under natural conditions one might expect the following sequence of events as a fine-grained, uncemented lime sediment becomes buried to increasing depths: (1) mechanical compaction at near room temperature with resultant loss of water but very little loss of hydrocarbons and related organic-matter substances down to porosities ranging between 30 and 40%; (2) chemical compaction concurrent with increasing temperature, resulting in expulsion of water containing dissolved hydrocarbons; this process would probably continue until porosities are reduced to about 5–10%; (3) expulsion of water and liquid hydrocarbons through microfracturing as porosities are further reduced by chemical compaction to about 2–5%; and (4) eventually a depth would be reached beyond the oil-generation zone at which hydrocarbons still remaining in the rock would be thermally cracked to gas condensate and gas.

The proposed sequence of events is approximate and would apply mainly to those rocks which had never experienced early marine or subaerial cementation. Such early-formed rocks would resist mechanical and chemical compaction more effectively than the uncemented, fine-grained sediments discussed here.

ACKNOWLEDGMENTS

Appreciation and thanks are given to Sr. Carlos Lubeck for the hydrocarbon measurements and Barbara Lidz for editorial assistance. We thank James Palacas and Robert Halley for encouragement and critical discussions, as well as J. H. Hudson for aid during sampling.

REFERENCES CITED

Bush, P.R., 1970, A rapid method for determination of carbonate carbon and organic carbon: Chemical Geology, v. 6, p. 59–62.

Dunnington, H.V., 1954, Stylolite development postdates rock induration: Journal of Sedimentary Petrology, v. 24, p. 27–49.

——, 1967, Aspects of diagenesis and shape change in stylolitic limestone reservoirs: London, Elsevier, Seventh World Petroleum Congress Proceedings, v. 2, p. 339–352.

Hunt, J.M., 1979, Petroleum geochemistry and geology: San Francisco, W.H. Freeman, 617 p.

Shinn, E.A., and D.M. Robbin, 1983, Mechanical and chemical compaction in fine-grained shallow-water limestones: Journal of Sedimentary Petrology, v. 53, p. 595–618.

Shinn, E.A., et al, 1977, Limestone compaction—an enigma: Geology, v. 5, p. 21–24.

Carbonate Sedimentary Rocks and the Origin of Heavy Crude Oils

Irving A. Breger[1]
U.S. Geological Survey
Reston, Virginia

Questions concerning the origin and constitution of so-called heavy crude oils have multiplied immensely with escalating interest in recovery of these oils. Although defining the term *heavy crude oils* presents a major and as-yet-unsolved problem, substances such as naturally occurring *tars, asphalts,* and *bitumens* are all considered to be expressions of heavy crude oils and are so treated in this presentation. Much evidence indicates that some heavy oils are residues of biodegraded conventional oils or are conventional oils from which the light ends have been stripped. On the other hand, such observations do not readily explain the origin of all heavy oils, among which are those found at great depth where neither biodegradation nor evaporative stripping is a likely mechanism.

The association of vanadium and sulfur with heavy crude oils also suggests that many such oils may have been formed by mechanisms other than those thought to be basic to the origin of conventional oils. In particular, evidence has recently been accumulated to show that carbonate rocks may be source beds for many heavy crude oils and that mechanisms leading to the direct formation of heavy oils are probably unlike those by which conventional oils are formed. Moreover, oil formed in carbonate sequences may have accumulated in source-bed reservoirs if conditions of porosity and permeability were appropriate.

[1]Now deceased.

Recurrent Appearance of Source-Rock Facies in Cretaceous to Eocene Carbonate Series of Tunisia

B. Tissot and R. Pelet
Institut Français du Pétrole
Rueil Malmaison, France

B. B. Furollet and J. L. Oudin
Compagnie Française des Pétroles Total
Paris, France

Oil production in Tunisia is obtained from a carbonate series of mainly shallow-water facies comprising Middle to Upper Cretaceous, Paleocene and Eocene formations interbedded with the reservoirs. Several organic-rich beds with good petroleum potential appear in the series; however, they are of limited geographic extent, which results in poor chronostratigraphic correlation from well to well. The deposition of these beds seems to have depended on the local appearance of anoxic conditions in an otherwise constant environment. The most plausible model, consequently, is a shallow, consistently oxygen-depleted basin with eutrophic planktonic growth in which anoxic conditions could occur in slightly depressed topographic hollows that varied both in place and time.

The character of the organic matter of these beds exhibits features that correspond well with the model suggested. The chloroform extract, rich in NSO (nitrogen, sulfur, and oxygen) compounds and especially in resins, generally shows a strong, even predominant, n-alkane distribution and a high phytane-pristane ratio. In some interesting but uncommon cases this picture is reversed, however. The chemically isolated, chloroform-insoluble fraction, studied by Rock-Eval pyrolysis, exhibits characteristics typical of type II organic matter that originates from marine planktonic remains. All these features show little variation with geologic age. Correspondingly, the oils in the reservoirs are all of similar composition independent of reservoir age. The composition of the oil is compatible with the concept of generation from the organic beds previously considered. Consequently, a single oil type appears to have been generated by distinct source beds corresponding to the recurrent appearance of the same organic facies.

Geochemistry of Crude Oils and Crude-Oil–Source-Rock Correlations in Four Carbonate Basins

Jacques Connan
Société Nationale Elf–Aquitaine, Pau, France

Georges Hussler and Pierre Albrecht
Université Louis Pasteur, Strasbourg, France

Crude oils and rock samples from carbonate basins in France (Aquitaine), Guatemala, Iraq, and Tunisia/Libya were analyzed by the usual methods of organic geochemistry. The studies were based mainly on computerized gas chromatography–mass spectrometry, with special emphasis on mass fragmentography. Several classes of biological markers were used for the purpose of characterizing the environments of deposition and for correlating crude oils with source rocks. The following specific features commonly were observed in the oil and rock samples from these carbonate formations: (1) the predominance of normal alkanes with even carbon numbers, usually more pronounced in the rocks than in the oils; (2) a ratio of pristane to phytane mostly below 1; (3) the presence in significant concentrations of the higher homologs of the hopane triterpanes, up to C_{35} (more important in rocks than in oils); (4) usually small amounts of steranes (C_{21}–C_{23} and C_{27}–C_{29} ranges), with predominance of non-rearranged structures; and (5) commonly an unusually high concentration of individual isoalkanes (C_{23}) and cyclohexylalkanes (C_{21}).

These molecular criteria partly reflect an important microbial contribution, which is usually more pronounced in highly anoxic environments. Furthermore, they have been used successfully in correlations among crude oils and between crude oils and their source rocks. The results agree with other geochemical and geological data.

Carbonate Source Rocks for Six Million Barrels of Oil per Day—Zagros Fold Belt, Southwestern Iran

R. Burwood
Petrofina UK, Ltd., Epsom, Surrey, England

The giant fields of the Zagros fold belt of southwestern Iran fall within the greater context of the Arabian–Iraq–Persian basin and contain cumulative recoverable reserves estimated at 87 billion barrels of oil and 514 trillion cubic feet of gas.

The regional geology of the area comprises a wedge of Paleozoic to Holocene sediments, 10–15 km (6–9 mi) or more in thickness, supported on a mobile Eocambrian salt economic basement. Forming a classic carbonate–evaporite sequence, the succession contains prolific source–reservoir combinations and effective seals of integrity. Located on the eastern subducting boundary of the Arabian plate, hydrocarbon generation has been largely controlled by the Neogene Zagros orogenic event. A number of potential source sequences have been recognized. With one exception, all are typically of an organic-rich, argillaceous, lime-mudstone litho-textural type.

Most of the giant oil accumulations in Asmari (Oligocene–Miocene) and Bangestan Group (Upper Cretaceous) reservoirs have a common provenance in the Kazhdumi Formation of Early Cretaceous (Albian) age. The Pabdeh (Paleogene), lower Garau–Gadvan (Lower Cretaceous, Neocomian), and Sargelu (Middle Jurassic) formations contribute less to overall reserves, having been the source for selected subordinate Asmari, Bangestan, and Khami Group (Upper Jurassic–Lower Cretaceous) reservoirs only. Gas in Permian–Triassic reservoirs has a provenance in Ordovician–Silurian siliciclastics.

Index

A reference is indexed according to its important, or "key" words.

Three columns are to the left of a keyword entry. The first column, a letter entry, represents the AAPG book series from which the reference originated. In this case, ST stands for Studies in Geology Series. Every five years, AAPG will merge all its indexes together, and the letter ST will differentiate this reference from those of the AAPG Memoir Series (M) or from the AAPG Bulletin (B).

The following number is the series number. In this case, 18 represents a reference from AAPG Studies 18. The third column lists the page number of this volume on which the reference can be found.

† = titles * = authors